CERAMIC LASERS

Until recently, ceramic materials were considered unsuitable for optics due to the numerous scattering sources, such as grain boundaries and residual pores. However, in the 1990s the technology to generate a coherent beam from ceramic materials was developed, and highly efficient laser oscillation was realized. In the future, the technology derived from the development of the ceramic laser could be used to develop new functional passive and active optics.

Co-authored by one of the pioneers of this field, the book describes the fabrication technology and theoretical characterization of ceramic material properties. It describes novel types of solid lasers and other optics using ceramic materials to demonstrate the application of ceramic gain media in the generation of coherent beams and light amplification. This is an invaluable guide for physicists, materials scientists and engineers working on laser ceramics.

AKIO IKESUE is the President of World-Lab. Co., Ltd. He is also an executive scientist at SCHOTT AG, Germany, and an invited professor at ENSCP (Ecole Nationale Supérieure de Chimie de Paris).

YAN LIN AUNG is a Senior Researcher at World-Lab. Co., Ltd. His research focuses on opto-ceramics.

VOICU LUPEI is a Professor at the National Institute for Laser, Plasma and Radiation Physics in Bucharest, Romania. His research interests are in photonic materials, quantum electronics processes in doped laser materials, and applications of solid-state lasers.

CERAMIC LASERS

AKIO IKESUE

World-Lab. Co., Ltd, Japan

YAN LIN AUNG

World-Lab. Co., Ltd, Japan

VOICU LUPEI

*National Institute for Laser, Plasma and Radiation Physics,
Romania*

CAMBRIDGE
UNIVERSITY PRESS

CAMBRIDGE
UNIVERSITY PRESS

University Printing House, Cambridge CB2 8BS, United Kingdom

One Liberty Plaza, 20th Floor, New York, NY 10006, USA

477 Williamstown Road, Port Melbourne, VIC 3207, Australia

314-321, 3rd Floor, Plot 3, Splendor Forum, Jasola District Centre, New Delhi - 110025, India

103 Penang Road, #05-06/07, Visioncrest Commercial, Singapore 238467

Cambridge University Press is part of the University of Cambridge.

It furthers the University's mission by disseminating knowledge in the pursuit of education, learning and research at the highest international levels of excellence.

www.cambridge.org
Information on this title: www.cambridge.org/9780521114080

© A. Ikesue, Y. L. Aung, V. Lupei 2013

First published 2013

A catalogue record for this publication is available from the British Library

Library of Congress Cataloging in Publication data
Ikesue, Akio, 1958–
Ceramic laser materials / authors, Akio Ikesue, Yan Lin Aung, Voicu Lupei.
pages   cm
Includes index.
ISBN 978-0-521-11408-0 (hardback)
1. Laser materials.   2. Ceramic materials.   I. Title.
TA1677.I44   2013
621.36´6 – dc23   2012042974

ISBN 978-0-521-11408-0 Hardback

Contents

Preface

Research and development on various types of solid-state laser has been carried out since the creation of the first ruby laser by Dr. T. H. Maiman in 1960 [1]. Examples of laser technologies include continuous wave (CW) lasers using Nd:YAG (neodymium-doped yttrium aluminum garnet) single crystals at room temperature developed by Dr. Geusic in 1964 [2], and tunable and ultra-short pulse lasers using Ti:sapphire single crystals developed by P. F. Moulton in 1982 [3]. These lasers are still being applied in industrial applications.

On the other hand, it is said that people started to use porcelain (ceramic ware) as standard tableware in the days before Christ. In the USA, modernization of ceramic technology started in the 1950s. Ceramics have now become an essential part of technologies that support industries such as electronics and engineering etc., and their industrial use has expanded greatly from year to year.

The interaction of ceramics with optics originated from the success that was achieved by Dr. R. L. Coble of MIT (Massachusetts Institute of Technology) in the late 1950s [4]. He succeeded in fabricating alumina ceramics including fewer residual pores by controlling the microstructure. This material exhibited translucency (i.e., diffused transmission, not in-line transmission throughout the specimen), and was applied as a discharge tube in a high-pressure sodium vapor lamp, thus contributing to the advancement of lighting technology. Research and development of various types of translucent and transparent ceramics then occurred around the world, but the optical quality of the translucent ceramics was much inferior to that of commercial single crystal and glass materials. In addition, there were no practical applications of these ceramics at that time, and most of the translucent ceramics were only of interest for laboratory research.

The development of transparent ceramic materials for applications in laser oscillation is an ultimate aim of materials technology. It is also a dream of materials scientists in the field of ceramics to develop ceramic (polycrystalline) materials which can be used for laser oscillation with a performance equal to or better than that of conventional single crystal materials. The first two attempts were from the USA, where laser oscillation using Dy:CaF$_2$ ceramics under cryostatic conditions was reported in 1964 [5], and pulse laser oscillation using Nd:ThO$_2$–Y$_2$O$_3$ ceramics was reported in 1973 [6]. Both succeeded in producing laser oscillation using ceramic materials, but the laser performance was very poor even though the ceramic samples were fabricated using the most advanced technology at that time. Almost two decades passed until the next attempt was made in 1995 (by the author) to produce efficient laser oscillation using ceramic materials [7], selecting Nd:YAG, which is a technologically and industrially important material in the field of solid-state lasers. It was demonstrated that polycrystalline Nd:YAG ceramics can be used for CW laser generation with an efficiency comparable to that of commercial single crystal lasers. Microchip lasers [8] and single mode laser oscillation [9] were achieved using heavily doped Nd:YAG ceramic, and the excellent performance of ceramic lasers was demonstrated. The importance of ceramic lasers began to achieve recognition worldwide from 2000, and nowadays research and development of ceramic lasers is being carried out around the world. Moreover, the development of new optical materials in addition to Nd:YAG ceramics was initiated by applying the technology to make highly transparent ceramics.

The ceramic laser changed completely the common concept of the conventional solid-state laser. It was possible to fabricate polycrystalline ceramic laser gain media with extremely low scattering and optical homogeneity, which cannot be obtained with commercial single crystals. In addition, ceramic technology permits the fabrication of composite laser elements with extremely complicated structures to improve the laser performance [10], and the fabrication of fiber laser elements to generate high beam quality lasers. In this sense, ceramic lasers can provide a broad range of design flexibility for the enhancement of laser performance compared to traditional single crystal lasers. New laser materials, which are difficult to obtain using the conventional melt-growth technology, such as Sc_2O_3 and Y_2O_3, can be fabricated by advanced ceramic processing, and high laser performance from such sesquioxide ceramics has been confirmed [11]. It is predicted that ceramic lasers will play a major role in solid-state lasers in the future, and will have applications in many industrial fields.

It is interesting that the author successfully produced a highly efficient ceramic laser even though he was not an expert in the field of ceramics or lasers at that time, but was engaged in the research and development of refractory materials. The readers may question (1) why the author focused on the development of ceramic lasers and (2) how the author managed to achieve technological success. Regarding the first question, it is normally the dream of a researcher to make a breakthrough in the development of the most difficult materials theoretically and technologically, and this was the best opportunity for the author to realize a contribution to ceramic technology. Regarding the second question, to be honest, at that time the author did not actually know about the special (strict) requirements (demand characteristics) of laser materials, and that is why he could perform such a reckless exploration of laser generation using ceramic materials. Therefore, even an engineer who does not know the materials technology may well open the door to new technologies if he or she has a challenging spirit and can keep the dream alive as an engineer. "Challenging nothing for fear of failure might be the biggest risk for a scientist."

In this preface, the development of ceramic lasers has been described briefly. In this book, a wide range of topics is described, from the fabrication technologies of ceramic lasers, which are expected to be used in the development of future solid-state lasers, to their performance.

References

[1] T. H. Maiman, Stimulated optical radiation in ruby, *Nature (London)* **187** (1960) 493–494.
[2] J. E. Geusic, H. M. Marcos, and L. G. van Uitert, *Appl. Phys. Lett.* **4** (10) (1964) 182–184.
[3] P. F. Moulton, *Opt. News* **8** (6) (1982) 9.
[4] R. L. Coble, *Am. Ceram. Soc. Bull.* **38** (10) (1959) 501.
[5] S. E. Hatch, W. F. Parson, and R. J. Weagley, *Appl. Phys. Lett.* **5** (1964) 153.
[6] C. Greskovich and J. P. Chernoch, *J. Appl. Phys.* **44** (10) (1973) 4599–4605.
[7] A. Ikesue, T. Kinoshita, K. Kamata, and K. Yoshida, *J. Am. Ceram. Soc.* **78** (4) (1995) 1033–1040.
[8] I. Shoji, Y. Sato, T. Taira, A. Ikesue, and K. Yoshida, Highly Nd^{3+}-doped YAG ceramic for high power microchip, CGCT-1, 2000/Abstract Book, T-C-05 (2000), pp. 345–355.
[9] A. Ikesue, Yan Lin Aung, T. Taira, T. Kamimura, K. Yoshida, and G. L. Messing, Progress in ceramic lasers, *Mater. Res. Annu. Rev.* **36** (2006) 397–429.
[10] A. Ikesue and Yan Lin Aung, Synthesis and performance of advanced ceramic lasers, *J. Am. Ceram. Soc.* **89** (6) (2006) 1936–1944.
[11] A. Ikesue, K. Kamata, and K. Yoshida. Synthesis of transparent Nd-doped HfO_2–Y_2O_3 ceramics using HIP, *J. Am. Ceram. Soc.* **79** (2) (1996) 359–364.

Acknowledgement

This research and development effort was started when the author was working in a refractory manufacturing industry (Kurosaki Harima Corporation) as an engineer. Support from the members of Kurosaki Harima Co. for research up to the present date is greatly appreciated. The authors are very thankful to collaborators around the world for their supportive collaboration: Professor Y. Iwamoto (Nagoya Institute of Technology), Associate Professor T. Kamimura (Osaka Institute of Technology), Associate Professor T. Taira (Institute of Molecular Science) and Associate Professor T. Yamamoto (Tokyo University) from Japan, and Dr. A. Lupei (Romania) and Dr. Yvonne Menke (SCHOTT AG, Germany) from Europe, and Dr. R. Shori (UCLA), Dr. J. Sanghera (Naval Research Laboratory) and Dr. M. Dubinsky (Army Research Laboratory) from the USA. Also, the authors greatly appreciate support and useful advice from Dr. Joan Fuller, Dr. K. Jata and Dr. K. Goretta of AFOSR (Air Force Office of Scientific Research)/AOARD (Asian Office of Aerospace Research and Development) for the development of the "high power density laser."

1

Introduction

1.1 Research background

There are many scientific, military, medical and commercial laser applications which have been developed since the invention of the laser in the 1960s [1–11]. It is clear that the laser is essential for optical communication, nowadays part of the optoelectronics industry. This industry is growing continuously and rapidly in scale and is comparable to the automobile, electronics, and the semiconductor industries. Lasers can be classified generally into four groups depending upon the type of gain medium: (1) semiconductor lasers, (2) gas lasers, (3) liquid (dye) lasers, and (4) solid-state lasers. (1) Semiconductor lasers are used as optical sources for optical communication, and read/write systems in recording media (optical disks, magneto-optical disks etc.). (2) One gas laser is the He–Ne laser which is used as an optical source for measurements in basic research because of its low cost. Excimer lasers are used as optical sources for etching in CVD (chemical vapor deposition), and for lithography. CO_2 lasers have many industrial applications such as laser cutting and welding. (3) Dye lasers use an organic dye as the gain medium. The wide gain spectrum of available dyes allows these lasers to be highly tunable, or to produce very short duration pulses on the order of a few femtoseconds. (4) Typical solid-state lasers are the glass lasers applied in nuclear fusion reactors, and YAG lasers used in mechanical processing and medical applications etc.

The discoveries of the ruby laser by Maiman in 1960 [1, 12] and of the YAG laser by Guesic *et al.* in 1964 [2] led materials scientists to develop a variety of solid-state laser materials including crystalline and non-crystalline materials. For instance, garnet materials such as GGG ($Gd_3Ga_5O_{12}$), GSGG ($Gd_3Sc_2Ga_3O_{12}$) and YSGG ($Y_3Sc_2Ga_3O_{12}$) [13–20], fluoride materials such as CaF_2 and MgF_2 [21], alexandrite materials [22–24], synthetic forsterite materials, and YVO_4 [25–29], and amorphous materials such as silicate glasses containing Rb, K and B [30, 31] were successfully developed. The basic requirements of the physical properties of solid-state laser materials are a high thermal conductivity, high chemical stability, and excellent machining properties. In addition, the laser gain materials must have a large cross-section of stimulated emission (σ) and long fluorescence lifetime (τ). Moreover, the laser threshold of the materials must be as low as possible, and stable laser oscillation in CW (continuous wave) or pulsed mode must be obtainable with a high energy conversion efficiency. The material that satisfies these demand characteristics is YAG, which occupies the majority of the solid-state laser market. It is not likely that there will be new materials in the future that can exceed the performance of YAG materials in terms of physical properties and laser characteristics. However, almost all of the materials used for solid-state lasers, including YAG, are single crystals produced by the melt-growth process. Initially, single crystals were produced by the Verneuil process, also called flame fusion [2], but because of quality issues in terms of laser amplification efficiency and laser beam quality, crystals grown by the Cz (Czochralski) process are now being used in industrial applications [32]. However, it takes about one month to grow an undoped YAG single crystal with a relatively large diameter, and it takes longer to grow an Nd:YAG single crystal. In addition, the optically homogeneous portion of the single crystal ingot which is suitable for laser oscillation is very limited because of the intrinsic core and facets. Therefore, the productivity and output characteristics are still unsatisfied from an industrial point of view.

Therefore, it was considered that if it were possible to produce YAG laser materials using ceramic technology, the above technical issues of single crystal YAG materials could be solved, but a weak point was that the grain boundaries in ceramics can disturb the efficient amplification of the coherence beam in the materials. This weak point was reflected in the results achieved from $Dy:CaF_2$ ceramics in 1964 and from $Nd:ThO_2-Y_2O_3$ ceramics in 1974. Although laser oscillation was demonstrated using these ceramic materials, the resulting beam quality and lasing efficiency were not sufficient for the development of ceramic materials as laser gain media in the future. However, this negative opinion changed abruptly with the successful demonstration of a highly efficient polycrystalline Nd:YAG ceramic laser at room temperature by the author in 1995. The author demonstrated that ceramic materials can show laser performance equivalent to that demonstrated for single crystal laser gain media and, moreover, ceramic lasers have potential which cannot be realized with conventional single crystal lasers. Nowadays, the optical quality of polycrystalline ceramic laser gain media is comparable to that of the high quality commercial single crystal counterparts, and further technological progress is expected in the future.

1.2 Technical problems of melt-growth single crystals

As described above, single crystal YAG materials are mainly used in solid-state lasers and almost all of the YAG laser gain media are produced by the Cz method (melt-growth process). Transition metal elements (Cr, Ti etc.) or lanthanide rare earth elements (Nd, Tm, Ho etc.) are doped as laser active ions in the YAG host material. Among these laser active ions, Nd is a four-level laser system, with narrow spectral width and high quantum efficiency. Therefore, Nd-doped YAG has become the most important laser material.

Figure 1.1 shows the principle of laser action in the Nd^{3+} ion. The Nd^{3+} ion has three fluorescent emissions: (1) 0.9 μm ($^4F_{3/2} \rightarrow {}^4I_{9/2}$), (2) 1.06 μm ($^4F_{3/2} \rightarrow {}^4I_{11/2}$) and (3) 1.3 μm ($^4F_{3/2} \rightarrow {}^4I_{13/2}$), and the radiative transition possibilities of each state are 0.25, 0.60 and 0.15, respectively [33, 34]. Emission (2) has the highest energy efficiency, and

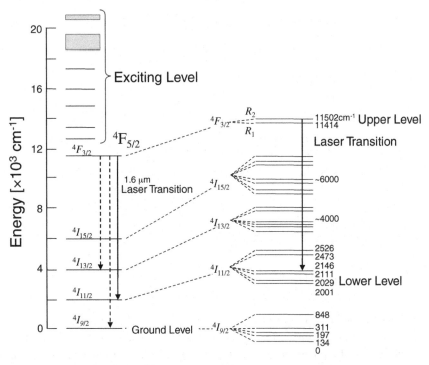

Figure 1.1 Energy diagram for Nd^{3+} ions in YAG crystal.

normally Nd:YAG crystals are used for 1.06 μm generation. To obtain laser action in
Nd:YAG materials, first the electrons of Nd ions in the ground state are excited with a
xenon or krypton lamp (white light source from a discharge flashlamp) so that the electrons
are pumped up to energy levels higher than $^4F_{3/2}$. More recently, a semiconductor laser (LD,
laser diode) with a certain wavelength has been used to pump the ground state electrons
directly to the $^4F_{5/2}$ band. Accordingly, these excited electrons which are in the upper levels
decay rapidly to the $^4F_{3/2}$ level (non-radiative transition), and then when these electrons are
transferred to $^4F_{13/2}$, $^4F_{11/2}$ and $^4F_{9/2}$, three fluorescent emissions (radiative transitions) ($\lambda =$
0.9, 1.06 and 1.3 μm) occur. Then the fluorescent emission is amplified in a set of mirrors
(the resonator), and finally an intensified light with a single wavelength (monochromatic
light) is emitted. In the Nd system, when electron transitions occur from the upper levels
to the $^4F_{3/2}$ level on exciting with a lamp source, the transitions are non-radiative and the
energy is released as heat. Therefore, effective laser oscillation cannot be achieved using
a lamp excitation system. (However, a high output power can be achieved using a lamp
excitation system because the lamp can provide very high input power to the laser gain
medium.) In the LD excitation system, laser diodes of wavelengths 808 nm or 885 nm
are used which can pump the ground state electrons directly to upper levels, $^4F_{5/2}$ or $^4H_{3/2}$
levels. Therefore, there are no excited electrons which undergo non-radiative transitions as
seen in the lamp excitation system, and efficient laser oscillation can be achieved. However,

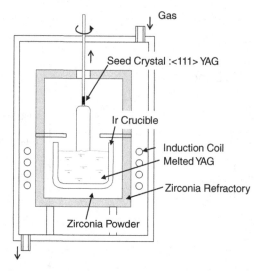

Figure 1.2 Schematic diagram of a YAG single crystal grown by the Czochralski method.

it was difficult to obtain high power LDs, and low power models were also very expensive. Therefore, the lamp excitation system was common for solid-state lasers. In the twenty-first century, the price of high performance LDs is becoming more reasonable, and recently a highly effective compact and high power (kilowatt range) Nd:YAG laser system with LD pumping has been commercialized. But, a technical issue of the LD excitation source is that the wavelength of the LD is shifted when its temperature is increased during laser operation. It is necessary to control the temperature of the main body of the LD and the spectral width of the emission band of the LD.

As described in Section 1.1, the Cz method is commonly used to produce YAG single crystals (see Figure 1.2). High purity raw materials (over 4N, 99.99 mass%) of Y_2O_3, Al_2O_3 and Nd_2O_3 are used. They are weighed in the YAG stoichiometric ratio, and blended. Then the blended powders are pressed into powder compacts and sintered. This relatively dense Nd:YAG sintered body is put into an Ir (iridium) crucible, and melted by high frequency induction heating at over 1950 °C. Then a YAG seed crystal is placed on top of the YAG melt, and the growing crystal is continuously pulled very slowly at a rate of 0.2 mm h^{-1} with a rotation speed of 10–30 rpm. Normally, a YAG seed crystal with an orientation of ⟨111⟩ is used because it has the largest surface energy, but in some cases, ⟨110⟩ and ⟨100⟩ seeds are also used. When Nd ions are doped into a YAG crystal, Nd replaces Y sites in the garnet structure. However, the ionic radius of the Nd ion is much larger than that of the Y ion, and it is well known that Nd hardly dissolves in YAG crystal. The segregation coefficient of the Nd ion in YAG crystal (the ratio of the Nd concentration in the crystal to the Nd concentration in the melt) is very small, about 0.2 [35]. Therefore, a highly doped Nd:YAG sintered body is prepared in advance to produce a YAG melt with 2–3 times higher Nd concentration than the target Nd concentration. (For instance, suppose the target composition is 1 at.%

Nd:YAG, then sintered bodies of 2 or 3 at.% Nd:YAG are prepared in advance, and they are melted and grown to obtain a crystal of 1 at.% Nd:YAG.) Although, high concentration Nd:YAG melts are prepared in advance, it is not certain that the grown crystal will have a high Nd concentration with homogeneous doping. When the Nd concentration in YAG is higher than 1 at.%, many inclusions (light scattering sources) are generated in the crystal, and it cannot be used as a laser crystal. Therefore, the concentration of Nd ions is limited to 1 at.% in YAG single crystals for laser applications. Even a 1 at.% Nd:YAG crystal ingot includes optically inhomogeneous parts such as the core and (211) facets from the pulling axis towards the rim area [35]. Accordingly, only a part (the optically homogeneous part) of the grown crystal can be used as a laser gain medium. Another technical issue with melt-growth crystals is that the Nd concentration varies slightly in the longitudinal direction of the ingot. This is because when the crystal is pulled, the concentration of Nd ions in the grown crystal is significantly lower than that in the melt. Thus, Nd ions are concentrated in the melt as the crystal growth progresses. Therefore, the Nd concentration in the grown crystal at the early stage of crystal growth (near the seed crystal) is slightly lower compared to the middle stage and terminal stage (the end of the ingot) along the longitudinal direction of the crystal ingot. As a result, this process produces a crystal rod which has a change in refractive index at each end. This is the technical limitation of the conventional melt-growth method.

A polarized image of approximately 1 at.% Nd:YAG single crystal (ingot), a Schlieren image and X-ray tomography image are shown in Figure 1.3(a), (b) and (c), respectively [35]. All of the images (especially the cross-sectional images) show crystal defects (such as core and facets) in the crystal ingot. A commercial Nd:YAG slab and polarized images (cross-nicol) are shown in Figure 1.3(d) and (e, f), respectively. The commercial Nd:YAG single crystal has very high transparency, and it looks optically very homogeneous to the naked eye (under natural light). However, when it is observed under a polarizer, lamella shaped facets can be seen across the length direction of the slab, suggesting that even the commercial high quality single crystal materials include such optically inhomogeneous components.

Any optical defects remaining in a laser gain medium can significantly reduce the light amplification efficiency and beam quality. Therefore, the demand characteristics for laser materials are necessarily very strict. Regarding quality control in commercial single crystal production, the thermal stress and refractive index distribution of the grown Nd:YAG ingot are inspected using a polarizer and interferometry. Optically homogeneous parts of the ingot are cut off and polished, and then the numbers of optical scattering centers in the samples are detected using a laser system. Finally, the most optically homogeneous part is selected as a high quality laser gain medium.

The optical loss of commercial high quality Nd:YAG single crystal is in the region of 0.2% cm^{-1}, and the quality has improved significantly compared with the crystal grown by the Bernoulli process in 1964. In fact, the melt-growth single crystal technology has reached its limits, and it is extremely difficult to improve further the quality of current single crystal materials.

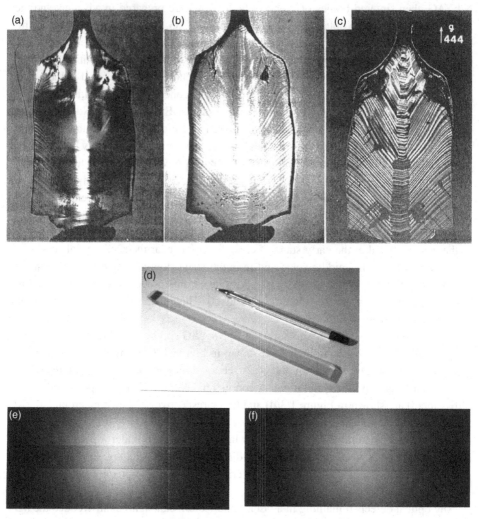

Figure 1.3 Cross-section of an Nd:YAG single crystal imaged by (a) polarizing light, (b) the Schlieren method (interferometry image), and (c) X-ray tomography. (d) Appearance of a commercial single crystal Nd:YAG slab, and (e, f) polarized images (open and cross-nicol) under a polarizer.

To grow a YAG crystal, a very expensive Ir crucible and growth unit is necessary. What is worse is that the growth rate is very slow. Normally it takes about one month to grow an ingot. Other economic issues are the very high initial costs of the fabrication equipment and running costs for a long delivery time (electricity, and recovery costs for the Ir crucible etc.). In addition, the yield percentage of high quality laser crystal is very low. There are also technological issues. Only a small cross-section rod can be achieved. The diameter of the ingot is limited to 5–6 inches, and a laser rod with approximately 10 mm diameter by 150 mm length or a laser slab of approximately $6 \times 20 \times 150$ mm^3 can be removed from the ingot. Therefore, large gain media for the future development of high power lasers cannot be produced by the current melt-growth crystal technology.

As described above, generally YAG has many advantages compared to other laser materials. However, the most important part in the laser system is the Nd:YAG material, and there are still many unsolved economic issues (including yield percentage) and technological issues. It is technologically almost impossible to achieve a breakthrough in the current single crystal growth process, and even a partial solution to the problems appears far away.

1.3 Technical problems of ceramics

Regarding translucent ceramics, Dr. R. L. Coble first developed translucent alumina ceramics in 1959 [4, 36], and then this material was used as a discharge tube in a high pressure sodium vapor lamp. Polycrystalline ceramics were thought to be opaque until Coble demonstrated experimentally that ceramics can transmit light when the bulk density is increased to very near the theoretical density by reducing the number of residual pores during the sintering process. Later, by controlling the purity, particle size and homogeneity of the starting raw materials, and by improving the traditional sintering process according to sintering theory, it was possible to manufacture high purity and high density ceramics with controlled microstructure. With a history of over 20 years, translucent and transparent ceramics composed of simple oxides such as MgO, Y_2O_3 and ZnO and multiple oxides such as ZrO_2–Y_2O_3 and spinel (MgO–Al_2O_3) and recently PLZT ((PbLa)(ZrTi)O_3), SBN ((SrBa)Nb_2O_6) using the electro-optical effect, Gd_2O_2S:Pr and (YGd)O_3:Eu for X-ray scintillators and highly thermally conductive ceramics (AlN) have been developed [37–42], and some of them have been commercialized.

However, up to now translucent ceramics have been developed for applications requiring only thin shapes, and have not been produced for high optical quality applications. Therefore, there was no published account of an optical ceramic with high optical quality until the transparent polycrystalline Nd:YAG ceramics developed by the author in 1995. In fact, the optical quality of conventional high quality translucent ceramics, which include many grain boundaries, was significantly inferior to that of commercial single crystals. For that reason, most of the published papers discussed mainly the optical quality and applications of thin transparent ceramics. For instance, in the case of alumina, which belongs to the hexagonal crystal system, the transmittance of 1 mm thick normal sintered ceramic is considerably lower than that of hot-pressed translucent alumina ceramic or single crystal sapphire (see Figure 1.4) [43]. Since the thickness of all materials was set at 1 mm in this experiment, when the transmittance is plotted with the thickness of each sample in accordance with the Lambert–Beer law, it can be seen that the slopes of the lines for single crystal and polycrystalline ceramics are considerably different, i.e., the scattering characteristics of the materials are different. The slope of the lines gives the optical attenuation coefficient (optical loss). This value was significantly larger even in the cases of transparent MgO or spinel ceramics which belong to the cubic crystal system. For these reasons, translucent

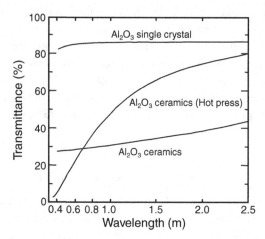

Figure 1.4 Relationship between measured wavelength and in-line transmittance for sapphire single crystal, hot-pressed Al₂O₃ ceramic, and Al₂O₃ ceramic produced by normal pressure sintering.

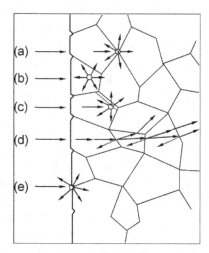

Figure 1.5 Illustration of optical scattering in polycrystalline ceramics due to various microstructural defects: (a) grain boundary, (b) residual pore, (c) secondary phase, (d) birefringence, and (e) surface roughness.

ceramics were mainly used for applications requiring a thin slice shape in the technology of the twentieth century.

Figure 1.5 illustrates the microstructure and scattering sources of normal translucent ceramics [44]. The ceramic is composed of many fine grains (crystallites) with randomized crystal orientations. Scattering sources in normal ceramics are (a) grain boundaries, (b) residual pores, (c) inclusions and secondary phases, (d) birefringence and double reflections, and (e) surface roughness. However, the fundamental differences in microstructure between single crystal materials and ceramic materials which affect the scattering

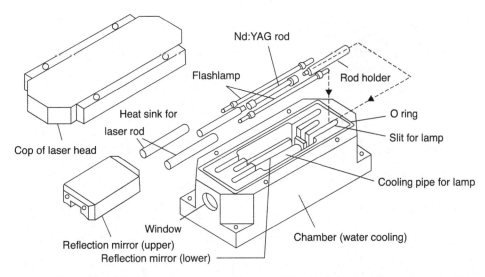

Figure 1.6 Schematic diagram of a commercial laser oscillator using a single crystal gain medium and flashlamp excitation.

characteristics are (a) the present of grain boundaries, and (b) the quantity by volume of residual pores. Accordingly, it is considered that these scattering sources significantly influence the optical quality of ceramic materials.

On the other hand, it is true that laser crystals with very low scattering loss can have a very high laser amplification efficiency. In fact, ultra low loss single crystals still cannot be produced using the current crystal technology. The optical loss of high quality commercial single crystals is around 0.3–0.2% cm^{-1} (some latest data show less than 0.2% cm^{-1}), and this quality can offer relatively efficient laser generation from commercial single crystals. A schematic diagram of a commercial laser oscillator using a single crystal gain medium and flashlamp excitation is shown in Figure 1.6 [45]. External energy (from a flashlamp or semiconductor laser) is input into the laser gain medium (doped with fluorescent ions) in order to excite the electrons continuously from the ground state to laser levels. Then, the emitted fluorescent beams are reflected and amplified between two mirrors. The laser beam quality and amplification efficiency are severely affected by the numbers of scattering sources in the laser gain medium.

Previously reported papers on translucent ceramics have discussed few details of the optical characteristics (optical constant), and there have been almost no reported papers which have described the scattering loss, condition of double refraction and homogeneity of the refractive index in ceramic materials. It can be considered that the scattering loss of the translucent or transparent ceramics developed in the past was very high, and they could not meet the requirements of laser materials.

However, one researcher, Dr. Greskovich of GE Co., developed a transparent Nd-doped 10%ThO$_2$–Y$_2$O$_3$ ceramic, and succeeded in producing laser oscillation at room

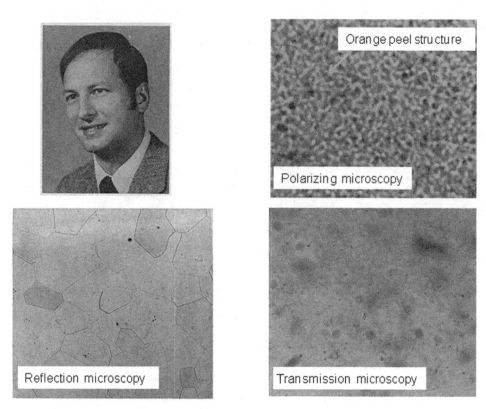

Figure 1.7 Picture of Dr. C. D. Greskovich and microstructure of Nd:ThO_2–Y_2O_3 ceramic.

temperature for the first time using a polycrystalline ceramic gain medium in 1973. (About 10 years before this, successful laser oscillation was achieved using Dy:CaF_2 ceramic under cryostatic conditions, but the details of the physical properties of the materials were not described.) Dr. Greskovich succeeded in fabricating transparent Nd:ThO_2–Y_2O_3 ceramic by sintering low sinterability Y_2O_3 powder, with 10% ThO_2 as a sintering aid, at 2200 °C for about 100 hours.

Figure 1.7 shows reflection microscopy and transmission microscopy images of the 1% Nd:ThO_2–Y_2O_3 ceramic together with a picture of Dr. Greskovich [6, 46, 47]. Although only a few residual pores are observed on the surface in the reflection micrograph, many internal defects (such as voids and grain boundary phases) can be observed in the transmission micrograph. An optically inhomogeneous structure, called an "orange peel" structure, can be seen with the naked eye. Although it was reported that the number of residual pores in the Nd:ThO_2–Y_2O_3 ceramic reached 10^{-5}–10^{-7} (10–0.1 ppm) in 1973, the optical loss of the material was lower than several percent per centimeter. Because of this optical quality, pulse laser oscillation could be produced by flashlamp excitation at room temperature, but the slope efficiency was as low as 0.1%. This technology was the best achieved in the

1970s, and the success of Dr. Greskovich encouraged materials scientists. However, there was no challenger who could compete with this achievement for about 20 years.

Compared to normal optics for active and passive applications, the optical quality of the single crystals required for laser application is significantly higher because light amplification occurs in the gain medium, and the amplified light is passed repeatedly through the gain medium. Therefore, even a small amount of optical loss of the material may have a significant effect on laser oscillation. In the past, various types of translucent or transparent ceramic materials have been developed. But it can be assumed that the scattering losses of those materials were very large compared to the materials reported by Dr. Greskovich. It was difficult to produce single crystal material with optical quality sufficient for laser applications, and it was far too difficult to produce conventional translucent ceramics, which include many crystal defects such as grain boundaries, residual pores and inclusions etc., which could meet the requirements for laser applications. For that reason, conventionally it was doubted that ceramic materials could be used as laser gain media in the future of solid-state lasers. If a technology existed which could remove microstructural defects perfectly (except grain boundaries) from ceramics, perhaps ceramic materials could be used as optical materials in laser applications. This problem was not solved until the author demonstrated high efficiency laser oscillation from a polycrystalline ceramic material. Details of this development are described in this book.

1.4 Purpose of this research

Normally the output power of a solid-state laser is comparable to that of a gas laser (e.g., CO_2 laser), and solid-state laser oscillators are very compact compared to gas lasers. There are many disadvantages to using gas lasers. For instance, no waveguide system is available in CO_2 gas lasers. However, in the case of Nd:YAG lasers, the oscillation wavelength is about 1 μm, and they can be used for optical fiber transmission systems. In addition, a high output power can be achieved from a compact solid-state laser system.

The most important material for solid-state lasers is Nd:YAG, and the optical quality of the Nd:YAG single crystal must be as high as possible, as mentioned before. In the case of ceramic processing, powders (the powder composition is very close to or the same as the target composition) prepared by a wet process such as the alkoxide method, co-precipitation method, or homogeneous precipitation method etc. have generally been used to obtain high performance ceramic materials. Using such wet processed powders, there are many reports of advanced ceramics with very high homogeneity and density or with transparent quality. However, the fabrication process is very complicated, and even the ceramics prepared from wet processed powders did not have optical quality comparable to that of the single crystal counterparts. The wet process was invented to overcome the problems generally seen in the solid-state reactive sintering process, for example, the formation of microstructural defects due to inhomogeneous blending of the powders, remaining unreacted phases, compositional variation among grains, and lack of uniformity of the grains. In the wet process, homogenization of the raw powder materials is performed at the formation of the host composition (e.g., garnet) step and calcination, in advance of the sintering process. In principle, therefore, it is very effective for improving homogeneity during the sintering process.

However, from another technical point of view, it can be considered that the wet process was invented to avoid the above described problems of the dry process because the properties of raw powder materials for the conventional dry process (solid-state reaction method) were not good enough for the production of optical quality ceramics. Whether this speculation is true or not hinges upon the fact that high optical quality ceramics can be produced by the dry process, using good quality raw powder materials. In this research, therefore, we focused on the solid-state reaction method without using synthesized raw powders prepared by the wet process. If the dry process can overcome the above mentioned general technical issues, it will become a very important industrial process for the economic production of optical ceramics because the powder preparation process is very simple and the preparation of powders with various compositions is very easy. Accordingly, the dry process was selected for this study. The main objectives of this study are as follows:

(1) to produce a transparent polycrystalline Nd:YAG ceramic of optical quality by the solid-state reaction method;
(2) to evaluate the optical quality of the ceramics produced, especially the laser oscillation characteristics;
(3) to clarify the effect of the microstructure of the ceramics on the laser performance;
(4) to explore new functionalities unique to ceramics (technologies that cannot be obtained using conventional single crystals);
(5) to figure out the technical problems involved in the fabrication of laser ceramics and to provide a means for solving the problems.

1.5 Outline of the book

This study is based on the above five objectives, and comprises eleven chapters. Outlines of the chapters are given here.

Chapter 2 introduces the main principles of solid-state lasers. After a brief description of the interaction of electromagnetic radiation with quantum systems and of laser emission, the three major components of the laser, the active laser material, the pumping system and the laser resonator, are presented, with emphasis on their roles and characteristics. The flow of excitation inside the pumped laser material is examined and it is shown that the competition of laser emission with other de-excitation processes, such as luminescence and non-radiative processes that generate heat, is determined by specific characteristics of these three parts of the laser. It is thus determined that the development of solid-state lasers, including optimization of the efficiency, use of new wavelength ranges, diversification of the temporal regime of emission, and scaling to very high power or energy, requires specific qualities of the laser material. From examination of the actual state of the art it is concluded that the traditional laser materials, based on doped crystals or glasses, cannot support the necessary development of lasers and new solutions must be looked for. Polycrystalline materials produced by ceramics techniques could be very promising in this development. The ceramic laser materials combine some of the advantages of glasses (large size, high optical uniformity and so on) with the qualities of crystals, adding new features such as enhanced compositional versatility, enhanced doping concentrations, tight control of the doping profile, possibility of producing composite materials and so on. Various classes of ceramic materials are described and the main directions of research are outlined.

In Chapter 3, the sintering behavior of raw powder materials, Y_2O_3 and Al_2O_3 powders prepared for this study and corresponding commercial powders used as starting materials for the synthesis of transparent YAG ceramics, are compared. In addition, YAG sintered bodies were prepared using these powder sources and their sintering behaviors were also investigated. Based on these results, the reasons why transparent ceramics were not produced successfully by the conventional dry process are investigated. Because a normal sintered body includes many residual pores which cause severe internal scattering, it is necessary to reduce the amount of residual pores to an extremely low level. To produce pore-free ceramics, it is necessary to use raw powder materials with good sinterability in the dry process. One more important point is the processing technology, especially the process of making the powder compact. This chapter describes a powder granulation method and the properties of the granulated powders for the fabrication of laser ceramics. A method for determining the packing condition of the granulated powders in the powder compacts and the packing condition in relation to the compacting pressure is described.

The HIP process has become important for the production of high optical quality YAG ceramics and large scale laser gain media. In this study, a HIP machine was used to reduce the porosity, one of the optical scattering factors in ceramics. Argon gas was used as a pressure medium. The applied pressure ranged from 9.8 to 196 MPa, and the treatment temperature ranged from 1500 to 1700 °C. The results achieved by a capsule-free HIP treatment are described. Argon gas, the pressure medium, was introduced into the Nd:YAG ceramics via grain boundaries, and the formation of pores due to argon gas was observed in HIP samples. As a result, the number of residual pores increased after HIP treatment. The disadvantages of the HIP process were determined qualitatively and quantitatively. Approaches to solving the problem of pore formation in the sintered body and a fabrication technology for optical ceramics with extremely low scattering are described.

Chapter 4 describes sesquioxides such as Sc_2O_3, Y_2O_3, and Lu_2O_3 etc. which are promising new materials for laser gain media. However, their melting points are higher than 2400 °C, and a phase transition point exists just below the melting point. For these reasons, it is very difficult to produce sesquioxide laser crystals by the conventional melt-growth process. In the ceramic process, it is possible to produce transparent sesquioxide materials because sintering occurs at temperatures very much below the melting point and phase transition point. In addition, ceramic technology allows the fabrication of large scale media, composite gain media, and advanced gain media with controlled absorption and emission spectral linewidth. Therefore, it is possible to create new functionalities such as wavelength-tunable lasers and short pulse lasers. These materials are unique to the ceramic technology, and there are many new functionalities to be discovered.

Chapter 5 describes the synthesis of Nd:YAG ceramics heavily doped with 0.3–4.8 at.% Nd ions, their optical characteristics and their laser performance. Generally, only about 1 at.% Nd ions can be dissolved homogeneously in YAG single crystal grown by the Cz method. If the doping concentration of Nd ions exceeds 1 at.%, segregation of Nd occurs, the optical quality decreases, and the material is not suitable for use as a laser gain medium. In the case of polycrystalline YAG ceramics, it was confirmed that a greater amount of Nd ions dissolved homogeneously compared to the melt-growth Nd:YAG single crystals. From the spectroscopic point of view, heavily doped polycrystalline Nd:YAG ceramic (more than 1 at.% Nd ions) has a new functionality as an advanced laser material. One of its applications is as a microchip laser and its laser performance (slope efficiency) was better than that of commercial Nd:YAG single crystals. In addition, single longitudinal mode laser oscillation was realized without using an etalon, mode selector, from a microchip ceramic laser.

In this chapter, the effect of impurities (Si added as a sintering aid) on the dissolution of Nd ions in YAG ceramics is described, and we discuss why a large amount of Nd ions can be dissolved homogeneously in YAG ceramics compared with single crystal YAG. The main difference between single crystals and ceramics in terms of microstructure is the presence or absence of grain boundaries, but a small amount of Si is also added in Nd:YAG ceramics. It was identified for the first time that the addition of a small amount of Si leads to more homogeneous dissolution of Nd ions in the YAG ceramics. The traditional theory says that Nd can hardly replace Y in the YAG lattice because of the large difference in ionic radius of the atoms. In this study, it was assumed that the lattice distortion, occurring when Nd^{3+} replaces Y^{3+} in the YAG lattice, was compensated for by the addition of Si, and as a result, a large amount of Nd could be doped homogeneously into the YAG lattice. In an actual experiment, it was confirmed that 7.2 at.% Nd was doped homogeneously into YAG ceramic using Si as a sintering additive.

In Chapter 6 we investigate the sources of optical scattering in polycrystalline ceramics, and summarize the results of fabrication techniques for very reliable high quality Nd:YAG ceramics. Single crystals have been utilized traditionally as the laser gain medium because they do not have the problem of grain boundary scattering. This advantage has led to the use of single crystals for optical applications. In this chapter, we focus on the pore volume of Nd:YAG ceramics and the amount of grain boundaries, and evaluate the amount of optical scattering quantitatively. In addition, the correlation between laser characteristics and optical scattering was investigated. It was verified that the number of residual pores and grain boundary phases greatly affects the laser oscillation characteristics. It was also realized that the laser oscillation characteristics do not depend upon the number of grain boundaries. Recent data demonstrated that the scattering coefficient of high quality laser ceramics at the laser oscillation wavelength regions is lower than that of commercial laser single crystals, and a breakthrough occurred in the optical performance of ceramics. The laser performance of ceramics with extremely low scattering has reached very close to the quantum oscillation efficiency, and ceramic materials have almost reached the ideal state for application in optics. In addition, using laser tomography, the scattering conditions of ceramics and commercial laser single crystals were investigated. Results of laser damage tests, which can be used in the future development of high power lasers, are also described.

In Chapter 7, the design flexibility of the ceramic technology is demonstrated with examples of different shapes of laser gain media such as composites and fibers. The laser performances are also described. Recently, a new type of composite consisting of two or more crystals having the same crystal structure but different compositions has been developed using a diffusion bonding technology in the USA. However, it has been pointed out that the composite is very weak and shows a lack of long-term durability attributable to the thermomechanical problems of the bonded interface of the composite. In this chapter, the problems of conventional bonding technology are described, along with approaches to solving those problems with the help of ceramic technology. Moreover, recently the author has succeeded in fabricating single crystals using a solid-state sintering process. Using this process, with a seed crystal, it is possible to fabricate bulky crystals as well as spherical single crystals several tens of micrometers in diameter. This technique is expected to find application in quantum devices using microsphere resonators. This chapter describes the composite formation technology which is essential for the creation of new functionalities in laser technology, and the fabrication of single crystals by sintering which can produce laser materials with ideal optical performance.

In Chapter 8, the current status of ceramic lasers is summarized. Nowadays this area is of great technological value, and research and development of ceramic lasers is very active not only in Japan but also in Asia, especially China, America and Europe, and it is expected that ceramics will become major materials for solid-state lasers in the future. The development of the $Dy:CaF_2$ ceramic in 1964 and of the $Nd:Y_2O_3$–ThO_2 ceramic in the USA led the way to the development of polycrystalline ceramic materials for application as solid-state lasers. However, the laser oscillation was a pulse laser operation produced under cryogenic conditions which was technologically and theoretically demanding. The oscillation efficiencies were extremely low, clouding the future of ceramic gain media for solid-state lasers. Since the successful development of high performance Nd:YAG ceramic lasers, the potential functionalities of ceramic laser materials, which cannot be produced using conventional melt-growth technology, have been realized successfully, and nowadays ceramic laser materials form a technological paradigm for solid-state lasers. Japan has led the way in ceramic laser technology from 1995 until today, but successful laser oscillations using ceramic gain media have also been reported from the USA, Europe and China. Thanks to the successful development of ceramic laser gain media, the development of high performance optical ceramics has become popular, and new types of transparent ceramic materials have been reported: for example, high refractive index materials such as zirconia and BNZT for ceramic lenses, Pr- or Ce-doped LuAG materials for radiation (X-rays or gamma-rays) detecting scintillators, and spinel ceramic materials for ultraviolet, infrared and visible light transmission windows. Nowadays, the development of high level optical ceramics is taking place worldwide because of their high marketability. This chapter explains the development of optical ceramics on a world scale, including ceramic lasers.

In Chapter 9, the technology and science described in Chapters 1–7 are brought together. Moreover, the role and the future of ceramics in laser technology are described. The fundamental idea of the ceramic laser is based on the fabrication of an ideal ceramic with defect-free microstructure. This basic idea will be applicable in other technological fields (e.g., electronic materials and structural materials etc.). Application technologies developed by the author, and their possible development in the future, are also described in this chapter.

Chapter 10 describes the main results in the spectroscopic investigation of doped transparent ceramics. It is shown that spectroscopic investigation has a double role, as a structural method for describing the variety, nature and structure of the various structural centers of the doping ions and the distribution of the doping ions in these centers, and as a method of characterizing the optical processes that determine laser emission. Thus, spectroscopic investigation provides the necessary information on the energy level structure and transition probabilities to enable mathematical modeling of the laser process and selection of suitable parameters for the laser material, pumping system and resonator design. Data on a large variety of spectroscopic studies on doped ceramics are presented and their implication in the structural properties of the ceramic and in modeling of the laser processes is discussed.

Chapter 11 describes the main factors that influence laser emission and heat generation in doped ceramics. It is shown that a proper exploitation of the properties of laser ceramics requires optimization of the pumping scheme for optimum utilization of the assorted excitations for laser emission and for a reduction of heat generation. The structure–properties–functionality relation for ceramic material is discussed and modalities to upgrade traditional laser materials or to tailor new materials are discussed, together with new schemes for power and energy scaling.

References

[1] T. H. Maiman, Stimulated optical radiation in ruby, *Nature (London)* **187** (1960) 493–494.
[2] J. E. Geusic, H. M. Marcos, and L. G. van Uitert, *Appl. Phys. Lett.* **4** (10) (1964) 182–184.
[3] P. F. Moulton, *Opt. News* **8** (6) (1982) 9.
[4] R. L. Coble, *Am. Ceram. Soc. Bull.* **38** (10) (1959) 501.
[5] S. E. Hatch, W. F. Parson, and R. J. Weagley, *Appl. Phys. Lett.* **5** (1964) 153.
[6] C. Greskovich and J. P. Chernoch, *J. Appl. Phys.* **44** (10) (1973) 4599–4605.
[7] A. Ikesue, T. Kinoshita, K. Kamata, and K. Yoshida, *J. Am. Ceram. Soc.* **78** (4) (1995) 1033–1040.
[8] I. Shoji, Y. Sato, T. Taira, A. Ikesue, and K. Yoshida, Highly Nd^{3+}-doped YAG ceramic for high power microchip, CGCT-1, 2000/Abstract Book, T-C-05 (2000), pp. 345–355.
[9] A. Ikesue, Yan Lin Aung, T. Taira, T. Kamimura, K. Yoshida, and G. L. Messing, Progress in ceramic lasers, *Mater. Res. Annu. Rev.* **36** (2006) 397–429.
[10] A. Ikesue and Yan Lin Aung, Synthesis and performance of advanced ceramic lasers, *J. Am. Ceram. Soc.* **89** (6) (2006) 1936–1944.
[11] A. Ikesue, K. Kamata, and K. Yoshida. Synthesis of transparent Nd-doped HfO_2–Y_2O_3 ceramics using HIP, *J. Am. Ceram. Soc.* **79** (2) (1996) 359–364.
[12] T. H. Maiman, *Phys. Rev. Lett.* **4** (11) (1960) 564–566.
[13] K. Maeda and Y. Fujii, *Oyo Butsuri (Bull. Appl. Phys. Jpn.)* **56** (1987) 1651.
[14] B. Struve and G. Huber, *J. Appl. Phys.* **57** (1988) 45.

[15] Y. Fujii, *Ceramics*, **21** (1) (1986) 26–31.
[16] T. Yokoyama, *Ceramics*, **23** (5) (1988) 461–463.
[17] G. Huber, E. W. Duczynski, and K. Petermann, *IEEE J. Quantum Electron.* **24** (6) (1988) 920–923.
[18] P. F. Moulton, J. G. Manni, and G. A. Rines, *IEEE J. Quantum Electron.* **24** (6) (1988) 960–973.
[19] J. A. Caird, M. D. Shinn, T. A. Kircoff, L. K. Smith, and R. E. Wilder, *Appl. Opt.* **25** (23) (1986) 4305–4320.
[20] M. Yamaga, *Oyo Butsuri (Bull. Appl. Phys. Jpn.)* **58** (2) (1988) 225–236.
[21] D. Welford and P. F. Moulton, *Opt. Lett.* **13** (1988) 975.
[22] K. Moriya, K. Yamagishi, and Y. Anzai, *CLEO'86 Technical Digest* (1986), paper THK19.
[23] J. C. Walling, O. G. Peterson, H. P. Jessen, R. C. Morris, and E. W. O'Dell, *IEEE J. Quantum Electron.* **16** (1980) 1302.
[24] R. Scheps, B. M. Gately, J. F. Myers, J. S. Krasinski, and D. F. Meeler, *Appl. Phys. Lett.* **56** (1990) 2288.
[25] T. J. Carring and C. R. Pollack, *OSA Proc. Advanced Solid State Lasers* (1991), vol. 10, p. 72.
[26] S. Amano, *Rev. Laser Eng.* **20** (9) (1992) 723–727.
[27] A. Sugioka et al., *Laser Sci. Prog. Rep. IPCR* **12** (1990) 124–126.
[28] P. F. Moulton, *J. Opt. Soc. Am.* **B3** (1) (1986) 125–132.
[29] H. Tsuiki, T. Masumoto, K. Kitazawa, and K. Fueki, *J. Appl. Phys.* **21** (7) (1982) 1017–1021.
[30] M. Leding, E. Heumann, D. Ehrt, and W. Seeber, *Opt. Quantum Electron.* **22** (1990) S107–S122.
[31] E. Snitzer, *Oyo Butsuri (Bull. Appl. Phys. Jpn.)* **34** (3) (1965) 164–175.
[32] M. Umino and M. Abe, *Optronics Mag.* **40** (1985) 93–97.
[33] L. G. DeShazer and L. G. Komai, *J. Am. Opt. Soc.* **55** (1965) 940.
[34] T. Kushida, H. M. Marcos, and J. E. Geusic, *Phys. Rev.* **167** (1967) 289.
[35] G. Adachi, K. Shibayama, and T. Minami, *New Technology for Advanced Materials*, Kagaku-Dojin, Kyoto (1987), p. 1830.
[36] R. L. Coble, U.S. Patent 3026210 (1962).
[37] G. D. Miles, R. A. J. Sambell, J. Rutherford, and G. W. Stephenson, *Trans. Br. Ceram. Soc.* **60** (1967) 619.
[38] R. W. Rice, *J. Am. Ceram. Soc.* **54** (1971) 205.
[39] F. W. Valdiek, *J. Less-Common Met.* **13** (1967) 530.
[40] K. Nagata, Y. Yamamoto, H. Igarashi, and K. Okazaki, Powder and Powder Metallurgy Society of Japan, Proceedings of Spring Meeting (1982), p. 120.
[41] C. Greskovich and K. N. Wood, GE Rep., No.72CRD (1972), p. 243.
[42] C. Grescovich, D. Cusano, D. Hoffman, and R. Rinder, *Bull. Am. Ceram. Soc.* **71** (7) (1992) 1120–1130.
[43] K. Nakano, *Electronics*, February (1985).
[44] K. Miyauchi and G. Toda, *Opto-ceramics*, Gihodo, Tokyo (1984).
[45] T. Hirai, *Jitsuyo-Laser-Gijyutsu (Practical Laser Technology)*, Kyoritsu, Tokyo (1987).
[46] C. Greskovich and K. N. Woods, *Ceram. Bull.* **52** (5) (1973) 473–478.
[47] C. Greskovich and J. P. Chernoch, *J. Appl. Phys.* **45** (10) (1974) 4495–4502.

2

Solid-state laser processes and active materials

Lasers constitute the main component (the optical range) of quantum electronics, the science of generation, amplification, the control of the properties (wavelength, temporal regime, power range) of radiation by the processes of interaction of electromagnetic radiation with quantum systems. The word LASER is an acronym standing for Light Amplification by Stimulated Emission of Radiation. The driving mechanism that determines the amplification of radiation in a laser is the stimulated emission in an ensemble of identical quantum systems (the active medium) with inversion of populations between a high energy level and a lower energy level that act as emitting and terminal energy levels for the emission, in a resonator that grants the necessary conditions for amplification and extraction of the amplified radiation beam and gives it a modal structure. A very important class of lasers is the solid-state lasers, where the quantum systems are embedded in a transparent solid material. The first laser was a solid-state laser based on the ruby crystal [1] and during their longer than fifty years history lasers have undergone tremendous scientific and technological development. Many textbooks discuss the basic physical properties and processes that determine the characteristics of laser materials and processes and the physics and engineering of solid-state lasers [2–20]. They contain extensive lists of relevant literature which will not be repeated here. Based on these books and on other relevant literature, this chapter discusses some of the major characteristics of laser materials and processes based on doped transparent polycrystalline materials produced by ceramic techniques.

2.1 Interaction of quantum systems with electromagnetic radiation (radiation absorption and emission processes in quantum systems)

2.1.1 Elementary processes

According to the analysis of Einstein [21], when interacting with an ensemble of identical quantum systems (of density n_0) with two energy levels E_1 and $E_2 > E_1$, an electromagnetic radiation of quantum $h\nu = E_2 - E_1$ can induce two types of processes, absorption from the level E_1 to E_2 and stimulated emission from E_2 to E_1, which compete with the spontaneous emission from E_2. These processes are shown schematically in Figure 2.1. In the absence of this interaction the densities of the populations of the two levels at temperature T are

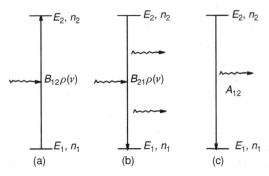

Figure 2.1 Absorption and emission processes in a two-level quantum system: (a) stimulated absorption, (b) stimulated emission and (c) spontaneous emission.

related by the Boltzmann distribution law $n_2/n_1 = (g_2/g_1)\exp[-(E_2 - E_1)/kT]$, where g_1 and g_2 are the degeneracies of the two energy levels. Interaction with the electromagnetic field induces variation of the population densities described by the rate equation

$$\frac{dn_2}{dt} = -\frac{dn_1}{dt} = W_{12}n_1 - W_{21}^{(st)}n_2 - W_{21}^{(sp)}n_2, \qquad (2.1)$$

such that at any moment of time the sum of the populations remains equal to the total population n_0, i.e. $n_1 + n_2 = n_0$. In Eq. (2.1), W_{12}, $W_{21}^{(st)}$ and $W_{21}^{(sp)}$ denote the rates of absorption, stimulated emission and spontaneous emission, respectively. If the action of the external electromagnetic field is suddenly stopped, the variation of the population n_2 is determined by the spontaneous emission only and is described by an exponential decay law with the spontaneous radiative emission (luminescence) lifetime $\tau_{rad} = (W_{21}^{(sp)})^{-1}$, i.e.

$$n_2(t) = n_2(0)\exp(-t/\tau_{rad}). \qquad (2.2)$$

The rates of the stimulated (induced) processes depend on the energy density $\rho(v)$ of the electromagnetic field, $W_{12} = B_{12}\rho(v)$ and $W_{21}^{(st)} = B_{21}\rho(v)$, while the rate of spontaneous emission $W_{21}^{(sp)} = A_{21}$ is constant and $\tau_{rad} = (A_{21})^{-1}$. The Einstein coefficients A_{21}, B_{12} and B_{21} are related by

$$A_{21} = \frac{8\pi v^2 h v}{c^3}B_{21} \qquad \text{and} \qquad B_{21} = \frac{g_1}{g_2}B_{12}.$$

Whereas in the spontaneous emission there is no phase relation between the quanta emitted by the various individual excited quantum systems of the ensemble, in the case of stimulated emission the emitted quantum of each member of the ensemble is in phase with the quantum that stimulates its emission. Stimulated emission can be induced both by photons of the incident resonant electromagnetic field and by photons already emitted in the spontaneous and stimulated emission processes. When the exciting electromagnetic field is coherent, the beam of stimulated emission quanta from the various members of the ensemble preserves the coherence; however, if the incident beam is non-coherent, the beam of stimulated photons will have a very low degree of coherence.

At low incident energy density $\rho(\nu)$ the de-excitation of the level E_2 by spontaneous emission (luminescence) dominates completely the stimulated emission. The spontaneous emission lines of the isolated quantum systems have small but finite width $\Delta\nu$ determined by the lifetime broadening (the natural linewidth), such that $\tau_{rad}\Delta\nu = (2\pi)^{-1}$, which influences uniformly all the identical quantum systems from the ensemble and the lines have Lorentzian shape. However, in reality these quantum systems are not completely free and isolated and thus their quantum states are influenced by additional interactions that induce broadening of the lines. The broadening of the lines can be homogeneous, when it influences all the quantum systems from the ensemble identically, or it can be inhomogeneous, when these additional interactions induce different shifts of the energy levels of the various quantum systems. If these shifts are not resolved spectrally, they result in broadening of the lines, although large perturbations could induce resolved effects. The homogeneously broadened optical lines (absorption and emission) are symmetric around the central frequency ν_0 and have Lorentz shapes

$$g_L(\nu) = \frac{2}{\pi\,\Delta\nu}\left[1 + \frac{4(\nu - \nu_0)^2}{(\Delta\nu)^2}\right]^{-1}, \tag{2.3}$$

where $\Delta\nu$ is the full width at half maximum (FWHM). In the case of inhomogeneous broadening the lines can be symmetric, with Gaussian shapes

$$g_G(\nu) = 2\left(\frac{\ln 2}{\pi}\right)\frac{1}{\Delta\nu}\exp\left[-\frac{4(\nu - \nu_0)^2}{(\Delta\nu)^2}\ln 2\right], \tag{2.4}$$

or asymmetric. Generally these two broadening mechanisms act together and sometimes symmetric broadening combining the Lorentzian and Gaussian shapes (Voight profiles) can be observed. Since the lineshapes are normalized, the linewidth $\Delta\nu$ and the peak value $g(\nu_0)$ are related, $g_L(\nu_0)\Delta\nu = 2/\pi \approx 0.637$, and $g_G(\nu_0)\Delta\nu = 2(\ln 2/\pi)^{1/2} \approx 0.939$. Under the action of narrow-band radiation all members of the ensembles with homogeneously broadened lines respond identically, whereas in the case of absorption lines with dominant inhomogeneous broadening the excitation is selective and involves systems whose peak frequency is near the exciting frequency and are confined in the spectral region delineated by the homogeneous broadening.

In real systems the characteristics of transitions are determined by the quantum states (energy levels, wave functions) and by interaction with the electromagnetic field (selection rules and transition probabilities for the various components – electric multipole and magnetic dipole – of the field). The optical transitions are investigated experimentally by static (absorption, emission spectra) or dynamic (emission decay) or combined (time-resolved spectroscopy) spectroscopic methods. The Einstein coefficients are seldom used in practice, and parameters more relevant to collection or interpretation of the experimental results are employed.

- The line strength S characterizes the intensity of the spectral lines and is related to the transitions between the quantum states for the various components i of the electromagnetic field (electric dipole, electric quadrupole, magnetic dipole),

$$S = \sum_i |\langle \varphi_1 | P_i | \varphi_2 \rangle|^2, \tag{2.5}$$

where φ_i are the wavefunctions of the states 1 and 2, and P_i is the operator of interaction. The line strength can be measured directly from the experimental absorption spectrum.
- The oscillator strength f is a dimensionless parameter that establishes the relation between the strength of transitions in a quantum system and the theoretical strength of a single electron in a harmonic oscillator model and is related to the line strength by

$$f_{12} = \frac{8\pi^2 m_e \nu}{3e^2 h g_1} S_{12}, \tag{2.6}$$

where m_e and e are the electron mass and charge and ν is the frequency of transition. The oscillator strengths for the downward and upward transitions are related by $f_{21} = (g_1/g_2) f_{12}$. The line strength and oscillator strength can be related to the Einstein coefficients,

$$A_{21} = \frac{8\pi^2 e^2 \nu^2}{mc^3} f_{12} = \frac{(8\pi^2)^2 \nu^3}{3hc^3 g_1} S_{12}. \tag{2.7}$$

- The cross-section expresses the ability of an individual quantum system to absorb or emit radiation and its dimension is that of an area (usually cm^2) per quantum system. The cross-section reflects the spectral composition of the optical process, $\sigma(\nu, \nu_0) = \sigma_0 g(\nu, \nu_0)$, where the peak cross-section σ_0 corresponds to the maximum ($\nu = \nu_0$) of line and, since the lineshape is normalized, $\int \sigma(\nu, \nu_0) d\nu = \sigma_0$. The absorption and emission cross-sections are connected to the Einstein coefficients by the Fuchtbauer–Ladenburg relations

$$\sigma_a \equiv \sigma_{12}(\nu, \nu_0) = \frac{1}{c} h\nu B_{12} g(\nu, \nu_0) \tag{2.8}$$

$$\sigma_e \equiv \sigma_{21}(\nu, \nu_0) = \frac{1}{c} h\nu B_{21} g(\nu, \nu_0) = \frac{c^2}{8\pi \nu^2} A_{21} g(\nu, \nu_0) = \frac{c^2}{8\pi \nu^2 \tau_{rad}} g(\nu, \nu_0) \tag{2.9}$$

and a relation of reciprocity between these two cross-section holds,

$$\int \sigma_{12}(\nu, \nu_0) d\nu = (g_2/g_1) \int \sigma_{21}(\nu, \nu_0) d\nu. \tag{2.10}$$

The emission cross-sections corresponding to the center of the emission line (the peak cross-sections) are related to the characteristics of the line and to the radiation lifetime: for the Lorentz lineshapes

$$\sigma_{21}(\nu_0, \nu_0) = \frac{c^2}{4\pi^2 \nu_0^2 \Delta\nu} A_{21} = \frac{c^2}{4\pi^2 \nu_0^2 \Delta\nu} \frac{1}{\tau_{rad}}, \tag{2.11}$$

whereas for Gaussian lines

$$\sigma_{21}(\nu_0, \nu_0) = \frac{c^2}{4\pi \nu_0^2 \Delta \nu} \left(\frac{\ln 2}{\pi} \right)^{1/2} \quad A_{21} = \frac{c^2}{4\pi \nu_0^2 \Delta \nu} \left(\frac{\ln 2}{\pi} \right)^{1/2} \frac{1}{\tau_{rad}}. \quad (2.12)$$

2.1.2 The absorption of radiation

By traveling through the ensemble of quantum systems (the active medium) the energy density of a resonant electromagnetic beam will be modified by these interactions. The evolution of energy density $\rho(\nu_e)$ after traveling a distance x inside this medium, neglecting the effects of spontaneous emission, is described by the equation

$$\frac{d\rho(\nu_e)}{dx} = -\rho(\nu_e) B_{12} h \nu_e \left(n_1 - \frac{g_1}{g_2} n_2 \right), \quad (2.13)$$

for which the solution is

$$\rho(\nu_e, x) = \rho_0(\nu_e) \exp\left[-h\nu_e B_{12} \left(n_1 - \frac{g_1}{g_2} n_2 \right) \frac{x}{c} \right]. \quad (2.14)$$

In the case of weak incident energy density $n_2 \approx 0$, and there is a net attenuation of the beam by absorption. Equation (2.14) then takes the known Beer–Lambert form

$$\rho(\nu_e, x) = \rho_0(\nu_e) \exp[-\alpha(\nu_e)x], \quad (2.15)$$

where $\alpha(\nu_e)$ is the absorption coefficient,

$$\alpha(\nu) = \frac{h\nu B_{12} n_1}{c} = n_1 \sigma_{12}(\nu). \quad (2.16)$$

For a broadened absorption line of shape $g(\nu, \nu_0)$ the peak absorption cross-section ($\nu_e = \nu_0$) can be related to the linewidth of the absorption line,

$$\sigma_{12}(\nu_0, \nu_0) \propto 1/\Delta \nu. \quad (2.17)$$

2.1.3 Amplification of radiation by stimulated emission

According to this picture, the absorption from the external electromagnetic field resonant with the difference between the two energy levels induces its attenuation, whereas the stimulated emission amplifies this field and the balance between these two processes at any moment of time is determined by the instantaneous populations of the levels on which the transitions originate: the population of the lower level for absorption and that of the upper level for stimulated emission. Under the action of the external resonant field on the two-level system, the population of the upper level cannot be larger than that of the lower level, and the absorption dominates the stimulated emission. As a consequence, no net amplification of the incident electromagnetic field can be obtained. Moreover, as mentioned above, if the incident electromagnetic radiation is non-coherent, then the stimulated emission radiation will have a very low degree of coherence.

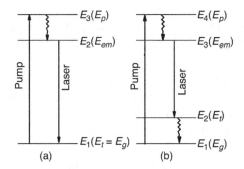

Figure 2.2 The three-level (a) and four-level (b) laser schemes.

However, inversion of populations between levels E_2 and E_1 can be obtained under the action of an external electromagnetic field by de-coupling the absorption from this field and the stimulated emission in three-level systems. In such systems the strong external electromagnetic radiation is absorbed efficiently into an upper level E_3 (E_p) from which the excitation decays rapidly to the level E_2 (E_{em}) that gives stimulated emission terminating on the ground level E_1 ($E_t \equiv E_g$), as shown schematically in Figure 2.2(a). If the external field is strong enough, the population of the excited level E_2 can be larger than that of the ground level E_1. This modality for creating inversion is called pumping. In the three-level (3L) scheme a spontaneous emission quantum between the levels E_2 and E_1 can be amplified by stimulated emission in an avalanche process and this grants a very high degree of coherence to the amplified beam since all the photons from the amplified beam have identical phase and propagate in the same direction. In order to dominate the global spontaneous emission, conditions for efficient amplification of the stimulated emission must be created, i.e. a long path for amplification, and this can be obtained by placing the quantum system in a resonator that recirculates the largest part of the stimulated emission radiation through the active medium and outcouples the rest of this beam. In the case of optical transitions, a Fabry–Perot interferometer consisting of two parallel mirrors, one fully reflecting (the end mirror, $R_2 \cong 1$), the other with a small transmission $T = -\ln R_1 \approx 1 - R_1$ (the exit mirror) at the wavelength of emission, can serve as resonator. Moreover, by its mode structure, the resonator determines the spectral characteristics of the laser radiation, the beam divergence, diameter and transverse energy distribution.

Since in the three-level quantum generators the stimulated emission transition terminates on the ground level, inversion of the population is difficult to achieve and very high pumping rates are necessary. This shortcoming can be avoided in an alternative approach, the four-level (4L) scheme (Figure 2.2(b)), in which the terminal level E_t of the stimulated emission is placed well above the ground level E_g such that $E_t - E_g \gg kT$, and its thermal population is negligible. Additionally, the excitation reaching this level must decay very fast to the ground level in order to avoid self-saturation of the laser emission. Intermediate cases, when the terminal level is close to the ground level ($E_t - E_g$ of the order of kT) and its thermal population causes partial reabsorption of emission can also exist: these are called quasi-three-level (Q3L) schemes due to the partial reabsorption of emission, although several authors prefer the name quasi-four-level (Q4L) scheme since the terminal level is not the ground state. In order to avoid stimulated emission on other transitions, in all these schemes the reflectivities of the laser mirrors must correspond to the wavelength of the desired laser emission.

A quantum generator of radiation using stimulated emission is then composed of three major parts: (i) the active material that contains the ensemble of quantum systems whose quantum properties determine the ability to create inversion of the population and to amplify the stimulated emission, as well as the major characteristics of this emission (wavelength, gain spectrum and so on); (ii) the system which produces the inversion of population between the emitting and the terminal energy levels (the pumping system); (iii) the resonator which provides the conditions for feedback and outcouples part of the beam of radiation amplified by stimulated emission. It is obvious that the starting point in designing a quantum generator is the active material. The other two parts, the pumping system and the laser resonator, must enable the most efficient exploitation of the properties of this material; moreover, optimization of the quantum generator of the radiation will require correlated optimization of these three parts. Fundamentals and engineering of laser emission are described in many textbooks, such as [2–11]; these textbooks also provide lists of the relevant literature, which will not be repeated here.

The stimulated emission can be considered as a negative absorption that increases when the beam travels repeatedly inside the pumped laser material due to the reflections on the mirrors with reflectivity $R_2 \cong 1$ and $R_1 < 1$. In order to reach gain at any given geometrical point inside the pumped laser material placed in the resonator, the stimulated emission beam must travel a round trip in the resonator. During this round trip the coherent radiation is amplified with a gain dependent on the length of the laser rod according to $\exp(2gl_{rod})$. However, taking into account that part of the laser beam is outcoupled through the mirror R_1, the global round-trip gain is reduced to $G = R_1 \exp(2gl_{rod})$. Similar to the absorption coefficient α, the emission coefficient equals $n_2\sigma_e$; however, taking into account the possible reabsorption of the amplified beam due to a residual population of the terminal level, the net effect on the amplified beam can be characterized by an effective gain coefficient $g = \Delta n\sigma_e$ with $\Delta n = n_2 - n_1$.

The laser material and the resonator can introduce additional losses at the wavelength of laser radiation, such as parasitic absorption, scattering, diffraction, residual Fresnel reflection and so on, and these are coined in a loss coefficient L. The amplified beam becomes intense enough to overcome the residual losses and the outcoupling loss when

$$R_1 \exp(2gl - L) \geq 1; \tag{2.18}$$

the strict equality in this equation defines a threshold condition for laser emission,

$$2gl = -\ln R_1 + L; \tag{2.19}$$

usually the reflectivity R_1 is large enough and $\ln R_1 = \ln(1 - T) \approx -T$ where T is the transmission of the exit mirror, and

$$2gl \approx T + L. \tag{2.20}$$

Below the threshold, the losses block the stimulated emission flux and a small-signal gain coefficient g_0 can be defined; immediately above the threshold, the stimulated emission is amplified according to the gain g_0 but for higher excitation the amplification of the light beam continues until the gain saturates for the inversion of the population determined by the actual pump intensity. The value of the saturated gain is determined by the pumping conditions and by the characteristics of the laser material and the design of the resonator and thus Eq. (2.18) establishes the connection between the output and input power by taking into account all these characteristics. The gain threshold condition defines the population inversion and the emission cross-section necessary to overcome the actual losses of the resonator: a larger emission cross-section determines a smaller threshold population inversion and thus the laser starts to oscillate for frequencies around the central value ν_0 of the function $g(\nu, \nu_0)$ and the gain profile will be considerably narrower than the width of this function. Nevertheless, pumping high above the threshold determines a broader gain profile. Since the peak value $g(\nu_0, \nu_0)$ of the function $g(\nu, \nu_0)$ is inversely proportional to the linewidth $\Delta\nu$, and the Einstein coefficient A_{21} is inversely proportional to the emission lifetime, the quantum systems with narrow emission lines and long lifetime will show a lower laser threshold.

Although the first laser was a three-level system, the ruby laser [1], very soon the advantages of four-level lasers became evident. The quantum systems with discrete energy levels able to give emission in the optical range (ultraviolet, visible, infrared) with wavelengths from hundreds of nanometers to tens of micrometers can be free atoms (such as in the He–Ne laser) or ions (such as Ar^+), free molecules (CO_2, CO, N_2 and so on), excimer noble gas halide molecules (ArF, KrF, XeF, KrCl, XeCl), ions of elements from the transitional groups (elements with incomplete inner electronic shells) doped in transparent solids, optically active defects in solids (color centers), solutions of organic molecules (dyes), and a special category is the semiconductors, where the optical transitions take place between collective electronic states (the conduction and valence bands). Efficient laser emission was demonstrated in all types of laser materials; the present book refers to lasers based on transparent solids doped with laser active ions, called solid-state lasers.

The phase coincidence of the photons in the stimulated emission process gives high temporal and spatial coherence to the amplified beam: it is monochromatic and propagates with reduced divergence. These properties open a very broad area of applications, based on the quality of laser radiation as a carrier of energy or information, as a highly monochromatic source for excitation or as a frequency etalon, as well as primary radiation for non-linear optical processes.

2.2 Solid-state lasers

2.2.1 Laser active materials

The solid-state laser material consists of a crystalline or amorphous (glass) transparent material doped with the laser active ions whose spectroscopic properties enable efficient excitation and laser emission schemes [18, 19].

2.2.1.1 Laser active ions

The quantum electronic states of a free multielectron atom are determined by a complex chain of interactions [12–16] reflected in the Hamiltonian

$$H = H_0 + H_{e-e} + H_{s-o}, \qquad (2.21)$$

where H_0 includes the electrostatic interaction between the electrons and nucleus and the movement of electrons (kinetic part), H_{e-e} describes the interactions between the electrons, and H_{s-o} describes the coupling of the orbital and spin moments of the electrons. Since the orders of magnitude of these interactions differ greatly, their effect can be treated by the perturbation theory. The eigenstates of the central-symmetric part of the Hamiltonian H_0 and of H_{e-e} are the electronic configurations consisting of electronic shells that contain electrons with the same main and orbital quantum numbers n and l. Similar to the hydrogen atom, the wavefunctions of the electrons can be written as a product between a radial and an angular part. Each electronic configuration is characterized by a definite energy: the differences in energy between the various electronic configurations are very large (typically of the order of 10^5 cm^{-1}) and only the configuration with the lowest energy (the ground configuration) is populated. The laser active ions used for solid-state lasers belong mainly to elements of transitional groups, i.e. of groups with incomplete inner electronic shells of electrons d (single electron orbital quantum number $l = 2$) or f ($l = 3$): the most popular in laser practice are the elements of groups 3d (iron group, $n = 3, l = 2$) and 4f (rare earths, RE or lanthanides, $n = 4, l = 3$). The 3d group contains ten elements with ground electronic configurations $1s^2\, 2s^2\, 2p^6\, 3s^2\, 3p^6\, 3d^n\, 4s^2$, with n running from 1 for ^{21}Sc to 10 for ^{30}Zn, whereas the 4f group contains 14 elements with ground configuration $1s^2\, 2s^2\, 2p^6\, 3s^2\, 3p^6\, 3d^{10}\, 4s^2\, 4p^6\, 4d^{10}\, 5s^2\, 5p^6\, 4f^n\, 6s^2$ (for two of these elements a 4f electron is replaced by a 5d electron), with n running from 1 for ^{57}La to 14 for ^{70}Yb. The excited electronic configurations of these elements are usually placed above the optical range. The radius of the 3d electronic shell in the 3d elements is smaller than that of the shell 4s; similarly, in the case of the 4f elements the radius of the 4f shell is smaller than those of the 5s, 5p and 6s shells, and thus the incomplete electronic shells in the free atoms from these groups are inner shells.

When introduced in a solid matrix the 3d elements lose the less bound 4s and one or more 3d electrons to form ions of various valence states and thus 3d becomes the outer shell; the 4f elements lose the 6s electrons and a 4f electron to form the dominant 3+ valence state, but 4f remains the inner shell, being screened from the exterior by the closed 5s^2 and 5p^6 shells. In the free ions with weak or moderate spin–orbit interaction (this includes the 3d and 4f elements) the asymmetric part of the electron–electron interaction H_{e-e} couples the individual spin and orbital moments of the electrons from each electronic shell into the total orbital and spin moments $\vec{L}$ and $\vec{S}$ characterized by the quantum numbers L and S and determines the splitting of the electronic configurations into new electronic states, the spectral terms, labeled by ^{2S+1}L (L takes the label S, P, D, F, G for $L = 0, 1, 2, 3, 4$ respectively). The L and S quantum numbers of the filled electronic shells are zero and thus the electronic configurations are not split into terms and transitions are only possible between the electronic configurations, at wavelengths usually outside the optical range. However, the incomplete electronic shells $3d^n$ and $4f^n$ discussed above enable several L or S quantum numbers and the corresponding spectral terms ^{2S+1}L are separated by energy gaps in the optical range, usually several thousands cm^{-1}. Thus, the ground configuration $3d^3$ of Cr^{3+} enables the ^4F, ^4P, ^2G, ^2H, ^2P, ^2D and ^2F spectral terms, the ground configuration $4f^3$ of Nd^{3+} contains nine spectral terms (^4I, ^4F, ^2H, ^4S, ^2G, ^4G, ^2D, ^2P, ^4D), whereas the ground configuration $4f^{13}$ of Yb^{3+} contains a unique spectral term, ^2F. The positions of the spectral terms can be expressed by three Racah parameters (A, B and C for d elements or E^1, E^2 and E^3 in the case of f elements), which are functions of the Slater integrals F^0, F^2, F^4 and F^6 of the electrostatic interaction between the radial parts of the corresponding states.

Subsequently, the spin–orbit interaction characterized by the parameter ς couples the moments $\vec{L}$ and $\vec{S}$ in the total angular momentum $\vec{J}$ (Russell–Saunders coupling), and the various modalities of coupling can be characterized by a chain of J quantum numbers between $L + S$ and $|L - S|$; this determines the splitting of each spectral term ^{2S+1}L in several energy manifolds $^{2S+1}L_J$. For instance the spectral term ^4I of Nd^{3+} is split into four manifolds, $^4I_{9/2}$, $^4I_{11/2}$, $^4I_{13/2}$ and $^4I_{15/2}$, the term ^4F is split into four manifolds, $^4F_{3/2}$, $^4F_{5/2}$, $^4F_{7/2}$ and $^4F_{9/2}$, whereas the term ^2F of Yb^{3+} is split into two energy manifolds, $^2F_{7/2}$ and $^2F_{5/2}$. The wavefunctions of the electrons in these manifolds would then be characterized by a set of quantum numbers (n, l, L, S, J); these manifolds have $(2J + 1)$ residual degeneracy according to the angular momentum projection quantum numbers M_J and this degeneracy can be raised completely or partially by the action of static electric or magnetic fields external to the atom. The energy gaps between the energy manifolds are several thousands cm^{-1}, smaller than those between the spectral terms, and increase towards the end of the 4f series due to the increased spin–orbit coupling parameter ς. In the case of ions with a large number of spectral terms, the energy manifolds belonging to different spectral terms can interpenetrate. The first excited electronic configuration of the trivalent rare earth ions is $4f^{n-1}5d$, usually in the deep ultraviolet range, although for the beginning of the series (the Ce^{3+} ion, ground configuration $4f^1$) the first excited configuration $5d^1$ can enter in the optical range.

A periodic perturbation of frequency v, such as an electromagnetic field, would induce transitions between the two states a and b of the free ions, with transition probabilities

$$W_{a \to b} \propto |\langle nlLSJ|A|n'l'L'S'J'\rangle|^2 g(E_a - E_b - hv), \qquad (2.22)$$

where A is the interaction operator. This perturbation can be expanded in a Taylor series and in the case of the electromagnetic field the main contribution to the interaction comes from the electric dipole and quadrupole moments, and from the magnetic dipole. The interactions with the laser active ions are specific for each of these moments and can be described by tensor operators with components C_{kq}. For instance, in the case of the electric dipole interaction the tensor operator is a first-rank tensor (vector) with components $-iz$, $(1/2)(x \pm iy)$; the operator for the electric quadrupole interaction is a second-rank tensor with components $(6)^{-1/2}(3z^2 - r^2)$, $z(x \pm iy)$ and $(1/2)(x \pm iy)^2$, whereas the magnetic dipole operator is a first-rank tensor with components $-(1/2c)(\vec{L} + 2\vec{S})$. The tensor operators of the various interactions with the electromagnetic field determine selection rules for the various quantum numbers that characterize the states involved in transition for each type of these moments.

- The matrix elements of the electric dipole operator are zero (the transitions are forbidden) between states with the same parity $(-1)^l$ and this imposes the selection (Laporte) rule $\Delta l = \pm 1$; additionally, the Laporte rules select the allowed electric dipole transitions between the spectral terms or energy manifolds, $\Delta S = 0$, $\Delta L = 0, \pm 1$ except for $L = 0 \rightarrow L' = 0$ and $\Delta J = 0, \pm 1$, except for the transition $J = 0 \rightarrow J' = 0$.
- For the electric quadrupole transitions $\Delta l = 0, \pm 2$, $\Delta S = 0$, $\Delta L = 0, \pm 1, \pm 2$ except for $L = 0 \rightarrow L' = 0$ and $\Delta J = 0, \pm 1, \pm 2$ when $J + J' > 1$, except the transitions $J = 0 \rightarrow J' = 0$, $J = 1/2 \rightarrow J' = 1/2$ and $J = 1 \rightarrow J' = 0$.
- For the magnetic dipole transitions $\Delta l = 0$, $\Delta S = 0$, $\Delta L = 0$, $\Delta J = 0, \pm 1$, except for transition $J = 0 \rightarrow J' = 0$.

These selection rules restrict severely the possible optical transitions for the free $3d^n$ or $4f^n$ ions. Generally, the electric dipole transitions are much stronger (several orders of magnitude) than the electric quadrupole and magnetic dipole transitions.

2.2.1.2 The transparent host material

The transparent materials that serve as hosts for the laser active ions are simple or complex halide, oxide or chalcogenide compounds of ions with complete electronic shells, such as Ca^{2+}, Sr^{2+}, Al^{3+}, Ga^{3+}, Sc^{3+}, Y^{3+}, Lu^{3+}, La^{3+}, although the $4f^7$ Gd^{3+} ion, whose energy levels do not allow transitions in the visible and near infrared, can also be used. These materials can contain a unique or several such cationic species and the densities of the cationic sites are very high, more than 10^{22} cm^{-3}. The relative doping concentrations of the laser active ions with respect to the substituted cations can range from tenths to tens of percent or even to full substitution, and thus a solid-state laser material can host very large and stable absolute concentrations (10^{19} cm^{-3} to 10^{22} cm^{-3}) of such laser ions.

The transparent materials used as host matrices can be crystalline or amorphous materials. The crystalline materials, either single crystals or polycrystals, are characterized by well-defined (anionic and cationic coordination numbers, cation–anion distances and bond angles) crystallographic sites. In the case of single crystals, the structure of the whole body has a defined symmetry (from the 32 crystallographic groups) and can be reproduced by translation of an elementary structure (cell), whereas for the polycrystalline materials, composed of crystalline grains of random orientation, these properties are valid for each grain. In the case of polycrystals, additional defective crystallographic sites can occur at grain boundaries, their proportion with respect to the normal (volume) sites being determined by the depth of the grain boundary and by the ratio (surface/volume) of the grains. Partial disordering of the crystals is obtained when specific crystallographic sites can be occupied by several species of cations, sometimes of different valences, but in proportion to maintain electrical neutrality. The structures of the cationic sites in amorphous materials, such as glasses, are not well defined: they are slightly different from each other and show a distribution (usually assumed to be Gaussian) around a majority center.

The role of the host material is not limited to that of support for a large concentration of laser active ions: the additional static interactions with the constituents of the host material or with other doping ions of the same or of different species can modify in an essential way the quantum states of the doping laser ion and its ability to interact with the electromagnetic radiation. As in the case of the free quantum systems, the most relevant information on quantum states of the doped solids is obtained from optical spectroscopy. These additional interactions will then determine the absorption and emission properties of the doping ion (wavelength, linewidth, transition probabilities). Moreover, since the doping ion and the other constituents of the lattice oscillate around their equilibrium positions, the static interactions are modulated by these specific vibrational modes of various symmetries and this influences the lineshape, structure, position and the probability of optical transitions as well as the dynamics of emission. Additional host structure-dependent effects on the dynamics of populations of the excited levels can be introduced by the transfer of excitation caused by the electrostatic and superexchange interactions between the doping ions.

2.2.1.3 The static effect of the host material on the electronic structure of the laser active ions

2.2.1.3.1 Crystal field splitting of the energy levels The sites occupied by the doping ions in the crystalline materials have well-defined local point symmetry that is not necessarily identical to the global symmetry of the host. The ionic radius of the doping ion can differ from that of the substituted cation, leading to local distortion of the lattice: usually this distortion preserves the original local point symmetry group of the site, although it can alter the ratio of the various structural parameters. When the doping ions are well isolated in the crystalline lattice their properties will be identical.

The neighbors of the doping ion in transparent solids determine additional static interactions (electrostatic, exchange, covalence, polarization and so on) that compete with the interactions specific to the free laser ion [13–20]. The covalence of the cation–anion bonds can influence the interelectronic interactions and modify the Racah parameters and the energies of the spectral terms ^{2S+1}L (the nephelauxetic effect). Moreover, the complex action of the ensemble of neighboring anions and cations on the electronic structure of the doping ion (the crystal field interaction) competes with the spin–orbit coupling and can be described by the phenomenological crystal field potential, whose symmetry is determined by the local point symmetry

$$V_c = \sum_{k,q} B_{kq} C_{kq}. \tag{2.23}$$

The crystal field interaction determines the partial or total removal of the residual degeneracy of the states of the free ion leading to the splitting of the energy states (Stark splitting) as well as to the mixing of wavefunctions. The radial crystal field parameters

$B_{kq} = \langle r^k \rangle A_{kq}$, where $\langle r^k \rangle$ is the average value of the kth power of the ionic radius of the doping ion and A_{kq} describe the effect of the neighboring ions, are a measure of the strength of the interaction and C_{kq} are tensor operators that reflect its symmetry.

Calculation of the effect of the crystal field on the states of the free ions implies calculation of matrix elements between the free-ion states $|nlLSJM_J\rangle$. The matrix elements of the tensor operators C_{kq} can be related to those of the components U_{kq} of the normalized (unitary) operators $U^{(k)}$,

$$\langle nlLSJM_J|V|n'l'L'S'J'M'_J\rangle = \sum_{k,q} B_{kq} \langle nlLSJM_J|C_{kq}|n'l'L'S'J'M'_J\rangle$$

$$= \sum_{k,q} (-1)^l [(2l+1)(2l'+1)]^{1/2} \begin{pmatrix} l & k & l' \\ 0 & 0 & 0 \end{pmatrix}$$

$$\times B_{kq} \langle nlLSJM_J|U_{kq}|n'l'L'S'J'M'_J\rangle .$$

The matrix elements of the tensor operator components U_{kq} are related to those of the unitary operator $U^{(k)}$ by

$$\langle nlSLJM_J|U_{kq}|n'l'S'L'J'M'_J\rangle = (-1)^{J-M_J} \begin{pmatrix} J & k & J' \\ -M_J & q & M'_J \end{pmatrix} \langle nlSLJ\|U^{(k)}\|n'l'S'L'J'\rangle$$

(Wigner–Eckart theorem). The Wigner 3-j and 6-j symbols from these relations are found in tables and they simplify greatly the calculation of the effects of the crystal field interaction on the energy levels of the doping ions. The 3-j symbol

$$\begin{pmatrix} l & k & l' \\ 0 & 0 & 0 \end{pmatrix}$$

imposes the selection rule for the k value in the crystal field potential, $k \le l+l'$. According to this rule the effect of the crystal field potential inside a given electronic configuration ($l = l'$) or between configurations of the same parity $(-1)^l$ will be different from zero only for the crystal field terms of even k and their number is determined by the value of l: for the 3d electrons $k \le 4$ and for 4f electrons $k \le 6$. The term with $k = 0$ is omitted since it determines equal shift for all levels. By contrast, the odd terms in the crystal field potential can have non-zero matrix elements between electronic configurations of opposite parity and mix their states. Thus, the configurations $4f^n$ and $4f^{n-1}5d$ of the rare earth ions can be connected by crystal field components with $k = 1, 3, 5$ and 7; however, such components are possible only in the case of doping ions in sites with inversionless local crystal field symmetries.

For each value of k the parameter $q \le k$ is selected by the local symmetry and this parameterization restricts greatly the summation in Eq. (2.24). In the presence of a symmetry axis of order p the parameter q can take the value of any multiple of p allowed by k, for instance, a second-order symmetry axis will determine $q = 0, 2$ and 4 for 3d electrons and 0, 2, 4 and 6 for 4f electrons. Thus, when acting inside the ground configuration 4f, for the cubic point symmetry groups $k = 4$ and 6, $q = 0$ and 4, for the tetragonal groups $k = 2, 4$

and 6, $q = 0$ and 4, for the trigonal groups $k = 2, 4$ and 6, $q = 0, 3$ and 6, for the rhombic groups $k = 2, 4$ and 6, $q = 0, 2, 4$ and 6, whereas for the 3d ions k is limited to 4.

The crystal field interaction modifies the quantum states of the doping laser ions. The point (local) symmetry groups of the sites occupied by these ions are sub-groups of the three-dimensional rotation group $R(3)$ that characterizes the symmetry of the free ions, which are systems with angular momentum. The crystal field interaction determines the partial breaking of the degeneracy of the angular momentum states, i.e. the splitting of the free ion energy levels (Stark splitting). The characteristics of splitting (the number of Stark levels and their remanent degeneracy) are determined by the decomposition of the irreducible representation of the $R(3)$ group for each angular momentum in the irreducible representations of the point symmetry group, as well as by the symmetry transformations with respect to the time inversion (Kramers degeneracy). This crystal field splitting is of paramount importance for solid-state lasers since the pump absorption and laser emission take place by transitions between the Stark levels.

For the 3d ions, where 3d is the outer electronic shell, the crystal field interaction is stronger than the spin–orbit coupling and thus it acts on the spectral terms ^{2S+1}L (intermediate crystal field) and induces the partial breaking of their $(2L + 1)$ degeneracy, resulting in crystal field (Stark) levels labeled according to the irreducible representations of the crystal field simple point group: in a crystal field of octahedral symmetry, the ^{2S+1}P term ($L = 1$, degeneration $2L + 1 = 3$) remains in the triplet state $^{2S+1}T_1$, the ^{2S+1}D term ($L = 2$) is split in a doublet ^{2S+1}E and a triplet $^{2S+1}T_2$, the ^{2S+1}F term ($L = 3$) is split into two triplets ($^{2S+1}T_1$ and $^{2S+1}T_2$) and a singlet $^{2S+1}A_2$, the term ^{2S+1}G ($L = 4$) is split into a singlet $^{2S+1}A_1$, a doublet ^{2S+1}E and two triplets ($^{2S+1}T_1$ and $^{2S+1}T_2$) and so on. The lower symmetry components of the crystal fields can split further, the triplets into doublets and singlets.

The strong crystal field interaction alters the energy level scheme of the 3d ions considerably, and is specific to each ion in each particular host: some of the Stark levels show weak dependence on the strength of the crystal field, whereas other levels are strongly influenced by the ratio between the crystal field strength and the electron–electron interaction parameters and this can modify strongly the relative positions of the various levels depending on the crystal field strength. This situation is reflected by the calculated diagrams of the energy levels of the various 3d ions depending on the crystal field strength for various cubic coordinations (Tanabe–Sugano diagrams). Thus, in the case of a Cr^{3+} ion (ground configuration 3d^3) in a site with regular octahedral coordination, the positions of the crystal field levels plotted versus the crystal field splitting parameter $10Dq$ between the ground $^4A_2$ and the first excited level $^4T_2$ originating from the ground spectral term ^4F, normalized to the Racah parameter B, show that the energy of the level $^4T_2$ increases linearly with the crystal field strength, whereas the ^2E level originating from the excited term ^2D shows a weak dependence at low crystal field strength then remains almost unchanged, and for $(10Dq/B) \approx 2$ these two levels intersect. The practical values of crystal field parameters are in this region, so for the crystals where $(10Dq/B) < 2$ (weak octahedral field) the first excited level is $^4T_2$, whereas for $(10Dq/B) > 2$ (strong octahedral field) this will be ^2E.

The weak spin–orbit coupling in the 3d ions splits these Stark levels further into spin–orbit levels, but this splitting is small, several cm^{-1} to a few tens of cm^{-1}.

For the rare earth ions, where the 4f shell is inner, the crystal field interaction is weaker than the spin–orbit coupling, and breaks partially the M_J degeneracy of the $^{2S+1}L_J$ manifolds. The residual degeneracy of the Stark components of J manifolds of the rare earth ions with odd number of 4f electrons (Kramers ions) in a solid is 2 or 4, whereas for ions with an even number of such electrons (non-Kramers ions) the residual degeneracy could be 1, 2 or 3, the largest value in both cases being possible only in cubic symmetry. In non-cubic symmetries each J manifold of the Kramers ions is split into $(2J+1)$ doubly degenerated Stark levels, whereas for non-Kramers ions the triplet states are split into doublets and singlets or only into singlets. Since the crystal field interaction for the 4f ions is weaker than the spin–orbit coupling, the splitting of the Stark levels is in most cases smaller than the gap between the manifolds J in the free ion and they show as groups of levels around the original positions of the manifolds from which they originate.

Accurate *ab initio* calculation of the crystal field parameters B_{kq} is very difficult and usually they are measured by fitting the experimental optical absorption and emission spectra. Such fitting will also provide new values for the free ion parameters that reflect the influence of the interactions in crystals (nephelauxetic effect, crystal field) on the positions of the spectral terms and energy manifolds. The calculated effect of the electrostatic interaction, which is the major contribution to crystal field splitting, indicates that the parameters A_{kq} are proportional to the electric charge of the doping and of the surrounding ions and depend strongly on the distance to the ions that produce the crystal field, $A_{kq} \propto R^{-(k+1)}$; strong distance dependence is shown by the other interactions that contribute to the crystal field potential too. Because of this dependence, the main contribution to the crystal field comes from the first anionic coordinating sphere around the doping ion although it can also be influenced sizably (especially the crystal field terms with lower k) by the ions from several other cationic and anionic coordination spheres around the doping ion. The ensemble formed by the doping laser ion and all the anionic or cationic constituents of the crystalline lattice that influence appreciably the crystal field is called the structural center.

2.2.1.3.2 Normal and perturbed centers Any disturbance in the structure or composition of the surrounding of the doping ion in the structural center (departure from ideal composition of the host material, statistical ensembles of doping ions in near lattice sites, co-doping ions, accidental impurities, vacancies) can modify the crystal field and the structure of the Stark levels of the center, leading to modification of its optical spectra. These perturbations are caused by the difference in ionic size and/or charge of the perturbing source with respect to the normal situation. Due to the discreteness of the crystalline lattice and to the dependence of the crystal field on distance and sometimes on the angle of perturbation with respect to the crystallographic axes, each type of perturbing source in the vicinity of the doping ion can determine a specific chain of discrete perturbations, whereas farther perturbations can determine a continuous distribution of the crystal field. Each of these discrete perturbations determines a specific structure of Stark levels and optical spectrum, and thus the variety of perturbations caused by a given type of perturbing source induces several individual optical spectra even in crystals with unique substitution site, i.e. multi-center structure of the global optical spectra. In many cases several kinds of such perturbing sources can coexist, each of them inducing specific perturbations. Since the concentrations of the perturbed centers are usually much smaller than that of the unperturbed center, the multicenter structure of crystals with unique doping sites is manifested in the optical spectra by specific structures of spectral satellites around the main line corresponding to the unperturbed center. A particular type of perturbed center is that determined by the mutual crystal field perturbations inside the ensembles of doping ions in near lattice sites: the most usual satellites in this case correspond to the pairs formed by the central laser ion with another laser ion from the first (nearest-neighbor, n.n.) or the second (next-nearest neighbor, n.n.n.) coordination sphere of available lattice sites (satellites M_1 and M_2). Investigation of the relative intensities of these satellites is a meaningful tool for assessment of the distribution of the laser ions in the host lattice.

2.2.1.4 The distribution of the doping ions and structural defects in the laser materials

The distribution of the perturbing sources in the crystalline lattice and the possibility of the existence of attractive or repulsive forces from the doping ion determine the probability of having such perturbators in its vicinity. In principle there are two classes of distributions: (i) correlated distributions, when the doping and the pertubing centers are coupled by additional attractive or repulsive forces; (ii) non-correlated distributions, in the absence of such interactions. The distribution of the doping ions and of defects can be modeled either numerically, by Monte Carlo generation of the doped lattice, or using analytical models.

The most usual cases of correlated distributions of the perturbing centers are those when the valence of the doping ion is different from that of the substituted cation and it becomes necessary to compensate for this difference. This can be accomplished either by local intrinsic lattice defects such as anionic or cationic vacancies, interstitial anions or cations, or by co-doping with charge-compensating ions; in special cases, particularly at high doping concentrations, severe restructuring of the lattice can take place. The additional attractive forces can determine migration of the charge compensating ions close to the doping ion; this can be facilitated by the fabrication technique and by the thermodynamics of the process, particularly by the temperature and duration.

In the case of non-correlated placement there are several models for the distribution, for example the following.

- Uniform distribution models, with two variants (i) the discrete uniform placement (the average-distance) model, which assumes that the doping ions are placed in the lattice at the same distance, $R = N_d^{-3}$, where N_d is the density of the doping ions, and (ii) the continuous uniform distribution model, which assumes that the density of doping ions $\rho(x, y, z)$ is the same at each geometrical point of the lattice. None of these models accounts for the crystallographic structure of the host. Moreover, at low doping concentrations the average distance is much larger than the distance between the nearest available sites and thus no ensembles of doping ions in near lattice sites such as pairs would be possible.

- Discrete equiprobable distribution models, which account for the fact that the doping ions or the defects are placed at random in the actual available discrete lattice sites i, with probability p_i equal to the relative concentration of the perturbing sources C, regardless of i. The doping ions or defects can be considered as placed on coordination spheres around each doping ion, the characteristics (radius, number of sites m) of these spheres being determined by the crystalline structure of the host. In the case of low enough concentrations of perturbing sources, the probability of having n occupied sites in a coordination sphere around the laser ion is given by the binomial law,

$$P_{nm} = \frac{m!}{n!(m-n)!} C^n (1-C)^{m-n}, \qquad (2.24)$$

and the relative concentration of perturbing sources is

$$C_{nm} = C P_{nm} = \frac{m!}{n!(m-n)!} C^{n+1} (1-C)^{m-n}. \qquad (2.25)$$

The probability of occurrence of a nearest-neighbor (n.n.) pair consisting of (doping ion–perturbing source) can be calculated by setting $n = 1$ in Eqs. (2.24) and (2.25), leading to $P_{1m} = mC(1-C)^{m-1}$ and $C_{1m} = mC^2(1-C)^{m-1}$: such pairs can occur at very low concentrations of perturbing sources and are favored by tight packing of the available sites (large m).

The resolution of the multicenter structure in the optical spectra is determined by the ratio between the line broadening and the difference between the lines of the various centers; the reduction of homogeneous broadening at low temperatures increases the resolution. Usually only the perturbations produced by centers from the near (mostly from the first) coordination spheres are resolved and determine the spectral satellite structure, whereas the non-resolved effect of farther perturbing sources leads to inhomogeneous, sometimes asymmetric, broadening of the lines. The measured relative intensities of the spectral satellites and of the main (unperturbed) spectral lines can then be compared with the calculated occurrence probabilities. In the absorption spectra, the intensity of each nm line is proportional to the relative concentration of the perturbing center and to its line strength S_{nm}, i.e. $I_{nm}^{(abs)} \propto C_{nm} S_{nm}$. The relative intensities of the various lines, specified by n in the frame of given m are then

$$I_{nm,rel}^{(abs)} = \frac{I_{nm}^{(abs)}}{\sum\limits_n I_{nm}^{(abs)}} = \frac{P_{nm} S_{nm}}{\sum\limits_n P_{nm} S_{nm}}.$$

For the optical transitions where the transition probabilities are not influenced by the perturbation that determine the satellite structure, $I_{nm,rel}^{(abs)} = P_{nm}$, i.e. the relative intensities of the satellites give a direct measure of their occurrence probabilities P_{nm} and any departure from this relation when P_{nm} is calculated within the random distribution model would be indicative of correlation of placement for the perturbing centers. Investigation of the satellite structure of the optical spectra can thus become a very meaningful tool for investigation of the microstructure of laser material and of the distribution of the doping ions and defects.

As will be made evident in a forthcoming section, the distribution of the doping ions influences the global dynamics of emission in the presence of de-excitation processes determined by the interactions between these ions. Any distribution model must predict in a consistent way all these properties, so in order to reveal its limits and to avoid any ambiguities, a structural model deduced from experiment must be verified by its applicability to other types of experiments.

The distribution of the doping ions and defects can be altered in regions of strong disturbance of the structure of the host material, such as the surfaces or interfaces (grain boundaries in polycrystalline materials), where strongly distorted structural centers can be formed or where strong segregation of the doping ions can take place: this can determine new lines in the optical spectra and modification of the relative intensities of the centers corresponding to ensembles of doping ions in near lattice sites. The extent of these effects in a global measurement will be determined by the depth of the disturbed region and by the ratio (surface/volume) as well as by the intensity of the segregation process. However, such effects can be evidenced and characterized by high spatial resolution experiments.

2.2.1.5 Transition probabilities in the optical spectra of the rare earth ions

Introduction of laser ions in inversion-symmetry sites which select the even k crystal field parameters in Eq. (2.23) preserves the Laporte rules for the electric dipole transitions inside the pure electronic configurations 3d or 4f. However, in crystal fields without inversion the crystal field potential can contain odd k terms that can mix the ground state with the excited configurations of opposite symmetry. Although, because of the large energy gap between the electronic configurations compared with the crystal field strength this mixing is quite weak, it can induce electric dipole optical transitions between the electronic states inside the ground configuration. Such a situation can arise either when the original symmetry of the site occupied by the laser ion lacks inversion, or in the case of ions substituted in sites with inversion, whose crystal field symmetry is lowered by near defects of the crystalline lattice, as discussed above. Lowering of symmetry can also be induced by the vibrations of the crystalline lattice and the effect on the transition probabilities is more pronounced for ions with enhanced sensitivity to the bonding in the crystal; the problem of these transitions is very complex and is beyond the scope of this book. In the case of the 4f ions the dynamic effects are weak and the induced electric dipole transitions are determined mainly by the low symmetry components of the crystal field.

An accurate treatment of the effect of configuration mixing on the intensities of transitions between the various Stark levels of the 4f ions in crystals would be excessively complex and approximate approaches have been used. According to the Judd–Ofelt theory [22, 23], the odd crystal field mixing of the ground $4f^n$ electronic configuration with electronic configurations of opposite parity, such as the first excited $4f^{n-1}5d$ configuration, determines induced electric dipole transitions between the energy manifolds J and the line strength of all these inter-manifold transitions can be described using only three intensity parameters Ω_λ, with $\lambda = 2, 4$ or 6,

$$S^{ed} = \sum_{\lambda=2,4,6} \Omega_\lambda |\langle 4f^n \alpha J \| U^{(\lambda)} \| 4f^n \alpha' J' \rangle|^2 \tag{2.26}$$

where $U^{(\lambda)}$ is the unitary tensor operator of rank λ and

$$\Omega_\lambda = (2\lambda + 1) \sum_{k,q} |A_{kq}|^2 \Phi^2(k, \lambda)(2k + 1)^2. \tag{2.27}$$

In this equation $\Phi(k, \lambda)$ describe the mixing of the electronic configurations of opposite parity. Induced electric dipole transitions are allowed when $|1 - \lambda| \le k \le |1 + \lambda|$ and this imposes selection rules for the rank k of the odd crystal field terms that determine each of the Judd–Ofelt parameters: for $\lambda = 2$, $k = 1$ or 3, for $\lambda = 4$, $k = 3$ or 5, and for $\lambda = 6$, $k = 5$ or 7. This determines specific sensitivity of the Judd–Ofelt parameters to the distance between the laser ion and its neighbors: since $A_{kq} \propto R^{-(k+1)}$, then $\Omega_\lambda \propto R^{-2(k+1)}$. Thus, the high order parameters ($\lambda = 4$ or 6) are determined mainly by the nearest neighbors in the crystal, whereas the low order parameter Ω_2 also feels the effect of the farther ions. The reduced matrix elements $\langle 4f^n \alpha J \| U^{(\lambda)} \| 4f \alpha' J' \rangle$ can be calculated and are collected in tables [18] and the Judd–Ofelt parameters can thus be determined from the experimental line intensities of the absorption spectra. By accounting only for the mixing of the J states from two electronic configurations, the Judd–Ofelt theory neglects the particularities of the transitions between the Stark levels and this introduces a degree of approximation that can reach 20%. The line strengths calculated with these parameters can then be used for calculation of the spontaneous emission Einstein coefficients $A_{JJ'}^{ed}$ for the transitions from a J manifold to any lower manifold J' and of the radiative lifetime τ_{rad} of the manifold J,

$$\tau_{rad} = \left(\sum_{J'} A_{JJ'} \right)^{-1} \tag{2.28}$$

as well as of the emission branching ratios for emission from the level J to any lower level J',

$$\beta_{JJ'} = \tau_{rad} A_{JJ'} = A_{JJ'} \bigg/ \sum_{J'} A_{JJ'}. \tag{2.29}$$

The magnetic dipole transitions between the manifolds of the 4f ions are allowed only with the selection rules $\Delta J = 0, \pm 1$ regardless of the crystal field symmetry and their line strength is

$$S^{md} = \left(\frac{eh}{2mc} \right)^2 |\langle 4f^n \alpha J \| \vec{L} + 2\vec{S} \| 4f^n \alpha' J' \rangle|^2 \tag{2.30}$$

and with few exceptions they are much weaker than the induced electric dipole transitions. There are cases when the electric and magnetic dipole transitions are allowed simltaneously, and both contributions to the observed line intensities must be accounted for. These considerations are valid for the global intensities of the $J \rightarrow J'$ transitions in the case of optically isotropic crystals. However, these intensities are distributed among the transitions between the Stark levels of these manifolds. Moreover, the observed intensities of the lines are also influenced by the fractional thermal population (Boltzmann) coefficient of the actual Stark level from which the transition originates. In the case of anisotropic crystals, which are outside the scope of this book, the polarization effects should be accounted for too.

As discussed previously, direct comparison of the measured relative global intensities of the absorption lines of the perturbed centers with the occurrence probabilities of the various configurations of the coordination spheres, calculated for various models of distribution, is only meaningful for transitions whose probabilities are not modified appreciably by perturbation. Since the perturbing sources are usually placed on the cationic coordination spheres, i.e. outside the nearest anionic coordination sphere, they will influence mainly the crystal field terms with weaker dependence on distance (low k), i.e. the Judd–Ofelt parameter Ω_2, whose effect on intensity can be eliminated by selecting electronic transitions with $\langle 4f^n \alpha J \| U^{(2)} \| 4f\alpha' J' \rangle \cong 0$: in the case of Nd^{3+} the transitions connecting the level $^4F_{3/2}$ with any of the ^4I_J manifolds or the ground level $^4I_{9/2}$ with other ^4I_J manifolds fulfill this condition and the line strengths S_{nm} do not depend on n. However, strong perturbations, such as those caused by charge difference, can influence the other Judd–Ofelt parameters (Ω_4 and Ω_6) and then the line strengths for the various centers might show differences.

2.2.1.6 Interactions with the lattice vibrations

The discussion of the quantum states of the laser materials in the preceding sections was based on the assumption that the positions of the ions constituting the laser materials are fixed in space. However, all the constituents of the lattice exercise vibrations of various symmetries around the equilibrium positions, which can influence the quantum states of the doping ions. The vibrational state of a crystal can be regarded as a superposition of normal modes of vibration for which the lattice vibrates uniformly with the same frequency ν_k. The normal vibration frequencies of a lattice can define quasi-particles of energies $h\nu_k$, called phonons. Thus, each lattice can be characterized by a discrete phonon spectrum corresponding to the various types of vibrations, and the energies of phonons depend on the mass of the constituents of the lattice, on the strength of bonds between these constituents and on the symmetry. The energies of phonons are in the range 300–400 cm^{-1} for fluorides, $\sim$500–600 cm^{-1} for sesquioxides, 700–800 cm^{-1} for garnets, and can reach 1200 cm^{-1} in crystals with molecular groups such as the borates. The phonons propagate inside the crystal; the low frequency phonons are called acoustical phonons, whereas those with high energy are called optical phonons and each of these phonon waves can have transverse and longitudinal mode structure. The densities of phonons of various energies (occupation number) are dependent on temperature according to the Bose–Einstein statistics,

$$\langle n(T) \rangle = [\exp(h\nu/kT) - 1]^{-1} \tag{2.31}$$

and this can induce specific temperature dependence for all the processes involving phonons.

These vibrations can be accounted for by introducing additional terms for the vibrations and for their coupling with the electronic states (electron–phonon interaction) in the Hamiltonian describing the state of the doping ion. These additional ("dynamic") interactions are usually weaker than the pure electronic ("static") interactions. The wavefunctions of the doping ions can then be written as products of functions describing the static situation and functions describing the dynamics of nuclei (Born–Oppenheimer states). These coupled states influence the optical spectroscopic properties of the doping ions [18, 19], as listed below, the effect being much stronger for the 3d ions than for the 4f ions.

- The appearance of additional lines (vibronics) corresponding to simultaneous absorption or emission transitions of the electrons and of one or more phonons. The manifestation of vibronics depends on the electronic structure of the doping ion and on the nature of the transition. In the case of 3d elements the vibronic lines can dominate the pure electronic (zero phonon) lines, especially for the spin allowed transitions, and usually the optical spectra are discussed in terms of the configuration-coordinate model. The vibronic optical bands of 3d ions can be very broad, several hundreds of nanometers. For such cases the maximum absorption and emission cross-sections connected to a specific electronic transition do not correspond to the zero phonon line but to the peaks of the vibronic absorption and emission bands and are shifted considerably (to higher energies for absorption and to lower energies for emission) from the zero phonon line. The Stokes shift between the peak of the absorption and that of the emission band of 3d ions is very large and there is reduced superposition between these bands. In the case of rare earth 4f ions the weaker electron–phonon interaction determines sizable vibronic transitions only at the ends of the 4f electronic series, as weak satellites whose shift from the pure electronic (zero phonon) lines corresponds to the energies of the optical phonons.
- Raman scattering of incident electromagnetic radiation: although the largest part of an incident narrow band electromagnetic radiation non-resonant with an electronic transition travels inside a transparent material without any change in frequency, a tiny fraction, below 10^{-6}, can undergo a positive or negative shift in frequency caused by the exchange of energy with a special category of optical phonons, the Raman active phonons. The dynamic polarization $P(t)$ of a material under the action of an external electromagnetic field E can be writted as $P = \sum_i \chi^{(i)} E^i$, where $\chi^{(i)}$ is the ith order susceptibility; the Raman shift is determined by the third-order susceptibility $\chi^{(3)}$ which can be non-zero in both optically isotropic and anisotropic materials. The characteristics of the Raman active phonons are determined by the symmetry of the material: in crystals with inversion symmetry the phonons that are active in the infrared absorption are inactive in a Raman process and vice versa, whereas in low symmetry crystals some phonons can be active in both processes. In Raman active laser crystals the emitted radiation can show weak additional structures of Raman satellites which can be used for stimulated emission at frequencies shifted from the pure electronic transition.
- Temperature dependent homogeneous broadening of the absorption and emission lines. This broadening determines the reduction of the absorption and emission cross-sections.

- Thermal shift of the optical lines.
- Redistribution of intensity and shift of lines in the case of resonance between the crystal field splitting and the phonon energies. Such an effect shows in the spectra of the RE^{3+} ions at the end of the series (Tm^{3+}, Yb^{3+}) but is less important in the case of Nd^{3+}.
- Non-radiative de-excitation of the excited levels to the nearest lower levels by transferring the energy to one or several phonons that can bridge the gap ΔE between these two electronic levels. Usually a phonon of high energy ($h\nu_{eff}$) is the most efficient in this de-excitation and the non-radiative de-excitation rate W_{nr} depends on the order $p = \Delta E/h\nu_{eff}$ of the phonon de-excitation process and on the phonon occupation number $n(T)$ according to the energy gap law

$$W^{nr} = C[n(T)+1]^p \exp(-\alpha\Delta E), \tag{2.32}$$

where the parameters C and α depend on the host material but not on the electronic transition and $n(T)$ induces the temperature dependence of the multiphonon de-excitation rate,

$$W^{nr}(T) = W_0^{nr} \left[\frac{\exp(h\nu_{eff}/kT)}{\exp(h\nu_{eff}/kT) - 1} \right]^p, \tag{2.33}$$

where $W_0^{nr} = C\exp(-\alpha\Delta E)$. The electron–phonon interaction does not alter the exponential character of the emission decay but reduces its lifetime,

$$\frac{1}{\tau_f} = \frac{1}{\tau_{rad}} + W^{nr}. \tag{2.34}$$

When the energy gap is very large compared with the effective phonon energy (high phonon order), W^{nr} is small and in many instances it can be neglected compared to τ_{rad}^{-1}. However, for low order processes W^{nr} can dominate τ_{rad}^{-1} completely and practically all the energy corresponding to the gap ΔE is transferred to vibrations. Thus, the pump level in a laser scheme is chosen so as to de-excite either directly or by a chain of successive steps via intermediate levels to the emitting level by low order ($p = 1$) electron–phonon processes and a similar rule applies to the gap between the terminal laser level and the ground state. By contrast, in order to prevent parasitic de-excitation by low order electron–phonon processes, the gap between the emitting level and the nearest lower energy level must be large compared with the phonon energy.

- Thermalization of the close energy levels: the lattice vibrations determine thermal Boltzmann distribution of the populations of the Stark energy levels of a manifold whose global population is non-zero. Thermalization can also involve levels whose emission processes are governed by different selection rules or have different emission lifetimes. In such a case the emission at a given wavelength from the group of thermalized levels will be influenced by the individual emission characteristics of all these levels, in a proportion determined by the thermalization process, i.e. by their relative positions and by temperature.

Besides these effects on the static and dynamic spectroscopic properties of the doping laser ions, the phonons also influence other properties of laser materials such as the thermal properties, especially the heat conduction.

2.2.1.7 The width of spectral lines

The width of the absorption and emission lines influences both the pump absorption capabilities and the laser emission process. The shape and linewidth of the spectral lines of the doped solids are determined by the common action of the temperature dependent electron–phonon interaction that determines the homogeneous broadening and the temperature independent distribution of the crystal field that determines the multicenter structure and the inhomogeneous broadening. Particularly large distributions of crystal fields occur in the case of compositionally disordered crystals with multiple occupancy of the cationic sites around the doping ion with host cations of various size and/or valence. In this case the abundances of several centers can be similar and the optical spectra contain broad and asymmetric bands with unresolved peaks and shoulders that are envelopes of the lines corresponding to the various structural centers and thus the linewidth can be engineered by a proper choice of the composition of the material.

2.2.1.8 Interactions between the doping ions: non-radiative energy transfer

2.2.1.8.1 Elementary energy transfer processes The doping ions in solids can interact with each other by electrostatic and exchange coupling. When one or both interacting ions are in excited states, they can exchange excitation by non-radiative processes induced by these static interactions; this process is called energy transfer (ET) [24, 25], the ion that loses energy is called the donor (D) and the ion whose energy is increased by transfer is called the acceptor (A). The donor and acceptor can be of the same or different species. A resonant donor–acceptor energy transfer, in which the donated and the accepted energy quanta are fairly equal, preserves to a good approximation the sum of the initial energies of the donor and acceptor and thus the final state of the acceptor is determined by the amount of excitation lost by the donor and by the initial state of the acceptor. In many instances, phonon assistance can help to correct the resonance mismatch. The energy transfer can take place regardless of the initial energy of the acceptor: when this is the ground level, the final state of the acceptor cannot be higher than the initial state of the donor even when the donor is completely de-excited by transfer; however, in most cases the final state of the donor is not the ground level, and thus the final state of the acceptor is below the initial excitation of the donor and generally a down-conversion of excitation takes place, as shown in Figure 2.3(a). When the initial state of the acceptor is an excited level, in many cases its final level can be higher than the initial energy level of the donor and thus upconversion of excitation (ETU) takes place, Figure 2.3(b). In special cases, determined by the energy level scheme of the acceptor, it can be excited successively by several donors, thus reaching a very high energy final state.

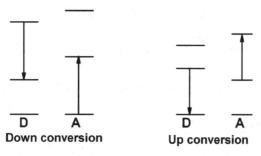

Figure 2.3 Direct donor–acceptor processes: (a) down-conversion and (b) upconversion.

Although energy transfer by interionic interactions is a coupled process of non-radiative de-excitation of the donor and excitation of the acceptor, it involves one-step transitions between the initial states of the donor and acceptor ions under the static ion–ion interactions. The energy transfer rate W_{DA} is then determined by the matrix element of the coupling interaction $H^{(i)}$ between the wavefunctions corresponding to the initial E_i and to the final E_f energy states of these two ions, $\langle \varphi_i(D)\varphi_i(A)|H^{(i)}|\varphi_f(D)\varphi_f(A)\rangle$. Since the energy transfer reduces the population of the initial energy level of the donor, it competes with other de-excitation processes in the quenching of donor emission. When the final state of the acceptor ion is metastable and can produce laser emission, the energy transfer from the donor acts as a sensitizer of emission.

The electrostatic coupling between donor and acceptor can be expressed as coupled multipolar contributions and the selection of the multipolarity for the energy transfer process is determined by the multipolarities of the transitions $E_i \rightarrow E_f$ for both ions. It can be of dipole–dipole (d-d), dipole–quadrupole (d-q) or quadrupole–quadrupole (q-q) type. Each such multipolarity of interaction is characterized by specific dependence of the ET rates W_{DA} on the distance R_{DA} between donor and acceptor:

$$W_{DA}^{(s)} = \frac{C_{DA}^{(s)}}{R_{DA}^s},$$

$$(2.35)$$

where $s = 6, 8$, or 10 for the d-d, d-q or q-q interactions respectively, and $C_{DA}^{(s)}$ is the microparameter of interaction

$$C_{DA}^{(s)} \propto \frac{1}{\tau_{rad}(A)} \int \sigma_e(D)\sigma_a(A)dv.$$

$$(2.36)$$

The superposition integral of the donor emission and acceptor absorption in this equation is determined by the electronic structure of the donor and acceptor ions and by the characteristics (selection rules, line strengths and lineshapes) of the transitions involved in transfer: these selection rules will thus select the possible ion–ion interactions responsible for transfer.

The ET rate in the case of exchange (superexchange) D–A coupling is

$$W_i^{ex} = \frac{1}{\tau_{rad}} \frac{2\pi}{\hbar} K^2 \exp\left(-\frac{2R_i}{L}\right) \int \sigma_e(D)\sigma_a(A)d\nu = \frac{1}{\tau_{rad}} \exp\left[\gamma\left(1 - \frac{R_i}{R_0}\right)\right] \quad (2.37)$$

with $\gamma = 2L^{-1}R_0$, where L is the effective Bohr radius (for RE^{3+} ions this is of the order of several hundredths or tenths of nanometers) and R_0 is the penetration depth of interaction (tenths of nanometers), i.e. the D–A distance at which the transfer rate is equal to the radiative rate. When R_i is larger than R_0 the rate W_i^{ex} drops strongly with distance. The exchange and the high-multipolarity electrostatic interactions are short-distance processes and determine high transfer rates to the nearest acceptors, whereas the d-d interaction can still be sizable at larger D–A distances.

The excitation can also be transferred completely from an excited donor to an unexcited ion of the same type, which ends in an excited state similar to the intial state of the excited donor, $E_{t,A(D)} \cong E_{i,D(D)}$. When the concentration of donors in the ground state is high, this process repeats with other donor ions and efficient migration of excitation at quite large distances from the original excited donor ion by diffusion or by hopping mechanisms can take place inside the system of donors, without altering the initial energy. This excitation can be further lost by transfer to an acceptor (trap) placed at quite large distance from the original donor, contributing to enhanced loss of excitation. The ion–ion interaction responsible for migration is usually d-d interaction.

2.2.1.8.2 The effect of direct donor–acceptor transfer on donor emission dynamics

As discussed above, the energy transfer rate depends on the donor–acceptor distance. In real systems the number of doping ions is very large and they can be placed at different distances from each other. The ensemble of doping ions able to be excited by an external source and to transfer subsequently the excitation to other ions defines the donor system of global relative concentration C_D with respect to the available substitution sites; normally only a fraction of these ions, $r(0)C_D$, is excited by the incident radiation. The acceptors in the laser materials can be the laser ions themselves (self-quenching systems) or foreign ions, and the ensemble of ions that can act as acceptors can be characterized by the global relative concentration C_A; nevertheless, the effective concentration of acceptors depends on the nature of the transfer process and on the intensity of excitation. Thus, in the case of down-conversion ET, in a system where the acceptors are of different species than the donors, their concentration can be considered constant at any moment of decay. However, in the case of self-quenching, the concentration of acceptors increases during decay from the initial value $C_A(0) = [1 - r(0)]C_D$ to C_D; obviously, for weak excitation in these systems $C_A \equiv C_D$. According to the crystalline structure of the host and the acceptor doping concentration, the acceptor ions in the direct donor–acceptor transfer can occupy only j of the specific available sites i relative to the positions of the donor ions. Due to the distribution of the acceptor ions in the crystalline lattice, each donor ion is surrounded by a particular configuration of acceptors.

In the presence of a unique acceptor ion placed in the jth lattice site with respect to the donor, the survival probability of the donor in its excited state at time t after a very short excitation pulse is reduced to $\exp(-t/\tau_D)\exp(-W_j t)$, whereas the whole particular configuration of surrounding acceptors modifies it to $\exp(-t/\tau_D)\prod_j \exp(-W_j t)$, corresponding to an exponential decay with rate $\left(1/\tau_D + \sum_j W_j\right)$, where τ_D is the lifetime of the donor in the absence of transfer. Since each donor is surrounded by a particular configuration of acceptors, its decay rate will be different from that of all the other donors. The survival probability at time t in the excited state in the presence of transfer for the whole ensemble of donors would require summation of survival probabilities of all the individual donors. However, since the number of donors is extremely large and the particular configurations of acceptor ions around each donor ion are not known, an exact calculation is not possible, although generation of the probable acceptor configurations in a Monte–Carlo simulation can enable the numerical analysis of decay. Nevertheless, the very large number of donors in a decay experiment makes reasonable the replacement of summation over all the donor ions by an averaging procedure over all sites i available to the acceptors around the donor sites. This averaging procedure depends essentially on the model of distribution of the acceptors at the available lattice sites. The global effect of ET can then be expressed with an energy transfer function $P(t)$, such that the survival probability in the excited state for the whole ensemble of donors in presence of transfer is $\exp(-t/\tau_D)\exp[-P(t)]$. Since the averaging over the whole ensemble of donors involves a large number of exponential decays with different lifetimes, the transfer function $P(t)$ for the D–A transfer is usually non-linear in time and the decay of the excited states of donors is non-exponential. Obviously, the transfer function $P(t)$ will be influenced by the acceptor distribution model involved in the averaging procedure.

- A very popular approach in discussing the effect of energy transfer on the evolution of the populations of the energy levels in the doped materials is the characterization of the process by a constant rate W_{DA} for the whole ensemble of doping ions. Such an approach is possible for down-conversion self-quenching within the model of *average donor–acceptor distance*, assuming that the rate of transfer is determined by the average distance, $R_{av} \propto (N_A)^{-3}$, where N_A is the absolute concentration (the density) of acceptors, i.e.

$$W_{DA} = \frac{C_{DA}}{R_{av}^s} = C_{DA}(N_A)^{s/3},\tag{2.38a}$$

and for d-d interaction it shows a quadratic dependence on the acceptor density,

$$W_{DA}^{d-d} = C_{DA}(N_A)^2.\tag{2.38b}$$

This model will determine exponential decay in the presence of transfer [26], with decay rate

$$\tau_e^{-1} = \tau_D^{-1} + W_{DA} = \tau_D^{-1} + C_{DA}(N_A)^{s/3};\tag{2.39a}$$

for d-d interactions this becomes

$$\tau_e^{-1} = \tau_D^{-1} + C_{DA}(N_A)^2.\tag{2.39b}$$

- In the *uniform continuous distribution model* for the acceptor ions the direct transfer function for the various multipolar interactions is [24, 25, 27]

$$P(t) = \Gamma \left(1 - \frac{3}{s}\right) \frac{C}{C_0} \left(\frac{t}{\tau_D}\right)^{3/s} = \gamma^{(s)} t^{3/s} \tag{2.40a}$$

where $\Gamma(x)$ is the Euler function. For d-d interaction

$$P(t) = \gamma^{d-d} t^{1/2} \tag{2.40b}$$

(the Forster–Dexter energy transfer function), with

$$\gamma^{d-d} = \frac{4}{3} \pi^{3/2} N_A (C_{DA})^{1/2}. \tag{2.40c}$$

This transfer function is linear in the acceptor concentration but non-linear in time and determines non-exponential decay of the donor excited state over the entire temporal range, with functional form determined by the multipolar interaction (the parameter s). The description of the decay of the donor using the uniform continuous distribution model is more realistic than for the average-distance model, but still has major shortcomings: it is restricted to a unique type of multipolar interaction, it does not reflect the actual crystalline structure of the host, and it predicts a finite acceptor density at the donor position that would determine faster decay at early times.

- Within the *discrete distribution models*, which account for the fact that the acceptor ions are placed at discrete lattice sites i with probability p_i the transfer function can be written

$$P(t) = \sum_i \ln[1 - p_i + p_i \exp(-W_i t)]; \tag{2.41}$$

in the case of completely random (equiprobable) distribution [28, 29] the probabilities p_i are the same for any available site and equal to the relative acceptor concentration C_A, and

$$P(t) = \sum_i \ln[1 - C_A + C_A \exp(-W_i t)]. \tag{2.42a}$$

Assuming that the available acceptor sites i are placed on l coordination spheres around donors, each with m_l sites, and assuming isotropic energy transfer, the transfer function can be rewritten

$$P(t) = \sum_l m_l \ln[1 - C_A + C_A \exp(-W_l t)], \tag{2.42b}$$

where W_l is the transfer rate to any of the acceptor ions placed on sphere l. The rates W_i (or W_l) involve all the donor–acceptor interactions responsible for ET, $W_i = \sum_s W_i^{(s)} + W_i^{(ex)}$.

The decay of the excited population of the donor in the presence of ET can then be written

$$n_D(t) = n_D(0) \exp\left(-\frac{t}{\tau_D}\right) \exp[-P(t)]. \tag{2.43}$$

Analysis of decay with the transfer functions (2.42) is not simple. However, within defined temporal ranges this can be approximated by simpler functional forms [30, 31].

- For the temporal range when $W_l t$ is small, the transfer function is linear in time,

$$P(t) \approx \sum_l m_l C_A W_l t = W_{lin} t, \qquad (2.44)$$

and the decay is quasi-exponential, with lifetime linearly dependent on the acceptor concentration. The temporal range t_1 of validity of this approximation is determined by the fastest W_i rate from summation and is inversely proportional to the energy transfer microparameter, $t_1 \propto r_{min}^s C_{DA}^{-1}$ and can range from microseconds to several tens of microseconds.

- For longer times, $P(t)$ can be approximated with the Inokuti–Hirayama function (2.39) for continuous distribution.

When a strong short range interaction couples the donors and the nearest k acceptors (usually from the first coordination sphere), such that the transfer rate to these acceptors W_i' is much larger than the transfer rates W_i to any other acceptors, the sum in Eq. (2.42a) can be broken into two parts:

$$P(t) = \sum_1^k \ln[1 - C_a + C_A \exp(-W_i' t)] + \sum_{i > k} \ln[1 - C_A + C_A \exp(-W_i t)]. \quad (2.45a)$$

The first term in this equation determines a very fast decay at early times, followed by a much slower decay determined by the second term. For very large transfer rates W_i', the first term can be replaced by $\sum_1^k \ln(1 - C_A)$, which at low acceptor concentrations equals kC_A and

$$P(t) \approx kC_A + \sum_{i > k} m_i \ln[1 - C_A + C_A \exp(-W_i t)]. \qquad (2.45b)$$

The first term will determine a very fast drop in donor excitation at early times, equal to kC_A, which could pass unnoticed in a low resolution decay experiment, followed by the decay described by the second term, which could show the linear part described by Eq. (2.44) but with the summation in W_{lin} only for $i > k$ and which evolves gradually to the continuous approximation decay. These simple dependences on time of the transfer function over restricted temporal ranges can be used for identification of the ion–ion interactions responsible for transfer and for estimation of the characteristic parameters.

The effect of energy transfer on the dynamics of the excited state population can be shown experimentally in an emission decay experiment with short pulse excitation. Usually these experiments are performed at low excitation intensities that give low $r(0)$. Such experiments show that of the three distribution models used for calculation of the transfer function $P(t)$, the discrete distribution is closest to the physical reality in crystals; moreover, its validity can be checked by other measurements such as the relative intensities of the spectral satellites corresponding to crystal field perturbations inside the statistical ensembles of doping ions in near lattice sites, especially of the M_1 pair satellite. A major advantage of this model is the possibility of analyzing the emission decay for the various structural centers in a multicenter structure, since each of these centers has a particular and well-defined structure of the environment. Obviously, the average-distance model has no physical reality and cannot accommodate the non-exponential decay or the pair satellites. It is also evident that the continuous distribution transfer function (Eq. (2.39)) describes satisfactorily the decay in crystals only after a certain lapse of time, and an arbitrary shift between the experimental and the calculated decay must be introduced to produce a fit, which could falsify the calculation of the emission quantum efficiency.

In the presence of two types of acceptors, A_1 and A_2, each in specific sublattices of sites i and respectively j, the transfer function $P(t)$ can be written [32] as the sum of the individual transfer functions $P_1(t)$ and $P_2(t)$, i.e.

$$P(t) = \sum_{i=1}^{N_1} \ln\{1 - C_{A1} + C_{A1} \exp[-W_{DA1}(R_i)t]\}$$

$$+ \sum_{j=1}^{N_2} \ln\{1 - C_{A2} + C_{A2} \exp[-W_{DA2}(R_j)t]\}. \qquad (2.46a)$$

If the two types of acceptors occupy the same type of sites the transfer function can be written

$$P(t) = \sum_{i=1}^{N} \ln\{1 - (C_{A1} + C_{A2}) + C_{A1} \exp(-W_{DA1}(R_i)t) + C_{A2} \exp(-W_{DA2}(R_i)t)\}.$$

$$(2.46b)$$

The two types of acceptors can be of different or the same species, but with different initial acceptor energy states, and in the latter case the transfer involves competing down-conversion and upconversion processes. For low acceptor concentrations the transfer function $P(t)$, Eq. (2.46b), can be approximated by the sum of the individual transfer functions to acceptors A_1 and A_2,

$$P(t) \approx P(t)_{A1} + P(t)_{A2}, \qquad (2.46c)$$

each of them having the form given by Eq. (2.42).

2.2.1.8.3 The effect of migration-assisted energy transfer on decay   The energy transfer between the donor ions which governs migration of excitation can be characterized by a microparameter C_{DD}. Similar to the down-conversion self-quenching, the concentration of acceptors $C_A^{(DD)}(t)$ in this process at any moment of time t is equal to the concentration of donor ions in the ground state, $C_A^{(DD)}(t) = [1 - r(t)]C_D$, where $r(t)$ is the fraction of donor ions in the excited state. At low excitation intensities, $C_A^{(DD)}$ can be considered independent of time and constant, $C_A^{(DD)} \approx C_D$; however, at high excitation it can be frustrated at early times of decay and it increases to C_D at the end of decay. In the absence of acceptors to quench the excitation, the total number of excited donors is not modified by migration and their de-excitation rate is not changed. However, when the crystal also contains acceptor ions, the migration of donors can facilitate the flow of excitation closer to these acceptors, enhancing the rate of de-excitation. In the case of migration of excitation by a hopping mechanism, specific to the rare earth ions, the additional migration-assisted de-excitation process can be characterized by an ensemble-averaged constant transfer rate for all donor ions [33], which is proportional to the relative concentrations of donors and acceptors,

$$\bar{W} = \bar{W}_0[1 - r(t)]C_D C_A. \tag{2.47a}$$

However, in the case of weak excitation, it can be considered constant,

$$\bar{W} \approx \bar{W}_0 C_D C_A. \tag{2.47b}$$

For a system with two types of acceptors (A_1 and A_2),

$$\bar{W} = \bar{W}(A_1) + \bar{W}(A_2),$$

with

$$\bar{W}(A_1) = \bar{W}_0(A_1)[1 - r(t)]C_D C_{A1} \quad \text{and} \quad \bar{W}(A_2) = \bar{W}_0(A_2)[1 - r(t)]C_D C_{A2}.$$

The direct and migration-assisted energy transfer act together over the entire temporal range of decay and the evolution of the excited state population after short pulse excitation can be written

$$n_D(t) = n_D(0) \exp\left(-\frac{t}{\tau_D}\right) \exp[-P(t)] \exp(-\bar{W}t). \tag{2.48}$$

The competition of direct and migration-assisted ET determines the shape of the decay: in the usual case of $C_{DA} > C_{DD}$ direct transfer dominates the behavior at short times, leading to departures from exponential, whereas the migration-assisted transfer dominates the end of decay, which becomes quasi-exponential. In the case of self-quenching, at low excitation the relative concentrations of the acceptors for both direct transfer and migration on donors are equal, $C_A^{(DD)} \equiv C_D = C_A = C$, and the direct donor–acceptor transfer function depends linearly on the doping concentration, $P(t) \propto C$, whereas the migration-assisted transfer rate depends on its square, $\bar{W} = \bar{W}_0 C^2$, and thus the doping concentration influences the relative importance of these two types of energy transfer. However, at high excitation intensities in the systems with self-quenching, the concentrations of the acceptors for migration and for down-conversion are diminished and both these transfer processes are frustrated.

2.2.1.8.4 Temporal evolution of the emission of the acceptor The decay function (2.48) is a solution of the equation of temporal evolution of the population n_D of the excited state of the whole ensemble of donors

$$\frac{dn_D}{dt} = -\frac{n_D}{\tau_D} - \frac{dP_{DA}(t)}{dt} n_D, \tag{2.49}$$

where $P_{DA}(t)$ accounts for both direct and migration-assisted transfer, whereas the evolution of the population of the energy level of the acceptor that is fed by transfer can be described by

$$\frac{dn_A}{dt} = -\frac{n_A}{\tau_A} + \frac{dP_{DA}(t)}{dt}n_D - \frac{dP_{AA}(t)}{dt}n_A, \tag{2.50}$$

where $P_{AA}(t)$ accounts for the possible energy transfer processes inside the system of acceptor ions. The general solution of this equation is [34]

$$n_A(t) = n_D(0)\exp\left[-\frac{t}{\tau_A} - P_{AA}(t)\right]$$

$$\times \int_0^t \exp\left\{\left(\frac{x}{\tau_A} - \frac{x}{\tau_D}\right) + [P_{AA}(x) - P_{DA}(x)]\right\}\frac{dP_{DA}(x)}{dx}dx \tag{2.51}$$

which does not have a simple analytical form. However, numerical calculation or analytical modeling with approximate models for the transfer function, such as a constant D–A transfer rate, predicts a complex temporal dependence of the population of the acceptor ion after a short pulse excitation of the donor, with an increase from zero to a maximum, followed by decay. The temporal evolution of the rise and decay portions, the moment of time t_{max} and the magnitude $n_A(\text{max})$ of the maximum in the acceptor population $n_A(t)$ are determined by the ratio of the total rate of de-excitation of the donor (intrinsic + energy transfer) to the total intrinsic rate of de-excitation of the acceptor, i.e. the rate obtained by pumping the acceptor directly. Additionally, $n_A(t)$ is proportional to $n_D(0)$ and thus depends on the excitation from an external source absorbed by the system of donor ions. The evolution of the rise as well as that of the decay part of $n_A(t)$ is determined by the strongest, respectively the weakest, global rate of donor or acceptor de-excitation process. The population $n_A(t)$ of an excited level of the acceptor by energy transfer from the donor can be utilized for emission from this level under pumping into the donor ion (the sensitization process). A very strong D–A transfer, which practically quenches the emission of the donor, will determine a temporal evolution of the acceptor emission similar to its intrinsic emission since in this case the moment of maximum acceptor population is $t_A(\text{max}) \cong 0$; however, the population of the acceptor n_A at $t = 0$ is now determined by the pump absorption by the donor ions, which can be much larger than is achievable by pumping the acceptor ion directly.

It is then obvious that the energy transfer processes in laser materials can have positive or negative effects. When the donor level is the laser emitting level, the self-quenching as well as the transfer to accidental impurities can reduce the laser efficiency. Moreover, if the final ET levels of the donor and acceptor ions are strongly coupled to the lattice vibrations, and when there is a ladder of closely spaced lower energy levels to the ground state, the whole initial excitation of the donor ion can be lost by low order electron–phonon processes and transformed into heat. On the opposite side, the ET sensitization can be used in practice for excitation of a weakly absorbing acceptor ion with good emission properties (activator) by transferring the excitation from a strongly absorbing donor ion (sensitizer).

2.2.1.9 Emission quantum efficiency

The emission quantum efficiency is defined as the fraction of excited ions that de-excite by radiative processes in the presence of non-radiative de-excitation:

$$\eta_{qe} = \frac{\int_{\infty}(I(t)/I(0))dt}{\int_{\infty}(I_{rad}(t)/I_{rad}(0))dt} = \frac{1}{\tau_{rad}}\int_{\infty}(I(t)/I(0))dt; \tag{2.52}$$

an effective emission lifetime $\tau_{eff} = \tau_{rad}\eta_{qe}$ can also be defined.

If the decay in the presence of non-radiative processes is exponential, with lifetime τ_{exp}, such as for electron–phonon interaction, the emission quantum efficiency equals $\eta_{qe} = \tau_{exp}/\tau_{rad}$. However, in the case of complex decay determined by the energy transfer (Eq. (2.48)) the calculation of η_{qe} will be influenced by the distribution model used in the calculation of $P(t)$.

2.2.1.9.1 Emission quantum efficiency in the presence of direct donor–acceptor transfer When the energy transfer is dominated by direct donor–acceptor transfer the emission quantum efficiency calculated within the distribution models described above is given by the following.

(1) With the *the average-distance model*

$$\eta_{qe} = \frac{1}{1+(N_A/N_0)^{s/3}} \tag{2.53a}$$

and

$$\tau_{eff} = \frac{\tau_D}{1+(N_A/N_0)^{s/3}}, \tag{2.53b}$$

where $N_0 = (C_{DA}^{(s)}\tau_f)^{-3/s}$ is a characteristic acceptor concentration that determines the reduction of the emission lifetime to half its value. For dipole–dipole interaction

$$\eta_{qe} = \frac{1}{1+(N_A/N_0)^2} \tag{2.54a}$$

and

$$\tau_{eff} = \frac{\tau_D}{1+(N_A/N_0)^2}, \tag{2.54b}$$

with $N_0 = (C_{DA}^{(s)}\tau_f)^{-1/2}$, i.e. η_{qe} and τ_{eff} in the case of direct D–A transfer by (d-d) coupling would diminish with the square of the acceptor concentration [26].

(2) In the case of *uniform continuous distribution* calculation of the emission quantum efficiency requires numerical methods. For dipole–dipole interaction [25]

$$\eta_{qe} = \frac{1}{\tau_D}(1 - \pi^{1/2}\exp(x^2)[1 - erf(x)]), \qquad (2.55)$$

where $erf(x)$ is the error function and $x = (1/2)\pi^{1/2}C_A/C_0$.

(3) For *discrete random distribution*, accurate calculation of the global emission quantum efficiency would require numerical calculation using the direct energy transfer function $P(t)$ (17) or (18); however, for $N_A C_{DA}$ values which are not too large (to the order of 2×10^{-19} cm^3 s^{-1}) an approximate analytical equation for the emission quantum efficiency can be calculated [35],

$$\eta_{qe} \cong \exp(-bC_A), \qquad (2.56)$$

where

$$b = \sum_i \frac{W_i}{\tau_D^{-1} + W_i}.$$

2.2.1.9.2 Emission quantum efficiency in the presence of direct and migration-assisted transfer In presence of direct and migration-assisted transfer (Eq. (2.48)) the emission quantum efficiency in the model of random discrete distribution can be written

$$\eta_{qe} \cong \frac{1}{1 + \tau_D \bar{W}} \exp\left(-\sum_i \frac{W_i}{\tau_D^{-1} + W_i + \bar{W}}C_A\right), \qquad (2.57)$$

which in the case of self-quenching becomes

$$\eta_{qe} = \frac{1}{1 + \tau_f \bar{W}_0 C^2} \exp\left(-\sum_i \frac{W_i}{\tau_D^{-1} + W_i + \bar{W}_0 C^2}C\right). \qquad (2.58a)$$

For large migration-assisted transfer rates $\bar{W}_0 C^2$ that determine the dominance of migration-assisted over direct transfer and luminescence de-excitation, η_{qe} can be approximated as

$$\eta'_{qe} \approx \frac{1}{1 + (C/C_0)^2}, \qquad (2.58b)$$

with $C_0 = (\tau_D \bar{W}_0)^{-1/2}$. This concentration dependence of η'_{qe} is similar to that predicted for direct D–A transfer in the average-distance model, Eq. (2.54), although the physical process and the critical concentration parameter are different, and this can cause confusion in the interpretation of experimental data. However, it must be stressed again that in the case of the discrete random distribution model this dependence on C is valid only at very high migration-assisted transfer rates $\bar{W}_0 C^2$, i.e. for materials with large $\bar{W}_0$ and/or C.

2.2.1.9.3 Relevance of emission quantum efficiency in multicenter systems The possibility of describing the emission decay for the various perturbed centers in a multicenter situation by taking into account the particular structure of the ensemble of acceptors for each center opens the possibility of calculating the emission quantum efficiency for each of these centers using the discrete placement model. Compared to the absorption spectra, where the intensities of the lines corresponding to the various centers are determined by the relative concentrations and depend on the line strength, in the case of steady-state lumines-cence emission under non-selective pumping, the intensities of the lines of the individual centers depend additionally on the emission quantum efficiency η_{qe}^{mn} and on the absorption efficiency η_a^{mn} for each type of structural center, $I_{mn}^{(em)} \propto C_{mn} S_{mn} \eta_{qe}^{mn} \eta_a^{mn}$. The dependence of $I_{mn}^{(em)}$ on η_{qe}^{mn} gives to the dynamics of emission a filtering role for information on the relative concentrations of the centers and on the models of distribution of the perturbing sources in the lattice. This could be particularly important in systems with self-quenching since the emission quantum efficiency of some of the perturbed centers determined by the statistical ensembles of ions in near lattice sites (pairs, triads and so on) can be so low as to make them unobservable in continuous wave excited luminescence.

2.2.1.10 Cooperative processes

In the previous sections the doping ions were considered as individual species, even in ensembles of near ions. However, when a strong interaction couples the two ions of a pair, the electronic structure can be radically altered and the pair can be considered as a special entity, a "dimer": if the electronic states of the interacting ions are $|i\rangle$ and $|j\rangle$, the states of the dimer will be $|ij\rangle$ and the energies are given, except for a small shift introduced by the coupling, by $E_{ij} \approx E_i + E_j$. These dimers show cooperative absorption and emission spectra [36, 37] whose shapes are determined by the convolution of the spectra of the isolated ions. The intensities of these spectra are usually very small and increase with the doping concentration, which determines the probability of occurrence of pairs, and the lifetime of cooperative emission is smaller than for the isolated ions. The cooperative processes can modify the dynamics of de-excitation from real levels of the isolated ions, which are placed in the vicinity of the dimer levels by ion-pair (dimer) energy transfer and can be very active both in self-quenching and in sensitization. The transfer function in the emission decay of a donor that transfers energy by d-d interaction to such a dimer (cooperative de-excitation) shows $t^{1/3}$ and n_A^2 dependence [38].

2.2.1.11 Evaluation of the spectroscopic parameters of interest for laser emission

Success in accommodating the laser emission schemes in the Stark energy level structures of the doping ions in solids depends on identification of suitable energy levels that fulfill the conditions for efficient excitation and emission. The structure of the energy levels, the shapes and widths of transitions and their cross-sections as well as the dynamics of emission are measured by static and dynamic optical spectroscopy. In most cases the Stark energy levels are not well isolated and are thermalized with nearby levels; since this process depends on temperature, the measured optical spectra provide the effective (not the absolute) transition cross-sections.

The parameters measured in a conventional spectroscopic experiment have a global character for the whole macroscopic volume traversed by the incident optical beam. Such an approach can give an accurate assessment in the case of homogeneous materials, without obvious local structural or compositional irregularities. Such disturbed micro-regions and their effect on the local spectroscopic properties can be investigated and characterized by spatially resolved methods of investigation in which the spectroscopic methods are coupled with microscopic methods that enable the accurate spatial confinement and localization of the excitation radiation and scanning of the excitation region over extended parts of the sample.

High resolution spectroscopy and emission decay under selective or non-selective excitation can be a very useful tool for investigation of the microstructure of laser materials [39]. Additionally, modeling of laser emission using spectroscopy enables selection of the characteristics of the laser material, pumping system and resonator design.

2.2.1.11.1 Absorption spectra The absorption spectra show the absorption transitions from the Stark levels that are thermally populated at the temperature of measurement. Such levels can belong to the ground manifold, leading to ground state absorption (GSA), or to higher energy manifolds populated by pumping (excited-state absorption, ESA). Absorption spectroscopy is a macroscopic method, the investigated samples have definite geometrical form and are well polished to avoid scattering, whereas the probe beam is unidirectional. This method enables the direct measurement of the effective absorption cross-sections of the various transitions by monitoring the attenuation of a low intensity beam (to avoid saturation or non-linear effects) in a sample of given size and doping concentration. The broad spectral composition of the incident beam enables the measurement of a wide range of optical transitions, depending on the energy or wavelength used. The radiation transmitted through the sample is scanned across the spectral range of interest with a high resolution dispersing element, usually a grating, and detected with a high sensitivity detector. High resolution of equipment is crucial for observation of the accurate shape of the sharp lines of the RE^{3+} ions and for evaluation of the cross-sections. Besides the structure of excited levels and absorption cross-sections, the absorption spectra provide information necessary for calculation of the Judd–Ofelt parameters and of the radiative lifetime as well as for evaluation of the emission cross-section by the reciprocity method.

2.2.1.11.2 Emission (luminescence) spectra Spontaneous emission (luminescence) spectroscopy can be used as a global or microscopic investigation technique, depending on the size of the exciting beam, using continuous wave or pulsed excitation, and the registration of spectra can be made in a stationary or time-resolved regime. The excitation can be done with broad-band sources, which determine absorption in several excited levels, or by narrow-band excitation in a specific level. For crystals with a multicenter structure, narrow-band excitation can be selective for each of these centers, or non-selective. As mentioned before, in contrast to the absorption spectra, where the intensities of the lines are detemined by the concentration of the absorbing centers and by the transition cross-section, the luminescence spectra show additional dependence on emission quantum efficiency. Moreover, the intensity of the emission lines can be altered by temperature, concentration and emission path dependent reabsorption or by energy transfer, as well as by the spectral sensitivity of the detector.

Because of the difficulties in collecting the whole omnidirectional emission, measurement of the emission cross-sections from luminescence spectra can be very inaccurate. However, the emission cross-sections can be derived from the absorption spectra using the Fuchtbauer–Ladenburg relation and the reciprocity method [40]. The equation [41]

$$\sigma_e(\nu) = \sigma_a(\nu)\frac{Z_l}{Z_u} \exp\left[\frac{E_{zl} - h\nu}{kT}\right], \tag{2.59}$$

relating the emission and absorption cross-sections between an upper (u) and a lower (l) manifold, each containing a number of Stark levels, is very useful. Z_l and Z_u are the partition coefficients for the two manifolds (the sum of fractional thermal population coefficients for all the Stark levels of each manifold) and E_{zl} is the energy gap between the lowest Stark levels of these manifolds.

Microscopic emission spectroscopy under laser excitation coupled with confocal microscopy can give very high spatial resolution, well below micrometer level, and can reveal differences (spectral composition, emission quantum efficiency and so on) in emission between very confined regions. These differences can then be related to differences in microstructure or microcomposition in these regions. However, accurate mapping and interpretation of the observed spatially resolved spectra is very difficult because additional differences can be induced by the excitation and by the detection process itself, as well as by differences in other properties such as the refractive index. Moreover, unambiguous correlation of these emission micro-properties with the global emission properties of the laser material is a difficult task.

2.2.1.11.3 Excitation spectra The excitation spectra are recorded by monitoring the variation of the emission intensity at a defined wavelength when tuning the excitation source over various ranges of absorption. Thus, the excitation spectra can be useful in selection of the most efficient pumping wavelength.

2.2.1.11.4 Emission decay The characteristics of the various processes that determine the dynamics of emission and the effects of external factors (temperature, doping concentrations) can be evaluated by monitoring the emission decay. Traditionally two methods of excitation are employed, very short pulse excitation and the measurement of decay after a square pulse excitation. At low doping concentration the decay is exponential and the lifetime is often assimilated to the radiative lifetime, although it can be shortened by the electron–phonon processes. The complex decay in the presence of energy transfer imposes drastic requirements on the measurements.

- In the experiments with short pulse excitation the duration of the exciting pulse must be shorter and the temporal resolution of the detection system must be better than any possible modification of the decay by the energy transfer.
- Registration and analysis of the decay over the entire temporal range, with special stress on the early-time behavior, is necessary for a meaningful characterization of de-excitation processes.
- In the case of multicenter structures, special care is needed in the section of the wavelength of the excitation and detection for measurement of global or of individual decay of these centers.
- The measurement based on square pulse excitation provides the correct decay only when this is exponential but cannot be used for investigation of the early-time departures from exponential induced by energy transfer. This decay contains contributions from ions excited at different moments inside this pulse. For most of these the fast evolution at early times is already completed at the moment when the excitation pulse ends and thus their contribution to the global decay will be limited to the slower part at the end of the individual evolution of their population, leading to smearing of the fast initial evolution and artificial lengthening of the measured decay.
- Special care should be taken to avoid the reabsorption of the emission since it re-circulates part of the emitted radiation back into the sample, lengthening the decay and deforming its shape. Since the reabsorption is dependent on the doping concentration and on sample size, it can smear the effect of the concentration dependent energy transfer terms in the rate equation.
- In the case of high emission cross-sections special care is needed to avoid accidental spontaneous emission amplification by internal reflections.

Accurate characterization of the energy transfer parameters from the decay measurements is crucial for subsequent calculation of the emission quantum efficiency. Experimentally the emission quantum efficiency can be measured by monitoring either the total radiation of the donor ion or the various effects of the heat generated by the joint action of the energy transfer and of electron–phonon interaction. Such measurement is difficult since it involves tedious calibration and utilization of parameters of the material or of the system of measurement that are not accurately known. Nevertheless, a check of the validity of the measured emission quantum efficiency with the calculated values based on the energy transfer parameters inferred from the emission decay is necessary.

2.2.1.11.5 Time-resolved spectroscopy In addition to conventional stationary excitation, luminescence spectra can also be obtained by short pulse excitation. Then, by registering the emission spectra in narrow temporal windows at various delay times after the exciting pulse, the emission spectra can reveal the particularities of emission for the various structural centers.

2.2.1.11.6 Raman spectroscopy When used as a macroscopic method, Raman spectroscopy provides global information on the structural centers and spectral data of importance for Raman shifted laser emission. Moreover, in the case of spatially resolved spectroscopy, the Raman spectra can reveal important local differences in the structure or composition that can be related to the results obtained by spatially resolved luminescence spectroscopy or other microstructural methods.

2.2.2 The optical pumping process

The laser material is processed to a well-established shape (cylindrical rod, plate, thin disk and so on) with highly polished plane ends. The size and shape of this component and the doping concentration or profile are determined by several factors, such as the need to absorb the excitation power, the need to limit the power (energy) per unit transverse surface of the beam (the fluence) to below the damage threshold of the active component and of the other optical components of the resonator, the need for efficient removal of the heat generated by non-radiative de-excitation and so on. The laser rod is optically pumped transversely or longitudinally with respect to the laser axis to produce the necessary population of the emitting level. The most usual optical pump sources are noble gas (Xe or Kr) arc discharge lamps or diode lasers, although laser or solar pumping can be very attractive in special cases. The efficiency of the pumping process can be characterized by the partial efficiencies of various steps.

- The transformation of the electrical power (energy) P_e supplied to the pump source into optical energy P_o of the pump source is characterized by the efficiency $\eta_{e-o} = P_o/P_e$.
- The transfer of the optical power of the pump source to the laser material is defined by the transfer efficiency $\eta_t = P_i/P_o$, where P_i is the power incident on the laser material.
- The absorption of the incident pump power by the laser material is characterized by the absorption efficiency η_a. Absorption from a monochromatic pump source obeys the Lambert–Beer law, i.e. after a path l inside a material with absorption coefficient $\alpha = n_1\sigma_a$ the radiation of intensity $I(0)$ is attenuated to $I(l) = I(0)\exp(-\alpha l)$ and the absorption efficiency is $\eta_a = 1 - \exp(-\alpha l)$; however, in the case of broad emission sources superposition with the absorption spectrum of the material must be taken into account.

2.2.2.1 Optical pump sources

2.2.2.1.1 Arc lamps Arc lamps can be operated in continuous wave or pulsed (hundreds of microseconds to several milliseconds) regime and have an emission spectrum extending from the ultraviolet to the near infrared, so they can be absorbed simultaneously into several absorption levels of the laser ions. The arc lamps are excited electrically and the input electric power per centimeter of discharge depends on the diameter of arc and in the CW regime is limited to hundreds of watts per centimeter, the lifetime of the lamp being hundreds to thousands of hours. In the pulsed regime the excitation energy is limited by the explosion energy, whose value is influenced by the pulse duration, being in the region of a hundred joules per centimeter, depending on diameter, whereas the lifetime (number of shots before 50% degradation) is determined by the operating point (percentage of the explosion energy) and by the repetition rate, and can range from thousands to billions of shots. The efficiency of transformation of electrical power into optical radiation η_{e-o} of arc lamps can be high, in the range 45–60%. The total power (or energy) of the lamp depends on its length: intense pumping would require long lamps. The constructive characteristics of these lamps impose processing of the laser material into rods whose length is adapted to that of the lamp, and the pumping configuration is transversal.

Utilization of arc lamps for pumping solid-state lasers has several major disadvantages. (i) The poor superposition of the emission spectrum with the absorption spectrum of many important laser materials limits utilization of the lamp emission, sometimes to only 10–20%. (ii) The short-wavelength portion of the emission spectrum can induce permanent or transient optically active defects in the pumped laser materials; in order to reduce this part of the incident pump radiation and transform it into more useful radiation, special spectral-converting or filtering glasses are used as envelope material for the lamps. (iii) Although the lamp and the laser material are placed in reflecting or diffusing enclosures, the omnidirectional emission of the lamps influences the efficiency of transfer η_t of the pump radiation to the laser material. (iv) The need for long laser components in the case of high power (energy) lasers excludes lamp pumping for laser materials that cannot be produced in large sizes. (v) The large quantum defect in lamp pumped infrared lasers leads to considerable wastage of the absorbed energy in non-radiative processes, inducing strong heating that limits severely the repetition rate of the laser. (vi) The size of the laser is large, the construction is complex, and maintenance and replacement of parts is difficult.

2.2.2.1.2 Diode lasers The diode lasers used for pumping solid-state lasers are broad area (wide-stripe) electrically excited semiconductor lasers whose emission is determined by recombination of electrons and holes in p–n or p–i–n heterostructures. The wavelength of emission is determined by the energy gap between the conduction and the valence bands of the semiconductor material and can be controlled, in specific ranges, by composition. Usually, ternary or quaternary III–V materials are used, such as InGaN active region on GaN or SiC substrate for blue emission in the range 380–470 nm, AlGaInP/GaAs for red emission (635–670 nm), AlGaAs/GaAs for 720–850 nm, AlInGaAs/GaAs for 840–900 nm, InGaAs/GaAs for 900–1100 nm, and InGaAsP/GaAs for 1000–1650 nm. The most powerful diode lasers are in the near infrared range: they have narrow emission bands (typically 2–4 nm, although 0.5 nm has been reported) at selected peak wavelengths corresponding to specific absorption lines of the laser materials. The peak emission wavelength is temperature sensitive (typically a few tenths of nanometers per degree kelvin, although under optimized conditions, for example using volume Bragg gratings, the stability can be improved by about an order of magnitude) and in many situations this imposes temperature stabilization. The optical–electrical efficiency η_{e-o} is high, above 50%, and in several cases values of 70% or attempts to reach over 80% have been reported. Usually the beam quality is quite poor, with high divergence, and the beam is elliptical and polarized. The diode lasers can be operated in CW or quasi-CW regime, with controlled duty cycle (repetition rate and pulse duration) and with peak power larger by about an order of magnitude than in the CW regime. The brightness is high but the power of individual lasers is quite small (several watts), however, they can be assembled in linear bars with CW power of tens of watts, or two-dimensional arrays of quite high global power (hundreds to thousands of watts). To increase the pump power further, the arrays of diodes can be configured in lattices of desired size and shape. The lifetime of the CW diode lasers is several tens of thousands of hours so CW arrays of diode lasers can compete with CW arc lamps for pumping. However, although the lifetime of pulsed diode lasers is high, the energies of the pulses cannot compare with those of the high energy flashlamps. The quality of the beam can be improved by coupling the diode laser to an optical fiber; the coupler contains correcting optics and a polarization scrambler and thus the radiation delivered by the fiber can be circular and symmetric, with a low degree of polarization. Fiber delivery (to several meters) can decouple spatially the pumping diode from the solid-state laser.

Diode laser pumping has several major advantages over lamp pumping: (i) laser emission of high efficiency with respect to the electrical power consumed, many (sometimes >10) times larger than for lamp pumping; (ii) increased utilization of the pump power and of absorption efficiency by matching the pumping wavelength to a strong absorption line close to the emitting level; (iii) considerably better utilization of the absorbed pump power due to the much smaller quantum defect, manifest in enhanced laser emission and reduced heat generation that enable higher beam quality, simplification of the cooling system and extended power scaling capability; (iv) the laser diode pumped lasers are compact and rugged and high power mobile units can be constructed; (v) easy maintenance and replacement; (vi) large variety of pumping configurations (transverse, longitudinal, brazing incidence and so on); (vii) high brightness pumping that enables efficient transfer (η_t to 85–98%) of very high pump power to small volumes of laser material and considerable reduction of the laser volume as well as construction of powerful lasers based on efficient materials that cannot be produced in large sizes; (viii) construction of new types of lasers such as fiber, waveguide, ring, corner-pumped, brazing incidence lasers and so on.

2.2.2.1.3 Laser pumping Laser pumping enables high intensity excitation in a very narrow bandwidth, in various regimes. Using tunable lasers the pump wavelength can be tuned with high accuracy to the absorption lines of the pumped material. Generally, due to the reduced efficiency of many pump lasers, the global efficiency of such laser pumped lasers can be quite low; however, in several cases this approach can be an acceptable (or unique) solution. Thus, compared with diode laser pumping, laser pumping has much greater wavelength stability over an extended power range and in many cases it can offer resonant pumping in wavelength ranges where efficient diode lasers are not available and the high brilliance enables efficient pumping for laser emission with reabsorption. Moreover, in many cases laser pumping can give the lowest quantum defect, i.e. minimal heat generation. Examples of such lasers are the Tm laser-pumped Ho laser, the Ti:sapphire laser pumped by the second harmonic of the 1 μm Nd or Yb laser and so on. Laser pumping with a tunable laser, particularly Ti:sapphire, is a very useful tool for investigation of the emission properties of various laser materials at different pump wavelengths.

2.2.2.1.4 Solar pumping There are several other pumping possibilities, one of the most challenging being solar pumping. The Sun is a practically non-exhaustible energy source: approximately 1.8×10^{14} kW energy is intercepted by the Earth, but only about 60% of this reaches the surface. The density of energy on the Earth's surface depends on location, $\sim$900 kWh m^{-2} in Europe and $\sim$2440 kWh m^{-2} in Africa. The solar radiation reaching the Earth is practically uni-directional and can be concentrated by Fresnel lenses and transferred with high efficiency to a laser material. The extraterrestrial solar emission spectrum can be considered as continuous over the ultraviolet, visible and near infrared ranges, and resembles black-body radiation at 5800 K, although the spectrum of radiation reaching the Earth's surface shows dips caused by absorption of the atmosphere. This spectrum has relative intensity of 0.8 to 1 between 410 and 720 nm, and efficient utilization of this radiation requires laser materials with good absorption properties in this range.

2.2.2.2 Transfer of pump radiation to the laser material

The pump radiation has definite size, shape and spatial distribution, omidirectional in the case of lamps and quasi-unidirectional, with specific intensity profile and polarization, in the case of lasers, diode lasers or solar pumping. This radiation is transferred to the laser material by various methods using reflection (profiled reflectors), or diffusion in the case of extended omnidirectional sources, or refraction (focusing optics) in the case of directional sources. Sometimes optical devices to modify the polarization state of the pumping system are coupled to the delivery system and the process is characterized by the pump transfer efficiency η_t defined above. Moreover, the transfer of pump radiation can modify the transverse distribution of the incident radiation. In order to increase the transfer, reflection of the pump radiation on the polished surface of the laser material in the case of longitudinal pumping must be reduced by an antireflection (AR) coating.

2.2.2.3 Absorption of the pump radiation in laser materials

According to Eq. (2.16) the absorption coefficient in a quantum system is determined by the population of the level from which the absorption originates and by the cross-section of the transition. In the case of solid-state laser materials the absorption transitions utilized in pumping originate from the Stark levels i of the ground manifold, with energies E_i ($E_1 = 0$). These Stark levels are thermallized, i.e. their population densities with respect to the global population of the ground manifold n_g are determined by the fractional thermal population coefficient, $n_i = f_i n_g$, with

$$f_i = \frac{\exp(E_i/k_B T)}{\sum_i \exp(E_i/k_B T)}.$$

The absorption coefficient is dependent on temperature, $\alpha_i = \sigma_i f_i n_g = \sigma_{i,eff} n_g$, both because of f_i and because of possible modification of σ_i induced by the thermal broadening and shift of the lines. A very important issue in optical pumping is the spectral matching of the pump emission and material absorption: if the pump emission band is narrower than the width of the absorption band and homogeneous broadening dominates, all the pump radiation can be used, whereas if the pump emission band is wider than the absorption, some of the pump radiation can leave the material without being absorbed.

2.2.2.4 Distribution of the pump radiation inside the laser material and the pump rate

The propagation of the pump radiation inside the (undoped) laser material can modify its intensity distribution, depending on the optical properties of the material (particularly the refractive index), its shape and size and the state of the surface (roughness). Finally, the intensity of pump radiation along the direction of propagation inside the laser material will be influenced by the absorption properties. Calculation of the actual distribution $r(x, y, z)$ would require numerical techniques, although in many instances approximate analytical distributions can be used.

In the case of diode laser longitudinal (end) pumping the transverse distribution of the absorbed pump beam, in the plane wave-front approximation, is determined by the transverse profile of the pumping beam and by the absorption inside the laser material. For materials with uniform doping and with matching pump and absorption profiles,

- in the case of an elliptical Gaussian pumping beam,

$$r(x, y, z) = \frac{2}{\pi w_p^2(z)l} \exp\left(-\frac{2(x^2 + y^2)}{w_p^2(z)}\right),$$

(2.60)

where w_p is the waist of the beam;
- for top-hat beams, specific to fiber-coupled diode lasers,

$$r(x, y, z) = \frac{1}{\pi w_p^2} \frac{\alpha \exp(-\alpha z)}{1 - \exp(-\alpha l)} \Theta\left(w_p^2 - x^2 - y^2\right)$$

(2.61)

where $\Theta(w_p^2 - x^2 - y^2)$ is the Heaviside step function.

In such cases, the pump radiation attenuates along its path inside the material according to the Beer–Lambert law but preserves its transverse profile. However, the distribution of the pump radiation in the laser material can be altered by non-uniform doping, either in the transversal plane, to favor low mode laser emission, or along the laser axis, to grant uniform longitudinal pump distribution or uniform heat generation. In the case of crystalline laser materials the multi-segmented rod approach can be used to control the pump absorption along the laser axis.

The spectral composition and transverse distribution as well as the dependence on time (continuous wave or pulsed) of the pump radiation together with the absorbing properties of the laser material, determine the dependence of the pump rate R per unit volume on the position (x, y, z) inside the laser material and on time, $R(x, y, z, t) = R_0(t)r(x, y, z)$

where $r(x, y, z)$ is the normalized pump distribution function $\int_V r(x, y, z) dV = 1$ and the total number of photons absorbed per unit time and unit volume is

$$\int_V R(x, y, z, t) dV = R_0(t) = \eta_a \eta_p \frac{P_{in}(t)}{h\nu_p V},$$

where $P_{in}(t)$ is the incident pump power. This non-uniform pumping rate induces a specific spatial distribution of the laser ions excited into the emitting level $n_{em}(x, y, z)$.

2.2.2.5 The pump level efficiency

A major criterion in selection of the laser pump level E_p is that all the absorbed pump energy should de-excite completely by very fast low order electron–phonon relaxation to the emitting level E_{em}. This assumes the existence of a ladder of closely spaced intermediate energy levels between the pump and the emitting level, with gaps of the order of the host lattice phonon energy; in many cases the pump level can be placed directly above the emitting level. In such cases the pump level efficiency η_p, which expresses the fraction of the photons absorbed into the emitting level that relax to the emitting level, is expected to equal unity. Despite fulfillment of this condition, in several important laser materials, such as Nd-doped yttrium aluminum garnet (YAG), the presence of additional de-excitation of the excited levels above the emitting level ${}^4F_{3/2}$ by coupling with high-phonon impurities (hydroxyl), or with color centers caused by the presence of hydroxyl, which circumvents the emitting level, has been claimed [42] in order to explain the observed aleatory variation of the emission quantum efficiency in YAG crystals with similar doping concentration. The doping ions coupled to such impurities would become inactive for laser emission and are called "dead sites." Although the presence of hydroxyl near the doping ion would be expected to perturb strongly the crystal field, leading to the appearance of specific optical spectra, no such additional lines were observed. It was also demonstrated that for diode laser pumping the observed non-radiative de-excitation can be fully explained by self-quenching of the emission. Moreover, investigation of the pump saturation effects in strongly absorbing materials, or comparison of the measured heat generation with estimations based on the emission quantum efficiency calculated with the energy transfer parameters, indicate clearly that in the case of Nd laser materials $\eta_p \approx 1$.

2.2.2.6 The quantum defect

Under monochromatic pumping of wavelength λ_p the quantum of pump radiation is larger than that of the emitted radiation and the difference is called the absolute quantum defect. In the case of lasers the emission wavelength is well established (λ_l) and for the four-level laser, the absolute quantum defect

$$\Delta_{qd}^{(abs)} = \Delta E_{pump} - \Delta E_l = (E_p - E_g) - (E_{em} - E_t)$$

is composed of two parts, the upper (superior) quantum defect $\Delta_{qd}^{(abs,up)} = E_p - E_{em}$ and the lower (inferior) quantum defect $\Delta_{qd}^{(abs,low)} = E_t - E_g$. Whereas the lower quantum defect is a characteristic of the emission scheme, the upper quantum defect is introduced for reasons connected with efficient pump absorption. In a three-level laser scheme the lower quantum defect is missing. The ratio between the absolute quantum defect and the pump quantum, which expresses the fraction of the pump energy quantum found in the laser quantum, defines the relative quantum defect,

$$\Delta E_{qd}^{(rel)} = \Delta E_{qd}^{(abs)}/E_p = 1 - \eta_{qd}^{(l)},$$

where $\eta_{qd}^{(l)} = h\nu_l/h\nu_p = \lambda_p/\lambda_l$ is the quantum defect (Stokes) ratio. However, in the case of luminescence, transitions from the emitting level E_{em} to several lower energy levels $E_t^{(f_i)}$ are possible, with branching ratios β_i; an average luminescence emission wavelength $\bar{\lambda}$ can be conventionally defined as

$$\bar{\lambda} \equiv \left(\frac{\bar{1}}{\lambda}\right)^{-1} = \left[\sum_i \beta_i \left(\frac{1}{\lambda_i}\right)\right]^{-1},$$

and the corresponding Stokes ratio is

$$\eta_{qd}^{(f)} = \lambda_p \left(\frac{\bar{1}}{\lambda}\right) = \lambda_p/\bar{\lambda}.$$

The average luminescence wavelength determines a fictitious luminescence terminal energy level $E_t^{(f)} = E_{em} - hc(\bar{1}/\lambda)$. In the case of broad band or polychromatic pumping a fictitious average pump wavelength $\bar{\lambda}_p$ can be defined by taking into account the convolution of the pump emission and laser material absorption spectra; this wavelength will then define an average quantum defect.

2.2.3 The laser resonator

2.2.3.1 The quality factor of the laser resonator

Optical resonators are characterized by the quality factor Q, defined as the ratio between the energy stored in the laser cavity and the energy lost per round-trip cycle. For a laser resonator of length l_{res} containing a laser rod of length l_{rod} and with refractive index n, which define an effective optical length $l_{eff} = l_{res} + (n-1)l_{rod}$, the quality factor $Q = 2\pi\nu_l\tau_c$ is related to the photon lifetime $\tau_c = (t_r/\varepsilon)$, where $t_r = 2l_{eff}/c$ is the photon round-trip time and $\varepsilon = -\ln R_1 + L$ defines the global losses. Consequently, the quality factor and thus the regime of laser emission can be controlled by optimizing l_{eff} and the losses ε.

2.2.3.2 The modal structure of optical resonators

The laser resonators are characterized by specific transversal and longitudinal oscillation modes, defined by the variation of the electromagnetic field perpendicular or along the laser axis, and this modal structure determines the characteristics of the laser emission. The transversal modes of the laser resonator determine the characteristics of the laser beam (energy distribution, diameter, divergence), whereas the longitudinal modes determine the spectral characteristics of the laser; consequently, the resonator gives spatial and temporal coherence to the laser beam.

The intensity distribution in a transversal (mn) mode TEM$_{mn}$, where m and n designate the number of peaks in the x and y directions perpendicular to the resonator axis z is given by

$$\rho_{mn}(x, y, z) = I_0 \left[H_m \left(\frac{\sqrt{2}x}{w(z)} \right) \exp \left(-\frac{x^2}{w^2(z)} \right) \right]^2 \times \left[H_n \left(\frac{\sqrt{2}y}{w(z)} \right) \exp \left(-\frac{y^2}{w^2(z)} \right) \right]^2$$

$$= \rho_0 \varepsilon(x, y, z) \tag{2.62}$$

where H_m and H_n are the Hermite polynomials. The total number of photons in a resonator mode is

$$q = \frac{1}{h\nu} \int_V \rho(x, y, z) dV = \frac{\rho_0}{h\nu} V_{eff},$$

where

$$V_{eff} = \int_{V_1} n\varepsilon(x, y, z) dV + \int_{V_2} \varepsilon(x, y, z) dV$$

accounts for the fact that the energy density inside the laser material volume V_1 is n times larger than in the volume of the rest of the resonator V_2, n being the refractive index of the laser material.

Of great practical importance are the lasers operating in the fundamental transversal mode TEM$_{00}$ whose profile is described by a Gaussian lineshape: they have the lowest divergence Θ and are characterized by the lowest beam parameter product $w(0)\Theta$. For these lasers the far-field beam waist increases with distance z according to $w(z) = (w(0)^2 + \Theta^2 z^2)^{1/2}$. The higher order TEM$_{mn}$ mode laser beams show m maxima in the x direction and n maxima in the y direction: their divergence is larger than for the TEM$_{00}$ mode and it can be characterized by a beam quality number M^2 such that $w(0)\Theta = M^2(w(0)\Theta)_{Gauss}$.

The existence of longitudinal modes is caused by the length l of the resonator being much larger than the wavelength of emission. The longitudinal modes are closely spaced in frequency, $\Delta\nu = c/2l$, so each transverse mode can accommodate several longitudinal modes. The spectral gain profile of the stimulated photons encompasses several longitudinal oscillation modes of the laser resonator and thus its spectral profile will consist of several discrete frequencies dephased with respect to each other.

2.2.3.3 Superposition of the pump and laser mode volumes

Only the excited laser ions placed inside the laser mode volume participate in the laser emission process and this selectivity is characterized by the superposition integral of pump radiation and laser mode volumes, $\eta_v = \int_V r(x, y, z)\varepsilon(x, y, z) dV$. The actual wavefront of pump radiation precludes an exact analytical calculation of this superposition integral; however, several approximations such as plane-wave or average waist along the whole rod have been used to calculate analytical equations for this integral. Approximation of the volume superposition η_v by the superposition of the transverse profiles (x, y) of the pump and laser beam has also been used. When the laser mode volume is larger than the pump volume at any point (x, y, z), the superposition integral is considered equal to unity. However, when the pump volume is larger than the laser mode volume, the superposition integral is smaller than unity and the excited ions that are outside the laser mode volume de-excite by luminescence and non-radiative processes. For instance, in case of monomode TEM$_{00}$ emission in a transversally pumped laser rod, the value of η_v can be quite low, in the range 0.3–0.4, but this can be increased to 0.8 in the case of multimode emission.

2.3 The flow of excitation inside the laser material

2.3.1 The steps in the flow of excitation inside pumped laser material

In real laser materials the laser level schemes are embedded into the more complex actual energy level scheme of the laser active ions and several of the remaining energy levels can introduce additional paths of de-excitation of the laser emitting level, offering conditions for parasitic radiative or non-radiative processes, such as multiphonon relaxation or self-quenching by energy transfer inside the system of active ions. The relation between the different de-excitation processes has been discussed in the literature [43]. A somewhat different approach [44], which enables a more convenient connection with the measured emission parameters and description of the spatial distribution of these effects, will be used here.

The flow of excitation in a four-level laser material is shown schematically in Figure 2.4. For sake of simplicity, only the down-conversion self-quenching by energy transfer between an excited ion and a non-excited ion (cross-relaxation on the intermediate level E_{cr}) is considered and it is assumed that below the level E_{cr} there is a ladder of closely spaced energy levels such that all excitation fed into this level relaxes to the ground level by electron–phonon interaction. The flow of excitation inside the laser material defined in Figure 2.4 corresponds to the following steps.

- *The first step* (**I**) is the absorption of pump radiation from the ground level E_g to the pump level E_p (the process (**1**)) and its efficiency is given by the absorption efficiency, $\eta(1) = \eta_a$.
- *The second step* (**II**) is the de-excitation of the pump energy level E_p. An efficient laser scheme requires fast non-radiative de-excitation of this level to the emitting level E_{em} (process (**2**)); however, sometimes de-excitation by other processes that circumvent the emitting level are possible (process (**3**)). The efficiencies of these processes relative to the absorbed excitation are $\eta(2) = \eta_p$ and $\eta(3) = 1 - \eta_p$.
- *The third step* (**III**) is de-excitation of the emitting level E_{em}. In a laser pumped above threshold there are two categories of excited ions: (i) ions that participate in laser emission (excited ions pumped above threshold in the region of superposition of the pump and laser mode volumes), which de-excite by stimulated emission (process (**4**)), and (ii) ions

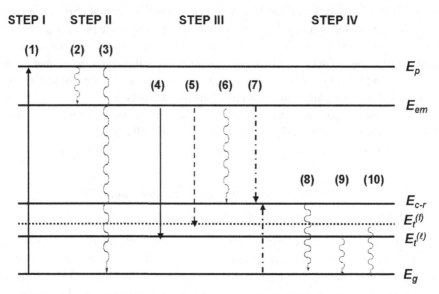

Figure 2.4 The flow of excitation in a four-level system embedded in a complex energy level scheme.

that do not participate in this process and can de-excite by three competing processes of luminescence (process (**5**), characterized by the emission quantum efficiency η_{qe}), multiphonon relaxation (process (**6**), with efficiency η_{mp}) or energy transfer (process (**7**), efficiency η_{et}). Obviously, $\eta_{qe} + \eta_{mp} + \eta_{et} = 1$. These two categories of excited ions are delineated by the laser emission efficiency η_l, which is determined by the characteristics of the laser emission process. With respect to the absorbed excitation, the relative efficiencies of the four processes of de-excitation of the emitting level are $\eta(4) = \eta_p \eta_l$, $\eta(5) = \eta_p(1 - \eta_l)\eta_{qe}$, $\eta(6) = \eta_p(1 - \eta_l)\eta_{mp}$ and $\eta(7) = \eta_p(1 - \eta_l)\eta_{et}$.

- *The fourth step* (**IV**) involves the de-excitation processes (**8**), (**9**) and (**10**) of the terminal levels E_{cr}, $E_t^{(l)}$, and $E_t^{(f)}$ for the processes (**4**) to (**7**). For most trivalent rare earth ions the energy levels below these terminal levels are dense enough to grant efficient non-radiative de-excitation of these terminal levels to the ground state E_g by electron–phonon interaction. The relative efficiencies of these processes with respect to the absorbed excitation are $\eta(8) = \eta_p(1 - \eta_l)\eta_{mp} + \eta_p(1 - \eta_l)\eta_{et} = \eta_p(1 - \eta_l)(1 - \eta_{qe})$, $\eta(9) = \eta_p \eta_l$, $\eta(10) = \eta_p(1 - \eta_l)\eta_{qe}$.

2.3.2 Manifestation of the de-excitation processes in pumped laser material

The de-excitation processes manifest as radiative processes (laser and luminescence) or as heat generation by non-radiative processes (multiphonon relaxation or energy transfer followed by multiphonon relaxation). Each of these processes can be characterized by the energy ΔE dissipated in the process and by its fraction $\eta = \Delta E / E_p$ of the excitation energy; these characteristics can be expressed with the partial efficiencies defined above.

Laser emission (process (4))

- The energy dissipated in the process

$$\Delta E^{(l)} = \eta_l \eta_p \left(E_{em} - E_t^{(l)} \right).$$ (2.63)

- The fractional laser emission coefficient

$$\eta^{(l)} = \frac{\Delta E^{(l)}}{E_p} = \frac{\eta_l \eta_p \left(E_{em} - E_t^{(l)} \right)}{E_p} = \eta_l \eta_p \eta_{qd}^{(l)}.$$ (2.64)

Luminescence emission (process (5))

- The energy dissipated in the process

$$\Delta E^{(f)} = (1 - \eta_l) \eta_p \eta_{qe} \left(E_{em} - E_t^{(f)} \right).$$ (2.65)

- The fractional luminescence emission coefficient

$$\eta^{(f)} = \frac{\Delta E^{(f)}}{E_p} = \frac{(1 - \eta_l) \eta_p \eta_{qe} \left(E_{em} - E_t^{(f)} \right)}{E_p} = (1 - \eta_l) \eta_p \eta_{qe} \eta_{qd}^{(f)}.$$ (2.66)

Non-radiative processes, which contribute to heat generation (processes (2), (3), (6), (7), (8), (9), (10))

- The energy dissipated in the process

$$\Delta E^{(h)} = E_p - \eta_p \eta_l \left(E_{em} - E_t^{(l)} \right) - \eta_p (1 - \eta_l) \eta_{qe} \left(E_{em} - E_t^{(f)} \right).$$ (2.67)

- The fractional non-radiative de-excitation coefficient

$$\eta^{(nr)} = \frac{\Delta E^{(h)}}{E_p} = 1 - \eta_l \eta_p \eta_{qd}^{(l)} - (1 - \eta_l) \eta_p \eta_{qe} \eta_{qd}^{(f)};$$ (2.68a)

obviously, $\eta^{(l)} + \eta^{(f)} + \eta^{(nr)} = 1$.

Equation (2.68a) shows the contribution to non-radiative de-excitation of the excited ions that participate in lasing (second term on the right side) and of those that do not lase (last term). The non-radiative de-excitation includes electron–phonon relaxation and energy transfer: when the final levels of the donor and acceptor de-excite by electron–phonon relaxation, all excitation of the donor ion is transformed into heat, and in this case the fractional non-radiative de-excitation coefficient is equivalent to the heat generation (thermal load) coefficient η_h which expresses the fraction of absorbed power transformed into heat in a laser [45], i.e.

$$\eta_h \equiv \eta^{(nr)} = 1 - \eta_l \eta_p \eta_{qd}^{(l)} - (1 - \eta_l) \eta_p \eta_{qe} \eta_{qd}^{(f)}.$$ (2.68b)

Equation (2.68b) shows that a high pump level efficiency η_p, large emission quantum efficiency of the emitting level and small quantum defect are desired to limit heat generation. By expressing explicitly the dependence of η_h on η_l, η_{qd} and η_{qe}, Eq. (2.68b) is instrumental in explaining the differences in heat generation between the lamp and diode laser pumped lasers, or between the different diode laser pumping wavelengths, as well as between different doping concentrations. The heat loading coefficient can be particularized for two extreme cases:

- heat generation by the excited ions that do not participate in lasing

$$\eta_h^{(f)} = 1 - \eta_p \eta_{qe} \eta_{qd}^{(f)};$$ (2.69)

- heat generation by the ions that participate in lasing

$$\eta_h^{(l)} = 1 - \eta_p \eta_{qd}^{(l)}.$$ (2.70)

Equation (2.70) shows that for the ions that de-excite by stimulated emission the heat generation is independent of the emission quantum efficiency; in the case of laser materials with self-quenching this results in the independence of heat generation by these ions of the doping concentration and, when $\eta_p = 1$, $\eta_h^{(l)}$ equals the relative quantum defect $\Delta E_{qd}^{(rel)}$. However, for heat generation by the ions that do not de-excite by stimulated emission, the emission quantum efficiency η_{qe} and its dependence on concentration is also important (Eq. (2.69)). As discussed above, under conditions of low density of excited ions in the emitting level, the self-quenching is dominated by down-conversion cross-relaxation and the emission quantum efficiency is not dependent on the pump intensity, whereas in the case of high density of excited ions not participating in lasing, the onset of upconversion will introduce a pump intensity dependence of the emission quantum efficiency and of the heat loading parameter $\eta_h^{(f)}$. According to Eqs. (2.69) and (2.70), the equality $\eta_{qd}^{(l)} = \eta_{qe} \eta_{qd}^{(f)}$ implies $\eta_h^{(l)} = \eta_h^{(f)}$. However, $\eta_{qd}^{(l)} > \eta_{qe} \eta_{qd}^{(f)}$ leads to $\eta_h^{(l)} < \eta_h^{(f)}$; obviously, when $\eta_{qd}^{(l)} < \eta_{qe} \eta_{qd}^{(f)}$ the opposite situation holds. For each laser emission wavelength the crossing point corresponds to a defined emission quantum efficiency $\eta_{qe}^{cross} = \bar{\lambda}/\lambda_l$, i.e. to a well-defined doping concentration and is independent of the pump wavelength. This specific heat generation by the two classes of excited ions will then determine the global behavior of the laser material, in proportion with the laser emission efficiency η_l. In the absence of laser emission ($\eta_l = 0$) all excited ions de-excite by non-radiative processes and the global fractional heat generation of the pumped laser material corresponds to $\eta_h^{(f)}$ (Eq. (2.69)), whereas in the presence of laser emission Eq. (2.68) must be used and this will approach $\eta_h^{(l)}$ (Eq. (2.70)), i.e. the relative quantum defect limit, only for very efficient laser emission ($\eta_l \approx 1$).

This discussion of the heat generation is strictly valid for systems without reabsorption of emitted radiation. The reabsorption recirculates part of the emission back to the emitting level, although not all this recycled excitation participates in a new laser emission process and thus it acts as a loss mechanism that reduces the laser emission efficiency η_l. Reabsorption modifies the global heat generation

$$\eta_h = 1 - \eta_l \eta_p \eta_{qd}^{(l)} (1 - \gamma f^{(l)}) - (1 - \eta_l) \eta_p \eta_{qd}^{(f)} \eta_{qe} (1 - \gamma f^{(f)})$$ (2.71)

where $f^{(f)}$ and $f^{(l)}$ are the fractions of luminescence, respectively laser radiation, reabsorbed in the laser material and γ is a parameter that accounts for the whole flow of excitation caused by the reabsorbed radiation and can be approximated by

$$\gamma \approx \frac{(1 - \eta_l)\left(1 - \eta_{qe}\eta_{qd}^{(f_i)}\right) + \eta_l\left(1 - \eta_{qd}^{(l_i)}\right)}{1 - [f^{(f)}(1 - \eta_l)\eta_{qe} + f^{(l)}\eta_l]}. \tag{2.72}$$

In this equation $\eta_{qd}^{(f_i)}$, $\eta_{qd}^{(l_i)}$ are the quantum defect ratios for luminescence and laser emission respectively, when only the lower quantum defect (the gap between the terminal energy level and the ground level) is considered. Reduction of the laser emission efficiency η_l by reabsorption increases the relative contribution to heat generation of the ions that do not participate in lasing and leads to a larger global heat loading than that determined only by the quantum defect.

2.4 Laser emission processes

As discussed in Section 2.1.3, laser emission requires an amplification gain larger than the global optical losses inside the resonator and at the mirrors, $2gl \geq L - \ln R_1 R_2$. The temporal dependence of these losses determines the temporal regime of laser emission, whereas the wavelength dependence of the emission cross-section and of losses determines the laser wavelength. Since $g = \sigma_e \Delta n$, the general features of the laser emission can be described by a system of coupled rate equations for the difference in populations $\Delta n = n_{em} - n_t$ in the presence of stimulated emission under conditions of pumping and for the light flux that determines the stimulated emission rate.

Because the excitation fed into the pump level E_p decays rapidly to the emitting level E_{em}, the only populations of relevance for the laser process are those of the emitting level n_{em}, labeled conventionally by n_2, of the terminal level n_t (or n_1) and of the ground level n_g (or n_0): the populations n_2 and n_1 determine the inversion of population, whereas the population n_0 influences the pump rate and their sum equals the total density n_{tot} of laser ions. In the four-level lasers $n_1 \approx 0$ and the inversion of population $\Delta n \approx n_2$; thus $n_2 + n_0 \approx n_{tot}$ and since n_2 is usually very small, $n_0 \approx n_{tot}$. Thus, in the case of four-level lasers the equation of evolution of population inversion Δn coincides with that of population n_2 of the emitting level.

In the case of three-level lasers the population of the terminal level $n_1 \equiv n_0$ can be considerable, whereas the population of the emitting level must be very large, $n_2 \geq (1/2)n_{tot}$, and the equation of evolution of inversion population requires the account of the evolution of population n_1 too: since $n_2 + n_1 = n_{tot} = $ constant, at any moment of time $dn_1/dt = -dn_2/dt$.

2.4.1 The free generation regime

When the resonator optical losses and the outcoupling losses are constant in time and small enough to give a high quality factor Q, the laser operates in the free generation regime and the temporal regime of laser emission is determined only by the regime of pumping. The equation describing the evolution of the difference in populations Δn in the presence of pumping and of de-excitation by luminescence, non-radiative electron–phonon relaxation, energy transfer from the emitting level and stimulated emission in the case of a laser with free generation can be written

$$\frac{d\Delta n}{dt} = -\left(\frac{1}{\tau_f} + \frac{dP(t)}{dt}\right)[\Delta n + n_{tot}(\gamma - 1)] - \phi\sigma_e c \Delta n \gamma + R \qquad (2.73)$$

where R is the density of ions fed into the emitting level by pumping and γ accounts for the variation of Δn for each emitted photon: for the four-level lasers $\gamma = 1$ whereas for the three-level lasers $\gamma = 1 + (g_2/g_1)$. In this equation τ_f is the luminescence lifetime at low doping concentrations, and $P(t)$ is the energy transfer function defined in Section 2.2.1.8. In the general case the pumping term R, the population inversion Δn, and the photon flux ϕ connected to the radiation energy density ρ by $\phi = \rho/h\nu$, depend on the position of the emitting ions inside the laser material and on time. When the energy transfer is dominated by down-conversion (low population n_2) the energy transfer function $P(t)$ does not depend on the pumping intensity; however, if upconversion is present, it introduces in the transfer function a term dependent on the pumping rate, leading to an additional term non-linear in Δn in Eq. (2.73). The relative contribution of the stimulated emission to the de-excitation of the emitting level can be expressed by the stimulated emission efficiency

$$\eta_{se} = \phi\sigma_e c \left/ \left(\frac{1}{\tau_f} + \frac{dP(t)}{dt} + \phi\sigma_e c\right)\right. , \qquad (2.74)$$

which is not constant and depends on the stimulated emission photon number of the cavity mode: at low photon number η_{se} is low, but above laser threshold it approaches unity asymptotically.

The variation of the stimulated emission photon flux ϕ is described by the rate equation

$$\frac{d\phi}{dt} = \sigma_e c \int_V \phi \Delta n \, dV - \frac{\phi}{\tau_c} + S; \qquad (2.75)$$

the integration is done over the laser material volume V and S accounts for the contribution of the spontaneous emission along the laser axis to the light flux and is usually neglected.

2.4.1.1 Continuous wave laser emission

In the case of CW laser emission $d\Delta n/dt$ and $d\phi/dt$ are equal to zero and the pump rate R and the population difference Δn are independent of time. The effect of the non-radiative de-excitation processes on luminescence in the absence of stimulated emission can be described by the emission quantum efficiency η_{qe} and by the effective lifetime $\tau_{eff} = \tau_f \eta_{qe}$ and

$$\eta_{se} = \phi\sigma_e c \left/ \left(\frac{1}{\tau_{eff}} + \phi\sigma_e c\right)\right. .$$

2.4.1.1.1 Four-level systems As discussed above, in the four-level lasers $\gamma = 1$, $\Delta n \approx n_{em}$, $n_0 \approx n_{tot}$, and thus $R = W_p n_0$, W_p being the pumping rate. The CW laser emission is then described by

$$-\frac{\Delta n}{\tau_{eff}} - \phi \sigma_e c \Delta n + W_p n_0 = -\frac{1}{\eta_{se}} \phi \sigma_e c \Delta n + W_p n_0 = 0 \qquad (2.76)$$

$$\sigma_e c \int_V \phi \Delta n dV - \frac{\phi}{\tau_c} = 0. \qquad (2.77)$$

Because of the spatial dependence of the pumping rate, of population inversion and of photon flux, an accurate analytical solution of these equations is not possible; however, approximate analytical solutions based on average (effective) values can prove useful for global descriptions of emission. With these equations the saturated gain g in the presence of stimulated emission can be related to the small-signal gain g_0 determined by the spontaneous emission at the laser threshold,

$$g = g_0 \left(1 + \frac{c \sigma \phi}{W_p + \tau_{eff}^{-1}} \right)^{-1} \qquad (2.78a)$$

or, by using the relation between the intensity and the photon flux $I = c \phi h \nu$,

$$g = g_0 \left(1 + \frac{I}{I_s} \right)^{-1} \qquad (2.78b)$$

where

$$I_s = \frac{h \nu}{\sigma_e \tau_{eff}} \qquad (2.79)$$

is the saturation intensity. The intensity of the light beam inside the laser resonator is then

$$I = I_s \left(\frac{2 g_0 l}{L - \ln R_1} - 1 \right) \qquad (2.80)$$

and the output power of the laser, outcoupled through the mirror R_1, is

$$P_{out} = A \left(\frac{1 - R_1}{1 + R_1} \right) I_s \left(\frac{2 g_0 l}{L - \ln R_1} - 1 \right), \qquad (2.81)$$

where $A = V_{eff}/l$ is the effective average area of the laser beam inside the resonator.

Under CW pumping the inversion of population in a four-level system in the absence of laser emission is $\Delta n = n_0 W_p \tau_{eff}$ and the small-signal gain coefficient is $g_0 = \sigma_e n_0 W_p \tau_{eff}$. An effective rate of feeding ions in the emitting level per unit volume can be defined in relation to the absorbed and incident pump power by dividing the total number of photons absorbed per unit time by the pumped volume V, i.e.

$$W_p n_0 = \eta_p \eta_{qd}^{(l)} \eta_v \frac{P_{ab}}{h \nu_l V} = \eta_p \eta_{qd}^{(l)} \eta_v \eta_a \frac{P_{in}}{h \nu_l V}$$

and the effective small-signal gain is

$$g_0 = \sigma_e \tau_{eff} \eta^{(in)} \frac{P_{in}}{h\nu_l V},$$

with $\eta^{(in)} = \eta_p \eta_{qd}^{(l)} \eta_v \eta_a$. Then Eq. (2.81) becomes

$$P_{out} = \eta_{sl}^{(in)} \left(P_{in} - P_{th}^{(in)} \right), \tag{2.82}$$

where

$$\eta_{sl}^{(in)} = \left(\frac{-\ln R_1}{L - R_1} \right) \eta^{(in)} \approx \frac{T}{T+L} \eta^{(in)} \tag{2.83}$$

is the slope efficiency, and

$$P_{th}^{(in)} = \left(\frac{L - \ln R_1}{2} \right) \frac{A I_s}{\eta^{(in)}} \approx \left(\frac{T+L}{2} \right) \frac{A I_s}{\eta^{(in)}} \tag{2.84}$$

is the laser threshold. Equation (2.82) shows that above the laser threshold the output power is linear in incident power with slope determined by η_{sl}. By taking into account the definition of saturation intensity and of the parameter $\eta^{(in)}$, the characteristic laser parameters can be expressed with the partial efficiencies described earlier,

$$P_{th}^{(in)} = \frac{1}{2} \frac{A h \nu_l}{\eta_a \eta_p \eta_{qd}^{(l)} \eta_v \tau_{eff} \sigma_{eff}} (T + L), \tag{2.85}$$

$$\eta_{sl}^{(in)} = \eta_a \eta_p \eta_{qd}^{(l)} \eta_v \frac{T}{T+L}, \tag{2.86a}$$

where the emission cross-section of the laser transition σ_e is replaced by the effective cross-section $\sigma_{eff} = f_{em}\sigma_e$, in order to account for the fact that in many materials the emitting level belongs to a group of thermallized levels and f_{em} is its fractional thermal population coefficient. By defining an outcoupling efficiency $\eta_{oc} = T/(T+L)$ as the fraction of the outcoupling loss from the total losses of the resonator, the slope efficiency can be written

$$\eta_{sl}^{(in)} = \eta_a \eta_p \eta_{qd}^{(l)} \eta_v \eta_{oc}. \tag{2.86b}$$

The equation of the light flux indicates that during operation of a continuous wave four-level laser the inversion of population inside the laser mode volume does not depend on pump intensity and remains clamped at the value corresponding to the laser threshold

$$n_{th} = \frac{1}{2} \frac{T+L}{\sigma_{eff} l}.$$

However, when the pumped volume is larger than the laser mode volume ($\eta_v < 1$), the population n_2 can increase when pumping above threshold, on account of the excited ions placed outside the laser mode volume [46]. This can contribute to the onset of higher order mode laser emission at high pump power as well as to increased de-excitation by luminescence and non-radiative processes, favoring amplified spontaneous emission and enhanced heat generation.

An optical–optical efficiency of the four-level CW laser can also be defined,

$$\eta_{o-o}^{(in)} = \eta_{sl}^{(in)}\left(1 - \frac{P_{th}^{(in)}}{P^{(in)}}\right) = \eta_a\eta_p\eta_{qd}^{(l)}\eta_v\left(1 - \frac{P_{th}^{(in)}}{P^{(in)}}\right)\frac{T}{T+L}$$

$$= \eta_a\eta_p\eta_{qd}^{(l)}\eta_v\eta_{oc}\left(1 - \frac{P_{th}^{(in)}}{P^{(in)}}\right) \tag{2.87}$$

such that

$$P_{out} = \eta_{o-o}^{(in)}P_{in}. \tag{2.88}$$

The laser emission parameters P_{th} and η_{sl} can also be defined with respect to the absorbed pump power $P^{(a)} = \eta_a P^{(in)}$. The optical–optical efficiency expressed as absorbed power is related to the fractional laser emission coefficient defined by Eq. (2.64) by

$$\eta_{o-o}^{(a)} = \eta^{(l)}\frac{T}{T+L} = \eta^{(l)}\eta_{oc}; \tag{2.89}$$

for a laser material with negligible residual losses ($L \approx 0$), $\eta_{oc} \approx 1$ and $\eta_{o-o}^{(a)} \approx \eta^{(l)}$.

Alternatively, using the definition of the stimulated emission efficiency η_{se}, Eq. (2.74), the saturated gain can be related to the small-signal gain by

$$g = g_0\frac{I_s}{I}\eta_{se} \tag{2.90}$$

and the output power for the CW four-level laser becomes

$$P_{out} = \eta\frac{T}{T+L}\eta_{se}P_{in} = \eta\eta_{oc}\eta_{se}P_{in}. \tag{2.91}$$

Comparison with Eq. (2.82) shows that the stimulated emission efficiency can be related to the measured laser emission data,

$$\eta_{se} = \left(1 - \frac{P_{th}}{P}\right). \tag{2.92}$$

The product

$$\eta_l = \eta_v\left(1 - \frac{P_{th}}{P}\right) = \eta_v(1 - f_{th}) = \eta_v\eta_{se} \tag{2.93}$$

entering in the equation for the optical–optical efficiency

$$\eta_{o-o}^{(in)} = \left(\eta_a\eta_p\eta_{qd}^{(l)}\frac{T}{T+L}\right)\eta_l = \left(\eta_a\eta_p\eta_{qd}^{(l)}\eta_{oc}\right)\eta_l \tag{2.94}$$

delineates the fraction $(1 - f_{th})$ of ions excited into the emitting level inside the laser mode volume which participate in lasing, and it can be defined as the laser emission efficiency. Many papers consider η_l equal to unity immediately above threshold and usually no connection with the volume superposition or with the operating point P/P_{th} is made. Figure 2.5 shows that even when $\eta_v = 1$, η_l tends asymptotically to unity only at very

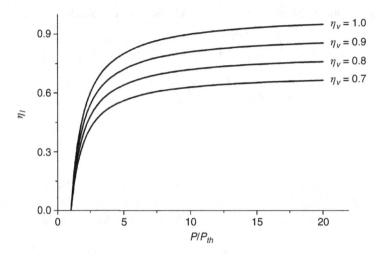

Figure 2.5 Laser emission efficiency versus pump power for various volume superposition efficiencies.

high pumping above threshold ($\eta_l = 0.9$ and 0.95 for $P/P_{th} = 10$ and respectively 20) and its value can be strongly reduced by a poor pump-laser mode volume superposition. This dependence is essential for a correlated assessment of the laser parameters and of heat generation during laser emission.

The residual losses L can be estimated [47] by comparing the emission threshold for a CW laser using extraction mirrors of various reflectivities R_1. With these values and with the known laser quantum defect the quantity $\eta_{qd}^{(l)}T/T + L$ can be calculated and any departure of the measured slope efficiency $\eta_{sl}^{(a)}$ from this quantity indicates a lower than unity $\eta_p\eta_v$ product; if $\eta_p = 1$, this could be indicative of poor volume superposition.

Several important conclusions can be drawn from Eqs. (2.85)–(2.87).

- A small-diameter laser beam favors lower threshold and this can be accomplished by focusing the pump beam or by reducing the transverse size of the laser rod (fiber or waveguide lasers).
- The threshold is reduced and the slope efficiency is enhanced for a lower quantum defect.
- The emission quantum efficiency η_{qe} and the effective emission cross-section σ_{eff} influence only the laser threshold, not the slope efficiency.
- A good overlap of the pump and laser beam determines improved laser parameters.
- Operating high above threshold improves the optical–optical efficiency.

There are two major classes of four-level laser materials. Either they have sharp zero–phonon transitions, usually based on rare earth ions, or they have broad emission lines based on vibronic transitions, specific to 3d elements. The emitting level can be thermallized with Stark levels from the same or near manifolds. For all transitions from the group of thermallized levels the effective lifetime τ_{eff} is the same, but the laser quantum defect ratio $\eta_{qd}^{(l)}$ and the emission cross-section σ_e are different. Moreover, since the fractional thermal population coefficients of these thermallized Stark levels depend on temperature, the effective emission cross-section σ_{eff} and, depending on the manifolds involved in thermallization, the effective lifetime τ_{eff} depend on temperature. Thus, the laser slope efficiency for the optical transitions originating from these thermallized levels will be influenced only by their individual quantum defect, although their laser threshold can differ strongly due to the differences in the effective emission cross-section.

The selection of the laser transition is determined by the spectroscopic properties of the active ion and by the spectral selectivity of the laser resonator as well as by the operating temperature. According to Eq. (2.85), in the case of a laser resonator without sharp spectral selectivity the lowest threshold will correspond to the wavelength λ that determines the largest product $\sigma_{eff}(\lambda)\eta_{qd}^{(l)}(\lambda)$ at the operating temperature and the laser will operate on this transition. Sometimes simultaneous emission on two or more transitions can be possible. Moreover, when changing the temperature, the laser emission wavelength follows the thermal shift of the selected optical transition, although sudden changes on other transitions can take place when the modification of thermal fractional coefficients f modifies the balance of the products $\sigma_{eff}\eta_{qd}^{(l)}$ of the various transitions. However, using spectrally selective elements inside the resonator (narrow-spectrum mirrors, gratings, prisms, etalons, filters, volume Bragg gratings and so on), it becomes possible to select any frequency for laser emission, regardless of $\sigma_{eff}\eta_{qd}^{(l)}$.

2.4.1.1.2 Three-level lasers In the case of three-level lasers the necessary inversion of population requires at least half of the laser ions in the emitting level. In this case the rate of feeding the emitting level is $R = W_p(n_{tot} - \Delta n)$ and implies a high pumping rate W_p to reach threshold. This requires laser materials with a large $\tau_{eff}\sigma_{eff}$ product and there are very few cases of pure three-level continuous wave solid-state lasers, although pulsed emission under the more powerful flashlamp pumping could be more attractive.

2.4.1.1.3 Lasers with partial reabsorption of laser emission When the terminal laser level is close to the ground level, its thermal population at the operating temperature cannot be neglected and partial reabsorption of the laser radiation can take place [46, 48]. This situation is specific to the lasers based on 4f ions, when the laser transition terminates on one of the excited Stark levels of the ground manifold. Such a laser scheme is usually termed a "quasi-three-level" (QTL) scheme, although many works prefer "quasi-four-level" scheme since the terminal laser level does not coincide with the ground level.

As mentioned before, the reabsorption recirculates part of the laser radiation back into the emitting level and thus its effect is complex: on the one hand it introduces an additional loss process for the stimulated emission, which reduces the gain, and on the other hand it introduces an additional "self-resonant laser pumping" process. Obviously, the recycled excitation does not return fully to laser emission and thus a net loss results. For such lasers the minimum fraction of the population of Nd ions $f_{min}n_0$ that needs to be in the upper laser level to give prevalence of laser emission over reabsorption, i.e. the transparency of the laser material to laser radiation, is $f_{min} = f_t/(f_t + f_l)$, where f_l and f_t are the thermal population coefficients of the emitting and terminal Stark levels of the ground manifold. However, because of the residual optical and outcoupling losses, the inversion at threshold must be larger than that imposed by f_{min}.

Reabsorption of laser radiation modifies the rate equation of the evolution of population inversion to

$$\frac{d\Delta n}{dt} = -\left(\frac{1}{\tau_f} + \frac{dP(t)}{dt}\right)[\Delta n + f n_{tot}] - \phi\left(\sigma_{eff}^e + \sigma_{eff}^{ra}\right)c\Delta n + R, \qquad (2.95)$$

where $\sigma_{eff}^e = \sigma_e f_l$ is the effective emission cross-section and $\sigma_{eff}^{ra} = \sigma_{ra} f_t$ is the effective reabsorption cross-section: in the case of these lasers $\sigma_e = \sigma_{ra}$ and the ratio $f = \sigma_{eff}^{ra}/\sigma_{eff}^e \equiv f_t/f_l$. The rate equation for the stimulated emission photon number q is similar to Eq. (2.75). In the case of CW quasi-three-level laser emission the threshold becomes

$$P_{th}^{(in)} = \frac{1}{2} \frac{Ah\nu_l}{\eta_a \eta_p \eta_{qd}^{(l)} \eta_v \tau_{eff} \left(\sigma_{eff}^e + \sigma_{eff}^{ra}\right)} (T + L + \delta_{ra}), \qquad (2.96)$$

where $\delta_{ra} = 2\alpha_{ra} l_{rod}$ is the additional reabsorption optical loss, with $\alpha_{ra} = \sigma_{eff}^{ra} n_0 = \sigma_e f_t n_0$; under conditions of low depletion of the ground level, $n_0 \approx n_{tot}$. Since the reabsorption recycles part of the laser emission back to the emitting level inside the laser mode volume above threshold, the laser slope efficiency will be similar to that of a four-level laser, Eq. (2.86). The enhanced threshold due to reabsorption makes it difficult to achieve a high operating point (P/P_{th}) and laser emission efficiency η_l, with a negative effect on the optical–optical efficiency and on heat generation.

Among the characteristics that determine reabsorption, a high emission cross-section is desired since this contributes to a reduced threshold, and the fractional thermal population coefficient can be controlled by temperature. On the other hand, in conventional single pass longitudinally pumped lasers, the length of the rod and the doping concentration influence the pump absorption. Under these circumstances several modalities to reduce reabsorption can be envisaged. Thus, the length of the rod can be decoupled from the pump absorption path by two modalities: a thin disk [49] of laser material can be pumped longitudinally by multipass of the pump radiation, or by transverse pumping. This approach also contributes to a reduction in residual losses L; additionally it enables very efficient dissipation of heat and tight control of the thermal field inside the material. The cooling of the laser material will reduce the fractional thermal population of the terminal laser level f_t and the reabsorption; this can also contribute to a more efficient dissipation of heat due to increased thermal conduction. Another possibility for obtaining good absorption of the pump radiation at very low doping concentration and low threshold with a small transverse area of the laser beam A, is utilization of long optical fibers.

In the case of systems with several possible emission lines to the various Stark components of the ground manifold, the reabsorption can act as a selective element for the lasing transition and this is not necessarily the one with the largest emission cross-section σ_{eff} as happens in the four-level systems. In certain cases and limits, the selection of the lasing transition could be influenced by the transmission T of the output mirror that modifies the relative balance among the losses and that of the losses and emission cross-section; unfortunately this choice can impose a high emission threshold. Nevertheless, the most effective way of utilizing the emission properties of these systems is transformation of as many as possible of the emission schemes into four-level laser schemes by cooling.

2.4.1.2 *Pulsed free generation laser emission*

The CW pump sources are limited to powers of the order of kilowatts to tens of kilowatts. Moreover, the CW regime is not always advantageous for specific applications. The free generation laser can also be pumped with pulsed radiation sources, such as flashlamps, with pulse duration of the order of several hundred microseconds to milliseconds, and with pulse energies of up to tens of kilojoules, i.e. with peak powers of the order of tens of megawatts. The free generation pulsed lasers can operate in isolated pulses or in trains of pulses (bursts) and this regime will also enable better dissipation of heat than the CW lasers. This pulsed laser emission is governed by the laser Eqs. (2.73) and (2.75), with proper account of the temporal dependence of the pump rate R which determines the duration of the laser pulse. For such lasers, the laser parameters (threshold, slope efficiency, output) are usually expressed in energy, not in power as in the case of CW emission. In the case of repetitive pulsing the average power of emission is also specified. The first laser [1], which employed a ruby crystal, was in fact a free generation flashlamp pumped three-level laser.

2.4.2 *Cavity loss controlled laser emission regimes*

2.4.2.1 *Q-switched laser emission*

The onset of laser emission (threshold) is determined by the balance between gain and resonator losses, which governs the population inversion. The level of losses determines the resonator quality factor Q: high Q implies modest intrinsic losses (excluding outcoupling) and thus the threshold population inversion is quite low; because the lifetime of the photons in the cavity τ_C is very short, the laser emission follows the temporal regime of pumping, whereas the power is limited below that of pumping. However, if the quality factor of the resonator deteriorates, the laser threshold will be raised very high and laser emission will be inhibited. With increased pump power the population of the emitting level can be increased much above the value corresponding to the laser threshold in the high Q resonator. On restoring the quality factor rapidly (switching), the photon flux in the resonator increases to very high values and almost all of the stored population inversion will be released in a very short pulse of peak power much higher than that of the pumping source. The Q-switched lasers can operate in isolated (single) pulse regime or in trains with the repetition rate limited by the need to restore the inversion of population to the low Q threshold value; additional limitation can be introduced by heat generation in the laser material in this regime of high population inversion storage.

The quality factor can be controlled by acting on the basic constituents of the laser resonator and/or by inserting in the resonator new types of components that can control the losses or propagation of the beam inside the laser material. These methods can be active or passive and sometimes mixed active–passive Q-switching can be used. The active Q-switching methods are based on external control of the temporal regime of the reflectivity, for example by tilting the mirror plane (spinning total reflection prism), or of the conditions of propagation (acousto-optic deflection, rotation of the polarization plane by electro-optic effect). Passive Q-switching is based on control of the absorption losses or of the reflectivity by inserting materials whose properties are dependent on the photon flux. Thus, the resonator absorption losses can be controlled by insertion in the resonator of materials that absorb strongly the increasing intracavity photon beam until the absorption saturates and the material becomes transparent: the high quality factor of the resonator is thus restored and intense pulses of very short duration are emitted. The emission reduces the intracavity laser beam intensity and thus the high absorption regime of the material is restored and the process repeats. Such saturable absorbers can be materials doped with ions with high absorption cross-section at the wavelength of the laser emission and with short lifetime, materials with color centers, semiconductors, nanosystems and so on. Such materials can also be used for passive modification of the reflectivity of the mirrors.

The Q-switching techniques can be employed with CW as well as with pulsed excitation. In the case of CW pumping with rate W_p the population of the emitting level in the absence of laser emission is $n_{em} = \tau_{eff} W_p n_0$ and the pump rate relates to the incident CW power P_i by $W_p = (P_i/V)\eta_a \eta_p$; thus, a high population in the emitting level will require a high incident pump power and high absorption efficiency. The power of the CW pump sources is limited, so the stored inversion is low and Q-switching of the CW pumped lasers can provide pulses of low energy, tens of nanoseconds in duration, at quite high repetition rates (tens of kilohertz with acousto-optic switching and hundreds of megahertz for passive switching), and with average power of tens of watts. Pulsed pumping can provide high pump intensities that feed a high population density into the emitting level: for a square pulse of duration t_p this equals $n_0 W_p t_p$, but due to luminescence and non-radiative de-excitation the maximum population of the emitting level reached at the end of the pump pulse is equal to

$$n_e(t_p) = n_0 W_p \tau_{eff}[1 - \exp(-t_p/\tau_{eff})] \qquad (2.97)$$

and increases with t_p and with τ_{eff}. With this equation, the fraction of excited ions available at the end of the pumping pulse (the storage efficiency) is

$$\eta_{st} = [1 - \exp(-t_p/\tau_{eff})](t_p/\tau_{eff})^{-1}; \qquad (2.98)$$

high storage efficiency can be obtained for short t_p and/or long τ_{eff}. As a compromise between high $n_e(t_p)$ and high η_{st}, it is customary to use $t_p \approx \tau_{eff}$.

The evolution of the inversion of population and of the photon flux under single pulse excitation in the presence of the Q-switch can be described by adaptation of the rate equations (2.73) and (2.75), taking into account that during the emission of the short pulse the pump rate and the fluorescence emission do not change. The additional time dependent loss $\zeta(t)$ introduced by the Q-switch must be accounted for in the calculation of the photon lifetime τ_c: the temporal evolution of this additional loss determines the characteristics of the Q-switched emission. Two major cases can be distinguished, fast and slow Q-switching: the electro-optic and mechanical Q-switches are typical examples of these two categories, respectively.

In the case of fast Q-switching the quality factor is switched almost instantaneously from the low to the high value, due to reduction of losses $\zeta(t)$ from the maximum value ζ_{max} when $t < 0$ to zero for $t \geq 0$. This determines the decrease in the inversion of population from the initial high value n_i to a very low value n_f, which is lower than the inversion at threshold n_{th} for free generation, accompanied by an increase in the photon flux to a very high value, which is reached at the moment when the declining population inversion reaches the value n_{th}, then declines to almost zero. These values of population inversion are related by the transcendental equation

$$n_i - n_f = n_{th} \ln \left(\frac{n_i}{n_f} \right). \tag{2.99}$$

The fast Q-switching determines generation of a single pulse whose energy and duration are given by

$$E_{out} = \frac{Ah\nu}{2\sigma_{eff}} \ln \left(\frac{1}{R_1} \right) \ln \left(\frac{n_i}{n_f} \right) \tag{2.100}$$

$$t_{Qsw} = t_r \frac{n_i - n_f}{n_i - n_{th} \left[1 + \ln \left(\frac{n_i}{n_{th}} \right) \right]} \tag{2.101}$$

where $t_r = 2l_{eff}/c$ is the round-trip time defined in Section 2.2.4.1 and l_{eff} is the effective resonator length. These parameters can be related [50] to the system parameters,

$$E_{out} = \frac{Ah\nu L}{2\sigma_{eff} \gamma} (z - 1 - \ln z) \tag{2.102}$$

$$t_{Qsw} = \frac{t_r}{L} \left(\frac{\ln z}{z[1 - a(1 - \ln a)]} \right) \tag{2.103}$$

where $z = 2g_0 l/L$ and $a = (z - 1)/(2 \ln z)$. The peak power of the Q-switched pulse will then be determined by the ratio (E_{out}/t_{Qsw}). In the case of high energy short-pulse pumping, very high peak power laser pulses reaching tens or hundreds of megawatts can be obtained (giant-pulse emission). Equation (2.103) shows that the pulse can be shortened by reducing the length of the resonator and this can be employed for the generation of high peak power pulses of moderate energy. The reflectivity of the output mirror influences the characteristic population inversion values n_i, n_f and n_{th} and an optimum reflectivity of the exit mirror, $R_1^{(opt)}$, dependent on the parameter z, can be defined: this reflectivity can be considerably lower than for free generation.

An extraction efficiency that expresses the fraction of stored energy that is found in the emitted Q-switched pulse can also be defined

$$\eta_{ext} = 1 - \left(\frac{1 + \ln z}{z} \right); \tag{2.104}$$

efficient extraction of the stored energy will then require a high (gain/loss) ratio z.

In the fast switching regime the moment of pulse generation is slightly delayed with respect to the switching. With slow temporal variation of loss $\zeta(t)$, for instance for the $\cos(\omega t)$ dependence specific to mechanical (prism spinning) switching, the evolution of population inversion can determine a large difference in time between the maximum of the photon flux (t_d) and the minimum of the variable loss, which influences the characteristics of laser emission: when $t_d \geq t_{min}$ the laser generates a single pulse, whereas for $t_d < t_{min}$ generation of multiple pulses occurs and this is accentuated at high pumping, limiting the possible pulse energy.

In the case of CW pumping the system can be Q-switched at a repetition rate, the population inversion undergoes cyclic variation and a train of pulses is emitted. The repetition rates can be very high, up to a hundred kilohertz, and the pulse duration is larger than for the pulse pumped electro-optic Q-switched lasers. Since the restoration of population inversion between pulses depends on the ratio between the repetition rate and the luminescent lifetime, the characteristics of pulse emission depend on the repetition rate. With increasing repetition rate the peak power of the pulses decreases, the pulse width increases and the average power increases then shows saturation behavior at a value approaching the power in the pure CW emission regime.

Theoretical modeling of the Q-switching process enables selection of the most suitable conditions, in conjunction with the properties of the laser material. The theory indicates that, although it reduces the free generation threshold, a high emission cross-section of the laser material is not favorable for population inversion storage and generation of high energy Q-switched pulses; moreover, it can facilitate the amplified spontaneous emission in directions outside the laser axis, leading to parasitic loss of inversion. However, a long luminescence lifetime τ_{eff} will facilitate population inversion storage. A major limiting factor for the performance (power level, repetition rate) of Q-switched lasers is the generation of heat facilitated by the increased population inversion storage. Suitably selected volume superposition, as well as a reduced quantum defect are then important factors of concern in the design of a high power Q-switched laser.

Because of its simplicity and because of the reduced length of the resonator, passive Q-switching using saturable absorbers (SA) is a preferred solution for many applications. A major characteristic of the SA is the modulation depth that defines the maximum possible modification of the losses in the process of switching. The intial transmission T_0 of the passive Q-switch is controlled by the cross-section σ_a of the saturable absorption transition, by the doping concentration C_{SA} and by length. Under pulsed pumping, in the presence of a photon flux of fluence E, the initial absorption coefficient α_0 of the SA is reduced to

$\alpha(E) = \alpha_0/[1 + (E/E_s)]$, where $E_s = h\nu/\sigma_a$ is the saturation fluence. A large absorption cross-section σ_a is thus necessary to grant high initial absorption and low saturation fluence; at the same time, although a short recovery time is desirable for the SA, it must be not be so short as to impede saturation. The final transmission of the SA can also be influenced by the presence of unsaturable losses. For several passive Q-switch materials the ground state saturable absorption can be accompanied by absorption from the excited state, which can reduce the efficiency of switching. The excited state lifetime of the SA must be vey short and the saturation fluence must be low. High heat conduction of the material is required since a large part of the absorbed energy is transformed into heat. The SA Q-switches can be used under CW pumping too: after a certain pumping rate the pulse duration and the peak power of the pulses do not increase further, although the repetition rate and the average power increase almost linearly with the pump power.

Very suitable saturable absorption properties are shown by ions such as Cr^{4+} in sites of tetrahedral symmetry for 1 μm lasers, V^{3+} in octahedral sites for 1.3 μm lasers, and Co^{2+} in tetrahedral sites for 1.2–1.6 μm lasers. Semiconductors or carbon nanotubes or graphene with very reduced selectivity to wavelength can be very useful. Of particular importance are the doped passive Q-switches based on the same host material as the active laser material. Since there is no sizable difference in the index of refraction for these two components, they can be expected to bond together or be produced as composite materials without any parasitic reflection at the interface. Such composite active-SA materials can enable considerable shortening of the resonator length and of the emitted pulse; this can be further reduced by deposition of the mirrors onto these components, opening the possibility of generating Q-switched pulses in the sub-nanosecond range.

2.4.2.2 Mode-locking

Shorter pulses, in the range of picoseconds or femtoseconds, can be generated by mode-locking (ML) techniques. In a free generation laser, the frequency gain profile covers a large number (to hundreds) of longitudinal modes, without any phase relation, and the interference of these modes of oscillation introduces strong and random fluctuations of the laser beam. If a fixed phase relation can be established between these modes of oscillation, the intensity of the spectral composition of emission will be controlled by the gain profile and a unique short pulse will be generated. Since this pulse travels inside the laser resonator of length l, the laser will generate such pulses with a repetition rate $c/2l$. The phase relation between the longitudinal modes of oscillation can be controlled (mode-locking) by passive or active methods. The passive mode-locking is based on the periodic modulation of the radiation using saturable absorption or the non-linear refraction index, whereas active mode-locking is obtained with acousto-optic modulation at a frequency corresponding to the distance between the longitudinal modes.

The saturable absorption at high pump intensity determines intensity dependent transmission. When the incident radiation is pulsed, the SA becomes transparent for the duration of the high intensity part at the center of the pulse, but will continue to absorb the low intensity radiation at the edges of the pulse. When placed in a laser cavity, such an absorber transmits the high intensity spikes inherent to the laser emission and the round-trip travel in the cavity will determine a train of $c/2l$ repetitive pulses. The SA must have very fast response time (picoseconds) and this can be found in various types of materials, such as organic dyes, semiconductors, transition metal doped solids, single-wall carbon nanotubes, graphenes and so on. The required modulation depth for mode-locking is quite low, and the saturation fluence of the SA must be low. Semiconductor SAs can be used to control the reflectivity of the mirrors by placing a thin slice between two reflecting surfaces to build a monolithic mirror whose reflectivity increases with the beam intensity (SESAM).

The duration of the mode-locked pulse is inversely proportional to the laser material gain bandwidth, $t_p \propto 1/\Delta \nu$, and the exact relation depends on the gain time-dependence profile. Thus, for Gaussian temporal dependence of the mode-locked pulse, $t_p = 0.44/\Delta \nu$; however, it was found that in the case of CW pumped mode-locked lasers the pulses can have the shape $I(t) = I(0) \, \text{sech}^2(t/\tau_p)$ and then $\tau_p \Delta \nu = 0.315$. Normallly the RE^{3+} emission linewidth is in the region of a hundred gigahertz, and thus the pulse duration is of the order of picoseconds; generation of ultra-short pulses, in the range of femtoseconds, would require broad emission bands, such as in the case of vibronic lasers or lasers with very strong inhomogeneous broadening.

Passive mode-locking can also be achieved by non-linear processes that do not rely on saturable absorption, such as the Kerr effect which enables selective focusing on the high power part of the incident pulse and generation of trains of short pulses. Kerr lensing is favored by laser materials with high non-linear refractive index; this can also be achieved by inserting into the laser cavity a piece of material with high non-linear optical properties. In many cases the Kerr non-linearity is too weak to start Q-switching and an additional fluctuation must be introduced by means such as SESAM. Kerr lensing mode-locking can enable generation of laser pulses with duration below the limit imposed by the emission linewidth by soliton-like interactions.

Low threshold mode-locked laser emission requires a large $\sigma_e \tau_{eff}$ product, so, by contrast to the Q-switched lasers, mode-locking is favored by a high emission cross-section of the laser material. The repetition rate in the case of ML lasers is very high, tens to hundreds of megahertz, and the pulse energy is very low, several nanojoules. Several SAs have properties suitable for both Q-switching and mode-locking, for instance, two different transitions (GSA and ESA) with different relaxation lifetimes from the upper states of the transitions (nanosecond and respectively picosecond) and usually in this case the passive mode-locking is accompanied by Q-switching and the emission consists of nanosecond trains of ML picosecond pulses. However, careful selection of the parameters allows control of the desired temporal regime: the generation of uniform, non-modulated ML pulses requires the elimination of parasitic Q-switching and this can be done by selecting the modulation depth (contrast), large for Q-switching and small for ML, by the initial transmission of the SA.

2.4.3 Laser amplifiers

The pulse energy from the Q-switched or mode-locked lasers can be increased subsequently by coherent optical amplification. The main element of an optical amplifier is a laser material pumped to high inversion in the absence of a reflecting system able to give enough gain to onset laser oscillation. An incident laser beam of frequency ν_l matching an emission line of the amplifier, can use the stored population inversion of the material to increase its energy. A major characteristic of the amplifier is the saturation fluence $E_s = h\nu/\sigma_e$, related to the stored energy per unit volume by $J = g_0 E_s$, where g_0 is the small-gain coefficient. Efficient extraction of the stored energy requires fluences similar to the saturation fluence of the material; it is obvious then that an amplifier material with large emission cross-section will not be favorable for the extraction of high energy pulses. A very important criterion in designing the optical amplifier is the density of power in the beam, which must be kept below the laser damage threshold for the various components of the amplifier system. This is typically around $10 \, \text{J cm}^{-2}$ for pulses in the nanosecond range, so saturation fluences in the vicinity of this value are recommended. Because of this limitation, the total energy that can be obtained from the amplifier will depend on the transverse section of the beam, for instance amplifiers in the kilojoule range require apertures of several tens of centimeters. An additional limitation of the emission cross-section is imposed by possible amplified spontaneous emission, especially for the large-aperture amplifiers. The amplification process is accompanied by non-radiative de-excitation processes that can induce intense generation of heat. This recommends amplifier materials with high emission quantum efficiency η_{qe} and pumping wavelengths to minimize the quantum defect. A high volume superposition of the pump and amplified beam volumes is also necessary.

There are several technical solutions for amplifiers and the choice is determined by the energy range of the laser beam. When the energy of the laser beam is high enough to approach saturation fluence, such as for the Q-switched lasers, the solution of choice is the master oscillator power amplifier (MOPA). Usually a single pass of the beam through the amplifying medium will be necessary, although sometimes a few passes can be more practical. However, for incident laser pulses of low energy, such as for the mode-locked lasers, the energy will be much below the saturation fluence and many passes of the radiation inside the amplifier will be necessary to extract efficiently the stored excitation (regenerative amplifier). In both MOPA and regenerative amplifiers, the pulse duration and the spectral width are determined by the oscillator, whereas the amplifier controls the energy. When pulses of very high energy are necessary, an amplifier chain consisting of regenerative amplifiers (preamplification stage) and MOPA (amplification stage) can be used and amplification factors up to 10^{10} can be obtained.

In the case of amplification of femtosecond pulses, the very short duration determines a very high peak power, beyond the limits acceptable for such amplifiers, and in this case a different approach, the chirped pulse amplification (CPA) [51] can be used. In a CPA chain, the temporal characteristics of the incident beam are initially changed by stretching it from the femtosecond range to nanoseconds, the pulse is then amplified in the normal way to high energy, and finally it is recompressed to the original femtosecond duration, resulting in very high peak power (TW-PW). Hybrid amplifiers using regenerative stages and PCA can be also used.

2.5 The spatial distribution of the de-excitation processes

2.5.1 Classes of excited ions

As discussed above, because laser emission is a threshold process and because of the spatial configuration of the pump and laser mode, not all the ions excited by pumping participate in laser emission. The proportion of excited ions participating in laser emission, luminescence and heat generation depends on the pumping and laser emission regime. In the case of CW laser emission, this proportion is determined by the laser emission efficiency $\eta_l = \eta_v(1 - f_{th})$ defined by Eq. (2.93). Accordingly, for the common case when the pumped volume is larger than the laser mode, three categories of excited ions can be conveniently defined [44].

- *Excited ions inside the laser mode volume that form the population inversion clamped at the laser threshold.* These ions de-excite by luminescence and non-radiative processes and the inversion at threshold is maintained by pumping. The fraction of these ions with respect to all the excited ions is $F(A) \equiv \eta_{th} = \eta_v f_{th}$.
- *Excited ions inside the mode volume pumped above the laser threshold*, with global fraction $F(B) \equiv \eta_{ath} = \eta_v(1 - f_{th}) \equiv \eta_l$.
- *Excited ions outside the laser mode*, with global fraction $F(C) \equiv \eta_{olm} = 1 - \eta_v$.

The existence of these three classes of excited ions, A, B and C, determines non-uniform spatial distribution of the de-excitation processes. The dependence on the operating point is particularly relevant in the case of longitudinal pumping, owing to attenuation of the pump radiation along the laser rod which can influence the relative proportions of these classes of excited ions.

2.5.2 Spatial distribution of laser emission

Laser emission is produced only by the excited ions B. For each of these ions the fraction of absorbed power in the laser emission equals $\eta_p \eta_{qd}^{(l)}$ and the total contribution of these ions to de-excitation (the fractional laser coefficient) is $\eta^{(l)}(B) = F(B)\eta_p \eta_{qd}^{(l)} = \eta_l \eta_p \eta_{qd}^{(l)}$, whereas $\eta^{(l)}(A) = \eta^{(l)}(C) \equiv 0$. The dependence of $\eta^{(l)}$ on η_l determines the dependence on the operating point P/P_{th} and thus the distribution of the absorbed power is a major factor.

2.5.3 Spatial distribution of luminescence

The excited ions that do not participate in laser emission (classes A and C) can de-excite by luminescence or non-radiative processes, according to the emission quantum efficiency

η_{qe}. The fraction of absorbed power transformed into luminescence for each of these ions is $\eta_p \eta_{qe} \eta_{qd}^{(f)}$ and thus the global contributions to the luminescence emission of the three classes of ions defined above are

$$\eta^{(f)}(A) = F(A)\eta_p \eta_{qe} \eta_{qd}^{(f)} = \eta_v f_{th} \eta_p \eta_{qe} \eta_{qd}^{(f)},$$
$$\eta^{(f)}(B) \equiv 0,$$
$$\eta^{(f)}(C) = F(C)\eta_p \eta_{qe} \eta_{qd}^{(f)} = (1 - \eta_v)\eta_p \eta_{qe} \eta_{qd}^{(f)};$$

and their sum,

$$\eta^{(f)}(A) + \eta^{(f)}(B) + \eta^{(f)}(C) = (1 - \eta_l)\eta_p \eta_{qe} \eta_{qd}^{(f)}$$

equals the fractional luminescence emission coefficient $\eta^{(f)}$ defined by Eq. (2.66).

The spatial distribution of the luminescence emission of the excited ions from class C is determined by the spatial superposition of the pump and laser volumes and when the former is larger ($\eta_v < 1$), this distribution in a plane perpendicular to the laser axis usually has annular profile, with the radius and height determined by the pump intensity. By contrast, for the ions of class A this distribution is flat, with the side profile corresponding to the profile of the laser mode and with height determined by the laser threshold, i.e. independent of the pump intensity. Summation of these two distributions, the first dependent on the pump intensity and the second independent of it, determines a new annular structure, with profile, radius and height dependent on the pump intensity, on the volume superposition and on the emission quantum efficiency η_{qe}. When the emission quantum efficiency is independent of the absorbed pump power (emission self-quenching determined only by down-conversion cross-relaxation) the luminescence power follows this spatial distribution: the fraction of luminescence emitted by the ions of class A will not depend on pump intensity, whereas that given by the ions of class C depends linearly on P_a. The presence of non-linear pump dependent de-excitation processes such as upconversion will modify this figure by smearing the maxima of the annular profile to a certain degree.

2.5.4 Spatial distribution of heat generation

All three classes of ions (A, B, C) contribute to heat generation: for each of the ions of classes A and C the fraction of absorbed power transformed into heat is given by Eq. (2.69) whereas for the ions of class B this is given by Eq. (2.70). The global contribution to heat generation for each of these classes of ions is

$$\eta_h(A) = F(A)\eta_h^{(f)} = f_{th}\eta_v \eta_h^{(f)} = f_{th}\eta_v \left(1 - \eta_p \eta_{qe} \eta_{qd}^{(f)}\right) \tag{2.105a}$$

$$\eta_h(B) = F(B)\eta_h^{(l)} = \eta_l \eta_h^{(l)} = \eta_v (1 - f_{th})\left(1 - \eta_p \eta_{qd}^{(l)}\right) \tag{2.105b}$$

$$\eta_h(C) = F(C)\eta_h^{(f)} = (1 - \eta_v)\eta_h^{(f)} = (1 - \eta_v)\left(1 - \eta_p \eta_{qe} \eta_{qd}^{(f)}\right). \tag{2.105c}$$

The sum $\eta_h(A) + \eta_h(B) + \eta_h(C)$ equals the fractional heat generation coefficient η_h given by Eq. (2.68) and for each class (i) of excited ions $\eta^{(l)}(i) + \eta^{(f)}(i) + \eta_h(i) = F(i)$. According to the definition of η_l, none of the fractional coefficients for the de-excitation processes

$\eta^{(l)}$ (Eq. (2.64)), $\eta^{(f)}$ (Eq. (2.66)) and $\eta^{(nr)}$ (Eq. (2.68)) is constant, but depends on η_v and on f_{th}. The heat generation coefficients given by Eqs. (2.105) are global parameters and, similar to the radiative processes, the spatial distribution of heat generation will be determined by the distribution of absorbed power and on these coefficients.

The thermal power $P_h = \eta_h P^a$ dissipated inside the various classes of ions under lasing conditions will be different: it is independent of the absorbed power above threshold and remains at the value corresponding to the inversion of population clamped at threshold for ions A, but will increase linearly, with different slopes, for the ions of classes B and C. Thus, the thermal power generated when the absorbed power P^a lies below the laser threshold ($P^a \le P^a_{th}$) is

$$P_{h,bth} = \eta_{h,bth} P^a = \eta_h^{(f)} P^a = \left(1 - \eta_{qe}\eta_{qd}^{(f)}\right) P^a, \tag{2.106}$$

whereas, using Eqs. (2.68) and (2.93), in the presence of laser emission ($P^a \, P^a_{th}$) it becomes

$$\begin{aligned}
P_{h,ath} = \eta_h P^a &= \left[1 - \eta_p\eta_{qe}\eta_{qd}^{(f)}(1 - \eta_l) - \eta_p\eta_{qd}^{(l)}\eta_l\right] P^a \\
&= \left(1 - \eta_{qe}\eta_{qd}^{(f)}\right) P^a - \left(\eta_{qd}^{(l)} - \eta_{qe}\eta_{qd}^{(f)}\right)\eta_v\left(P^a - P^a_{th}\right) \\
&= \left(1 - \eta_{qe}\eta_{qd}^{(f)}\right) P^a_{th} + \left[1 - \eta_{qe}\eta_{qd}^{(f)}(1 - \eta_v) - \eta_{qd}^{(l)}\eta_v\right]\left(P^a - P^a_{th}\right) \\
&= \eta_{h,bth} P^a_{th} + \eta_{h,ath}\left(P^a - P^a_{th}\right). \tag{2.107}
\end{aligned}$$

When the emission quantum efficiency η_{qe} is independent of excitation power, both heat load coefficients $\eta_{h,bth}$ and $\eta_{h,ath}$ are constant. These equations show that $\eta_{h,ath}$ can be larger or smaller than $\eta_{h,bth}$ depending on the sign of $(\eta_{qd}^{(l)} - \eta_{qe}\eta_{qd}^{(f)})$ and outline the very important role of η_v for the heat loading parameter above threshold $\eta_{h,ath}$: when $\eta_v = 1$, $\eta_{h,ath} = (1 - \eta_{qd}^{(l)}) = \eta_h^{(l)}$ but it decreases for smaller η_v. In the presence of non-linear quenching processes $\eta_{h,bth}$ becomes dependent on the absorbed power; however, $\eta_{h,ath}$ depends on P^a only when $\eta_v < 1$.

There is quite a large literature discussing the thermal effects in lasers and generally the distribution of the heat source is considered identical to the distribution of pumping radiation inside the laser material, i.e. no difference like that indicated by Eqs. (2.69) and (2.70) between the heat generated by the ions which participate and which do not participate in laser emission is taken into account. However, as discussed above, an accurate account of these differences shows that the distribution of the heat source can be different from that of the pumping field and is determined by the laser design (mismatch between the laser mode volume and the pump volume, laser emission wavelength, threshold), by the characteristics of the laser material (doping concentration) and of the pumping process (wavelength, pump intensity). Thus, whereas below the laser threshold the distribution of heat generation is similar to that of the absorbed power, on operating above threshold with a laser mode smaller than the pump distribution, the heat source has a (η_v and f_{th}) dependent annular structure that is more accentuated for low emission quantum efficiencies. The annular distribution of heat generation will be slightly different from that of luminescence emission since in the first case all three classes of excited ions contribute, whereas the second case involves only the ions of classes A and C. The pump dependent non-radiative de-excitation processes such as upconversion increase the fraction of ions of class A but they will also modify the spatial distribution of heat generation and of luminescence emission for ions of class C: the maximum of the annulus for the luminescence distribution will be reduced non-uniformly at the expense of the maximum heat generation distribution. As mentioned above, these considerations are strictly valid for CW laser emission and must be adapted for other pumping and emission regimes.

2.6 Thermal field inside the pumped laser material and thermal effects

2.6.1 Thermal field distribution inside the laser material

Heat generation by non-radiative processes in the pumped laser material is manifested as a rise in temperature. The spatial and temporal dependence of the pumping radiation and of absorption as well as that of the radiative and non-radiative processes determine a definite spatial and temporal configuration of the heat source $Q_t(x, y, z, t)$, i.e. a time dependent distribution of temperature gradients inside the laser material. However, in a solid material the actual configuration of the thermal (temperature) field $T(x, y, z, t)$ is different from that of the heat source and is influenced by the shape and size of the laser material and by its properties (density ρ, specific heat C_p, thermal conductivity K) which determine the transport of heat inside the laser material and its dissipation. The problem of thermal effects in laser materials has been discussed in many papers, reviews and textbooks such as references [7, 52, 53].

The distribution of the thermal field can be calculated with the differential equation

$$\rho C_p \frac{\partial T(x, y, z, t)}{\partial t} - K\nabla^2 T(x, y, z, t) = Q_t(x, y, z, t). \tag{2.108}$$

In this equation the material parameters (ρ, C_p, K) are usually considered constant over the temperature interval (usually several tens of degrees kelvin) determined by the heat generation and dissipation processes. In a stationary regime the first term of Eq. (2.108) cancels and the thermal field is connected to the configuration of the heat source only by the thermal conductivity,

$$\nabla^2 T(x, y, z) = -\frac{1}{K} Q_t(x, y, z). \tag{2.109}$$

However, in the case of non-stationary heat sources, such as in pulse pumping, the density and the heat capacity of the material become relevant since, together with the thermal conductivity, they determine the regime of increasing temperature under pumping and its decay after the pumping pulse. A thermal relaxation time at which the increase in temperature after the exciting pulse at the center of a laser rod of radius r_0 drops to $(1/e)$ of its value can be defined as $\tau_{th} = r_0^2/(4k)$, where $k = K/(C_p\rho)$ is the thermal diffusivity. When the interval between the pulses is well above this value, the rod temperature drops naturally to this value before excitation; however, for short intervals between pulses the heat builds inside the rod, reaching quasi-stationary values that can approach that of continuous wave pumping with similar power.

The configuration of the heat source Q_t depends on the configurations of the absorbed pump radiation and of the heat load parameter, i.e. $Q_t(x, y, z, t) = \eta_h(x, y, z)P_a(x, y, z, t)/V$, where V is the pumped volume, and the solution of the heat field equation is determined also by the conditions at the limit, such as the shape and size of the laser material. In many cases the laser material is cooled externally to dissipate part of the heat generated by pumping and this influences the boundary conditions. This problem has been much discussed in the literature: however, as a general rule, the heat load coefficient η_h is considered constant in space and thus the configuration of the heat source is automatically considered to be coincident with the pump distribution, and analytical solutions for various pumping configurations such as uniform transversal pumping, Gauss, super-Gauss or top-hat longitudinal pumping have been obtained. For instance, in the case of uniform pumping of a cylindrical laser rod of length L and radius r_0, the distribution of heat was calculated [53] as parabolic, with the maximum along the laser axis and with the difference in temperature between the center ($r = r_0$) and the periphery ($r = 0$) of the rod equal to $\Delta T = \eta_h P_a/(4\pi L K)$. Such a variation in temperature will induce a variable thermal gradient along the rod radius. The temperature at the periphery can be controlled by transferring the heat to a heat sink of temperature T_s (cooling). With the heat transfer coefficient H the efficiency of cooling can be expressed by $T(r = r_0) - T_s = P_h/(2\pi r_0 L H)$: a rod of large radius will dissipate the heat better. However, since the size of the doped rod is limited by other restrictions, artificial "thickening" of the rod by cladding with undoped laser material could be useful. In the case of longitudinal pumping the equations for the thermal field inside the laser material are more complex, but the temperature differences and gradients will again be inversely proportional to the heat conductivity.

2.6.2 Thermal effects

The thermal field can influence the state of the laser material through various effects such as thermomechanical effects and their influence on the optical properties, and through the thermal variation of different properties (optical, electron–phonon, spectroscopic, thermal).

2.6.2.1 Thermomechanical effects

The temperature gradients induced by the thermal field cause strain accompanied by stress inside the laser material; both these are tensors. The numbers of components ε_{ij} and respectively σ_{kl} are determined by the symmetry of the material, and are connected by Hooke's law $\sigma_{ij} = S_{ijkl}\varepsilon_{kl}$, where the Einstein summation rule is applied and the components of the tensor S_{ijkl} are determined by the mechanical properties of the material. For isotropic materials the number of components is restricted to three and the stress and strain parameters can be related using only two mechanical parameters, Young's modulus E and Poisson's ratio ν. The main directions of the stress components in a plane perpendicular to the laser axis are determined by the geometry of the laser material [54]. For cylindrical rods the approximation of plane strain is suitable and the radial, tangential and longitudinal directions are most relevant,

$$\sigma_r(r) = QS(r^2 - r_0^2) \tag{2.110a}$$

$$\sigma_t(r) = QS(3r^2 - r_0^2) \tag{2.110b}$$

$$\sigma_z(r) = 2QS(2r^2 - r_0^2) \tag{2.110c}$$

where $S = \alpha_l E[16K(1 - v)]^{-1}$ and Q is the total amount of heat deposited in the rod. The stress components $\sigma(r)$ can be positive or negative, leading to tension or compression. According to Eqs. (2.110), they change sign at a certain radial position: while at the center of the rod ($r = 0$) all these components are negative, with maximum value $\sigma_t = -2QS$, at the rod surface ($r = r_0$), $\sigma_r = 0$ and both σ_t and σ_z are positive, $\sigma_t(r_0) = \sigma_z(r_0) = 2QS$.

Another effect is thermal expansion: under uniform heating, this determines the increase in the volume V of the laser material, $\Delta V = V(T) - V_0(T_0) = V_0(T_0)[\alpha_V(T - T_0)]$ and similar equations hold for the increases in the linear sizes: for isotropic materials the linear thermal expansion coefficient is the same along all directions and $\alpha_l \approx (1/3)\alpha_V$, whereas for anisotropic materials the linear expansion coefficients have particular values along the main axes. Usually the linear expansion coefficient α_l decreases at low temperatures.

2.6.2.2 *The consequences of thermomechanical effects*

2.6.2.2.1 Thermal bulging In a rod with free end surfaces the non-uniform radial distribution of the thermal field can induce radially dependent thermal expansion, $l(r) = l_0\alpha_l[T(r_0) - T(r)]$. The larger temperature on the rod axis leads to bulging of the rod according to the profile of the thermal field distribution. This can be restricted if caps of undoped material are bonded at the ends of the rod (capped rods).

2.6.2.2.2 Thermophotoelastic effect The thermal strain and stress effects in the laser materials can modify the refractive index by the photoelastic effect. In a cylindrical rod of isotropic material the thermal strains modify the local refractive index in a plane normal to the laser axis from spherical to ellipsoidal, with the axes determined by the main directions of the strain tensor, leading to shifts given by

$$\Delta n_{r,t} = -\frac{1}{2}n_0^3 \frac{\alpha_l P_h}{K} C_{r,t} r^2,$$

where the photoelastic coefficients $C_{r,t}$ are complex functions of the Poisson ratio v. This local variation of the refractive indices along two main directions induces radially dependent birefringence

$$\Delta n_r - \Delta n_t = n_0^3 \frac{\alpha_l P_h}{K} C_B r^2.$$

Thus, in crystalline materials, even for cubic symmetry the thermal birefringence will be dependent on the orientation.

2.6.2.3 Thermo-optical effects: temperature variation of the refractive index

The temperature can induce positive or negative modification of refractive index $\Delta n = n(T) - n(T_0) = (dn/dT)(T - T_0)$; in the latter case this can compensate partly for the effect of thermal stress. The thermo-optical coefficient dn/dT usually decreases at lower temperatures.

2.6.2.4 Thermospectral effects

As discussed in Section 2.1.6, the electron–phonon interaction influences the static and dynamic spectroscopic properties of the laser materials and thus the temperature dependence of this process will induce a corresponding temperature dependence of these properties.

- Shift of absorption and emission lines due to the temperature dependent electron–phonon interaction. Additional shift can be caused by the effect of lattice thermal expansion on the static interactions (crystal field, nephelauxetic effect).
- Homogeneous broadening of the spectral lines and modification of the peak absorption and emission cross-sections.
- Modification of the fractional thermal population coefficients of the energy levels belonging to the same manifold: this would modify selectively the effective cross-sections, but the lifetime of all emission transitions will be the same.
- Thermallization with upper levels from manifolds with different emission cross-section and lifetime in the case of 4f ions or with levels with different selection rules for emission in the case of 3d elements; this could modify the linewidths, cross-sections and the emission lifetimes.
- Modification of the intensities of vibronic bands (3d ions) or satellites (4f ions).
- Modification of the dynamics of emission either by its influence on the electron–phonon de-excitation rate or by its influence on the energy transfer rates, by its effect on the superposition integral of donor emission and acceptor absorption spectra.

2.6.2.5 Thermo-thermal effects: temperature dependence of the heat capacity and thermal conductivity

The thermal conductivity determines the heat flow per unit area in a system with a temperature gradient. The carriers of heat can be free electrons, phonons or photons. In the case of laser materials the dominant carriers are the phonons and the thermal conductivity can be related to material properties by $K = \frac{1}{3}C_p v\delta$, where C_p is the heat capacity per unit volume, v is the velocity of phonons (equivalent to the speed of sound) and δ is the mean free path of the phonons. Of these, v shows reduced temperature dependence while C_p and δ are strongly dependent on temperature. At temperatures much below the Debye temperature, $\theta_D = h v_m / k$, of the material, where v_m is the maximum normal mode vibration frequency, the heat capacity $C_p \propto T^3$, whereas above $(0.1\text{–}0.2)\theta_D$ it evolves to a constant value. The mean free path δ is influenced by any disturbance of the lattice periodicity, such as boundaries of the body (which determine a specific path δ_b, dependent on the volume delineated by boundaries and on the quality of surfaces), doping ions or impurities (δ_i, dependent on the nature of these ions, especially on the difference in mass with respect to the substituted ion [54]), multiple occupancy of sites by ions of different species in the case of disordered crystals (δ_{dis}) and inelastic scattering of the acoustic phonons (δ_{ph}, dependent on the nature of these ions). The global mean free path accounts for all these processes, $1/\delta = \sum_k (1/\delta_k)$. At low temperatures the main limiting factor is the boundary condition and δ is almost constant. With increasing temperature towards $(0.1–0.2)\theta_D$, the effects of doping or disorder in reducing δ become apparent, but the main reduction is introduced by the enhanced inelastic phonon–phonon scattering determined by the increasing phonon occupancy number at higher temperatures, which determines a temperature dependence of the global mean free path described approximately by $\delta \propto \exp(-\theta_D/T)$ which evolves finally to a T^{-1} dependence. The temperature dependence of the heat conduction is then determined by the joint effect of the heat capacity and by the photon mean free path: at low temperatures it is dominated by the evolution of C_p and increases according to the T^3 law up to the region of $\sim(0.1–0.2)\theta_D$ where it reaches a maximum determined by the impurities or disorder, then drops according to the contribution from the phonon scattering. For the most common laser materials the region $(0.1–0.2)\theta_D$ corresponds to several tens of degrees kelvin.

2.6.3 Manifestation of the effects induced by heating on the laser emission

2.6.3.1 Thermal fracture

2.6.3.1.1 Mechanical properties of the laser materials The mechanical properties of the laser material must enable the material to be processed to high accuracy without deformation. The material must have the ability to resist accidental mechanical action without surface damage (hardness), the ability to resist high mechanical effort (tensile stress) as well as the ability to resist fracture under stress in the presence of a crack (fracture toughness). The fracture toughness is expressed by the plane strain stress intensity factor K_{1C}, in units of MPa m$^{1/2}$. A material with large K_{1C} is prone to ductile fracture, i.e. it would undergo plastic deformation and absorb considerable energy before breaking, whereas materials with small K_{1C} undergo brittle fracture, with small deformation and absorption of energy before breaking.

2.6.3.1.2 Thermal fracture of the laser materials The components of the thermal stress reach maximum values at the surface and at the center of the laser rod, but since at the surface the two stresses (σ_t and σ_z) combine vectorially, the maximum stress will be larger,

$$\sigma_{max} = 2^{1/2}\sigma_t(r_0) = \frac{2^{1/2}\alpha_l E}{8\pi K(1-v)}\frac{P_h}{l}.$$

This stress can reach the fracture limit of the material σ_{fr} (the tensile strength), leading to mechanical damage of the rod, and thus the heat generation per unit length must be limited to $P_h/l = 8\pi R_{fr}$. The parameter R_{fr} contains all the material parameters that limit the power absorbed by the material,

$$R_{fr} = \frac{K(1-\nu)\sigma_{fr}}{\alpha_l E},$$ (2.111)

and is called the thermal shock parameter since it characterizes the ability of the material to withstand large variations in temperature induced by the pumping regime. A large thermal shock parameter would be desired for high power lasers, and it is evident that a high thermal conductivity would be desired. The actual tensile stress is influenced by the state of the surface, so a proper treatment (reinforcement) of the surface can increase R_{fr}.

2.6.3.2 Thermal lensing

Since the thermal effects discussed above (thermo-optical effect, thermophotoelastic effect, thermal bulging) are linear with temperature, their spatial distribution will be determined by the configuration of the thermal field inside the pumped laser material. In a uniformly pumped cylindrical rod, and assuming constant thermal load coefficient, the parabolic radial dependence of temperature will determine the modification of the optical path of an optical beam parallel to the rod axis and wavefront distortion, similar to the effect of a thick lens [54], of dioptric power

$$D_{th} = \frac{1}{f_{th}} = \frac{P_h}{KA} \left[\frac{dn}{dt} + \alpha_l C_{r,t} n_0^3 + \frac{1}{l}\alpha_l r_0 (n_0 - 1) \right].$$ (2.112)

The departures from parabolic distribution of the thermal field can determine strong aspherical aberrations; such distortions can be caused by non-uniform pump absorption, non-uniform distribution of the heat load coefficient η_h discussed in Section 2.5.4, non-uniform laser material and so on, which must be taken into account for any particular case. Depending on the cause, the thermal distortion of the wave front can follow the temporal regime of pumping and laser emission or have rapid fluctuations. The thermal lensing effects can also be calculated for other shapes of laser materials and pumping configurations, and in all cases they are determined by the material parameters and are proportional to the ratio between the heat power and the thermal conductivity. Thermal lensing can be deleterious for laser emission since it influences the conditions of amplification and above a certain absorbed power the output laser power levels off and drops rapidly with further increase in pumping power. The "pure" thermo-optical lens determined by the parabolic distribution of the thermal field can be compensated for by placing in the resonator a lens of opposite dioptric power. However, compensation for the aspherical aberrations or for the rapidly fluctuating aberrations necessitates adaptive compensation. Additional power-induced lensing of an electronic nature, modification of the refractive index due to population of the excited levels, can further limit the useful absorbed power [56].

2.6.3.3 The effect on laser emission efficiency

The thermospectral effects can influence the pump absorption and the emission properties of the laser materials. A shift in the absorption lines can affect the matching with the radiation of narrow and wavelength stabilized pump radiation. Moreover, modification of the fractional thermal coefficients f as well as modification of the absorption cross-sections σ_ρ can influence the effective pump absorption cross-section σ_{eff}; similar effects can influence the emission properties, and sometimes the laser emission can jump from one transition to another. For the quasi-three-level lasers, cooling to cryogenic temperatures can reduce the reabsorption of laser emission by orders of magnitude. The thermal broadening can influence the ability of the laser material to generate ultra-short pulses or tunable emission.

2.6.3.4 Depolarization loss

Thermal birefringence can influence the polarization of the laser beam. As discussed in Section 2.6.2.2.2, the thermal birefringence depends on the heat power P_h and on the orientation of the crystal, even in the case of cubic symmetry. In polycrystalline materials, non-uniform depolarization effects inside the grains can lead to a granular structure of the laser beam or can limit the performance of electro-optic Q-switched lasers.

2.7 Performance scaling of solid-state lasers

The analysis of the laser emission processes reveals that an improvement of laser performance in several main directions such as increased efficiency, extension of the wavelength range, diversification of the temporal regime, power or energy scaling, requires a correlated approach for the three major parts of the laser, the laser material, the pumping system and the laser resonator design, in order to optimize the flow of excitation inside the pumped laser material. At the same time it is obvious that a major factor limiting scaling to higher power or energy is the thermal field determined by the heat generated by non-radiative de-excitation. Optimization of laser emission and reduction of heat generation can be achieved by working on the partial efficiencies of the various quantum processes defined in the preceding section. As mentioned, some of the material characteristics can influence several such partial efficiencies and thus figures of merit that characterize the whole effect of a given factor must be introduced, for instance the effect of the doping concentration on the laser threshold can be characterized by a spectroscopic figure of merit $\eta_a \eta_{qe}$. Scaling to higher power would be possible by increasing the laser emission for a given amount of heat generation, and in this case a spectroscopic figure of merit of significance could be the ratio $\eta^{(l)}/\eta_h$. Such figures of merit can be calculated for each particular laser material and pump condition. According to this task of improving the laser processes, specific requirements are set forth for each of these major parts of the laser.

2.7.1 Requirements for the laser material

The active material is the central part of the solid-state laser since it offers the conditions for inversion of population and amplification of radiation by stimulated emission, and the pumping system and laser resonator must be designed to allow optimal exploitation of these conditions. The physical processes that determine the properties of the laser active centers and of their interaction with the electromagnetic radiation determine a complex inter-relation between the composition, structure, properties and functionality of the laser material. In order to optimize this relation the laser material must fulfill a large variety of requirements, such as the following.

- *Characteristics of the fabrication process*: highly reproducible fabrication of laser materials of desired volume and shape, high technological yield, good use of raw material, reduced energy consumption, and reduced need for expensive equipment, components and auxiliary materials. The fabrication process should also enable good control of the composition.
- *Doping characteristics*: ability to incorporate the necessary concentrations of doping ions in the desired valence state and with the desired concentration profile.
- *Possibility of producing composite laser materials*: such as rods with cladding or undoped caps or monolithic multifunctional (lasing and passive Q-switching) materials.
- *Optical properties*: high optical transparency in the spectral regions of significance for the pumping process and for laser emission, high optical homogeneity, very low second-order refractive index, small variation of the refractive index with temperature.
- *Chemical properties*: high resistance to the ambient atmosphere, chemical, thermochemical and photochemical stability.
- *Mechanical properties*: high hardness, high stress fracture limit, high threshold for optical damage.
- *Thermal properties*: small thermal expansion, high thermal conductivity, high thermo-mechanical shock parameter, reduced or favorable temperature dependence of all the material properties that influence laser performance.
- *Microstructural properties*: unique and well-defined substitution sites of low symmetry in the case of monochromatic lasers and conditions for inhomogeneous broadening of lines in the case of tunable or ultra-short pulse lasers, substitution sites of valence similar to the desired valence of the doping ion, possibility of controlled charge compensation when the valences of the doping and substituted ions differ.
- *Spectroscopic properties*: the combination doping ion–host material must provide electronic energy levels and transition probabilities that enable the embedding of efficient laser emission schemes in the desired temporal regime:
  - good absorption of the pump radiation and the possibility of controlling the pump rate distribution at pump wavelengths that allow a high quantum defect (Stokes) ratio $\eta_{qd}^{(l)}$;
  - very efficient de-excitation of the pump level to the emitting level ($\eta_p \approx 1$);
  - long radiative lifetime τ_{rad} of the emitting level;
  - negligible electron–phonon de-excitation W^{nr} of the emitting level, but very efficient de-excitation of the pump level to the emitting level and of the terminal level to the ground state;
  - reduced self-quenching by down-conversion or upconversion energy transfer and high emission quantum efficiency η_{qe}, to give long τ_{eff};

- emission cross-section σ_e adapted to the temporal regime: the CW and ML lasers require high σ_e (above $\sim 10^{-20}$ cm^2) and large $\sigma_{eff}\tau_{eff}$ products, the Q-switched lasers require long τ_{eff} and moderate σ_{eff} to enable storage of population inversion, whereas in large aperture laser materials used for amplifiers the emission cross-section should be confined to the range $(2-6) \times 10^{-20}$ cm^2 to control the saturation fluence close to the damage limit of the optical components and to avoid efficient ASE;
- reduced parasitic de-excitation by excited state absorption from the emitting level of the pump or laser emission;
- reduced reabsorption of the laser emission.

2.7.2 Requirements for the pumping system

- Emission spectrum with good superposition on the absorption spectrum of the laser material, of particular interest being diode lasers with emission in strongly absorbing energy levels close to or resonant with the emitting level, to give the lowest possible quantum defect;
- Desired temporal regime (continuous wave, pulsed);
- High power or brilliance;
- Long lifetime and stable emission;
- Good electrical–optical efficiency;
- Possibility of high transfer efficiency to the laser material.

2.7.3 Requirements for the laser resonator

- Proper selection of the laser transition;
- Good control of outcoupling;
- Good control of the quality factor to enable the desired temporal regime of emission;
- Control of the mode structure and beam size;
- Control of the thermal field (dissipation of heat, external cooling);
- Technical requirements, for example, robust, compact, easy maintenance, easy replacement of parts.

2.8 The laser material

The laser material must be selected with suitable absorption and emission properties for the doping ion and proper structural, optical, mechanical and thermal properties for the host material. The general bases of selection are reviewed in this section and detailed spectroscopic and laser properties are discussed in Chapters 10 and 11.

2.8.1 Selection of the laser active ion

As discussed in Section 2.2.1.1, the main criteria for selection of the laser active ions are the energy level structure and the transition probabilities: it was determined that the most useful laser ions are those with ground configuration $3d^n$ or $4f^n$, embedded in crystallographic sites with low local symmetry, which does not correspond necessarily to the global symmetry of the host.

2.8.1.1 3d ions

The vibronic laser materials offer the best prospect for tunable or short pulse laser emission. The most popular system is the Ti^{3+}-doped trigonal Al_2O_3 (sapphire) crystal that allows tunable emission (several watts in the CW regime) over a range of about 230 nm around 800 nm, and generation of pulses as short as a few femtoseconds. Peak powers in the petawatt range are obtained in CPA systems based on Ti:sapphire. However, the applications of this laser are limited by the difficulties of pumping (second harmonic of 1 μm Nd or Yb lasers) and by the limited size of these crystals (melting temperature 2100 °C). Alternatives to this material are being sought and other 3d ions, especially chromium ions of different valences (2, 3 or 4) with suitable absorption and emission properties embedded in low symmetry sites of cubic crystals are very attractive. The spectroscopic and dynamic properties of these ions are very sensitive to the crystal field and thus the selection of the host material is of paramount importance.

According to Section 2.2.1, the Cr^{3+} ions in six-fold coordinated sites can have broad vibronic emission lines in the near infrared if the excited level $^4T_2$ is placed below the ^2E level: the spin-allowed emission to the ground $^4A_2$ level is strong and, at the same time, the closeness to the ^2E level grants enough thermallization to obtain a fairly long (60–100 μs) effective emission lifetime. Cr^{3+} has strong vibronic absorption lines in the visible suitable for direct lamp or diode laser pumping.

The Cr^{4+} ions in sites with tetrahedral coordination have strong absorption bands in the visible and 1 μm range (σ_a of the order of $(3.5-7) \times 10^{-18}$ cm^2) and very intense (σ_e in the range $(3-5) \times 10^{-19}$ cm^2) short lived (several microsecond) broad (over 200 nm) infrared emission in the 1350–1580 nm range, depending on host. These materials can be used either as passive Q-switches for the 1 μm Nd or Yb lasers or as laser materials for tunable or ultra-short pulse emission.

The Cr^{2+} ion tetrahedral sites have very strong (σ_a in the 10^{-18} cm^2 region) and broad (350 nm) absorption centered on 1750 nm and short ($\sim$5 μs), strong ($\sigma_e \sim 1.3 \times 10^{-18}$ cm^2) and broad (860 nm) emission centered around 2450 nm, corresponding to vibronic transitions between the levels $^5T_2$ and ^5E. The absorption and emission properties of Cr ions of different valences offer potential for use as laser emitting ions or as sensitizers for the weakly absorbing 4f ions.

The V^{3+} ions in octahedral sites show strong absorption in the 1.05 to 1.5 μm range (absorption cross-section $\sim$1.3 $\times$ 10^{-18} cm^2) suitable for Q-switching of 1.3 μm lasers, whereas Co^{2+} in weak field tetrahedral sites shows promise for Q-switching of 1.3 μm lasers.

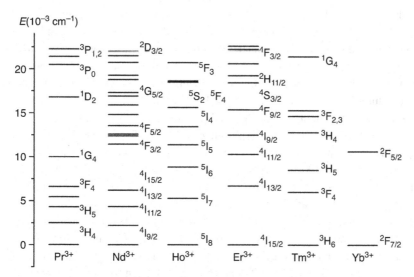

Figure 2.6 Energy manifolds for several RE^{3+} laser ions (based on data in YAG).

2.8.1.2 4f ions

A considerably larger selection is offered by the 13 RE^{3+} ions with ground electronic configuration 4f^n, with n running from 1 (Ce^{3+}) to 13 (Yb^{3+}). Practically all these ions have been made to lase and high-performance has been demonstrated for Pr^{3+} (4f^2), Nd^{3+} (4f^3), Tm^{3+} (4f^{10}), Er^{3+} (4f^{11}), Ho^{3+} (4f^{12}) and Yb^{3+} (4f^{13}). The structure of the lowest energy manifolds in the range of interest for laser processes for several RE^{3+} ions is shown in Figure 2.6: for all these ions, except Yb^{3+}, the energy level structure is fairly complex and in many cases allows laser emission on several inter-manifold transitions.

2.8.1.2.1 Pr^{3+} as laser active ion The most important laser transitions for Pr^{3+} are in the visible (blue, green, orange, red) or near infrared and originate from the manifold $^3P_0$ placed in the 20 500 cm^{-1} region. Although the lifetime of $^3P_0$ is short (few tens of microseconds) and influenced by self-quenching, the fairly large emission cross-section, especially for the red $^3P_0 \rightarrow {}^3F_2$ transition, makes this ion of interest for lasing. The main limiting factor for Pr^{3+} lasers is the absence of suitable pump sources and the weak absorption; in several cases special upconversion schemes (simple or sensitized photon avalanche) has been used, but the output was limited to watts. However, the recent development of blue GaN laser diodes could improve this situation.

2.8.1.2.2 The Nd^{3+} ion The ground configuration 4f^3 Nd^{3+} contains a complex structure of energy manifolds but only the manifold $^4F_{3/2}$, placed around 11 500 cm^{-1}, satisfies the conditions of metastability: the calculated radiative lifetime of $^4F_{3/2}$ depends on the host material and ranges between 80 and 500 μs. Below this metastable manifold there are four manifolds ^4I_J ($J = 9/2, 11/2, 13/2, 15/2$), originating from the same spectral term ^4I: the gap from $^4F_{3/2}$ to the nearest of these, $^4I_{15/2}$, is around 5500 cm^{-1} and it can be normally bridged (except for materials with unusually high phonon energies) only by very high order (p of 7–10) electron–phonon processes. Based on this, the radiative lifetime can be safely approximated with the measured luminescence lifetime at very low Nd concentrations and low temperatures. In the low symmetry crystal field each of these energy manifolds is split in $(2J + 1)/2$ Kramers doublets, labeled traditionally as R_i ($i = 1, 2$) for $^4F_{3/2}$, X_j ($j = 1$ to 7) for $^4I_{13/2}$, Y_j ($j = 1$ to 6) for $^4I_{11/2}$ and Z_j ($j = 1$ to 5) for $^4I_{9/2}$. The global crystal field splitting of the manifolds in various hosts varies within very large limits, from 0 to 250 cm^{-1} for $^4F_{3/2}$, 400 to 1000 cm^{-1} for $^4I_{15/2}$, 240 to 690 cm^{-1} for $^4I_{13/2}$, 190 to 525 cm^{-1} for $^4I_{11/2}$, and 280 to 880 cm^{-1} for $^4I_{9/2}$. The laser emission takes place on optical transitions between the Stark levels of the emitting and terminal manifolds and thus in case of the Nd^{3+} ion there are $2j$ possible optical transitions between the two Stark levels of the emitting manifold $^4F_{3/2}$ and the Stark levels of the lower manifolds. For all these transitions the effective lifetime τ_{eff} is the same, but the laser quantum defect ratio $\eta_{qd}^{(l)}$ and the emission cross-section σ_e are different; moreover, the temperature dependence of the fractional thermal population coefficients $f_{1,2}$ of the levels $R_{1,2}$ of $^4F_{3/2}$ induces temperature dependence of the effective emission cross-sections σ_{eff} and of the laser threshold. However, for all the laser transitions, the slope efficiency $\eta_{sl}^{(a)}$ will be influenced only by the particular quantum defect and it will be independent of temperature.

Laser emission can be obtained on many of these transitions in the 900 nm range (to the ground manifold $^4I_{9/2}$), with room temperature cross-sections for some transitions larger than 10^{-20} cm^2, in the 1 μm range (to the $^4I_{11/2}$ manifold, placed around 2200 cm^{-1}) with σ_e sometimes larger than 10^{-19} cm^2, and in the 1.3–1.45 μm range (to the $^4I_{13/2}$ manifold, placed around 4000 cm^{-1}) with σ_e larger than 10^{-20} cm^2; the emission to the manifold $^4I_{15/2}$, placed around 6200 cm^{-1} is very weak. According to Eq. (2.29), the branching ratios β_{ij} for the various emission transitions originating from an excited manifold can be calculated with the Judd–Ofelt parameters. In the case of the $^4F_{3/2}$ emission of Nd^{3+}, the quality factor $X = \Omega_4/\Omega_6$ is of particular relevance: for very small X the branching ratio $\beta_{3/2,11/2}$ dominates (0.66 for $X = 0$) but decreases with increasing X, whereas $\beta_{3/2,9/2}$ (0.17 for $X = 0$) increases, the crossing point for these two branching ratios being around $X \approx 1.15$. The branching ratio $\beta_{3/2,13/2}$ decreases from 0.17 at $X = 0$ to less than 0.1 for $X = 2$, whereas $\beta_{3/2,15/2}$ is small, especially for large X, and is usually neglected.

The high density of excited levels above the emitting level enables excitation with a broad selection of wavelengths in the visible and near infrared. All this excitation is relaxed to $^4F_{3/2}$ by a chain of very fast processes and this enables pumping of the Nd lasers with a large variety of optical sources (lamps, solar, diode lasers). Although some of these absorption transitions have quite large absorption cross-section in the range of the emission of Xe and Kr lamps, the general utilization of the emission spectrum of these sources is quite small, usually below 20%, and the average quantum defect of the 1 μm Nd lasers is very large, around 50–60%. For several laser materials, sensitization with ions strongly absorbing in the visible, such as Ce^{3+} or Cr^{3+}, was instrumental in increasing the utilization of lamp or solar radiation. A much larger absorption efficiency and lower quantum defect of Nd-doped materials can be obtained by diode laser pumping: traditionally this is done with powerful CW or QCW AlGaAs 800 nm diode lasers using the strong absorption into the excited level $^4F_{5/2}$ at around $12\,500$ cm^{-1}. Whereas the lamp pumping is little influenced by temperature, the absorption of the diode lasers shows temperature dependence, according to the fractional thermal population coefficient of the Stark level from which absorption originates and to the effects on the linewidths. The absorption efficiency can be managed by proper selection of the Nd doping concentration and of the path of pump radiation inside the laser material. The high power 808 nm CW diode lasers enabled construction of highly efficient Nd lasers reaching hundreds of kilowatts; however, the lack of high energy pulsed diode lasers leaves room for flashlamp pumping for very high energy pulse generation or amplification.

A major shortcoming of the complex electronic structure of Nd^{3+} is self-quenching of the emission, both by down-conversion and by upconversion. In the case of down-conversion an Nd^{3+} excited ion in the $^4F_{3/2}$ level transfers part of its excitation to an unexcited Nd^{3+}: at room temperature the process is dominated by the cross-relaxation process $(^4F_{3/2}, {}^4I_{9/2}) \rightarrow$ $(^4I_{15/2}, {}^4I_{15/2})$. In most Nd-doped materials the final level of the donor and acceptor is further de-excited by efficient low order electron–phonon processes to the ground state and thus the initial excitation of the donor ion is completely wasted to heat. The interaction between two excited Nd^{3+} ions can induce upconversion of excitation to an upper level, according to the cross-relaxation processes $(^4F_{3/2}, {}^4F_{3/2}) \rightarrow ({}^4I_{15/2}, {}^4G_{5/2})$ or $\rightarrow ({}^4I_{13/2},$ $^4G_{7/2})$ or $\rightarrow ({}^4I_{11/2}, {}^4G_{9/2})$ or $\rightarrow ({}^4I_{9/2}, {}^2P_{1/2})$. Due to the high density of levels above $^4F_{3/2}$, all the excitation transferred to the acceptor will be lost by electron–phonon interaction and the excited acceptor returns to the initial excited level $^4F_{3/2}$. From this picture it is evident that down-conversion is more wasteful for excitation since the unique excitation, of the donor ion, is completely lost, whereas in the case of upconversion only one of the two excitations of the initial states of the two ions involved in the process (donor and acceptor) is lost. Efficient upconversion assumes a high density of excited ions, so it is favored by a high pumping intensity in highly doped materials but it can be neglected when the density of excited ions is low. Moreover, it is evident that the upconversion prepares the pair of two interacting ions for further down-conversion so both these processes should be accounted for together when upconversion is present. As discussed before, self-quenching can increase the laser threshold and heat generation and thus the Nd concentration in the laser materials is selected by a compromise between its effect on the absorption and on the emission quantum efficiency.

The large variety of emission wavelengths, the extended possibilities of pumping, the high efficiency and the variety of temporal laser regimes, from CW to hundreds of femtoseconds, the very large power and energy range have made the Nd lasers one of the most popular class of lasers.

2.8.1.2.3 The Ho^{3+} ion The $4f^{10}$ ground electronic configuration of Ho^{3+} has several well separated levels that can give emission in the visible or infrared, the most important being the 2.1 μm emission $^5I_7 \rightarrow {}^5I_8$ with cross-section around 10^{-20} cm^2 and lifetime in the 7 ms range. Since the visible and near infrared absorption lines of Ho^{3+} are weak, sensitization with Tm^{3+} or with Cr^{3+}–Tm^{3+} was traditionally used; however, efficient 2 μm laser emission with low quantum defect was demonstrated by pumping with a Tm^{3+} laser or with diode lasers directly into the emitting level $^5I_7$. Excitation of visble emission can be made by various upconversion schemes.

2.8.1.2.4 The Er^{3+} ion The ground configuration $4f^{11}$ of Er^{3+} has a very rich electronic level structure, although several levels in the visible and near infrared are well separated and can give emission. At the same time, the accidental coincidence of the gaps between the various pairs of energy levels could facilitate a large variety of ET down-conversion or upconversion processes. Some of the low energy levels, such as $^4I_{11/2}$ (placed around 10 000 cm^{-1}) and $^4I_{13/2}$ (around 6500 cm^{-1}) have long radiative lifetimes, in the range 6–9 ms, although the electron–phonon interaction can reduce considerably the luminescence lifetime of the first level. Efficient emission can be obtained on the four-level transition $^4I_{11/2} \rightarrow {}^4I_{13/2}$, in the 2.7–2.95 μm range; unfortunately, the luminescence lifetime of the terminal level is larger than that of the emitting level, leading to rapid self-saturation of the laser emission. However, at high Er doping the migration-assisted ET upconversion $(^4I_{13/2}, {}^4I_{13/2}) \rightarrow (^4I_{9/2}, {}^4I_{15/2})$ can depopulate efficiently this level and recirculate part of its excitation to the $^4I_{9/2}$ level from which it relaxes back to the emitting level $^4I_{11/2}$. The effect of this upconversion is to a certain extent reduced by the upconversion from the level $^4I_{11/2}$, according to the scheme $(^4I_{11/2}, {}^4I_{11/2}) \rightarrow (^4S_{3/2}, {}^4I_{15/2})$ and a careful balance between these two upconversion processes must be controlled in order to produce highly efficient 3 μm laser emission.

The quasi-three-level transition $^4I_{13/2} \rightarrow {}^4I_{15/2}$, with emission cross-section in the 5×10^{-21} cm^2 range could be used for laser emission in the 1.55–1.65 μm range. Such laser emission could be useful as "eye-safe" radiation to replace the eye-dangerous 1 μm lasers or in telecommunications, since it matches the range of minimal absorption of glass fibers. The ET upconversion mentioned above could restrict laser emission from $^4I_{13/2}$ to diluted Er materials. Other transitions useful for Er^{3+} laser emission are in the visible (violet, blue, green, red) from various high energy levels; of particular interest could be the green or red emission from $^4S_{3/2}$, although efficient down-conversion ET processes could reduce its emission quantum efficiency.

A shortcoming of Er^{3+} is the weak pump absorption; however this can be circumvented by co-doping with sensitizer ions. Thus, sensitization with Cr^{3+} was useful in the case of lamp pumping, although it causes a large quantum defect. Excitation of the $^4I_{11/2}$ level by powerful 960 nm diode lasers is possible but it is not very efficient because of the low absorption; however, this can be improved by co-doping with Yb^{3+}, which has larger absorption and can transfer the excitation efficiently to Er^{3+}. This sensitized pumping could be useful for excitation of the 1.6 μm $^4I_{13/2}$ emission too, but its efficiency would be diminished by the emission or upconversion from $^4I_{11/2}$; however, it was found that $^4I_{11/2}$ can be rapidly de-excited to $^4I_{13/2}$ using the ET to other ions such as Ce^{3+}, whose absorption is resonant with this energy gap. The recent advent of 1.5 μm diode lasers could enable efficient direct pumping of the 1.6 μm Er lasers, with extremely low quantum defect. The upper energy levels for visible emission can be excited by ET upconversion and/or excited state absorption of the infrared pump radiation and the energy transfer from Yb^{3+} could be useful for sensitization of a variety of such upconversion processes.

2.8.1.2.5 The Tm^{3+} ion The energy level diagram of Tm^{3+}, ground configuration $4f^{12}$, with quite well separated levels enables emission in the visible and infrared, the most important being the quasi-three-level emission in the 2 μm range on transition from the first excited manifold $^3F_4$, placed in the region 5000 cm^{-1}, to the ground level $^3H_6$. Since Tm^{3+} is close to the end of the 4f series the electron–phonon interaction determines marked homogeneous broadening of the lines; the emission cross-section in this transition could reach several times 10^{-21} cm^2, but this is largely compensated for by the long lifetime of level $^3F_4$, 8–10 ms. Moreover, it was found that the $^3F_4$ level can be efficiently populated by quantum splitting of the excitation of the level $^3H_4$, placed around 12 500 cm^{-1}, by the cross-relaxation $(^3H_4, {}^3H_6) \rightarrow ({}^3F_4, {}^3F_4)$. A major difficulty for pumping Tm^{3+} is the weak absorption in the visible and infrared, and it was found that 3d ions such as Cr^{3+} or Fe^{3+} could be efficient sensitizers for population of the $^3H_4$ level under broad-band lamp excitation. However, with the recent development of diode lasers, direct excitation of $^3H_4$ or even of $^3F_4$ levels could become possible. The visible emission from Tm^{3+} can be excited by upconversion processes, particularly important being the Yb-sensitized upconversion in several steps that enables excitation of blue emitting levels under infrared excitation.

2.8.1.2.6 The Yb^{3+} ion The ground electronic configuration $4f^{13}$ of Yb^{3+} has only two manifolds, $^2F_{7/2}$ (ground) and $^2F_{5/2}$ (in the region of 10 000 cm^{-1}), which preclude emission self-quenching by cross-relaxation or by excited state absorption. Due to its position at the end of the lanthanide series, the electron–phonon interaction and crystal field effects for Yb^{3+} are stronger than for Nd^{3+} and thus Yb-doped materials have broader (homogeneous and inhomogeneous) emission lines and quite strong vibronic satellites. The global crystal field splitting of $^2F_{5/2}$ (three Stark levels) and of $^2F_{7/2}$ (four Stark levels) in various hosts can reach 600 cm^{-1}, and respectively 1200 cm^{-1}. The emission transitions are in the range 960–1060 nm; however, the thermal population of the lower Stark levels of $^2F_{7/2}$ could determine strong reabsorption that restricts the useful range for laser emission to 1025–1060 nm and thus the laser quantum defect is much lower (below 10%) compared with the 1 μm Nd lasers (around 24% for 800 nm diode laser pumping). Compared with the emitting level $^4F_{3/2}$ of Nd^{3+}, the level $^2F_{5/2}$ of Yb^{3+} has longer (3 to 5 times) lifetime but smaller (up to an order of magnitude) absorption and emission cross-sections.

Since Yb^{3+} has no absorption in the emission range of arc lamps, the development of Yb lasers only became possible in the last two decades, with the advent of efficient CW and QCW diode lasers in the wavelength range corresponding to the absorption spectrum of the $^2F_{5/2}$ level. The $^2F_{7/2}(1) \rightarrow {}^2F_{5/2}(1)$ absorption line (960–980 nm range), with the highest peak absorption cross-section is usually very sharp and shows strong homogeneous broadening: in the ordered crystals the 300 K FWHM could be of nanometer size, but at cryogenic temperatures it sharpens to a few tenths of nanometers. Diode pumping of this line would then require tight control of the peak wavelength and FWHM of the pump. Actually, a broader $^2F_{7/2}(1) \rightarrow {^2F_{5/2}(2)}$ absorption line in the 940 nm range is usually used for diode laser pumping, and its lower cross-section can be managed by technical solutions, such as fiber or thin-disk multi-pass laser configurations. For the bulk Yb lasers, quite high doping concentrations are necessary: however, this would favor the migration of excitation inside the system of Yb ions and easy access to various quenching traps, as well as Yb cooperative upconversion processes. It could also cause a severe reduction in heat conductivity, especially for materials where the mass of the Yb^{3+} ion differs strongly from that of the substituted cation.

In most Yb laser materials the $^2F_{5/2}(1) \rightarrow {^2F_{7/2}(3)}$ emission line has the highest peak cross-section; however, the reabsorption could act as a selective element for the emission transition under given conditions of temperature and transmission T of the exit laser mirror. Because of the considerably lower room temperature thermal population of the level $^2F_{7/2}(4)$ compared with $^2F_{7/2}(3)$, the reabsorption could determine (particularly at low T) a lower laser threshold for the much weaker 1.05–1.07 μm transition $^2F_{5/2}(1) \rightarrow {^2F_{7/2}(4)}$, which is closer to a four-level laser scheme, than for the quasi-three-level (1.03–1.05 μm) emission $^2F_{5/2}(1) \rightarrow {^2F_{7/2}(3)}$. However, at liquid nitrogen temperature the populations of the levels $^2F_{7/2}(4)$ and $^2F_{7/2}(3)$ become negligible and the Yb lasers behave as almost pure four-level lasers. Reduction of the electron–phonon interaction at low temperatures weakens the homogeneous broadening of the lines and increases strongly the peak absorption and emission cross-sections; at the same time it increases the relative contribution of the inhomogeneous broadening. Such inhomogeneously broadened Yb^{3+} emission lines, of very large line width, could be engineered in disordered laser materials.

Based on these qualities, Yb laser materials show promise for highly efficient, low-heat diode pumped laser emission in various regimes: the large product $\sigma_{eff}\tau_{eff}$ enables efficient CW emission, the long τ_{eff} favors efficient storage of inversion for Q-switched emission, whereas the broad emission lines favor mode-locked short pulse emission.

2.8.2 Selection of the host material

The characteristics of the laser emission process and of parasitic non-radiative de-excitation require careful selection of the laser material (active ion, host material) and of the doping characteristics. The doping ability is largely determined by the matching of the ionic radii of the doping and substituted cations. The Shannon ionic radii of the RE^{3+} ions in condensed state with anionic coordination 6 or 8 are $r_6 = 0.103$ to 0.115 nm and $r_8 = 0.113$ to 0.128 nm respectively, the largest values corresponding to the ions at the beginning and the lowest to the ions at the end of the 4f series [57]. These ions can substitute for host cations of similar ionic radius, such as Ca^{2+}, Sr^{2+}, La^{3+}, Gd^{3+}, Y^{3+}, Lu^{3+}, Sc^{3+} and so on, the degree of substitution being influenced by the conditions of fabrication (especially the thermodynamic regime) of the laser material. Doping with laser active ions can reduce

Table 2.1 *The Shannon ionic radii and atomic mass for several major laser active ions and host cations of laser materials*

Ion	Ionic radius (nm)			Atomic mass
	r_8	r_6	r_4	
Doping ions				
Ce^{3+}	0.1238	0.115		140.12
Pr^{3+}	0.1266	0.113		140.907
Nd^{3+}	0.1249	0.1123		144.24
Sm^{3+}	0.1219	0.1098		150.35
Ho^{3+}	0.1155	0.1041		164.930
Er^{3+}	0.1144	0.1030		167.26
Tm^{3+}	0.1134	0.1020		168.934
Yb^{3+}	0.1125	0.1008		173.04
Cr^{2+}		0.087		51.996
Cr^{3+}		0.0755		
Cr^{4+}		0.069	0.055	
Host cations				
La^{3+}	0.130	0.1172		138.91
Gd^{3+}	0.1193	0.1078		157.25
Y^{3+}	0.1159	0.1040		88.905
Lu^{3+}	0.1117	0.1001		174.97
Sc^{3+}	0.1010	0.0885		44.958
Ga^{3+}		0.076	0.061	69.72
Al^{3+}		0.0675	0.043	26.9815
Sr^{2+}	0.140			87.62
Ca^{2+}	0.126			69.72

drastically the thermal conductivity of the host material and a major additional criterion for selection of the host material is that the atomic mass of the substituted cation should be as close as possible to that of the doping ion. Table 2.1 gives the ionic radii of some of the most common laser active ions and of some host cations that can be substituted in the laser materials, together with the atomic mass. The doping ability is also influenced by the relation of the valence between the doping and the host cations; the differences in valence impose charge compensation and this can be achieved by lattice defects, such as vacancies or interstitials, by clustering of the doping ions or by co-doping with charge compensators. In several cases, heavy doping with ions of valence different from the host can determine severe restructuring of the crystalline lattice.

Obviously, other caracteristics of the host materials, such as mechanical properties, chemical stability, fabrication processing and so on can be determining factors in the selection of the host. The availability and cost of the raw material and the performance and cost of the fabrication process are of major importance too. As discussed before, the hosts that show the best prospect for solid-state lasers are large-gap dielectric materials such as simple, complex or mixed oxides and/or fluorides, that can have single crystal or polycrystalline structure, or amorphous materials (glasses). The fabrication technology can involve (quasi-) solid-state processes, such as used for polycrystalline materials produced by ceramics techniques, and solidification and crystallization of melts, such as used for glasses and respectively single crystals.

2.8.2.1 Single crystal laser materials

Single crystals have a regular structure that can be generated by translation of a basic structure (unit cell). The single crystals contain one or several types of identical cationic sites surrounded by defined polyhedrons of anions. The local symmetry of the cationic sites can differ from the global symmetry of the crystal, for instance cubic symmetry crystals such as the garnets or sesquioxides can offer substitution sites of low symmetry without inversion, which enable intense forced electric dipole transitions. Many low symmetry crystals have non-linear optical properties that allow the construction of integrated devices based on self-frequency conversion of the fundamental laser emission of the doping laser ion. The crystalline structure determines several other characteristics suitable for lasers, such as good mechanical and thermal properties. Most RE-doped laser crystalline materials are oxides such as garnets $A_3B_2C_3O_{12}$ (where $A = La$, Gd, Y, Lu; $B = Al$, Ga, Sc; $C = Al$, Ga, the most popular being the yttrium aluminum garnet $Y_3Al_5O_{12}$, YAG [58]), sesquioxides R_2O_3 (where $R = Y$, Lu, Sc), perovskites, niobates, tungstates, fluorophosphates, hexa-aluminates, and so on or fluorides (alkaline earth fluorides, trifluorides, complex fluorides), whereas simple oxides, such as Al_2O_3 or garnets, can host 3d elements.

The single crystals of laser materials can be produced by various techniques based on crystallization from melts with a crucible (such as the Czochralski technique, edge-defined film-fed growth EFG, micropulling, vertical or horizontal Bridgman growth, heat exchange method) or without a crucible (Verneuil, floating zone growth, laser heated pedestal growth LHPG), or crystallization from a flux and so on. A very important condition for avoiding large non-uniformities or phase separation is homogenization of the raw material and of the melt before crystallization: in the case of crucible-based methods, the high mobility of the ions in the melt or in solution as well as stirring of the melt during Czochralski growth contribute to further uniformity of the melt as well as to the manifestation of segregation or clustering processes during crystallization.

The performances of melt-growth technologies (uniformity, size of the crystal, maximum doping concentration, productivity) are largely determined by the thermal properties of the material such as the melting temperature and heat conductivity, by its composition and capacity for incorporating the doping ions as well as by the thermochemical or thermostructural properties. Obviously, the performance of the crystal growth equipment (size of the crucible, tight control of the thermal field and of mechanical displacement and so on) is also a determining factor. The segregation coefficient K_s, defined as the fraction of doping ions from the melt that is incorporated in the crystal at each moment of growth, and which is influenced by the difference in ionic radii between the doping and the host cations, leads to variation of the doping concentration along the grown crystal and limits its maximum value.

Large single crystals of garnets with diameters of several inches and lengths of 10–12 inches can be grown by the Czochralski method by pulling along the $\langle 111 \rangle$ direction. The process is slow (typically below 1 mm^{-1} h) and it takes several weeks to grow a large crystal. For systems with low segregation coefficient, such as Nd:YAG ($K_s \approx 0.18$) these crystals show coring, a central region with enhanced doping concentration that cannot be used for the fabrication of laser rods. The dimensional mismatch between the doping Nd^{3+} and the host cation Y^{3+} limits the maximum doping concentration of these Nd:YAG crystals to 1–1.2 at.% although crystals with small-size parts with up to 3–4 at.% Nd can be grown by methods based on thermal gradient. Moreover, the low K_s induces strong variation of the doping concentration along the crystal, severe enrichment of Nd in the melt, and limits severely the fraction of melt that can be used for growth. These factors limit the size of laser active components that can be fabricated from these crystals (maximum transverse section around 1 cm^2 and maximum length of 15–20 cm) and determine poor utilization of the raw material (below 10%). The high melting temperature of some oxides (1970 °C for YAG, 2420 °C for cubic sesquioxides) imposes the utilization of iridium or rhenium crucibles and a carefully controlled atmosphere during growth. High temperature melt-growth of many complex crystals is difficult because of parasitic temperature or atmosphere dependent thermophysical and thermochemical processes that can cause departure from the desired composition and structural defects.

Crystalline laser materials can also be deposited as thin films by various techniques (epitaxy, pulsed laser deposition, chemical vapor deposition and so on) onto lattice matched substrates. Such materials show considerable doping capabilities compared to bulk crystals. Guided structures can be fabricated in such films or in bulk materials by various techniques such as ionic bombardment, femtosecond laser writing and so on, for construction of lasers of interest in information technology. Attempts to grow crystalline fibers have produced promising results.

The remarkable variety and qualities of crystalline laser materials has contributed essentially to the development of laser physics and the technology and of applications. However, the production of single crystal materials is limited by various factors, for example the following:

- difficulties in growing crystals of large size, such as are needed for high power lasers (tens of kilowatt range) or for high energy large-aperture lasers or amplifiers necessary for inertial fusion;
- low technological yield;
- necessity of very tight control of the process over long periods of time;
- poor use of the raw material;
- severe limitations in composition of the host material and in the doping level and profile;

- limited optical quality, especially at high doping concentration;
- need for very expensive equipment and technological components;
- high energy consumption;
- the fabrication of multifunctional materials such as capped rods, combinations of active material–passive Q-switch and so on requires bonding of unifunctional crystalline parts.

2.8.2.2 Glasses

The sites available for doping of glasses can have quite large differences in the anionic coordination number, cation–anion distances and angles of bond, leading to inhomogeneous broadening of the lines, with corresponding reduction of the absorption and emission cross-sections. Moreover, the ability to incorporate high doping concentrations of rare earth ions is much enhanced. There is quite a large variety of laser glasses, most of them oxide (silicate, phosphate) glasses, but also chalcogenide or fluoride or mixed glasses, and these can be doped to high concentrations with trivalent rare earth ions. The temperature range required for the fabrication of glasses is much lower than that required for laser crystals and this enables dimensional scaling of the technological equipment. The glasses have extremely high optical quality and can be fabricated as large bulk bodies or as very long optical fiber lasers. Major shortcomings of the glasses are the low absorption and emission cross-sections as well as the poor mechanical and thermal properties, which preclude efficient or high power CW laser emission and reduces the possibility of scaling high energy pulsed bulk glass lasers to high average power. Interesting laser materials could be glass–ceramic composites in which part of the material is re-crystallized by thermal treatment to form nanocrystals of controlled composition, suitable for specific laser applications. This combines the possibility of drawing fibers, specific to the glasses, with the spectroscopic properties of doped single crystals.

2.8.2.3 Transparent polycrystalline laser materials produced by ceramic techniques

A very important class of laser materials, which combine and extend the advantages of the single crystals (well-defined crystalline structure, high thermomechanical qualities) and of the glasses (large size, high doping ability) are the transparent polycrystalline materials produced by ceramic techniques. These materials consist of tightly packed single crystal grains of uniform size and random orientation. The main factor limiting the optical transparency of these materials is the scattering (and possible absorption) of light due to microstructural defects (pores, inclusions, phase separation), and/or to differences in the refractive index between grains along the direction of propagation. The scattering depends on the relation between the wavelength of the light and the size of the scattering centers: in the optical range scattering centers of the order of a micrometer can be deleterious although those of several tens of nanometers would cause negligible scattering. Reducing the volume density of the scattering centers, particularly of the pores trapped at the junction of grains, to less than 1% would give a density of the polycrystalline material 99.99% of that of the bulk crystalline material.

The fabrication of transparent doped ceramics of high optical quality to meet the requirements for efficient laser emission is an extremely complex and challenging problem. These materials must conserve (or improve) the spectroscopic, mechanical and thermal properties of the corresponding single crystals.

There is a large variety of ceramic techniques that enable the fabrication of laser quality polycrystalline materials and they can be grouped in two classes: solid-state synthesis (reaction) or wet (soft) chemistry synthesis. The solid-state synthesis method was the first to produce transparent polycrystalline cubic oxide materials of laser quality. The elementary oxide materials and sintering aids are thoroughly milled and mixed in a ball mill, the slurry is spray dried to grains of tens of micrometers in size, the material is pressed in pellets and pre-sintered to synthesize the doped material. Compared with traditional ceramics techniques, which did not give high optical quality owing to the high density of pores trapped at grain boundaries or at corners, a crucial innovative step that enabled the control of the density of pores and fabrication of highly transparent materials was cold isostatic compression (CIP) at $\sim$200 MPa of these pre-sintered pellets [59, 60], combined with proper selection of a sintering aid that facilitates the migration of pores trapped at grain boundaries and the control of the grain growth. The purity, shape, size and uniformity of the starting raw material is of major importance. The grain growth of ceramics is governed by a combination of solid-state diffusion and melt controlled growth in the presence of the sintering aid at the temperatures necessary to produce a high degree of compaction of the grains and to facilitate the limitation of grain size after compaction. The solute drag effect at high doping concentrations can contribute to limitation of the grain size. The pellets are finally sintered in vacuum at temperatures of 1700–1750 °C in the case of YAG and cubic sesquioxides. The thermal regime of sintering should produce complete compaction of the ceramic and elimination of pores; however, above a certain temparature, where the compaction is almost complete, a strong increase in grain size is observed. In certain cases subsequent hot isostatic compression becomes necessary.

In the techniques based on wet chemistry [61], the soluble compounds (nitrates, sulfates, carbonates, chlorides and so on) of the desired cationic species are dissolved together with a precipitating agent, resulting in a precipitate which is then reduced by heating to the final compound. Techniques based on precipitation and reduction by combustion can also be employed. The precipitate is amorphous and is transformed by heat treatment to nanocrystalline powder. This technique enables the use of slip casting to produce the desired shape and size of the material. The final step is vacuum sintering at 1700–1750 °C, similar to the solid-state technique. In this technique the large surface energy of the nanocrystallites contributes to the grain growth (nanotechnology process), and reduces the need for a sintering aid. A hybrid approach, combining the use of nanocrystallites with isostatic compression, has also been reported.

In both these techniques the grains have very shallow grain boundaries (order of nanometers) and very low density of scattering pores. The mean grain sizes of the ceramics produced by solid-state synthesis are several tens of micrometers (coarse-grained ceramics) whereas for those produced by the nanotechnology process they are of the order of several micrometers (fine-grained ceramics). Since the orientation of the individual single crystal grains in these materials is random, the possibility of producing highly transparent materials by these ceramic techniques has been limited to materials with isotropic refractive index, i.e. to cubic materials (garnets, sesquioxides, fluorides).

Although there is a certain mobility of the cations in the ceramic grain growth process, this is much restricted compared with the melt-growth of crystals and its effect is confined to each grain: complete homogenization or the desired local and global composition and structure must then be carried out during the preliminary steps before sintering. This is particularly important for mixed host laser materials or for materials doped with ions that necessitate charge compensation.

In order to keep the ceramic grains at a small size, which is claimed to grant better mechanical properties and reduced doping segregation, a double sintering technique that allows full compaction by keeping the grain size to less than a micrometer without using a sintering aid has been developed for the cubic sesquioxides. The temperature is raised to the point where a high (85–95%) relative density is achieved, then it is lowered and kept for tens of hours at a value where compaction continues without grain size expansion, and the inherent residual losses are efficiently removed by subsequent hot isostatic pressing [62, 63]. An attempt to keep the size of the ceramic grains close to the initial size (30–40 nm) of the starting Nd:YAG nanopowders was undertaken by low temperature (450–600 °C) sintering under high pressure (2–6 GPa). Indeed, the grain size was kept low, but although the ceramics were optically transparent and dense (close to the theoretical limit) the extinction coefficient (absorption and scattering) was 5–6 times larger than for single crystals [64], which was attributed to residual nanopores. Moreover, reduced heat conduction in these ceramics can be expected because of the small grain size. Small grains can also be obtained by fast sintering techniques such as plasma arc sintering.

Ceramics with grain sizes of tens of nanometers (nanoceramics) of various materials have also been produced. Since the effect of birefringence on scattering is reduced, this would enable the production of fairly transparent ceramics of birefringent materials, but their reduced heat conduction precludes their use in laser techniques. A confusing situation was generated by the use by several authors of the denomination "nanoceramics" for the fine-grained ceramics (around micrometer size) produced by the nanotechnology process. However, it must be outlined again that these two classes of ceramics (true nanoceramics and fine-grained ceramics) are quite distinct.

The random orientation of the grains in the coarse- and fine-grained ceramics is largely determined by the random orientation of the material submitted for sintering, and this imposes a restriction to cubic materials. However, if the orientation of the particles in the green could be controlled, the fabrication of transparent ceramics of anisotropic materials would become possible.

In all ceramic techniques the size of the bulk material is controlled, within reasonable limits, by the characteristics of the technological equipment. This opens the possibility of producing very large size laser materials, comparable with those available for glasses.

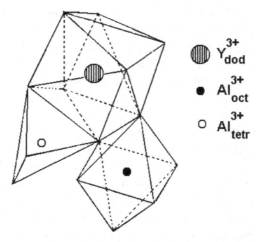

Figure 2.7 The YAG structure with the three cationic sites: dodecahedral, octahedral and tetrahedral.

2.8.3 Main classes and properties of transparent ceramic laser materials

2.8.3.1 Selected classes of cubic laser materials for transparent ceramics

2.8.3.1.1 Garnets The laser garnets have cubic symmetry $Ia3d$ and the chemical composition $A_3B_2C_3O_{12}$, where A are large trivalent ions such as La^{3+}, Gd^{3+}, Y^{3+}, Lu^{3+} or divalent ions (Ca^{2+}) surrounded by eight O^{2-} anions in the corners of a distorted cube with local symmetry D_2 (c-sites), B are smaller size ions such as Al^{3+}, Ga^{3+}, Sc^{3+}, Nb^{5+} and so on, surrounded by six O^{2-} ions that form a trigonally distorted octahedron of C_{3i} symmetry (a-sites), whereas C are small ions (Al^{3+}, Ga^{3+}) surrounded by a tetragonally distorted tetrahedron of oxygen atoms, with S_4 local symmetry (d-sites) (Figure 2.7). Based on ionic size, the doping RE^{3+} ions should substitute preferably for the large A cations. The garnets are compositionally ordered when each type of site is occupied by a unique cationic species, such as $Y_3Al_5O_{12}$ (YAG), $Gd_3Ga_5O_{12}$ (GGG), $Gd_3Sc_2Ga_3O_{12}$ (GSGG); however, compositional disorder can be induced when a site can be occupied at random by several ionic species, sometimes of different valence, such as $Y_3(Sc_x,Al_{1-x})_2Al_3O_{12}$ (YSAG) or $Ca_3(Li_xNb_{1-x})_2Ga_3O_{12}$ (CLNGG).

A major problem of Czochralski growth of garnet crystals is evaporation of part of the aluminum or gallium oxide and/or the shift of the congruent melting composition from the ideal stoichiometric composition, leading to an excess of A ions that can enter the octahedral sites normally occupied by the B ions [65]. The extent of such departures from ideal stoichiometry depends on the difference in ionic radii of the A and B cations and on the temperature of growth, being much lower in flux growth (growth temperature about 600 °C below melting temperature) than in the melt-grown crystals. In the case of YAG, the percentage of a-sites occupied by Y^{3+} ions (Y(a)) is ~1.75% and thus it can be stated that the Czochralski grown YAG crystals are in fact Y(a)-doped YAG. It would then be.

expected that the low fabrication temperature of the transparent ceramics would reduce considerably such departures from stoichiometry.

The ideal garnet structure of the YAG crystals was challenged by the EXAFS investigation of melt-grown crystals [66]: in order to explain several lines that did not fit the ideal garnet structure a large-scale inversion between the Y^{3+} and Al^{3+} in occupancy of octahedral and dodecahedral sites was assumed, which would reduce the symmetry from cubic to trigonal. However, no other experiment has confirmed this assumption. At the same time, as discussed in Section 2.2.2.5, the quality of YAG as laser material was questioned in references [42] by assuming the presence of large and uncontrolled concentrations of "dead" Nd sites coupled with hydroxyl or color centers that de-excite these Nd ions faster than relaxation to the emitting level, making them inactive for laser emission; again, no obvious independent proof of such dead sites exists.

RE-doped ordered or compositionally disordered garnet ceramics have been produced by a variety of techniques based on solid-state [59, 60] or soft synthesis [61]. The maximum doping concentration was much larger than in crystals, up to 9 at.% in the case of Nd:YAG. Comparative global spectroscopic investigation of the doped YAG ceramics and single crystals indicates that the main features of the optical spectra are similar over the entire range of concentrations available for comparison. Although the extension to higher doping concentration could induce specific modification of the optical spectra (inhomogeneous asymmetric broadening of lines, shifts induced by slight modification of the lattice parameter), these spectra preserve the general features of the diluted materials. The high resolution spectroscopy of coarse-grained Nd:YAG ceramics [67] revealed that the global satellite structure of the spectra induced by the statistical ensembles of doping ions are similar and retain fairly well the features specific to a random doping distribution, and that the emission decay of samples with up to 7 at.% Nd can be described with the energy transfer parameters derived from the measurements on low concentration (up to 2.5 at.%) crystals and ceramics. Nevertheless, the fabrication process and the granular structure of the ceramics can modify this picture locally, the main factor being the inherent dopant segregation at the grain boundaries and the diffusion of the doping ions or of ions from the sintering aid. Theoretical [68] and spatially resolved experimental investigation [69, 70] indicates doping concentration dependent differences in the spectroscopic properties between the region of the grain boundaries and the bulk of the ceramic grains. No precise correlation between the spectroscopic data and results obtained by microscopic methods of compositional analysis exists, and further investigation wiil be necessary. The effect of these perturbed regions on the global spectroscopic and laser properties will be influenced by the ratio (surface/volume) of the grains and by the depth of the affected region.

The mechanical properties of the YAG ceramics are improved with respect to the single crystals. The hardness of YAG crystals is 14.5 GPa and increases by $\sim 15\%$ in the case of ceramics; however, the fracture toughness increases from 1.8 MPa $m^{1/2}$ in the case of crystals to 4.3–8.7 MPa $m^{1/2}$ in ceramics, the largest values corresponding to the ceramics with larger grains [71]. Generally, in the case of ceramics it is considered that the tensile strength is related to the grain size by $\sigma \propto d^{-1/2}$ [72]. The thermal conductivity at room temperature of Nd:YAG coarse-grained ceramics (10.2 W m^{-1} K^{-1}) was similar to that of single crystals but that of fine-grained ceramic was smaller, 9.58 W m^{-1} K^{-1} [73]. Doping with laser ions could reduce the thermal conductivity, especially when there is a large difference in atomic mass between the doping and the host cations, the reduction being influenced by the doping concentration. At low temperatures the thermal conductivity of the YAG crystals increases, 46 W m^{-1} K^{-1} at $\sim$100 K, but for ceramics this is weaker, 38 W m^{-1} K^{-1} [74]; the effect is smaller for the fine-grained ceramics due to scattering of the phonons at the ceramic grain boundaries. The thermo-optic coefficient dn/dT decreases with temperature, from 7.8 $\times$ 10^{-6} K^{-1} at 300 K, to 0.9 $\times$ 10^{-6} K^{-1} at 100 K, whereas the linear thermal expansion coefficient α_L decreases from 6.2 $\times$ 10^{-6} K^{-1} to 2 $\times$ 10^{-6} K^{-1} over this temperature interval.

The potential of Nd:YAG ceramics with 1 at.% Nd in the construction of lasers was demonstrated very soon after the advent of these materials, for both types, with similar or improved laser parameters compared with the single crystals. A CW power of up to $\sim$1.5 kW was achieved [59, 61], and in a few years the range of Nd concentration showing laser emission was extended to 3.5% [75] and $\sim$7% [76]. The similarity of laser emission of the Nd:YAG ceramics and of crystalline lasers was subsequently demonstrated to 3.5 at.% Nd [77] and since then a very large variety of technological, structural, spectroscopic and laser emission results have been reported. Ceramic Nd:YAG laser components of very high optical quality and very large size (10 $\times$ 10 $\times$ 2 cm^3) have been instrumental in scaling the free generation heat capacity burst laser to the unprecedented value of 67 kW [78]. Various composite materials such as capped Nd:YAG rods, rods with radial or longitudinal composition gradient, rods clad with undoped material or with Sm^{3+}-doped material to suppress ASE, YAG composite materials with parts containing Nd^{3+} as laser ion and Cr^{4+} as passive Q-switch material and so on were also produced. Improvement of the mechanical and thermal properties of YAG ceramics compared with the single crystals as well as the favorable evolution of these properties and of several spectroscopic characteristics at low temperature gives promise for utilization of these ceramics in the construction of advanced (including cryogenic) lasers. The Yb^{3+}-doped ceramics are considered potential candidates for very high energy pulsed lasers (tens of hertz) for inertial laser nuclear fusion.

2.8.3.1.2 Cubic sesquioxides The cubic sesquioxides R$_2$O$_3$ (R = Gd, Y, Sc) have the Bixbyite structure, *Ia*3 space group, with two cationic sites surrounded by six oxygen ions, of C_2 and C_{3i} local symmetry, in the proportion 3:1 and the RE^{3+} ions can substitute at both these sites (Figure 2.8). The high density and the tight packing of cationic sites and the short cation–cation distances in these materials favor strong crystal field interactions and efficient energy transfer between the doping ions.

The first transparent ceramics of cubic sesquioxides (Nd:Y$_2$O$_3$) were produced by conventional sintering with ThO$_2$ sintering aid, and demonstrated pulsed lasing [79]. Improved optical quality was then demonstrtated in Y$_2$O$_3$ ceramics with HfO$_2$ sintering aid formed using hot isostatic pressing [60]. Subsequently, sesquioxide ceramics were

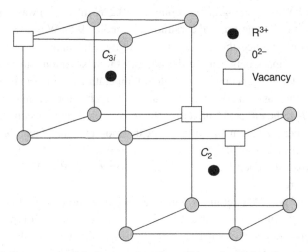

Figure 2.8 The R_2O_3 structure with the two crystallographic cationic sites C_2 and C_{3i}.

also produced by wet techniques [80]. The cubic sesquioxide ceramics show improved mechanical and thermal properties compared with the crystals [74, 81]. The thermal conductivity of the sesquioxides is larger than that of garnets, showing room temperature values of 17 W m^{-1} K^{-1} for Sc_2O_3, 12.5 W m^{-1} K^{-1} for Lu_2O_3 and 13.6 W m^{-1} K^{-1} for Y_2O_3 and increasing at low temperatures (52 W m^{-1} K^{-1} for Y_2O_3 at 92 K) [82].

These data show that the garnet and sesquioxide ceramics possess much better mechanical and thermal properties than the glasses (fracture toughness and thermal conductivity are larger by about an order of magnitude) and thus show unprecedented potential for scaling the power, energy or repetition rate for large-aperture solid-state lasers with special applications, such as inertial nuclear fusion [83, 84] or for generation of ultra-high peak power laser pulses [84, 85] for nuclear physics studies.

2.8.3.1.3 Cubic fluorides The alkaline earth fluorides MeF_2 (with $Me^{2+} = Ca^{2+}$, Sr^{2+}, Ba^{2+}) have cubic symmetry, $Fm3m$ space group; the Me^{2+} site has cubic symmetry and is coordinated by a regular cube of eight F^- ions. In order to maintain electrical neutrality the F^- cubes are alternately occupied by cations or empty. The large Me^{2+} cations can be substituted by large laser ions of f-electronic groups such as rare earths or actinides. The alkaline earth fluorides show very high optical qualities and thermal conductivity close to that of YAG. In fact, U^{3+}-doped CaF_2 was the second laser, demonstrated [86] soon after the advent of the first (ruby) laser.

The melting temperature of the MeF_2 compounds is quite low, so they can be easily grown by melt-based methods. Substitution of the Me^{2+} ion with RE^{3+} in MeF_2 implies the need for charge compensation: at low doping concentrations, typically below 0.1 at.%, the substitutional RE^{3+} ions can be considered isolated from each other and charge compensation can be done by interstitial F_i^- ions or by co-doping with substitutional alkaline ions M_s^+, such as Na^+, Li^+, or K^+. The charge compensation can be done in the vicinity of the doping RE^{3+} ion and in this case a chain of perturbations can be produced, leading to a low-symmetry multicenter spectral structure, or it can be done far from the doping ion and in this case the original local cubic symmetry is preserved. The position of the charge compensator F_i^- or M_s^+ with respect to the RE^{3+} ion can be denoted (n, l, m) in units of the fluorine cube edge. In the case of interstitial F^- charge compensation $(n + l + m) = 2k + 1$ ($k \geq 0$) and the nearest compensators reduce the cubic crystal field symmetry to tetragonal C_{4v} $(1, 0, 0)$, trigonal C_{3v} $(1, 1, 1)$, monoclinic $(2, 1, 0)$ and so on. For M^+ charge compensation $(l + m + n) = 2j$, with $j \geq 1$, and the nearest compensators reduce the crystal field symmetry to orthorhombic C_{2v} $(1, 1, 0)$, tetragonal C_{4v} $(2, 0, 0)$, and so on. The existence of such centers determined by F_i^- or Na_s^+ was first demonstrated by techniques such as electron paramagnetic resonance and the optical spectra confirm this variety. For instance, in the case of U^{3+} in CaF_2, it was shown by EPR that co-doping with various alkaline ions (Li^+, Na^+, K^+) induces a similar variety of perturbed centers, but whereas the characteristics of the centers compensated by $M_s^+(2, 0, 0)$ are quite similar (weak perturbation), the differences in the parameters of the centers with $M_s^+(1, 1, 0)$ charge compensation are quite large [87]. The absorption and emission cross-sections for the RE^{3+} in CaF_2 are not very large and high doping concentrations would be necessary for efficient laser emission. However, for large concentrations, a strong tendency of clustering of RE^{3+} ions, with involvement of the charge compensators, is observed and it was claimed that charge compensation by M_s^+ reduces this process. Increasing the doping concentrations above 1%, such as for Yb^{3+}-doped CaF_2, leads to strong restructuring of the fluorite structure: all doping ions coagulate in large hexametric ensembles involving six n.n. Yb^{3+} ions, with cubo-octahedron structure, without the need for external charge compensation [88]: a unique optical spectrum, with trigonal symmetry was thus obtained. Such clustering was not observed in the case of the large RE^{3+} ions or when $Me = Sr$ or Ba. In all the cases described above, the charge compensation process is determined by the mobility of the charge compensating and/or of the doping laser ion and this in turn is determined by the thermodynamic conditions of production and can be very different in the case of single crystals and ceramics.

CaF_2 single crystals of high optical quality with diameter in the range of 40 cm can be grown from melt (Kyropoulos technique); however, a major problem is the perfect cleavage that can limit their potential for high power lasers. Attempts to use the disordering induced by doping and/or of various mixed systems (such as solid solutions of fluorides of Ca and Sr) can be useful. The transparent polycrystalline materials can solve this problem; in fact, the first ceramic laser was based on $Dy:CaF_2$ ceramics produced by hot pressing [89]. A major problem that needs detailed investigation in the case of CaF_2 ceramic laser materials is control of the charge compensation process and its effects on the properties of the active centers. The typical solid-state reaction ceramic fabrication process starting from dispersed material limits the diffusion of the ions and thus mixing of the elementary compounds (host, doping) enabling sufficient homogenization and control of the charge compensation, leading to high spatial non-uniformity. This requires special preparation of the material that is to be sintered, such as wet synthesis of the doped material under conditions that will eliminate the possibility of oxygen contamination [90]. A modality to avoid this problem could be hot pressing ("hot forming") of single crystal material. This material shows improved mechanical and thermal parameters compared with single crystals, especially at cryogenic temperatures [91].

2.8.3.1.4 Other cubic materials There are several other classes of cubic materials that can host laser active ions or ions with saturable absorption for passive Q-switching. Examples of such materials are the cubic spinels, with the representative material $MgAl_2O_4$, which offers tetrahedral Mg^{2+} sites and octahedral Al^{3+} sites for substitution. A big advantage of this material is the high heat conductivity, $\sim 25\%$ higher than for YAG, that could facilitate heat dissipation when used as saturable absorber. The material is hard and very stable and the melting temperature is high, $2135\,°C$, making crystal growth difficult, but highly transparent ceramics are produced industrially for high resistance optical, mechanical or refractory components. Particularly attractive for use in lasers is Co^{2+}-doped $MgAl_2O_4$ [92] which shows strong absorption from 1.2 to 1.6 μm (absorption cross-section 3.5×10^{-19} cm^2 at 1.54 μm), with potential for passive Q-switching of 1.3 μm Nd or of 1.5 μm Er eye-safe lasers.

Transparent ceramics of ferroelectric lanthanum-modified lead zirconate titanate (PLZT) doped with RE^{3+} ions have potential for laser technology since they can offer additional electro-optic properties, such as the possibility of modulation of the emission of the doping ions with an external electric field [93]. PLZT has a pseudo-cubic perovskite structure ABO_3, with the 12 O^{2-} coordinated A sites occupied at random by Pb^{2+} or La^{3+} and the sixfold O^{2-} coordinated B sites occupied by Zr^{4+} or Ti^{4+}. The best optical properties correspond to the ratios (Pb/La) = 90/10 and (Zr/Ti) = 65/35. The presence of La^{3+} in the A sites implies the need for charge compensation, and this can be naturally accomplished by cation vacancies. Based on ionic size arguments, the large RE^{3+} could show a preference for the A sites, whereas the smaller RE^{3+} ions, such as Yb^{3+}, could prefer the B sites.

As discussed in Section 2.8.1.1, Cr^{2+} ions in tetrahedral coordination show potential for efficient tunable emission in the infrared. Very promising hosts for these ions are cubic II–VI semiconductor hosts, such as ZnSe. When doped with divalent 3d ions such as Co^{2+} or Fe^{2+}, these materials could be used as saturable absorbers for passive Q-switching of infrared lasers. Recent success in the fabrication of transparent ZnSe ceramic by hot pressing could be an important step for the development of such lasers or Q-switches [94].

2.8.3.2 Transparent ceramics of anisotropic materials

Many laser materials have a low symmetry structure and are optically anisotropic and thus the ceramic techniques discussed so far, which confer random orientation to the grains, will not result in transparent materials. Examples of such anisotropic materials are Al_2O_3 (sapphire) doped with Ti^{3+}, the main ultra-short pulse laser material, the Sr fluoroapatite (SFAP) doped with Yb^{3+} which shows good qualities for high repetition rate high energy amplifiers in inertial nuclear fusion [95], RE-doped double vanadates, perovskites or wolframates, numerous non-linear materials for frequency multiplication or mixing and so on. A method of potential impact on the fabrication of transparent ceramics of such low symmetry materials is orientation of the nanocrystalline grains of the material in external fields, particularly in magnetic fields, before sintering [96, 97]. This process can be more efficient in materials doped with ions with incomplete electronic shells (such as laser ions from the 3d and 4f groups) that possess paramagnetic properties: attempts to produce Nd- or Yb-doped SFAP laser ceramics resulted in materials of fairly good optical transparency [96]. Another possibility for ceramics of low symmetry materials could be selection of the spectral range where birefringence is zero, such as the 560 nm range for the trigonal oxysulfide La_2O_2S [91].

2.8.3.3 The main characteristics of the ceramic laser materials

The results obtained so far in the fabrication of transparent polycrystalline materials by ceramic techniques and in investigation of their properties, as well as in their application in the construction of solid-state lasers have been reviewed in the literature [75, 98–108] and several specific features of these materials have been revealed.

Characteristics of fabrication techniques

- Fabrication temperatures lower by 400 to 700 °C than those used for the melt growth of crystals.
- Very high technological yield.
- High reproducibility.
- Good control of the fabrication process.
- Large doping concentrations possible, for instance in the case of Nd:YAG ceramics of high optical quality, Nd concentrations up to 9 at.% have been produced.
- Controlled profile of composition or doping (homogeneous, transverse and/or longitudinal gradient, profiled, and so on) over the whole volume of the ceramic.
- Possibility of producing composite materials, containing parts with defined composition, such as laser rods clad or capped with undoped material, materials composed of a laser active part and a passive Q-switch part, Nd-doped material surrounded by ASE suppressing material and so on.
- Large bodies of size (plates of tens of centimeters on the edges and centimeters thick) and shape close to that of the final optical component can be produced.
- Further innovative potential for new materials and techniques, such as solid-state crystal growth, composites, single crystal ceramics, ceramic fibers and so on.

Structural properties

- Uniform grain sizes, of several micrometers to tens of micrometers, depending on the fabrication process.

- Low size and volume density of residual intergrain pores (order of ppm).
- Very shallow (order of micrometer) grain boundaries.

Functional properties – tailoring of new laser materials

- High compositional versatility.
- Good control of composition.
- Optical transmission spectrum similar to that of single crystals.
- Room temperature thermal properties similar or better than those of single crystals and a reasonable increase at low temperatures.
- Similar or improved mechanical properties with respect to single crystals.
- Superior mechanical and thermal properties compared with glasses.

Economical aspects

- Potential for reduced production cost.
- Very good use of raw material.
- No need for expensive components such as rare metal crucibles.
- Reduced energy consumption.

2.8.4 Further research and development in the optical spectroscopy and laser emission of doped ceramics

The characteristics of the transparent ceramics discussed above reveal the enormous potential for research and development of transparent ceramics and of their applications. These materials could have a major contribution to the development of basic science (physics of multielectron systems and of their interaction with radiation, optics, quantum electronics and laser physics, photonics, materials science and so on) and applied science (lasers, optical systems, scintillators and so on). At the same time, the complexity of the problems related to the fabrication of doped transparent ceramics with properties tailored to optimize, extend or generate new applications, as well as the early stage of knowledge of many aspects, determined by the novelty of the problems, demands a sustained research effort for the characterization and technological development of these materials.

Nevertheless, ceramic grain growth and the resulting granular structure as well as utilization of sintering aids could influence the microstructural, optical and spectroscopic properties of the doped ceramics and the properties of these materials should be carefully investigated:

- the variety, nature and structure of the centers formed by the doping ions;
- the quantum state of the doping ions (energy levels, transition probabilities, crystal field effects);
- the distribution of the doping ions at the available lattice sites, including the defective sites at the surface or induced by the sintering aid;

- the interactions between the doping ions and energy transfer processes;
- the segregation and diffusion of the doping ions at the grain surfaces and their local effects on the spectroscopic properties, energy transfer processes, heat generation and variation of the refractive index;
- correlation of the data collected by spatially resolved microscopic investigation with the global spectroscopic properties;
- the influence of the spectroscopic, optical and thermomechanical effects of granular structure of ceramics on the laser emission and on the characteristics of the laser beam.

The results and implications of these spectroscopic investigations are presented in Chapter 10 and the results and prospects for the construction of solid-state lasers based on ceramic laser materials is discussed in Chapter 11.

References

[1] T. Maiman, Stimulated optical radiation from ruby, *Nature* **187** (1960) 493–494.

[2] A. E. Siegman, *Lasers*, University Science Books, Sausalito, CA (1986).

[3] W. P. Risk, T. R. Gosnell, and A. V. Nurmikko, *Compact Blue Green Lasers*, Cambridge University Press, Cambridge (2003).

[4] W. T. Silvfast, *Laser Fundamentals*, Cambridge University Press, Cambridge (2004).

[5] M. C. Gupta and J. Belatto, eds., *The Handbook of Photonics*, second edition, CRC Press, Boca Raton, FL (2006).

[6] Y. Y. Kalisky, *The Physics and Engineering of Solid-State Lasers*, SPIE Press, Bellingham, WA (2006).

[7] W. Koechner, *Solid-State Laser Engineering*, sixth edition, Springer Science + Business Media, New York (2006).

[8] F. Trager, ed., *Springer Handbook of Lasers and Optics*, Springer Science + Business Media, New York (2007).

[9] R. Menzel, *Photonics: Linear and Nonlinear Interactions of Laser Light and Matter*, second edition, Springer, Berlin (2007).

[10] R. Ifflander, Laser systems, in *Landolt–Bornstein Group VIII, Advanced Materials and Technologies* (2008), Vol. 12, pp. 3–96.

[11] O. Svelto, *Principles of Lasers*, fifth edition, Springer Science + Business Media, New York (2010).

[12] G. M. Dieke, *Spectra and Energy Levels of Rare Earth Ions in Crystals*, Wiley-Interscience, New York (1968).

[13] B. R. Judd, *Operator Techniques in Atomic Spectroscopy*, McGraw Hill, New York (1963).

[14] B. DiBartolo, *Optical Interactions in Solids*, John Wiley, New York (1968).

[15] S. Hufner, *Optical Spectra of Transparent Rare Earth Compounds*, Academic Press, New York (1978).

[16] C. A. Morrison, *Crystal Field for Transition Metal Ions in Laser Host Materials*, Springer, Berlin (1992).

[17] F. Gan, *Laser Materials*, World Scientific, Singapore (1995).

[18] A. A. Kaminskii, *Crystalline Lasers: Physical Processes and Operating Schemes*, CRC Press, Boca Raton, FL (1996).

[19] R. C. Powell, *Physics of Solid State Laser Materials*, Springer, New York (1988).

[20] B. Henderson and R. H. Bartram, *Crystal Field Engineering of Solid-State Laser Materials*, Cambridge University Press (2000).

[21] A. Einstein, Zur Quantentheorie der Strahlung (On the quantum theory of radiation), *Z. Phys.* **18** (1917) 121.

[22] B. R. Judd, Optical absorption intensities of rare earth ions, *Phys. Rev.* **127** (1962) 750–761.

[23] G. S. Ofelt, Intensities of crystal spectra of rare earth ions, *J. Chem. Phys.* **37** (1962) 511–520.

[24] Th. Forster, Zwischenmolekulare Energiewanderung und Fluoreszenz, *Ann. Phys. (Leipzig)* **2** (1948) 55–75.

[25] D. L. Dexter, A theory of sensitized luminescence in solids, *J. Chem. Phys.* **21** (1953) 836–850.

[26] H. G. Danielmeyer, M. Blatte, and P. Palmer, Fluorescence quenching in Nd:YAG, *Appl. Phys.* **1** (1973) 269–274.

[27] M. Inokuti and F. Hirayama, Influence of energy transfer by the exchange mechanism on the donor luminescence, *J. Chem. Phys.* **43** (1965) 1978–1989.

[28] S. I. Golubov and Yu. V. Konobeev, Procedure of averaging in the theory of resonance transfer of electron excitation energy, *Sov. Phys. Solid State* **13** (1972) 2679–2682.

[29] V. P. Sakun, Kinetics of energy transfer in a crystal, *Sov. Phys. Solid State* **14** (1973) 1906–1914.

[30] A. G. Avanesov, B. I. Denker, V. V. Osiko, S. S. Pirumov, V. P. Sakun, V. A. Smirnov, and I. A. Shcherbakov, Kinetics of nonradiative relaxation from the upper active level of neodymium in a $Y_3Al_5O_{12}$ crystal, *Sov. J. Quantum Electron.* **12** (1982) 744–755.

[31] V. Lupei, A. Lupei, S. Georgescu, and W. M. Yen, Effects of energy transfer on quantum efficiency of YAG:Nd, *J. Appl. Phys.* **66** (1989) 3792 –3799.

[32] V. Lupei, Selfquenching of Nd^{3+} emission in laser garnet crystals, *Opt. Mater.* **16** (2001) 137–152.

[33] A. I. Burshtein, Concentration quenching of noncoherent excitation in solutions, *Sov. Phys. Usp.* **27** (1984) 579–606.

[34] V. Lupei, A. Lupei, and G. Boulon, On the characteristics of sensitized emission in laser crystals, *Phys. Rev. B: Condens. Matter* **53** (1996) 22–32.

[35] V. Lupei, and A. Lupei, Emission dynamics of the $^4F_{3/2}$ level of Nd^{3+} in YAG at low pump intensities, *Phys. Rev. B: Condens. Matter* **61** (2000) 8087–8098.

[36] D. L. Dexter, Cooperative optical absorption in solids, *Phys. Rev.* **126** (1962) 1962–1967.

[37] H. J. Schugar, E. I. Solomon, W. I. Cleveland, and L. Goodman, Simultaneous pair electronic transitions in Yb_2O_3, *J. Am. Chem. Soc.* **97** (1975) 6442–6450.

[38] T. T. Basiev, K. K. Pukhov, and I. T. Basieva, Cooperative quenching kinetics: computer simulation and analytical solution, *Chem. Phys. Lett.* **432** (2006) 367–370.

[39] V. Lupei, A. Lupei, C. Tiseanu, S. Georgescu, C. Stoicescu, and P. M. Nanau, High resolution optical spectroscopy of Nd in YAG – a test for structural and distribution models, *Phys. Rev. B: Condens. Matter* **53** (1995) 8–15.

[40] D. E. McCumber, Theory of phonon terminated optical masers, *Phys. Rev.* **134** (1964) A299–A304.

[41] S. A. Payne, L. L. Chase, L. K. Smith, W. L. Kway, and W. F. Krupke, Infrared cross-section measurements for crystals doped with Er^{3+}, Tm^{3+}, and Ho^{3+}, *IEEE J. Quantum Electron.* **28** (1992) 2619–2627.

[42] D. P. Devor and L. G. DeShazer, Evidence of Nd:YAG quantum efficiency on nonequivalent crystal field effects, *Opt. Commun.* **46** (1983) 97–102; D. P. Devor, R. C. Pastor, and L. G. De Shazer, Hydroxyl impurity effects in YAG ($Y_3Al_5O_{12}$), *J. Chem. Phys.* **81** (1984) 4104–4117; D. P. Devor, L. G. De Shazer, and R. C. Pastor, Nd:YAG quantum efficiency and related radiative properties, *IEEE J. Quantum Electron.* **25** (1989) 1863–1873.

[43] D. C. Brown, Heat, fluorescence and stimulated emission power densities and fractions in Nd: YAG, *IEEE J. Quantum Electron.* **34** (1998) 560–572.

[44] V. Lupei, Directions for performance enhancement and power scaling of the Nd lasers, *Prog. Quantum Electron.* (to be published).

[45] T. Y. Fan, Heat generation in Nd:YAG and Yb:YAG, *IEEE J. Quantum Electron.* **29** (1993) 1457–1459.

[46] W. P. Risk, Modeling of longitudinally pumped solid state lasers exhibiting reabsorption losses, *J. Opt. Soc. Am. B* **5** (1988) 1412–1423.

[47] D. Findlay and R. A. Clay, The measurement of internal losses in 4-level lasers, *Phys. Rev. Lett.* **20** (1966) 277–281.

[48] T. Y. Fan and R. L. Byer, Continuous-wave operation of a room-temperature, diode-laser-pumped, 946-nm Nd:YAG laser, *Opt. Lett.* **12** (1987) 809–811.

[49] A. Giesen and J. Speiser, Fifteen years of work on thin-disk lasers: results and scaling laws, *IEEE J. Sel. Top. Quantum Electron.* **13** (2007) 598–609.

[50] J. J. Degnan, Theory of optimally coupled Q-switched laser, *IEEE J. Quantum Electron.* **25** (1989) 214–220.

[51] D. Strickland and G. Mourou, Compression of amplified chirped optical pulses, *Opt. Commun.* **56** (1985) 219–224.

[52] W. A. Clarkson, Thermal effects and their mitigation in end-pumped solid-state lasers, *J. Phys. D: Appl. Phys.* **34** (2001) 2381–2395.

[53] S. Chenais, F. Druon, S. Forget, F. Balembois, and P. Georges, On thermal effects in solid-state lasers: the case of ytterbium-doped materials, *Prog. Quantum Electron.* **30** (2006) 89–153.

[54] W. Koechner, Absorbed pump power, thermal profile and stress in a cw pumped Nd:YAG crystal, *Appl. Opt.* **9** (1970) 1429–1434.

[55] R. Gaumé, B. Viana, D. Vivien, J. P. Roger, and D. Fournier, A simple model for the prediction of thermal conductivity in pure and doped insulating crystals, *Appl. Phys. Lett.* **83** (2003) 1355–1357.

[56] E. Anashkina and O. Antipov, Electronic (population) lensing vs thermal lensing in Yb:YAG and Nd:YAG laser rods and disks, *J. Opt. Soc. Am. B* **27** (2010) 363–369.

[57] R. D. Shannon, Revised effective ionic radii and systematic studies of interatomic distances in halides and chalcogenides, *Acta Crystallogr. A* **32** (1978) 751–767.

[58] J. E. Geusic, H. M. Marcos, and L. G. Van Uitert, Laser oscillation in Nd-doped yttrium aluminium, yttrium gallium and gadolinium garnets, *Appl. Phys. Lett.* **4** (1964) 182–184.

[59] A. Ikesue, T. Kinoshita, K. Kamata, and K. Yoshida, Fabrication and optical properties of high-performance polycrystalline Nd:YAG ceramics for solid state lasers, *J. Am. Ceram. Soc.* **78** (1995) 1033–1040.

[60] A. Ikesue, K. Kamata, and K. Yoshida, Synthesis of transparent Nd-doped HfO_2–Y_2O_3 ceramics using HIP, *J. Am. Ceram. Soc.* **79** (1996) 359–365.

[61] J. Lu, M. Prabhu, J. Song, C. Li, J, Xu, K. Ueda, A. A. Kaminskii, H. Yagi, and T. Yanagitani, Optical properties and highly efficient laser oscillation on Nd:YAG ceramics, *Appl. Phys. B* **71** (2000) 469–473.

[62] L. W. Chen and X. H. Wang, Sintering dense nanocrystalline ceramics without final-stage grain growth, *Nature* **404** (2000) 168–171.

[63] K. Serivalsatit and J. Ballato, Submicrometer grain-sized transparent erbium-doped scandia ceramics, *J. Am. Ceram. Soc.* **93** (2010) 3657–3662.

[64] R. Fedyk, D. Hreniak, W. Lojkowski, W. Strek, H. Matysiac, E. Grzanka, S. Gierlotka, and P. Mazur, Method for preparation and structural properties of transparent YAG nanoceramics, *Opt. Mater.* **29** (2007) 1252–1257.

[65] C. D. Brandle and L. R. Barns, Crystal stoichiometry of Czochralski grown rare earth gallium garnets, *J. Cryst. Growth* **26** (1974) 169–170.

[66] J. Dong and K. W. Lu, Noncubic symmetry in garnet structres studied using extended X-ray absorption fine structure spectra, *Phys. Rev. Condens. Matter B* **43** (1991) 8808–8821.

[67] V. Lupei, A. Lupei, S. Georgescu, T. Taira, Y. Sato, and A. Ikesue, The effect of Nd concentration on the spectroscopic and emission decay properties of highly doped Nd:YAG ceramics, *Phys. Rev. Condens. Matter B* **64** (2001) 092102.

[68] U. Aschauer and P. Bowen, Theoretical assessment of Nd:YAG ceramic laser performances by microstructural and optical modelling, *J. Am. Ceram. Soc.* **93** (2010) 814–820.

[69] M. O. Ramirez, J. Wisdom, H. Li, Y. L. Aung, J. Stitt, G. L. Messing, V. Dierolf, Z. Liu, A. Ikesue, R. L. Byer, and V. Gopalan, Three-dimensional grain boundary spectroscopy in transparent high power ceramic laser materials, *Opt. Express* **16** (2008) 5865–5973.

[70] W. Zhao, C. Mancini, P. Amans, G. Boulon, T. Epicier, Y. Min, H. Yagi, T. Yanagitani, and A. Yoshikawa, Evidence of the inhomogeneous Ce^{3+} distribution across grain boundaries in transparent polycrystalline Ce^{3+}-doped $(Gd,Y)_3Al_5O_{12}$ garnet optical ceramics, *Jpn. J. Appl. Phys.* **49** (2010) 02200.

[71] A. Kaminskii, M. S. Achkurin, R. V. Gainutdinov, K. Takaichi, A. Shirakawa, H. Yagi, T. Yanagitani and K. Ueda, Microhardness, and fracture toughness of Y_2O_3 and $Y_3Al_5O_{12}$ based nanocrystalline laser ceramics, *Cryst. Rep.* **50** (2005) 869–873.

[72] R. W. Rice, *Mechanical Properties of Ceramics and Composites: Grain and Particle Effects*, CRC Press, Boca Raton, FL (2000).

[73] T. Taira, RE^{3+}-ion-doped YAG ceramic lasers, *IEEE J. Sel. Top. Quantum Electron.* **13** (2007) 798–809.

[74] T. Y. Fan, D. J. Rippin, R. L. Aggarwall, J. R. Ochoa, B. Chann, M. Tilleman, and J. Spitzberg, Cryogenic Yb^{3+}-doped solid-state lasers, *IEEE J. Sel. Top. Quantum Electron.* **13** (2007) 448–459.

[75] I. Shoji, Y. Sato, S. Kurimura, T. Taira, A. Ikesue, and K. Yoshida, Optical properties and laser characteristics of highly Nd^{3+}-doped $Y_3Al_5O_{12}$ ceramics, *Appl. Phys. Lett.* **77** (2000) 939–941.

[76] V. Lupei, A. Lupei, N. Pavel, T. Taira, I. Shoji, and A. Ikesue, Laser emission under resonant pump in the emitting level of concentrated Nd:YAG ceramics, *Appl. Phys. Lett.* **79** (2001) 590–592.

[77] V. Lupei, N. Pavel, and T. Taira, Efficient laser emission in concentrated Nd laser materials under pumping into the emitting level, *IEEE J. Quantum Electron.* **38** (2002) 240–245.

[78] R. M. Yamamoto, C. D. Booley, K. P. Cutter, S. N. Fochs, K. N. LaFortune, J. M. Parker, P. H. Paks, M. D. Rotter, A. M. Rubenchik, and T. F. Soules, *SPIE Conf. 6552 on Laser Technology for Defense and Security III. Orlando, FL, USA*, April 2007, paper 655204.

[79] C. Greshkovich and J. P. Chernoch, Improved polycrystalline ceramic laser, *J. Appl. Phys.* **45** (1974) 4495–4502.

[80] J. Lu, J. F. Bisson, K. Takaichi, T. Uematsu, A. Shirakawa, M. Musha, K. Ueda, H. Yagi T. Yanagitani, and A. A. Kaminskii, $Yb^{3+}:Sc_2O_3$ ceramic laser, *Appl. Phys. Lett.* **83** (2003) 1101–1103.

[81] G. L. Bourdet, O. Casagrande, N. Deguil-Robin, and B. Le Garrec, Performances of cryogenic cooled laser based on ytterbium doped sesquioxide ceramics, *J. Phys. Conf. Ser.* **112** (2008) 032054.

[82] K. Petermann, G. Huber, L. Fornasiero, S. Kuch, E. Mix, V. Peters, and S. A. Basun, Rare earth doped sesquioxides, *J. Lumin.* **87–89** (2000) 973–975.

[83] J. Kawanaka, N. Miyanaga, T. Kawashima, K. Tsubakimoto, Y. Fujimoto, H. Kubomura, S. Matsuoka, T. Ikegawa, Y. Suzuki, N. Tsuchiya, T. Jitsuno, H. Furukawa, T. Kanabe, H. Fujita, K. Yoshida, H. Nakano, J. Nishimae, M. Nakatsuka, K. Ueda, and K. Tomabeki, New concept for laser fusion driver by using cryogenically cooled Yb:YAG ceramic, *J. Phys. Conf. Ser.* **112** (2008) 032058.

[84] E. A. Khazanov and A. M. Sergeev, Concept study of a 100 PW femtosecond laser based on laser ceramics doped with chromium ions, *Laser Phys.* **17** (2007) 1398–1403.

[85] E. A. Khazanov and A. M. Sergeev, Petawatt lasers based on optical parametric amplification: their state and prospects, *Phys. Usp.* **51** (2008) 969–974.

[86] P. P. Sorokin and M. J. Stevenson, Stimulated infrared emission from trivalent uranium, *Phys. Rev. Lett.* **56** (1960) 557–559.

[87] I. Ursu and V. Lupei, EPR of uranium ions, *Bull. Magn. Res.* **6** (1984) 162–224.

[88] V. Petit, P. Camy, J.-L. Doualan, X. Portier, and R. Moncorge, Spectroscopy of $Yb^{3+}:CaF_2$: from isolated centers to clusters, *Phys. Rev. Condens. Matter B* **78** (2008) 085131.

[89] S. E. Hatch, W. E. Parsons, and R. J. Weagley, Hot pressed polycrystalline $CaF_2:Dy^{2+}$ laser, *Appl. Phys. Lett.* **5** (1964) 153–154.

[90] A. Lyberis, G. Patriarche, P. Gredin, D. Vivien, and M. Mortier, Origin of light scattering in yttrium doped calcium fluoride transparent ceramics for high power lasers, *J. Eur. Ceram. Soc.* **31** (2011) 1619–1630.

[91] T. T. Basiev, Y. A. Demidenko, K. V. Dikelskii, P. P. Fedorov, E. I. Gorocheva, I. A. Mironov, Yu. V. Orlovskii, V. V. Osiko, and A. N. Smirnov, Optical fluoride and oxysulfide ceramics: preparation and characterization, in *Developments in Ceramic Materials Research*, ed. D. Rosslere, Nova. Science Publishing, New York (2007).

[92] K. V. Yumashev, Nonlinear optical properties and passive Q-switch performance of $Co^{2+}:MgAl_2O_4$ crystal, *Laser Phys.* **9** (1999) 525–632.

[93] A. S. S. De Camargo, C. Jacinto, L. A. O. Nunes, T. Catunda, D. Garcia, E. R. Botero, and J. A. Eiras, Effect of Nd concentration quenching in highly doped lead lanthanum zirconate titanate transparent ceramics, *J. Appl. Phys.* **101** (2007) 053111.

[94] S. Mirov, V. Fedorov, I. Moskalev, D. Martishkin, and C. Kim, Progress in Cr^{2+} and Fe^{2+} doped mid-IR laser materials, *Laser Photon. Rev.* **4** (2010) 21–41.

[95] S. Payne, L. D. DeLoach, W. Kway, J. Tassano, and W. F. Krupke, Laser, optical and thermomechanical properties of Yb-doped fluoroapatite, *IEEE J. Quantum Electron.* **30** (1994) 170–179.

[96] X. Mao, S. Wang, S. Shimai, and J. Guo, Transparent polycrystalline alumina ceramics with orientated optical axes, *J. Am. Ceram. Soc.* **91** (2008) 3431–3433.

[97] J. Akiyama, Y. Sato, and T. Taira, Laser ceramics with rare-earth-doped anisotropic materials, *Opt. Lett.* **35** (2010) 3598–3600.

[98] A. Ikesue, Polycrystalline Nd YAG ceramic lasers, *Opt. Mater.* **19** (2002) 183–187.

[99] J. Lu, K. Ueda, H. Yagi, T. Yanagitani, Y. Akiyama, and A. A. Kaminskii, Neodymium doped yttrium aluminum garnet ($Y_3Al_5O_{12}$) nanocrystalline ceramics – a new generation of solid state laser and optical materials, *J. Alloys Comp.* **341** (2002) 220–225.

[100] V. Lupei, A. Lupei, and A. Ikesue, Single crystal and transparent ceramic Nd-doped oxide laser materials: a comparative spectroscopic investigation, *J. Alloys. Comp.* **380** (2004) 61–65.

[101] A. Ikesue and Y. L. Aung, Synthesis and performance of advanced ceramic lasers, *J. Am. Ceram. Soc.* **89** (2006) 1936–1944.

[102] A. Ikesue, Y. L. Aung, T. Taira, T. Kamimura, K. Yoshida, and G. L. Messing, Progress in ceramic lasers, *Annu. Rev. Mater. Res.* **36** (2006) 397–429.

[103] A. A. Kaminskii, Laser crystals and ceramics: recent advances, *Laser Photon. Rev.* **1** (2007) 93–177.

[104] V. Lupei, Comparative spectroscopic investigation of rare earth doped oxide transparent ceramics and single crystals, *J. Alloys Comp.* **451** (2008) 52–55.

[105] V. Lupei, A. Lupei, and A. Ikesue, Transparent polycrystalline ceramic laser materials, *Opt. Mater.* **30** (2008) 1781–1786.

[106] A. Ikesue and Y. L. Aung, Ceramic laser materials, *Nature Photon.* **2** (2008) 721–727.

[107] V. Lupei, Ceramic laser materials and the prospect for high power lasers, *Opt. Mater.* **31** (2009) 701–706.

[108] S. G. Garanin, A. V. Dmitryuk, M. D. Mikhailov, A. A. Zhilin, and N. N. Rukavishnikov, Laser ceramic. Spectroscopic and laser properties, *J. Opt. Technol.* **78** (2011) 393–399.

3

Experimental technique: powder characteristics and the synthesis of optical grade ceramics, effects of sintering aids

The word "ceramics" originates from the Greek "Keramos," meaning pottery and porcelain, used as tableware since ancient times. Ceramics were considered to be fundamentally opaque because there was no technology available to control the raw materials, molding procedure, sintering method and so on. However, in the 1950s Dr. Coble at MIT figured out the reason why ceramic materials are opaque [1].

By observing the microstructure, he confirmed that residual pores in the material are the main scattering sources. He succeeded in minimizing the scattering by forming a better microstructure, controlling the growth rate of grains and the migration rate of residual pores during the sintering process. As a result, he achieved a translucent alumina ceramic, and the technology was later used by GE (General Electric Company) to produce the arc tube for high pressure sodium vapor lamps. This discovery stimulated the development of other types of transparent ceramics, classified as passive types such as spinel and MgF_2, and active types such as PLZT $((PbLa)(ZrTi)O_3)$, $(Pr, Ce)(GdY)_2OS$. Since the optical quality of the transparent ceramics at that time was not comparable with that of single crystals, they were hardly considered to be of practical use. A technology which could produce transparent ceramic of optical quality comparable to that of single crystals was strongly desired.

Section 3.1 describes the production of the major material used for solid-state lasers, Nd:YAG, using polycrystalline ceramic technology rather than the traditional melt-growth technology. In Section 3.2, HIP (hot isostatic pressing) technology is described as a promising method to produce large scale ceramic laser materials with higher optical quality. The problems of the HIP process and approaches for achieving high quality materials are discussed.

3.1 Introduction

The highest quality is required for the laser element in optical applications. This is because the deterioration of lasing efficiency or beam quality is caused mainly by the presence of fractional scattering or optical inhomogeneities in the optical element, since the excitation and laser beams travel from several tens to thousands of cycles in the laser gain medium.

It is not easy to satisfy the characteristics required, even using current commercial high quality single crystals. The first trials of laser oscillation using polycrystalline materials were carried out with laser oscillation using $Dy:CaF_2$ ceramics at cryogenic temperature in 1964 [2], and with pulse oscillation using $Nd:ThO_2-Y_2O_3$ ceramics at room temperature by Greskovich in 1974 [3]. Although the laser oscillation was successful, the lasing efficiency was extremely low, owing to the presence of many scattering sources in the materials.

Although YAG (yttrium aluminum garnet) is the major material used for solid-state lasers, there were no reports on the synthesis of transparent polycrystalline YAG ceramics until the 1980s because its synthesis is more difficult than that of the materials described above. In the 1980s, the groups led by de With [4, 5] and Sekita [6, 7] reported the synthesis of high density and transparent YAG ceramics using a wet process in which raw powders with the YAG composition or a precursor are prepared by pyrolysis of sulfate or urea in advance. In the solid phase reaction method, the intermediate phase formed during sintering, especially the $YAlO_3$ phase, inhibits sintering, and Al_2O_3 which coexists with $YAlO_3$ tends to remain as an impurity phase. Therefore, the wet process was introduced to produce more homogeneous raw powder compared to the solid phase method. The synthesis of high density YAG ceramics by a solid phase method was reported by the group of Haneda [8, 9]. Although commercial Al_2O_3 and Y_2O_3 raw powders were used as starting materials in their experiments, they concluded that the starting powder should be single phase YAG because volume expansion occurred on the formation of the intermediate phase, $YAlO_3$, during reactive sintering of a powder mixture of Y_2O_3 and Al_2O_3. Additionally, it is easy to understand how difficult it is to prepare high density YAG ceramics using a solid phase method because the relative density of the sintered body prepared by Haneda *et al.* reached only a maximum of 69%.

Given these challenges, it is quite certain that the wet process is preferable to the solid phase process for the preparation of YAG ceramics in principle. However, there is some doubt whether the raw materials used in the solid phase process were suitable. Since the development of translucent alumina ceramics by Coble, it has become easy to obtain high purity Al_2O_3 powder (with good sinterability) which approaches the theoretical density at low sintering temperatures.

However, there was a low demand for Y_2O_3 powder for marketing and sintering, and it is considered that the properties of Y_2O_3 powder as raw material were inferior to those of the Al_2O_3 powder available at that time. As can be noticed from the report by Greskovich, it took long hours of sintering to achieve high density Y_2O_3 ceramics at approximately 2200 °C [10, 11]. Y_2O_3 is an ionic crystal consisting of Y^{3+} and O^{2-}, and the diffusion coefficient (mass transfer rate) of each ion [12] is significantly larger than that of the Al^{3+}, Mg^{2+}, and O^{2-} ions in Al_2O_3 and MgO. Nevertheless, the temperature required to produce dense Y_2O_3 ceramics was much greater than the temperature required to produce translucent Al_2O_3 or MgO ceramics. Therefore, the sintering of Y_2O_3 powder as raw material was problematic. Previously, Y_2O_3 powder with a crystallite size of several micrometers (the exact size is unknown) was used for the preparation of dense Y_2O_3 ceramics; this is extremely disadvantageous for sintering compared with the primary particle sizes of

Table 3.1 *Impurities of high purity*
Al_2O_3 used in the present experiment

Impurity	Mass (ppm)
Na	<2
Si	13
Mg	5
Cu	1
Fe	7
Ni	<1

0.2∼0.5 μm [13, 14], or 0.05 μm [15, 16] for sinterable Al_2O_3 or MgO powders. The primary purpose of the experiment described in the present section is to ascertain the basic characteristics and sintering mechanism of the raw materials, Al_2O_3 and Y_2O_3 powders, used for producing transparent YAG ceramics in a solid phase process by comparing their sintering characteristics with that of commercial powders. The secondary purpose is to reveal the function of Si derived from TEOS (tetraethyl orthosilicate) which was added as a sintering aid to YAG ceramics. Finally, the manufacture of YAG ceramics with fewer structural defects (e.g., residual pores) and the methods used to evaluate their properties are discussed.

3.1.1 Sintering characteristics of Al_2O_3 and Y_2O_3 raw material powders

3.1.1.1 Sintering performance of Al_2O_3 powder

Table 3.1 shows the micro-analytical results obtained using ICP (inductively coupled plasma) for the impurities of Al_2O_3 powder used in the synthesis of YAG ceramics described in the present experiment. The detected impurities are Na, Si, Mg and so on, the total amount is less than 100 mass ppm.

Figure 3.1 shows the shrinkage behavior of green powder compacts at pressed 98 MPa using CIP (cold isostatic pressing) measured by TMA (thermal mechanical analysis) with a heating rate of 5 °C min^{-1}. The compacts were made from two types of Al_2O_3. One was prepared from alkoxide, with primary particle size approximately 400 nm and purity 99.99 mass%; the other was commercial powder manufactured by TM-10 of Taimei Chemical, prepared from pyrolysis of ammonium aluminum carbonate, with primary particle size approximately 200 nm and purity 99.99 mass%. There was no obvious difference in the shrinkage of these compacts. They started to shrink around 1000 °C and then the shrinking was almost complete around 1500 °C. The shrinkage of the powder compact derived from commercial powder was greater than that of the other sample. It is considered that the difference in packing density during the pressing step affected the shrinkage. This was caused by the difference in diameter of the primary particles. Additionally, the sintered

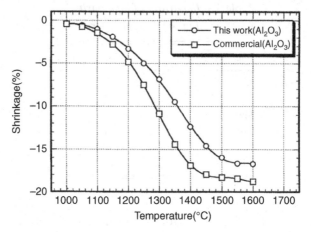

Figure 3.1 Sintering shrinkage of the present experimental powder compact and commercial powder compact heated at 5 °C min^{-1}.

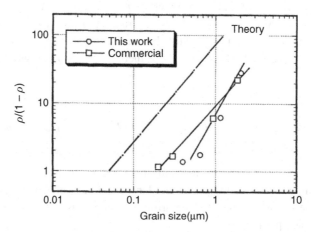

Figure 3.2 Relationship between log $\rho/(1-\rho)$ and log R for the experimental and commercial Al$_2$O$_3$ powders.

densities were determined by the Archimedes method for the two types of powder compacts sintered at 1600 °C for 1 hour. There was no significant difference in sintered density. The values were 99.6% and 99.7% of the theoretical density for the present experimental and the commercial Al$_2$O$_3$ powders, respectively.

Figure 3.2 shows a theoretical relationship between log $\rho/(1-\rho)$ and log R, where ρ is the relative density, R is the grain size, when the powder is packed ideally, and ideal densification takes place, reported by the group of Ikegami [17]. Data from the present experimental and commercial Al$_2$O$_3$ powders are also plotted. At the same density, the grain size of both Al$_2$O$_3$ powders was slightly larger than that of the theoretical calculation,

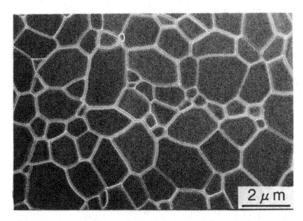

Figure 3.3 Surface of Al_2O_3 ceramic formed using the experimental powder sintered at 1600 °C for 60 min under oxygen atmosphere.

but the densification behavior was very similar to the theoretical calculation. The relative densities of both green powder compacts (before sintering) were 58% and 54%, respectively. It is reported that the rate of grain growth of pressed green compact depends on the bulk density [17]. Therefore, it can be considered that the packing of each power compact was not sufficient (over 60% of relative density is necessary), and the experimental plots are shifted to the grain growth side compared with the theoretical calculation.

Figure 3.3 shows a thermal etched surface of a polished specimen of the experimental Al_2O_3 powder sintered at 1600 °C for 1 hour. There are no pores in the observation area. The specimen consists of relatively uniform grains (around 2 μm).

The above result confirms that the Al_2O_3 powder used for the present experiment has good sintering performance, comparable to or better than the commercial powder used for the production of translucent Al_2O_3 ceramics.

3.1.1.2 Sintering performance of Y_2O_3 powder

Table 3.2 shows the analytical results obtained by ICP for the impurities of Y_2O_3 powder used for YAG synthesis described in the present experiment and after Chapter 3. The elements detected as impurities are Cr, Fe, and Si but their level is very low if Si is not considered as contamination, but rather as a sintering aid. Figure 3.4 shows the shrinkage behavior of compacts pressed at 140 MPa using CIP, measured by TMA at a heating rate of 5 °C min^{-1}. The compacts are made from two types of Y_2O_3. One is prepared from pyrolysis of basic carbonate, with primary particle size approximately 60 nm and purity 99.99 mass%; the other is commercial powder manufactured by Mitsubishi Chemical, prepared by the oxalate method, with primary particle size approximately 40 nm and purity 99.95 mass%.

The powder compact derived from the present experimental powder starts to shrink around 1200 °C and almost completes densification around 1600 °C. In contrast, the

Table 3.2 *Impurities of high purity* Y_2O_3
powder used in the present experiment

Impurity	Mass (ppm)
Sc, La	<2
Ce, Pr, Nd	<10
Sm	<7
Eu, Ho, Er, Tm	<0.5
Gd, Tb, Dy	<3
Mn, Zn, Ti	<0.5
Cu, Ni	<1
Cr	4
Fe	6
Si	25

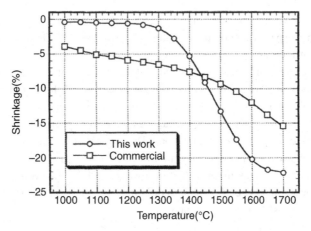

Figure 3.4 Sintering shrinkage of compacts derived from the experimental and commercial powders
heated at 5°C min^{-1}.

compact derived from commercial powder starts to shrink earlier but the shrinkage rate is
poor. At 1700 °C, the shrinkage of the commercial powder is approximately 7% smaller
than that of the experimental powder.

Figure 3.5 shows the relationship between $\log \rho/(1-\rho)$ and $\log R$ for the experimental
and commercial powder green compacts, as in Figure 3.2. Although the packing densities after pressing are almost the same, the commercial powder compact drifts along the
theoretical line as the densification becomes higher. This suggests that more grain growth
occurred than the requirement for densification. Since the relative density of the present
Y_2O_3 experimental powder compact is approximately 49%, the plot line shifts to the grain
growth side with respect to bulk specific gravity. This phenomenon is the same as found
for the above mentioned Al_2O_3 powder, and the reason would also be the same.

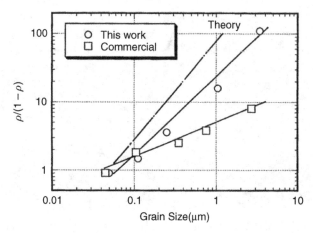

Figure 3.5 Relationship between $\log \rho/(1 - \rho)$ and $\log R$ for both the experimental and commercial Y_2O_3 powders.

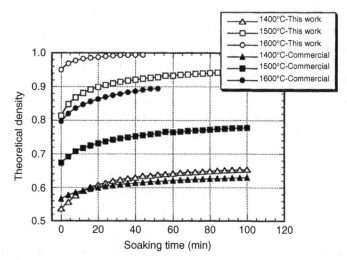

Figure 3.6 Relative densities of present experimental and commercial powder Y_2O_3 compacts sintered at 1400, 1500, and 1600 °C.

Figure 3.6 shows the relationship between soaking time and relative density of the experimental and commercial powder compacts heated at 1400, 1500, and 1600 °C. The difference in relative densities at each temperature is clear. The Y_2O_3–10ThO_2 ceramic developed by Dr. Greskovich was synthesized by the addition of ThO_2 to Y_2O_3, 99.99 mass% purity, and the relative density of the sintered body reached the theoretical density above 2000 °C [3, 18]. The present experimental powder reaches theoretical density at only 1600 °C (about 400 °C lower) without a radioactive sintering aid such as ThO_2. The commercial powder reached 90% of relative density at 1600 °C in this experiment, but in the previously cited

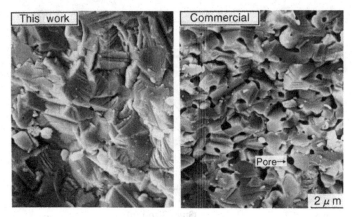

Figure 3.7 Fracture surface of Y_2O_3 ceramics formed using the present experimental and commercial powders sintered at 1600 °C for 100 min in oxygen atmosphere.

paper 1800 °C was necessary to achieve 90% of relative density. The conclusion is that the Y_2O_3 powder used for the present experiment requires a lower sintering temperature, about 200 °C lower than that of the traditional raw materials.

Figure 3.7 shows SEM images of fracture cross-sections of Y_2O_3 compacts sintered at 1600 °C for 100 min derived from the experimental and commercial powders. The sintered body derived from the commercial powder includes a lot of residual pores, but the sintered body derived from the experimental powder includes only localized pores. Therefore, it is clear that there is a difference in sintering performance between these powders from a microstructural point of view.

3.1.2 Production of YAG ceramics using the present experimental and commercial powders

Two combinations of powders were used to make the YAG ceramics. The first was a combination of the Al_2O_3 and Y_2O_3 powders used for the present experiment. The second was a combination of the commercial Al_2O_3 and Y_2O_3 powders. Figure 3.8 shows the shrinkage behavior of the two types of powder compacts (YAG composition) prepared from the experimental and commercial powders. (Details of the shrinkage behavior of the powder compacts are described in this chapter onwards.) Although the compact derived from commercial powder starts to shrink earlier than the compact derived from the present experimental powder, the shrinkage is almost completed at around 1400 °C. Furthermore, expansion occurred above 1400 °C, and then shrinkage due to densification was not observed. On the other hand, the shrinkage of the present experimental powder compact starts at 1200 °C, which is slightly higher than the equivalent temperature of the commercial powder. Although the shrinkage rate of the experimental powder compact decreased around 1400 °C, rapid densification occurred afterward. This obvious difference in the sintering behavior of the powder compacts was observed in the case of production for YAG ceramics.

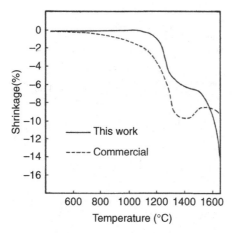

Figure 3.8 Sintering shrinkage of the present experimental and commercial YAG powder compacts heated at 5 °C min^{-1}.

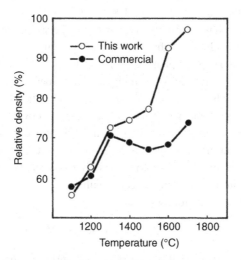

Figure 3.9 Relationship between relative density and heating temperature (no soaking time) under oxygen bleeding.

Figure 3.9 shows the density calculated by the Archimedes method for compacts of the present experimental and commercial powders, which were heat treated with a ramp of 5 °C min^{-1} up to the desired temperature then quenched at 20 °C min^{-1}. The commercial powder has a decrease in density at 1400 °C and starts to expand, then mild densification occurs at 1600 °C. The relative density of the sintering body at 1700 °C is around 70%. The present experimental powder reaches 97% of relative density at 1700 °C; no decrease in density is observed during the sintering process.

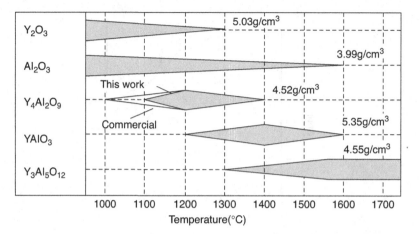

Figure 3.10 X-ray diffraction patterns of phase formation in YAG with increasing temperature.

Figure 3.10 shows XRD (X-ray diffraction) patterns of compacts of the present experimental and commercial powders at every 100 °C from 1000 to 1700 °C. The reticulate
range indicates detected phases formed qualitatively in the specimens.

The formation of new phases with increasing temperature is almost the same in both
compacts except that the $Y_4Al_2O_9$ phase appears at a lower temperature in the commercial
powder compared with the present experimental powder because shrinkage of the commercial powder compact also started at a lower temperature. Therefore, there is no obvious
difference in the solid reactivity between Y_2O_3 and Al_2O_3 powders inside the compacts
formed using the present experimental and the commercial powders. By the way, an expansion phenomenon was observed above 1400 °C for the sintered body using commercial
powder. The specific gravity of $YAlO_3$ formed during sintering as an intermediate phase
is 5.35 g cm^{-3}. This intermediate phase reacts with Al_2O_3 to give YAG single phase, of
specific gravity 4.55 g cm^{-3}, as follows;

$$3YAlO_3 + Al_2O_3 \rightarrow Y_3Al_5O_{12}(YAG). \tag{3.1}$$

At this transition point, theoretically 10% of volume shrinkage occurs by the formation
of a low density YAG phase (4.55 g cm^{-3}) from the high density phase (5.35 g cm^{-3}).
For this reason, the commercial powder causes a volume expansion of the compact during
sintering, which means that it is impossible to produce a high density sintered body using
the solid method. The same phenomenon was confirmed by Haneda's group.

Figure 3.11 shows the particle size distribution during sintering for the samples analyzed
in Figure 3.10. In the commercial powder, grain growth starts at 1200 °C then continues
moderately with increasing temperature. On the contrary, although in the experimental
powder grain growth hardly occurs below 1400 °C, rapid grain growth starts after that.
Shrinkage of the commercial powder accompanies grain growth simultaneously, whereas
that of the experimental powder does not occur until it reaches 6% at 1400 °C and only

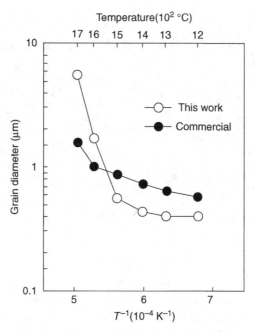

Figure 3.11 Change in grain diameter during sintering, on heating at 5 °C min^{-1}.

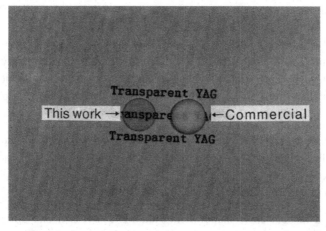

Figure 3.12 Appearance of mirror polished YAG ceramics prepared using the present experimental and commercial powders.

the solid phase reaction proceeds without obvious grain growth. Experiments to explain the sintering phenomenon in detail have not yet been conducted. It is concluded that the properties of Y_2O_3 powder are key to producing high density YAG sintered bodies.

Figure 3.12 shows the appearance of YAG sintered bodies derived from the present experimental and commercial powders. Both samples were sintered at 1750 °C for 5 hours

Figure 3.13 Fracture surface of powder compact (sintering powders) observed by SEM.

under 1.3×10^{-3} Pa of vacuum. The sintered body made using commercial powder is 92% of relative density and is not transparent, whereas that made using the experimental powder is close to the theoretical density and has perfect transparency and optical isotropy.

Figure 3.13 shows a SEM image of the compact derived from the present experimental powder at high magnification. Observation of the particles reveals the presence of approximately 500 nm secondary particles but it was not possible to determine the alignment because the Y_2O_3 particles were too small.

Figure 3.14 illustrates a model of sintered Y_2O_3 and Al_2O_3 based on the above mentioned information. Although the two types of powders have different primary particle size, crushing the powders using a pot mill gives primary, or close to primary, particles dispersed homogeneously in a slurry. By eliminating solvent from the slurry using a spray dryer, the two types of powders can be classified into the following alignment categories.

(1) Formation of an agglomeration of small Y_2O_3 particles deposited onto large Al_2O_3 particles. In other words, a hybrid structure of nuclear Al_2O_3 surrounded by Y_2O_3 having strong affinity.
(2) Formation of agglomerated Y_2O_3 and Al_2O_3 alternately.

It is considered in both cases that sintering among Y_2O_3 itself and diffusion between Y_2O_3 and Al_2O_3 takes place during sintering; only a change of phase formation would be observed, without grain growth. However, the solid phase reaction between Al_2O_3 and Y_2O_3 is hard to produce and it is difficult to change the phase formation actively when the sintering process follows alignment (2). Since the present experiment involves active phase change, the structure of the particles is considered to be alignment (1), i.e., close to the alignment illustrated in the figure. In this case, secondary particles having a hybrid structure are thought to form following the mechanism illustrated in the figure during sintering. First of all, the contacting surface between Al_2O_3 and Y_2O_3 forms $Y_4Al_2O_9$ by

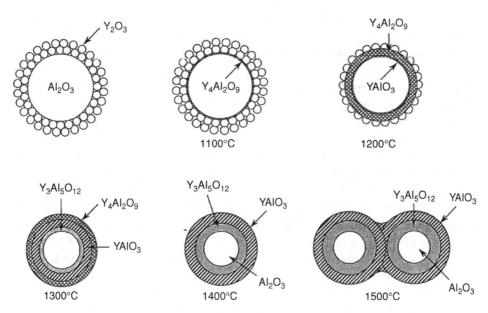

Figure 3.14 Sintering model of a hybrid powder process in the present experiment.

mutual diffusion of particles of Y_2O_3 deposited onto Al_2O_3 to give a uniform composition inside the agglomerated particles. At this time, the Y_2O_3 also sinters together. This process results in sintering shrinkage, of approximately 6%, owing to a slight shrinkage in size of the secondary particles with hybrid structure. Since this shrinkage is only 6%, SEM observation cannot detect the change in particle size. At 1200 °C, the solid phase reaction progresses and the thickness of $Y_4Al_2O_9$ as a boundary phase increases, retaining a small amount of Y_2O_3 on the shell of agglomerated particles. Moreover, the formation of $YAlO_3$ starts at the phase boundary between Al_2O_3 and Y_4AlO_9. Y_2O_3 disappears above 1300 °C, and the solid phase reaction

$$Y_4Al_2O_9 + Al_2O_3 \rightarrow 4YAlO_3$$

occurs. The $YAlO_3$ phase is obviously formed by consumption of $Y_4Al_2O_9$. At 1400 °C, the inside of the particles becomes only the $YAlO_3$ and Al_2O_3 hybrid structure. Below this temperature range, only the solid state reaction occurs and there is no accompanying grain growth. There is no difference in composition from a micro point of view of particle size. YAG single crystallization and abrupt sintering shrinkage, densification, begin at this temperature as a result of the solid phase reaction inside particles as a secondary stage, $3YAlO_3 + Al_2O_3 \rightarrow Y_3Al_5O_{12}$, and the sintering between particles having a hybrid structure begins.

Expansion, theoretically 10% of volume expansion, is not observed even though the low density YAG phase is formed via $YAlO_3$ and rapid densification occurs after a slight decrease in shrinkage rate using the present experimental powder. The reason is thought to be because the powder activity is retained until the densification stage due to a small increase in grain growth by the special sintering mechanism mentioned above. Additionally, it is thought that expansion is retained when YAG is formed because the agglomerated particles do not contact each other if a homogeneous reaction of secondary particles of the hybrid structure as in the assumed sintering mechanism occurs. Therefore, it is estimated that a dense YAG sintered body is produced without expansion by preserving the particle alignment.

Commercial powder with similar primary particle size of Al_2O_3 and Y_2O_3 underwent the same procedure. Considering the trend of phase formation shown by X-ray diffraction, secondary particles having similar orientation were formed. The difference is that the compact of commercial powder causes obvious grain growth due to homogenization of particles inside having the hybrid structure and also due to sintering between adjacently agglomerated particles simultaneously. It is assumed that sintering expansion is caused by a volume change when the YAG phase is formed from $YAlO_3$. The reason is the difference in sintering performance of Y_2O_3 in particular, during mass transfer, but the relationship between the synthesis conditions of the powder and sintering behavior is still not understood.

3.1.3 Effect of (tetraethyl oxysilicate) TEOS (Si) as sintering aid [19]

SiO_2 is a component which is also reported in the synthesis of YAG ceramics described by the groups of G. de With [4, 5] and Haneda [6, 7]. Its purpose is to eliminate Al_2O_3 and $YAlO_3$ impurity phases when producing transparent YAG ceramics. However, the impurity phase Al_2O_3 remains in the sintered body even when Si is added. In earlier reports, the control of properties using this element was not discussed although its effect in eliminating impurities was confirmed. The qualitative and quantitative data for the effect of Si in these ceramics have not been reported. Additionally, there is little discussion of whether Si can dissolve in the crystalline lattice of YAG host material or not, although Si is a main component of natural garnet. The addition of Si would cause another problem in the application of solid laser material if it forms impurity phases such as grain boundary phases, even though it is effective in eliminating impurity (intermediate) phases from YAG ceramics. In this section, we discuss the role of the pyrolysis of TEOS, producing Si as a sintering aid. Additionally, an experimental procedure was developed.

Figure 3.15 shows the result of relative density determined by the Archimedes method for pressed green compacts of YAG composition with added 320 and 930 mass ppm Si produced by heat treatment from 1500 to 1800 °C with a heating ramp of 5 °C and cooling ramp of 20 °C without dwell time. The pressed green compact without added Si shows linear densification without the expansion phenomenon described in Section 3.1.2. It was recognized clearly that adding Si to this powder improves the densification. However, the compact with added 320 mass ppm Si is only slightly prevented from reaching theoretical density and there is no difference between the compacts with varying amount of Si. All of the pressed green compacts, with or without Si, when sintered at 1750 °C for 10 hours achieved higher than 99.5% of relative density. From this, it is thought that one effect of Si

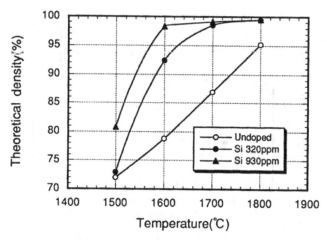

Figure 3.15 Relative density of YAG ceramics with varying amounts of Si and sintered at various temperatures.

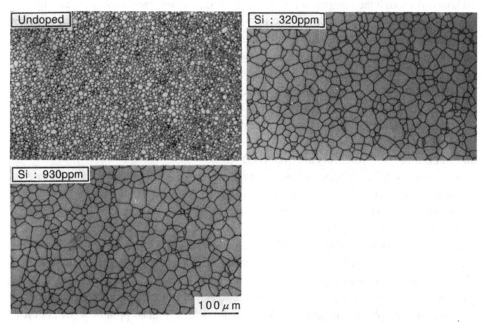

Figure 3.16 Micrographs by reflecting microscopy of the surface of YAG ceramics with varying amounts of Si after thermal etching.

is to increase the rate of densification, reducing the sintering temperature and shortening the sintering time.

Figure 3.16 shows micrographs taken with a reflecting microscope of pressed green compacts with no, 320 and 930 mass ppm of Si added before sintering at 1750 °C for

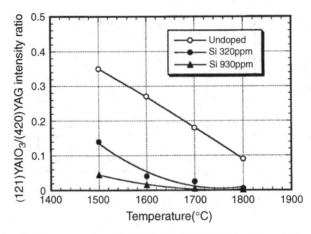

Figure 3.17 Diffraction intensity ratio of (121) $YAlO_3$/(420) YAG by X-ray diffractometry.

10 hours. Although the compact without added Si has a grain size which is smaller by one order of magnitude than the others, there is no obvious difference between the compacts with 320 and 930 mass ppm Si which have grain sizes of 40 and 60 μm, respectively. Therefore, Si promotes grain growth as a secondary effect.

Figure 3.17 shows the relative diffraction intensity of (121) and (420) plane indices for $YAlO_3$ which is an intermediate impurity phase in the compact after heat treatment from 1500 to 1800 °C, as in Figure 3.15. The compact without added Si shows a trend of decreasing impurity phase as the temperature increases. Although the other compacts with added Si show decreasing impurity phase with the amount of added Si, both compacts almost become YAG single crystal over 1600 °C.

Additionally, X-ray diffraction shows that the plain compact becomes YAG single crystal on sintering at 1750 °C for 10 hours. In this case, Si enhances extinction of the impurity phase. Therefore, a third effect of Si is to promote extinction of impurity phases.

Figure 3.18 shows the relationship between the amount of Si in the sintered body and the transmission of compacts sintered at 1750 °C for various times. The plain compact sintered for 1 hour has almost zero transmission, then the transmission increases gradually as the sintering time is extended. The compact with added Si reaches transmission 85% faster than the plain compact but the level of transmission attained is comparable. In Figure 3.19, a grain boundary phase is observed in the compact with 930 mass ppm Si added, but it is hard to say which sintered body, about 1 mm thick, shows better transmission.

Figure 3.19 is a qualitative measurement of an impurity phase, an Si compound, segregated at the grain boundary on setting the cooling rate to a very moderate level, 10 °C min^{-1}, down to room temperature, which is found in specimens with varied Si content after sintering at 1750 °C for 20 hours [20]. The segregation at the grain boundaries, the effect of added Si and the cooling rate are described in the following section. It is seen from this figure that Si dissolves homogeneously into the host material without segregation at grain

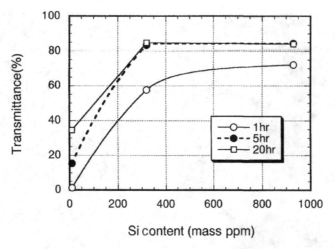

Figure 3.18 In-line transmittance of YAG ceramics with varying amounts of Si sintered at 1750 °C for 1, 5, and 20 h, respectively.

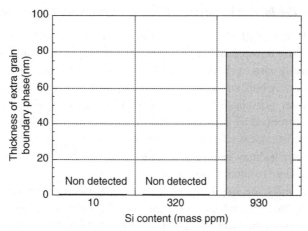

Figure 3.19 Effect of Si content in YAG ceramics on the thickness of the extra grain boundary phase observed by TEM.

boundaries when the concentration is less than approximately 300 mass ppm. Therefore, it is important to use the minimum amount of added Si, which will work effectively as a sintering aid and prevent segregation, for the production of transparent YAG ceramics.

Lastly, Si has atomic number 14 and does not have any electrons belonging to the d or f orbital which contribute to light emission. The main application of YAG is in solid-state lasers, so the sintering aid must not interfere optically with laser oscillation, for example it must not cause degradation of the wavelength or efficiency by interacting with the emitting ions. From this point of view, because Si is a light element it meets this purpose.

Figure 3.20 Granulated Y_2O_3–Al_2O_3, YAG composition, powders made by spray drying. Photograph (b) is a picture magnified ten times of the granulated powder in (a).

3.1.4 Production technology of low scattering loss YAG ceramics

Although suitable Y_2O_3 and Al_2O_3 powders and an appropriate amount of Si additive are required to synthesize transparent YAG ceramics by the solid method, finally it is necessary to produce relatively large volumes such as laser rods. CIP (cold isostatic pressing) is widely used for casting bulk sintered bodies. Powder for CIP is prepared by spray drying.

However, powder granulated using the spray dry method cannot be crushed by the pressing process described by the group of Uematsu and others [21–23]. If crushing of the particles is incomplete, local voids or defects may be formed, or the packing condition of the constituent particles may become non-uniform. Defects in the pressing compact are expected to cause large residual pores or inhomogeneous particles in the microstructure.

For optical ceramics, the amplified efficiency deteriorates seriously because a minimal structural defect in the ceramic bulk causes scattering, and it may be impossible to produce laser oscillation. Laser materials have very strict demand characteristics such as high in-line transmittance and almost zero optical loss. The maximum pore diameter in the Nd:YAG ceramics prepared in the present experiment was approximately 5 μm and the amount of residual volume determined by transmission microscopy in samples prepared by the author in the 1990s was only 1–2 vol. ppm [20, 24]. The relative density of ordinary ceramics ranges from 90% to 99% and that of Nd:YAG ceramics is 4% to 6% lower owing to defects and residual pores. Although the compact was cast by CIP using granulated powder in the present experiment, the amount of pores varied dramatically with the character of the granulated powder. In this section, the crushing behavior of granulated powder produced by spray drying is reported.

Figure 3.20 shows SEM images of granulated powder formed by mixing and spray drying Y_2O_3, Al_2O_3, and Nd_2O_3 powders of composition 1.1 at.% Nd:YAG. The diameter of granulated powder formed by spray drying is approximately 50 μm and the packing is very dense.

Figure 3.21 shows the crushing behavior of the compact obtained in Figure 3.20 on pressing from 2 to 20 MPa. The point of observation is the center of the compact. A carcass (defects) of granulated powder remains until 10 MPa, but a homogeneous structure is formed by unification of mixed granulated powders completely at 20 MPa.

Figure 3.22(a) shows the distribution of pore diameter measured by the mercury intrusion technique for the compact pressed at 140 MPa by CIP from the granulated powder of Figure 3.21. The pore diameter of the compact is in the range 25–40 μm, and there are no pores greater than this. Figure 3.22(b) shows the distribution of pore diameter of a compact pressed under the same conditions using non-optimized granulated powder. The distribution of the pore diameter is similar, between 25 and 40 μm, but coarse voids larger than a micrometer are detected. These voids will remain in the Nd:YAG sintered body as residual pores after sintering and become scattering centers. Therefore, the preparation of high quality pressed powder, of compacts without large voids, is extremely important for laser applications.

The pore diameter of the compact is governed by the packing condition of Y_2O_3 powder if the compact is composed of the powder listed in Tables 3.1 and 3.2. The relative density, based on 4.55 g cm^{-3} for YAG, after casting using powder of YAG composition, is around 50%. Additionally, the average particle diameter for Y_2O_3 powder is approximately 60 nm. When voids are present, in the eight coordination structure of particles of Y_2O_3, the packing density and pore size are 52% and 20–30 nm, respectively. The fact that the pore diameter of the compact determined by measurement agrees with the theoretical value calculated from the packing density indicates that the powder forms an eight coordination structure homogeneously. This result shows that the granulated powder is crushed almost completely when it is pressed in the cast, and ideal orientation of particles is maintained.

Figure 3.23 shows the relationship between CIP pressure and transmission of the sintered body, of thickness 3 mm, measurement wavelength 1000 nm, heat treated at 1750 °C for 20 hours. As depicted in Figure 3.21, the compact pressed at higher than 20 MPa using CIP, which is enough to crush the granulated powder, achieves theoretical transmission after sintering. This fact reveals that the pressed powder forms a good quality compact without large voids, and even at low CIP pressure it is possible to produce transparent sintered bodies with theoretical transmission. Pressing at higher than 20 MPa with CIP is thought to shorten the distance between the particles in the crushed granulated powder or change their orientation. Therefore, the formation of granulated powders is very important for the production of a homogeneous packing structure under low casting pressure, giving stable crushing behavior when fabricating YAG ceramics. It is proved that the powder in the compact retains the ideal packing structure even under pressing at a relatively low pressure.

However, pores in the range approximately 1–2 vol. ppm remain in the YAG sintered ceramics, although pressed powder with almost perfect particle arrangement is formed by

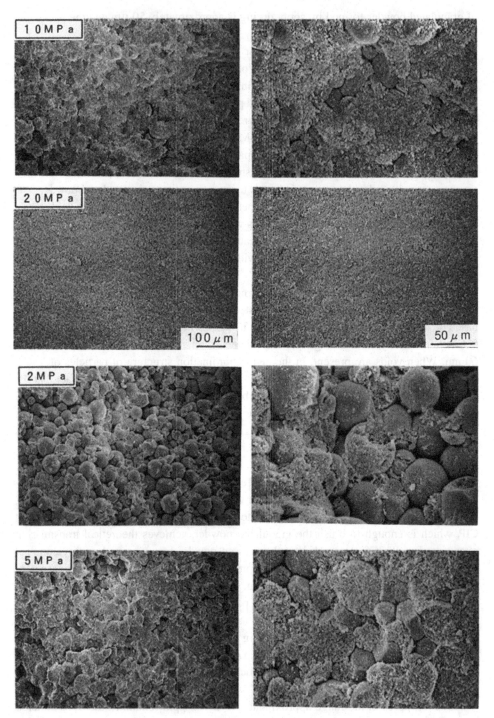

Figure 3.21 Fracture surfaces of powder compacts under 2 to 20 MPa of uniaxial pressing. (The figures on the right are the left figures magnified twice.)

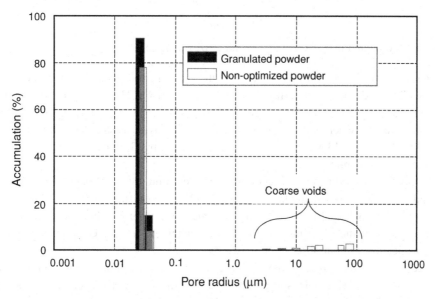

Figure 3.22 Pore size distributions of powder compacts obtained in the present experiment.

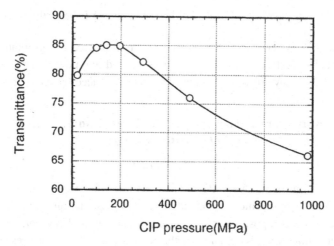

Figure 3.23 In-line transmittance of YAG ceramics sintered at 1750 °C for 20 hours under vacuum after pressing at 980 MPa with CIP.

casting. The residual pores are the result of localized defects, packing voids, caused by the casting process. This will be discussed in the following section. Since residual pores act as scattering centers in YAG ceramics, the establishment of a molding technology which produces fewer defects and HIP (hot isostatic pressing) is important for enhancing the laser oscillation efficiency.

3.1.5 Conclusion

The results obtained in this section are summarized as follows.

(1) There is no obvious difference between the present experimental powder and commercial powder in the sintering characteristics of Al_2O_3. However, the sintering performance of the present experimental Y_2O_3 powder is clearly superior to that of the commercial powder, and it reaches theoretical density at around 1600 °C. This temperature is 400 °C lower than the temperature used by Greskovich to produce Y_2O_3 transparent ceramics.

(2) In the case of producing YAG ceramics using the solid process, both the experimental and commercial powders change to the YAG single crystal phase via $YAlO_3$ as an intermediate phase. However, although commercial powder is difficult to sinter because of expansion, the experimental powder can change to transparent YAG ceramic by densification without observable expansion.

(3) The reason why the present experimental powder gives dense transparent YAG ceramics, as mentioned in (2), is the special agglomerated structure of the experimental powder and the difference in sintering behavior of Y_2O_3.

(4) The Si added as a sintering aid (A) promotes densification, (B) promotes grain growth, (C) eliminates impurities and (D) reduces the sintering time. However, excess Si, more than 900 mass ppm, easily forms an extra grain boundary phase with main composition Si.

(5) It is key that YAG ceramics with fewer pores are produced using granulated powder which is easily crushed at low pressure. It was found that powder with homogeneous packing is formed by crushing granulated powder produced by spray drying.

3.2 Microstructure and optical characteristics of Nd:YAG processed by HIP (hot isostatic pressing) [25]

3.2.1 Abstract

Small samples of transparent Nd:YAG ceramics produced by a solid method and vacuum sintering technology can achieve high efficiency laser oscillation which is comparable to that of the same composition of single crystal material. However, 2 vol. ppm of pores of several micrometers remains in the present Nd:YAG. The presence of pores in YAG ceramics causes light scattering. Minimizing the scattering leads to a decrease in leakage of the laser beam from the optical resonator, so it is extremely important to minimize the numbers of pores. The residual pore refractive index $n = 1.00$, which is very different from the refractive index of the host material; the refractive index of YAG is $n = 1.82$ [26].

Polycrystalline ceramics contain intrinsic microstructural defects, namely pores, inhomogeneous particle sizes or impurity phases. Often, HIP (hot isostatic pressing) or HP (hot pressing) are chosen as the sintering technology to minimize such defects. Currently, present sintering equipment is being applied to produce MgO, infrared transparent devices including spinel [27–29], electro-optical devices including PLZT [30], scintillator devices including Gd_2OS [31, 32], magnetic devices including Mn–Zn ferrite [33] and so on; the field of application is expected to expand.

In order to minimize the amount of pores in Nd:YAG ceramics, HIP was chosen because it allows use of a high temperature and pressure compared with HP; the maximum temperature and pressure are generally 1500–1600 °C and 20–30 MPa, respectively. Additionally, HIP can use various types of gases as pressing media, Ar gas for convenience, neutral or inert gases including N_2 and Ar, and oxidative gases including O_2. A detailed investigation was carried out to determine how the processing conditions of HIP, using Ar as pressurizing medium, affect the microstructure and optical characteristics of Nd:YAG.

3.2.2 Experimental methodology

This experiment was also conducted using 99.99 mass% high purity Al_2O_3, Y_2O_3, and Nd_2O_3 powders as mentioned before. These powders were weighed to give the composition 1.1 at.% Nd:YAG, and the slurry was prepared by the method described before. The slurry was prepared as 150–50 μm diameter spherical granulated powder using spray drying. The granulated powder was cast in a compact 20 mm in diameter and pressed at 10 MPa with a die, then it was pressed at 196 MPa using CIP. The prepared pressed green compact was sintered at 1600 °C for 3 hours under 1.3×10^{-3} Pa of vacuum, heat treated for 3 hours at a temperature in the range 1500–1700 °C under pressure in the range 9.8–196 MPa using capsule-free HIP. Specimens treated with HIP do not have sufficient transmission because the grain growth is incomplete, and require additional sintering at 1750 °C for 20 hours under vacuum. These are defined as standard specimens. The experimental details are almost the same as described in Section 3.1, excluding HIP treatment.

The microstructure of the specimens after sintering was analyzed by reflection microscopy, transmission microscopy, SEM, and EDX, energy dispersive X-rays. The specimens were 10 mm in diameter, 5 mm thick and polished to $R_a = 3$ nm on both sides in order to increase the transmittance. The transmittance of the specimens was measured by spectrophotometry at wavelengths in the range 200–1000 nm.

3.2.3 Experimental results

Details of the present experiment are described in Table 3.3. Samples A1–A3 had added 0.1, 0.5, and 1.0 mass% TEOS and were treated at 1700 °C under 196 MPa with HIP. Samples B1 and B2 had varied HIP temperatures from 1500 to 1600 °C while the amount of TEOS and the pressure were fixed at 0.5 mass% and 196 MPa, respectively. Sample C1 underwent HIP at a pressure of 9.8 MPa while the amount of TEOS and the HIP temperature were fixed at 0.5 mass% and 1700 °C, respectively. The experimental conditions allow the effects of the amount of TEOS, and the HIP temperature and pressure to be determined. The amount of TEOS was varied because it is expected to act as a liquid lubricant and fill

Table 3.3 *Specification of specimens in the present experiment*

Specimen	TEOS content (mass%)	HIP temperature (°C)	HIP pressure (MPa)
A1	0.1	1700	196
A2	0.5	1700	196
A3	1.0	1700	196
B1	0.5	1500	196
B2	0.5	1600	196
C1	0.5	1700	9.8

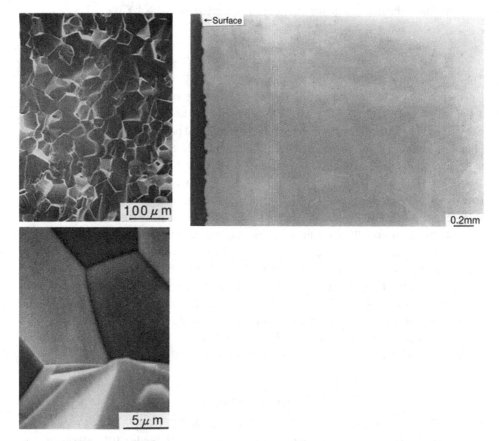

Figure 3.24 SEM and reflection micrographs of Nd:YAG ceramics as reference sintered at 1750 °C for 20 hours under vacuum.

structural defects under HIP pressure because SiO_2 formed by pyrolysis of TEOS resides at grain boundaries as a liquid during sintering.

Figure 3.24 shows a SEM fracture cross-section and reflection micrograph of the polished surface of a standard specimen sintered at 1750 °C for 20 hours under vacuum, under

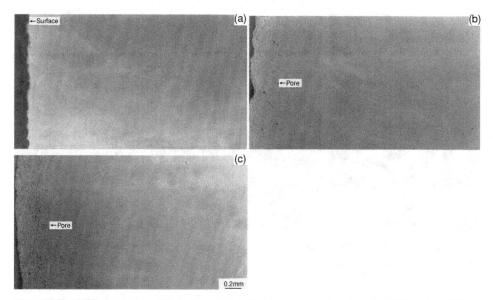

Figure 3.25 Reflection micrographs of specimen A2 after (a) pre-sintering, (b) HIP, and (c) HIP + vacuum sintering.

the same conditions as used for ordinary YAG ceramics. The average particle size and relative density are approximately 50 μm and 99.9997% counted by observation through a transmission microscope. Good transparent ceramic is formed, and pores and grain boundary phase are not detected in the SEM micrograph.

Figure 3.25 shows reflection micrographs before and after vacuum sintering following HIP treatment. Although it is impossible to see pores at this magnification in the specimen after pre-sintering, there are a lot of bubbles in the superficial layer. The number of bubbles in the specimen sintered under vacuum after HIP treatment is greater than in the specimen only treated by HIP.

Figure 3.26 shows SEM micrographs of the superficial layer and center of specimen A2 after HIP treatment. Although a lot of bubbles are observed at grain boundaries in the superficial layer, there are almost none at the center. However, when the inside of the specimen is observed at high magnification, it can be seen that there is both a very thin layer of grain boundary phase and bubbles at the grain boundary.

Figure 3.27 shows SEM micrographs of the superficial layer and center of specimen A2 after sintering under vacuum following HIP treatment. Although grain growth does not occur obviously at the superficial layer (compare with Figure 3.26), more bubble formation is seen, than compared with HIP treatment only, after heat treatment under vacuum. On the other hand, the average particle size is around 50 μm at the center of the specimen and the bubbles reside at the grain boundary phase. The specimen with added 0.1 mass% TEOS has especially small particles, while grain growth is saturated in the specimen with 0.5 mass% TEOS; both of these specimens are pore free, as shown in Figure 3.28.

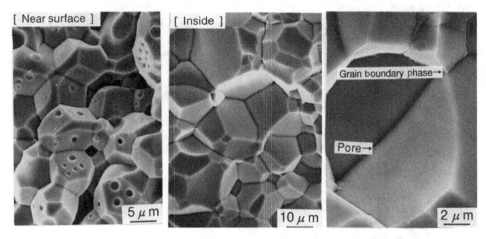

Figure 3.26 SEM micrographs of specimen A2, near the surface and inside, after HIP.

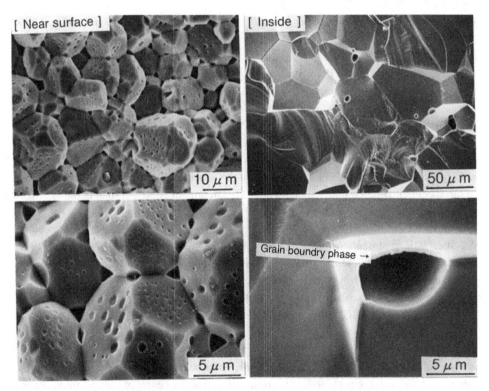

Figure 3.27 SEM micrographs of a superficial layer and the center of specimen A2 after vacuum sintering following HIP treatment.

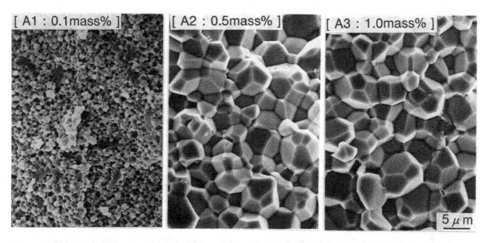

Figure 3.28 SEM micrographs of specimens A1, A2, and A3 pre-sintered at 1600 °C for 3 hours under vacuum.

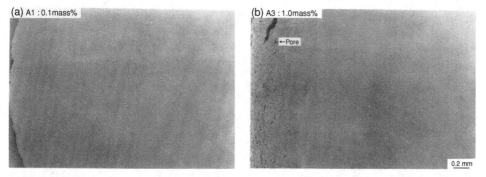

Figure 3.29 Reflection micrographs of (a) sample A1 with 0.1 mass% TEOS and (b) sample A3 with 1.0 mass% TEOS after vacuum sintering following HIP treatment.

Figure 3.29 shows reflection micrographs of specimens A1 and A3 with added 0.1 and 1.0 mass% TEOS after vacuum sintering following HIP treatment. A lot of bubbles emerge at the superficial layer of specimen A3 compared with specimen A2 shown in Figure 3.25 produced under the same conditions. However, only a few bubbles are observed at the superficial layer of specimen A1. It is concluded that increasing TEOS tends to increase the number of pores under the same HIP conditions.

Figure 3.30 indicates that the amount of TEOS added affects transmittance at 1000 nm; the specimens measured were B1, A2, and B2 with 0.1, 0.5, and 1.0 mass% TEOS sintered at 1750 °C for 20 hours under vacuum following HIP treatment at 1700 °C under 196 MPa, respectively. The transmittance of the specimens with 0.1 and 1.0 mass% TEOS is deteriorated due to incomplete grain growth and elimination of the intermediate phase $YAlO_3$. There is an increase in bubble formation with Ar gas following HIP treatment (compare

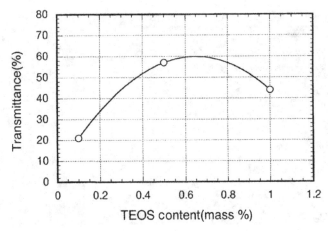

Figure 3.30 Dependence of in-line transmittance at 1000 nm on TEOS content added to the specimens.

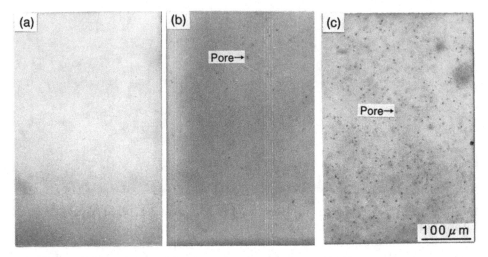

Figure 3.31 Transmission micrographs of (a) a reference sample after vacuum sintering only, (b) sample B1 HIPed at 1500 °C, and (c) sample A2 HIPed at 1700 °C.

with the specimen with 0.5 mass% TEOS). Since the transmittance of the sample with 0.5 mass% TEOS sintered under vacuum is 85%, it is clear that HIP treatment causes a deterioration in transmittance.

Figure 3.31 shows transmission micrographs of a standard specimen (0.5 mass% TEOS, only vacuum sintering), specimen A2 (0.5 mass% TEOS, HIP at 1700 °C under 196 MPa + vacuum sintering), and specimen B2 (0.5 mass% TEOS, HIP at 1500 °C under 196 MPa + vacuum sintering). After vacuum sintering the standard specimen has scarcely any pores, and it is seen that increasing HIP temperature increases the number of pores dramatically.

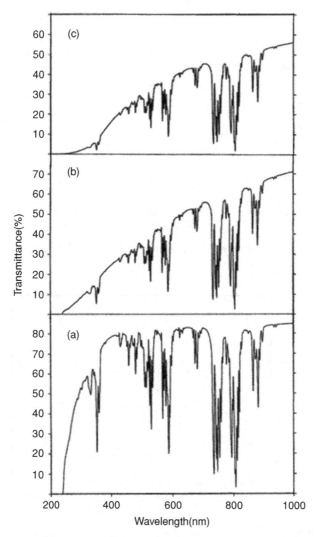

Figure 3.32 Transmission spectra for (a) a reference sample after vacuum sintering only, (b) sample B1 HIPed at 1500 °C, and (c) sample A2 HIPed at 1700 °C.

Figure 3.32 shows transmission spectra of the specimens, of thickness 5 mm, shown in Figure 3.31. The transmittance of the standard specimen is greater than 80% at background level, and deterioration of the transmittance is not observed in the short wavelength range. Although specimens B1 and A2 with bubbles caused by the HIP treatment have deteriorated transmittance over all of the measured range, light scattering in the short wavelength range increases, resulting in low transmittance. It is confirmed that the transmittance deteriorates as a result of light scattering caused by the presence of pores and

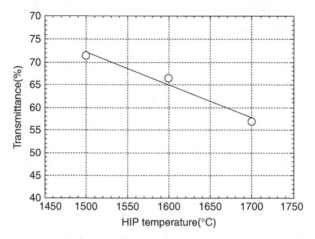

Figure 3.33 Dependence of in-line transmittance at 1000 nm on the HIP temperature.

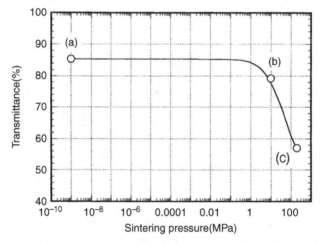

Figure 3.34 Dependence of in-line transmittance at 1000 nm on sintering pressure. (a) Reference sample after vacuum sintering only, (b) sample B1 HIPed at 1500 °C, and (c) sample A2 HIPed at 1700 °C.

that increasing the temperature of HIP treatment increases the formation of bubbles in the specimens.

Figure 3.33 shows the relationship between the temperature of HIP treatment and the transmittance measured at 1000 nm for specimens with 0.5 mass% TEOS treated under 196 MPa of HIP. It is found that the transmittance of the specimens degrades linearly depending on HIP temperature. This decrease in transmittance is caused by the amount of bubble formation in the specimens. Figure 3.34 shows the relationship between the transmittance at 1000 nm and the sintering pressure for the standard specimen only sintered

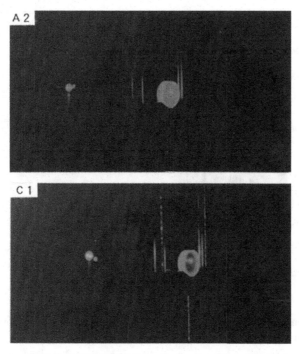

Figure 3.35 Appearance of 1.1 at.% Nd:YAG ceramics irradiated by a He–Ne laser beam (632 nm). Specimens A2 and C1 are HIPed at 1700 °C under 196 and 9.8 MPa of Ar, respectively.

under vacuum, sample C1 HIPed at 9.8 MPa, and sample A2 HIPed at 196 MPa, both sintered at 1750 °C for 20 hours under vacuum after HIP treatment. The transmittance of the specimen sintered under vacuum is the best and transmittance tends to decrease on HIP treatment at a pressure above 9.8 MPa. The amount of bubbles in the specimens are in the order standard specimen (vacuum sintering) → C1 (9.8 MPa) → A2 (196 MPa); increasing amounts of bubbles decrease the transmission spectra, depending on the HIP pressure.

Figure 3.35 shows the appearance of specimens A2 and C1 irradiated with a He–Ne laser (632 nm), which were HIPed at 1700 °C and 196 MPa and 1700 °C and 9.8 MPa under argon gas, followed by vacuum sintering. Significant scattering was seen in the specimen treated at higher HIP pressure. The main scattering sources are segregated secondary phases formed during the HIP process and bubbles generated after HIP treatment.

Figure 3.36 shows EDX analysis at two spots, a grain boundary (spot 1) and inside a grain (spot 2) of specimen A2 after HIP treatment. The picture shows the spot position and fluorescence X-ray shows Al, Y, Nd, and Si at the spot. The table below gives the analytical data. Although the diameter of the electron beam used for spot analysis is approximately 1 μm, the region of analysis is several micrometers from the center so that analytical data at the grain boundary refer to the grain boundary and the grain. Only Al, Y, and Nd as main

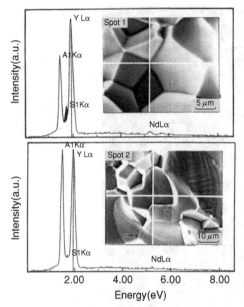

	Spot 1 (Grain boundary)	Spot 2 (Inner grain)
Al_2O_3	33.6	41.1
Y_2O_3	58.1	58.0
Nd_2O_3	3.2	0.9 (mass%)
SiO_2	5.1	Tr.

Figure 3.36 Spot analysis at a grain boundary by EDX of specimen A3. Spot 1 and spot 2 indicate a grain boundary and inner grain, respectively.

elements of Nd:YAG ceramics are detected inside the grain. On the other hand, a large amount of Si and Nd, greater than inside the grain, are detected at the grain boundary. The major composition at the grain boundary is thought to be SiO_2 and Nd_2O_3. However, residual Ar forming bubbles is not detected because it is below the detection limit of EDX. This composition at the grain boundary is not observed in the standard specimen sintered under vacuum shown in Figure 3.24, only in specimens after HIP treatment.

3.2.4 Effective utilization of HIP

HIP is a high pressure sintering technology in which enforced pressurizing with a gas medium is effective in producing dense ceramics. Although HIPing to 99–95% of relative density of a material gives dense sintered ceramics in general, usually the density of ceramics obtained by HIP treatment is much lower than that of the Nd:YAG ceramics pre-sintered in the present experiment. In the present experiment, a precise molding method and vacuum sintering technology led to elimination of residual pores to the ppm level. The utilization of HIP for the synthesis of a laser gain medium used to generate high power was investigated, but the results obtained were not successful. Several possible methods for solving this problem are introduced in this section.

Figure 3.37 shows the outline of an Nd:YAG ceramic encapsuled with platinum and the inside of the material analyzed by transmission and polarized microscopy. As previously mentioned, HIP treatment was carried out after the platinum capsule was attached to the

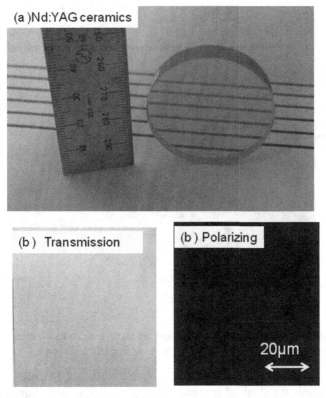

Figure 3.37 (a) Appearance and (b) microstructure of transmission and polarizing light of Nd:YAG ceramics after Pt encapsuled HIP.

specimen in order to prevent the pressurizing medium penetrating the specimen during HIP treatment. No scattering centers are detected because there are no residual pores in the material. Pores are formed by a foam of pressurized Ar gas when it is released from high pressure after the high pressurized gas has penetrated into material along grain boundaries. The capsule is very effective in inhibiting the penetration of Ar gas from the outside.

Figure 3.38 shows Nd:YAG ceramics produced using HIP at 1750 °C for 30 min under 196 MPa of Ar gas without a capsule. Since gas foam along with Ar gas penetration is observed at the surface of the specimen after HIP treatment, high quality ceramic was obtained on removing the surface with machining. The size of the obtained specimen is $7 \times 10 \times 70$ mm^3, and one residual pore around 3 μm in size was detected in the specimen. The amount of residual pores is 3 ppt, 3×10^{-12}. This value is extremely low, and allows the synthesis of extremely low scattering ceramics.

Figure 3.39(a) shows the lasing characteristic of 2.4% Nd:YAG ceramics produced using HIP with a capsule. The optical loss at 1064 nm reached 0.1% cm^{-1} at an extremely low scattering level. This result gives a slope efficiency of 58% even though the specimen does not have an AR coating on the surface. Since the slope efficiency improves by about 20%

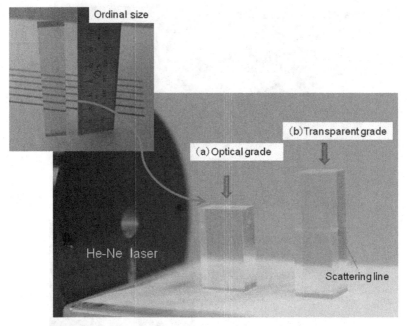

Figure 3.38 Scattering condition of Nd:YAG ceramics with (a) extremely low (optical grade) and (b) relatively large (transparent grade) porosity after capsule-free HIP treatment.

with an AR coating, the obtained specimen may approach the limit of quantum efficiency of 74% [34]. Similarly in Figure 3.39(b), the laser performance of 7% Yb:YAG ceramic pumped with a 940 nm LD is shown. The scattering level of this material is very low, and hence a very high oscillation efficiency (almost quantum efficiency) was achieved [35].

From the above results, inappropriate HIP conditions increase the residual pores in Nd:YAG ceramics, which results in a loss of transmittance. However, appropriate application of the HIP process can dramatically reduce the residual pores, which cause scattering, to approximately ppm–ppt levels. It contributes to an exponential improvement in optical properties. Using optimized HIP treatment, the lasing characteristic reached very close to 74%, the value calculated from quantum theory.

3.2.5 Discussion

3.2.5.1 Why the transmittance is not improved by HIP treatment

As mentioned in Section 3.2.1, other published reports [36] have shown that HIP is an effective means of eliminating the structural defects formed in ceramics. It is easy to produce translucent material using HIP even though the raw materials are usually difficult to sinter under vacuum or ambient atmosphere, for example Al_2O_3, MgO, PZT, PLZT, and so on.

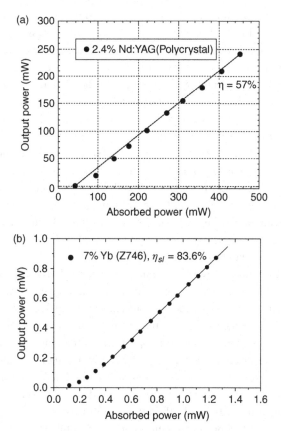

Figure 3.39 Laser performance of (a) Nd:YAG and (b) Yb:YAG ceramics with low optical loss.

A small amount of SiO_2 decomposed from TEOS is added as a sintering aid in the synthesis of Nd:YAG ceramics. SiO_2 as 300 mass ppm Si promotes densification and homogenization of Nd:YAG ceramics by forming an infinitesimal liquid phase at grain boundaries as soon as it reacts with the host material on an increase in temperature. Most of the SiO_2 formed during sintering finally disappears at grain boundaries because it diffuses into the host Nd:YAG at high temperature. This is confirmed in the next section by a change in lattice constant and observation of grain boundaries with SEM and TEM.

As shown in Figures 3.27 and 3.28, a grain boundary phase was observed in the Nd:YAG ceramics treated with HIP. This grain boundary phase is hardly detected by SEM (around ten thousand magnification) in the Nd:YAG ceramics which were simply sintered under vacuum. Additionally, spot analysis by EDX revealed that the main composition formed at grain boundaries in specimens treated with HIP are SiO_2 and Nd_2O_3. It is possible that the grain boundary phase is involved in bubble formation because the bubbles certainly are located at grain boundaries. The bubbles increase significantly with an increase in TEOS, HIP temperature, or HIP pressure. Pre-sintered material has SiO_2 at the grain boundaries.

It is considered that bubble formation occurred as a result of pressure release after Ar gas (pressurizing medium of HIP) dissolved into SiO_2 under high pressure.

Although Ar was not detected by EDX analysis in the grain boundary phase, a silicate phase formed after HIP treatment as shown in Figure 3.36. A previous hypothesis, which involves dissolving Ar gas into the grain boundary at the HIP step → pressure release → vaporizing of Ar gas dissolved into the grain boundary, is considered reasonable, especially given the presence or absence of the grain boundary phase and how the HIP conditions affect the amount of bubble formation. It is conceivable that Ar penetrates into the grain boundary phase through a superficial layer of the specimen as mentioned before. The volume of vaporized Ar gas expands by an order of magnitude even though a minimal amount of Ar gas is dissolved to a depth of less than a micrometer in the grain boundary phase. The amount of Ar dissolved in the grain boundaries in the superficial part of the Nd:YAG ceramic where many bubbles are observed with SEM can be estimated to be very small; EDX is not able to detect Ar in this region.

The above discussion has only mentioned adverse effects caused in Nd:YAG ceramics by the grain boundaries produced by HIP treatment. However, HIP can improve or eliminate structural defects such as voids and the pressure during molding can arrange the constituent particles in a way that cannot be corrected by ordinary sintering. However, Nd:YAG ceramic requires ideal crushing behavior of the granulated powder, and a packing structure with almost no defects in the pressed green compact. Since the amount of pores in Nd:YAG sintered bodies produced under vacuum sintering only and their average size are around 2 vol. ppm and 5 μm, respectively, the amount of structural defects formed in the pressed green compact before sintering and their size are considered to be very small. There is very little literature discussing statistically the relationship between the amount of structural defects and their size in the pressed green compact and after HIP treatment. In order to reduce the quantity of defects it is necessary to determine the relationship between the defect size and structural particle size in specimens produced by HIP. When the defect size in ceramics is substantially smaller than the structural particle size, although the defect can be filled by plastic change of the structural particles, it is difficult to produce materials with good thermal and mechanical properties. On the other hand, when the defect size in ceramics is substantially greater than the structural particle size, ceramics with no structural defects can be formed by filling the defects among the structural particles.

The process of filling defects, defect elimination by particle migration, can be supposed to follow the above process because the size of the original defects in Nd:YAG in the present experiment was 5 μm and the particle size before HIP treatment was adjusted to equal to or smaller than this. Although little defects residing in Nd:YAG ceramics might be eliminated by HIP treatment, the harmful effect produced by HIP was too great to notice this.

However, the technical problems of capsule-free HIP treatment have been revealed, and the optical quality can be improved dramatically by reducing the level of defects in the material to ppm level as described in Section 3.2.4. The success of HIP technology is dependent on the method used.

3.2.5.2 Gas formation in ceramics by HIP treatment

It is found that deterioration in the densification and transparency of Nd:YAG ceramics occurred by the penetration of Ar gas under HIP treatment as mentioned above. In other words, vacuum sintering for the production of Nd:YAG ceramics is considered suitable. A similar phenomenon was reported by Faile and Roy [37], and by Kwon and Messing [38] regarding inert and neutral Ar, solubility in glass and gas diffusion in liquid sintered ceramics treated with HIP, respectively. The solubility of Ar in silicate glass is expressed as follows:

$$C_i = C_g \exp(-A\gamma/RT), \tag{3.2}$$

where C_i and C_g are the gas concentrations in the atmosphere and in silicate glass, respectively, A is the surface area of bubbles formed by the penetration of gas, γ is the surface tension of glass, R is the gas constant, and T is temperature. The solubility of gas increases, depending upon the gas pressure and temperature linearly. It was found that inert gases such as Ar or nitrogen gas are dissolved into various types of glass by HIP. The gas composition in the glass forms a foam by pressure release on annealing at ambient pressure. The gas foam, Ar gas, and the grain boundary phase, silicate, can be observed after HIP treatment in Nd:YAG. The grain boundary phase is shown to contain additional dissolved Ar, confirming that the pressurizing medium is dissolving in the silicate glass. Ar dissolves into the grain boundary phase of composition mainly SiO_2 by HIP, and the optical quality deteriorates as shown by experiment. Additionally, once a foam of Ar gas builds up in Nd:YAG ceramics, it is considered that elimination by diffusion, mass transfer, is almost impossible because the atomic radius of Ar is relatively large. It is easy to see that elimination of the gas foam is difficult because many bubbles remain even when a specimen is heated at 1750 °C for 20 hours after HIP treatment.

Although the grain boundary phase in Nd:YAG ceramics can dissolve Ar by the pressurizing effect of HIP, the grain boundary phase containing Ar still remains in the Nd:YAG ceramic even when it is sintered at high temperature under vacuum. Why is it difficult to dissolve the grain boundary phase containing Ar in Nd:YAG crystal at high temperature? When Nd:YAG ceramic is annealed at low temperature after HIP treatment, Ar as a gas phase is formed along with the grain boundary phase as its solubility in the grain boundary phase is extremely low. Accordingly, the grain boundary phase containing Ar is gradually dissolved into the Nd:YAG crystal and only the Ar gas introduced by HIP remains. Although a lot of foam debris (residual pores) remains along with the grain boundary phase at the specimen surface, there is no grain boundary phase at this part judging from SEM observation. Since the diffusion distance of Ar from the periphery to the inner part of the specimen is very long, the amount of Ar gas foam is small due to the relatively small amount of Ar penetrating.

To utilize HIP successfully to produce high quality YAG ceramics, (1) the penetration of Ar gas must be blocked, and (2) the distance Ar gas is able to penetrate must be shortened.

3.2.6 Conclusion

HIP, mainly capsule free, which is a high pressure sintering method, was applied in order to reduce approximately 2 vol. ppm of pores in Nd:YAG ceramics produced by the solid phase method and vacuum sintering. The results are as follows.

(1) The transparency of Nd:YAG ceramics treated with HIP is much inferior to that of Nd:YAG ceramics sintered under vacuum only. The reason is bubbles of gas generated by the HIP treatment and the effect of the grain boundary phase.
(2) The scattering centers, the bubbles, increase depending on the HIP temperature and pressure, and the optical quality of the specimen deteriorates as the amount of SiO_2 in the specimen increases.
(3) The bubbles formed in the specimen by HIP treatment are caused by Ar gas dissolving in the grain boundary phase under pressure release.
(4) The optical properties are optimized by reducing the amount of residual pores in the Nd:YAG ceramic to ppt level; HIP treatment using a capsule and high temperature for short duration can solve the problems of HIP treatment.

References

[1] R. L. Coble, *Am. Ceram. Soc. Bull.* **38** (10) (1959) 501, US Patent No. 3026210.
[2] S. E. Hatch, W. F. Parson, and R. I. Weagley, *Appl. Phys. Lett.* **5** (1964) 153.
[3] C. Greskovich and J. Chernoch, *J. Appl. Phys.* **44**, (1073) 4599–4605.
[4] G. de With and H. J. A. van Diji, *Mater. Res. Bull.* **19** (1984) 1669–1674.
[5] C. A. Mudler and G. de With, *Solid State Ionics* **16** (1985) 81–86.
[6] M. Sekita, H. Haneda, T. Yanagitani, and S. Shirasaki, *J. Appl. Phys.* **67** (1) (1990) 453–458.
[7] M. Sekita, H. Haneda, and S. Shirasaki, *J. Appl. Phys.* **69** (6) (1991) 3709–3718.
[8] H. Haneda, M. Sekita, T. Mitsuhashi, S. Shirasaki, and T. Yanagitani, *Electronic-Ceramics, January 17–23* (1993).
[9] Technical Report of National Institute for Inorganic Materials, No. 49, pp. 38–49 (Research for Sintered Materials regard to Opto-Electronics).
[10] Y. Tsukuda and A. Muta, *Yogyo Kyokai-shi (J. Ceram. Soc. Jpn.)* **84** (1976) 585–589.
[11] C. Greskovich and K. N. Woods, *Am. Ceram. Soc. Bull.* **52** (1973) 473–478.
[12] W. D. Kigrey *et al.*, *Introduction to Ceramics* (Japanese edn.), Uchida Rokakuho Book, Tokyo (1980), p. 233.
[13] N. Yamamoto and H. Irokawa, *J. Illum. Eng. Jpn.* **69** (1985) 11–14.
[14] G. Toda, T. Noro, and A. Muta, *J. Powder Powder Metall. Soc. Jpn.* **21** (1974) 76–80.
[15] A. Ikesue, S. Matsuda, and S. Shirasaki, *Taikabutsu* **43** (1991) 451–460.
[16] E. Smethurst and D. W. Budworth, *Trans. Br. Ceram. Soc.* **78** (1995) 225–228.
[17] Technical Report of National Institute for Inorganic Materials, No. 11, p. 21 (Research for Magnesium Oxide).
[18] C. Greskovich and K. N. Wood, GE Rep., No. 72CRD, 243 (1972).
[19] A. Ikesue and K. Kamata, *J. Jpn.Ceram. Soc.* **103** (5) (1995) 489–493.
[20] A. Ikesue, K. Yoshida, T. Yamamoto, and I. Yamaga, Optical scattering centers in polycrystalline Nd:YAG laser, *J. Am. Ceram. Soc.* **80** (6) (1997) 1517–1522.

[21] K. Uematsu, *New Ceram.* **5** (1992) 47–50.

[22] K. Uematsu, J. Y. Kim, Z. Kato, N. Uchida, and K. Saito, *J. Ceram Soc. Jpn.* **98** (5) (1990) 515–516.

[23] K. Uematsu, FC Reports (Japan Fine Ceramics Association), **9** (5) (1991) 187–189.

[24] A. Ikesue and K. Yoshida, Scattering in polycrystalline Nd:YAG lasers, *J. Am. Ceram. Soc.* **81** (8) (1998) 2194–2196.

[25] A. Ikesue and K. Kamata, Microstructure and optical properties of hot isostatically pressed Nd:YAG ceramics, **79** (7) (1996) 1927–1933.

[26] A. Ikesue and K. Yoshida, Influence of pore volume on laser performance of Nd:YAG ceramics, *J. Mater. Sci.* **34** (1999) 1189–1195.

[27] M. W. Bebecke, N. E. Olson, and J. A. Pask, *J. Am. Ceram. Soc.* **50** (7) (1967) 365–368.

[28] E. Smethurst and D. W. Bundworth, *Trans. Br. Ceram. Soc.* **71** (2) (1972) 45–53.

[29] K. Hamano and S. Kanzaki, *J. Jpn. Ceram. Soc.* **85** (5) (1977) 225–230.

[30] F. W. Ainger, D. Appleby, and C. J. Kirk, *J. Mater. Sci. Lett.* **8** (12) (1973) 1825–1827.

[31] H. Yamada, A. Suzuki, Y. Uchida, M. Yoshida, H. Yamamoto, and Y. Tsukuda, *J. Electrochem. Soc.* **136** (1989) 2713–2713.

[32] N. Matsuda, *Bull. Ceram. Soc. Jpn.* **28** (2) (1993) 130–133.

[33] K. Hirota and E. Hirota, *Bull. Ceram. Soc. Jpn.* **18** (3) (1983) 190–197.

[34] A. Ikesue, Single crystal to polycrystalline materials with optical grade, *Appl. Phy. (Oyo-butsuri)*, **75** (5) (2006) 579–583.

[35] E. Pwalowski *et al.*, *SPIE Proc.*, Solid State Lasers, XIX, Technology and Devices, **7578** (2010) 757815–1.

[36] Y. Ito, H. Yamada, M. Yoshida, H. Fujii, G. Toda, H. Takeuchi, and Y. Tsukuda, *J. Jpn. Appl. Phys.* **27** (1988) L1371.

[37] S. P. Faile and D. M. Roy, *J. Am. Ceram. Soc.* **49** (12) (1966) 638–643.

[38] Oh-Hun Kwon and G. L. Messing, *J. Am. Ceram. Soc.* **72** (6) (1989) 1011–1015.

4

Synthesis of polycrystalline ceramic lasers (RE-doped sesquioxides)

There are many hundreds of solid-state media in which laser action has been achieved, but relatively few types are in widespread use. Of these, the most common type is the neodymium-doped YAG laser, which is mainly used as an industrial laser. This host material has a garnet structure. Recently, in addition to oxide host materials such as YVO_4, fluoride host materials such as YLF ($YLiF_4$) have been reported. However, garnet structure laser materials are still usually used as the major gain media in solid-state lasers.

Generally, lanthanide rare earth elements can be used as laser active ions. However, some rare earth oxides such as Sc_2O_3, Y_2O_3, and Lu_2O_3 etc. do not show emission. These oxide materials are promising laser host materials but it is very difficult to grow single crystals of these oxides by the conventional melt-growth method because of their high melting temperature. However, recently, advanced sintering technology has enabled the fabrication of polycrystalline ceramic laser materials for use as laser gain media in solid-state lasers. One of the advantages of sintering technology is that materials can be produced below the melting temperature without melting the materials – that is a solid-state reaction occurs between the starting raw powder materials. This has opened the way to fabrication of high melting point materials and incongruently melting materials, which could not be produced by the conventional melt-growth technology.

In this chapter, technical issues in the fabrication of rare earth oxide single crystals using conventional melt-growth technology are described, and we discuss why such new materials are necessary for future laser technology. Regarding the fabrication process of polycrystalline rare earth oxide laser gain media, Y_2O_3 ceramic materials will be described as a typical example. In addition, experimental results on $Er:Sc_2O_3$ and $Nd:Y_2O_3$ ceramic laser gain media prepared by the author will be described, together with results on the optical properties and laser oscillation performance of rare earth oxide laser materials reported by other groups.

4.1 Current status of single crystal technology

A sesquioxide is an oxide containing three atoms of oxygen with two atoms (or radicals) of another element. Its chemical formula is expressed as RE_2O_3. Many sesquioxides contain the metal in the +3 oxidation state and the oxide ion. Normally, when the RE site is replaced

160

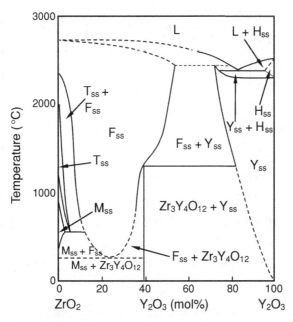

Figure 4.1 Phase diagram for the ZrO_2–Y_2O_3 system.

with lanthanide rare earth elements such as Sc, Y and Lu, these sesquioxides can be host materials. By doping laser active ions such as Nd, Yb and Er ions into these sesquioxide host materials, they can be used as laser gain media.

In fact, RE_2O_3 materials have been investigated for many years as use as laser gain media, as well as garnet materials such as YAG. But their melting points are higher than 2300 °C and it is very difficult to fabricate such single crystals by conventional melt-growth technology. For example, when growing such sesquioxide crystals by the Bridgman technique or Czochralski process, the crucible must be made of high temperature material which can be used to melt the sesquioxide raw materials [2, 3]. Therefore, it is very difficult to resolve the technical and cost problems with the current melt-growth technology. In order to solve the cost problem, the Verneuil process was introduced because it does not require a crucible to grow a single crystal. However, the optical quality of the grown crystals was very poor and they could not be used as laser crystal materials. In addition, large scale crystals cannot be grown by this process [4–6].

A phase diagram of the ZrO_2–Y_2O_3 system is shown in Figure 4.1. As seen from the diagram, the melting point of Y_2O_3 is very high (about 2430 °C). Another technical issue is the phase transition point which exists near 2200 °C, at which the hexagonal system changes to a cubic system. Therefore, when the melt is crystallized, stresses can remain in the crystal materials.

The phase diagram of the lanthanide rare earth oxides is shown in Figure 4.2 [7]. When Y_2O_3 is heated up, it changes from a cubic to a monoclinic crystal system (B-type) around

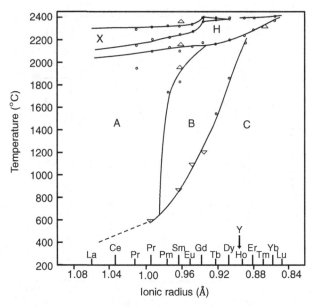

Figure 4.2 Phase diagram for various lanthanide rare earth oxide materials.

1800 °C. Around 2200 °C, it changes to a hexagonal crystal system (H-type). Finally, it starts to melt at over 2432 °C. As for the other rare earth oxides which have a smaller ionic radius than Y, generally the phase transition occurs at a relatively lower temperature compared to Y_2O_3. Therefore, it is necessary to fabricate such materials at temperatures below these phase transition temperatures, which means that it is very difficult to fabricate these ceramic materials even by solid-state sintering. For example, as seen in the phase diagram, transparent rare earth oxide ceramics such as Ce_2O_3 and La_2O_3 must be prepared below 400 °C. Given such conditions, these materials are difficult to produce using both crystal melt-growth technology and ceramic process technology.

4.2 Requirements for sesquioxide ceramic lasers

In this section, a general overview of the crystal structure of sesquioxide materials and why these materials are required for new types of lasers are discussed based on their physical properties.

4.2.1 Crystal structure and typical physical properties

Generally, RE_2O_3 (RE implies Sc, Y and lanthanide rare earth elements) takes a type of crystal structure which is called "Bixbyite" (see Figure 4.3), and its space group is $Ia3$ (T_h^7). Two lanthanide (RE^{3+}) sites in the Bixbyite structure are represented in the figure. RE atoms occupy two crystallographically independent sites, $8a$ and $24d$, whereas all the

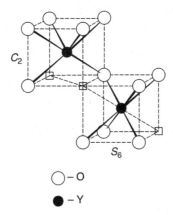

Figure 4.3 Illustration of the Bixbyite structure.

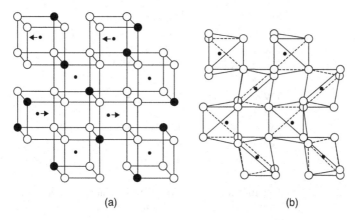

(a)   (b)

Figure 4.4 Crystal structure of Y_2O_3: (a) MO_2 florite structure, (b) C type M_2O_3 structure.

O atoms are located at normal sites, 48e [8]. So, another explanation is to say that the Bixbyite structure is formed when some of the O atoms are removed from a fluorite (CaF_2) structure, which belongs to the space group $Fm3m$ (O_h^5).

Figure 4.4(a) shows the crystal structure of CaF_2 (MO_2). Large circles (both white and black) represent oxygen atoms, and small black circles represent metallic ions. Black circles (some of the oxygen atoms) are removed from this structure and metallic ions are moved in the directions of the small arrows. Then the remaining oxygen atoms are rearranged as shown in Figure 4.4(b). This structure is called a C type M_2O_3 structure, and it is normally seen for medium to heavy rare earth sesquioxides. Y_2O_3 also takes this structure.

As a representative of the Bixbyite structure, the physical properties of Y_2O_3 are compared with those of $Y_3Al_5O_{12}$ and $YLiF_4$ in Table 4.1 [9].

Table 4.1 *Physical properties of Y_2O_3, $Y_3Al_5O_{12}$ and $YLiF_4$ [9]*

Property	Y_2O_3	$Y_3Al_5O_{12}$	$YLiF_4$
Crystal system	Cubic	Cubic	Tetragonal
Structure type	Bixbyite	Garnet	Scheelite
Space group	$T_h^7 - Ia3$	$O_h^{10} - Ia3d$	$C_{4h}^7 - I4_1/a$
Y^{3+} site symmetry	C_2 and C_{3i}	D_2	S_4
Y^{3+} site density	2.681×10^{22} cm^{-3}	1.388×10^{22} cm^{-3}	1.396×10^{22} cm^{-3}
Lattice constant	10.607 Å	12.002 Å	$a = 5.167$ Å
			$c = 10.733$ Å
Melting point	2410 °C	1940 °C	1650 °C
Debye temperature	460 K	560 K	279 K
Phonon energy	560 cm^{-1}	700 cm^{-1}	450 cm^{-1}
Specific heat	0.170 J g^{-1} K^{-1} at 100 K	0.137 J g^{-1} K^{-1} at 100 K	–
	0.360 J g^{-1} K^{-1} at 200 K	0.456 J g^{-1} K^{-1} at 200 K	–
	0.455 J g^{-1} K^{-1} at 300 K	0.626 J g^{-1} K^{-1} at 300 K	0.791 J g^{-1} K^{-1} at 300 K
Thermal conductivity	40 W m^{-1} K^{-1} at 100 K	55 W m^{-1} K^{-1} at 100 K	–
	20 W m^{-1} K^{-1} at 200 K	23 W m^{-1} K^{-1} at 200 K	–
	13 W m^{-1} K^{-1} at 300 K	14 W m^{-1} K^{-1} at 300 K	1.7 W m^{-1} K^{-1} at 300 K
Thermal expansion	4.2×10^{-6} K^{-1} at 100 K	4.2×10^{-6} K^{-1} at 100 K	$a = 13.3 \times 10^{-6}$ K^{-1}
	6.0×10^{-6} K^{-1} at 200 K	6.0×10^{-6} K^{-1} at 200 K	$c = 8.3 \times 10^{-6}$ K^{-1}
	7.5×10^{-6} K^{-1} at 300 K	7.5×10^{-6} K^{-1} at 300 K	both axes at 300 K
Density	5.04 g cm^{-3}	4.56 g cm^{-3}	3.98 g cm^{-3}
Mohs hardness	6.8	8–9	4–5
Index of refraction	1.8892 at 1 μm	1.8192 at 1 μm	1.4362 (*o*) at 1 μm
			1.4708 (*e*) at 1 μm

4.2.2 Thermal conductivity

Initially, these sesquioxide materials were very attractive as laser materials because it was expected that their thermal conductivity would be superior to that of YAG materials although their Mohs hardness is less than that of YAG crystal [10]. For this reason, these sesquioxide materials were considered to have high thermal shock properties for use as high power laser host materials. However, recent results indicated that the thermal conductivities of these materials are almost equivalent to those of YAG materials as shown in Table 4.1.

Table 4.2 *Thermal conductivity of YAG and sesquioxide materials*

	YAG	Y_2O_3	Sc_2O_3	Lu_2O_3
Undoped k (W m^{-1} K^{-1})	10.1	12.8	15.5	12.2
3 mol% Yb^{3+} k (W m^{-1} K^{-1})	7.6	7.4	6.4	10.8
$\Delta k/k$ (%)	−25	−42	−59	−11

In addition, it was verified that the thermal conductivity of Bixbyite materials can be decreased by doping of rare earth ions (laser active ions) in the host materials (see Table 4.2). In particular, when 3% Yb ions were doped in Sc_2O_3, the thermal conductivity at room temperature was reduced by about 60%. On the other hand, when 3% Yb ions were doped into YAG, the thermal conductivity was reduced by only about 25% [11]. The thermal conductivity of the 3% Yb:YAG was 7.6 W m^{-1} K^{-1}, which is close to the thermal conductivity of sesquioxide materials [12]. In the case of 6% Yb:YAG, for which the concentration is equivalent to 3% Yb:Y_2O_3, the thermal conductivity was reduced by only 32%.

However, it is noteworthy that in the case of Lu_2O_3 host materials doped with rare earth ions, the thermal conductivity was only slightly reduced. One reason for this is that the mass of the doping rare earth element is similar to that of the host Lu. For example, the thermal conductivity of 3% Yb doped Lu_2O_3 was reduced by only 11%, compared with undoped Lu_2O_3 (12.2 W m^{-1} K^{-1}). For comparison, based on the same reason, the thermal conductivity of 50% Yb doped LuAG ($Lu_3Al_5O_{12}$) was reduced by only 10%, i.e., the value 7.8 W m^{-1} K^{-1} for undoped LuAG decreased to 7.1 m^{-1} K^{-1} [13], which is lower than that of the Lu_2O_3 system materials.

4.2.3 Concentration quenching

Generally, concentration quenching is seen in sesquioxide laser materials doped with rare earth fluorescent ions. When a large amount of fluorescent ions are doped into the host sesquioxide material, the radiative quantum efficiency is decreased. In the case of Nd:Y_2O_3, when the Nd concentration was 0.1 at.% in Y_2O_3, the fluorescence lifetime was 271 μs. However, when the Nd doping was increased to 1 at.%, the fluorescence lifetime decreased to 158 μs, which is a reduction of about 42%. (For comparison, the fluorescence lifetime of 2 at.% Nd:Y_2O_3 was 77 μs) [14]. It was considered that the concentration quenching occurs mainly as a result of electric dipole–dipole interaction between the doped fluorescent ions [15]. On the other hand, in the case of 1 at.% Nd:YAG, the fluorescent lifetime was decreased by only 20% [16].

However, in the Bixbyite structure (see Figure 4.3), the rare earth atoms (RE) share the oxygen atom (RE–O–RE). Therefore, the distance between these two RE atoms becomes shorter and the effect of electric dipole–dipole interaction between these fluorescent ions becomes larger. The lattice constant of Y_2O_3, is about 10 Å, and Sc and Lu, which have a smaller ionic radius than Y, also have smaller lattice constants than Y_2O_3. Accordingly, it can be said that concentration quenching occurs easily in sesquioxide system fluorescent materials. In contrast, the distance between two RE atoms in YAG is not as small as in sesquioxides, and the lattice constant of YAG is about 12 Å, which is larger than the value for the sesquioxides. For this reason, it is considered that concentration quenching scarcely occurs in YAG compared to the Bixbyite system.

To explain the decrease in fluorescent emission efficiency due to dipole–dipole inter-action, it is necessary to take into account how the fluorescence relaxation occurs in the various electron levels. For instance, we can look at the fluorescence process in the case of the most popular laser active ion Nd^{3+}. The ground state ($E = 0$ cm^{-1}) of Nd^{3+} is $^4I_{9/2}$, and when it is excited with a light source the ground state electrons are trans-ferred to upper energy levels and then they move to a metastable state at $^4F_{3/2}$ ($E \sim$ 12 000 cm^{-1}). When these electrons transfer further from this level to lower energy levels, normally fluorescence relaxation occurs and finally the electrons are returned to the ground state $^4I_{9/2}$ again. However, the energy level $^4I_{15/2}$ ($E \sim 6000$ cm^{-1}) exists between the metastable state $^4F_{3/2}$ and the ground state $^4I_{9/2}$. Therefore, when an inter-action between excited Nd ions and non-excited Nd ions occurs, a transition state $^4I_{9/2}$, $^4F_{3/2} \to {}^4I_{15/2}, {}^4I_{15/2}$ may occur. This means that quenching occurs when electron transition takes place via the intermediate level $^4I_{15/2}$ due to interactions between Nd ions in the $^4I_{9/2}$ state and $^4F_{3/2}$ state. When electrons accumulate in the $^4I_{15/2}$ state, Nd ions do not produce fluorescence emission and are in a quenching condition. Due to dipole–dipole interaction, a transition state $^4I_{9/2}$, $^4F_{3/2} \to {}^4F_{3/2}, {}^4I_{9/2}$ (interionic reciprocal state of Nd ions) can also occur. This process is not quenching. It is called a diffusion process in which the excitation density is decreased, and the fluorescent intensity per unit volume is also decreased. The magnitude of the quenching process in the $Nd:Y_2O_3$ system was measured and compared with the Nd:YAG system. The rate of the quenching process in Y_2O_3 was 3.7×10^{-39} cm^6 s^{-1} and it was about 20 times higher than that in YAG (1.8×10^{-20} cm^6 s^{-1}) [17].

Accordingly, from the viewpoint of concentration quenching, Nd ions may not be appropriate with Bixbyite structure laser host materials. In the Yb system, for example, this kind of intermediate energy level does not exist, and the concentration quenching problem does not occur. Conversely, in the Ho system, an intermediate energy level ($^4I_7$) exists, and an electron transition state ($^4I_8$, $^4I_5 \to {}^4I_7, {}^4I_7$) can occur. However, this state produces fluorescence emission, and it can probably intensify the quantum efficiency.

4.2.4 Spectrum characteristics

In Nd-doped YAG materials, normally the laser transition $^4F_{3/2} \to {}^4I_{11/2}$ occurs and a wavelength of 1 μm is emitted. However, in the case of Nd-doped Bixbyite materials, the most common transition is $^4F_{3/2} \to {}^4I_{9/2}$ which emits a wavelength of 0.9 μm. Fluorescence spectra for $Nd:Y_2O_3$ and Nd:YAG are shown in Figure 4.5. The laser transition $^4F_{3/2} \to$ $^4I_{9/2}$ is considered a three-level system, and lasing wavelengths of 914 nm and 946 nm are available. These wavelengths are suitable for application to ozone measurement [18, 19].

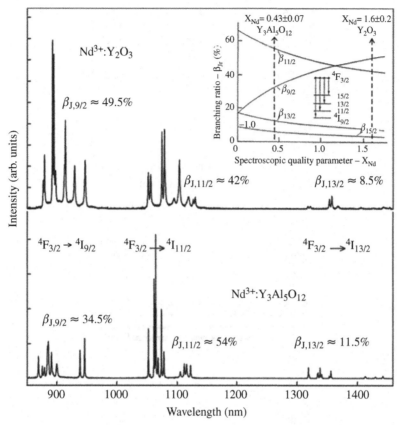

Figure 4.5 Fluorescence spectra for Nd-doped Y_2O_3 and YAG. The Japan Society of Applied Physics, copyright 2011 [19].

The Yb-doped laser materials have a three-level system. The energy levels consist of only two states: the ground state $^2F_{7/2}$ and excited state $^2F_{5/2}$. Therefore, the fluorescence is not as complicated as in the Nd system. When an electron transition occurs from $^2F_{5/2} \rightarrow {}^2F_{7/2}$, a broad fluorescent light of approximately 1 μm wavelength is generated. The Yb system can be focused to give a high efficiency laser, with a quantum efficiency estimated to be higher than 90%. Absorption and emission spectra of Yb-doped Y_2O_3 and Sc_2O_3 are shown in Figure 4.6. Compared to Yb:YAG, Yb-doped Bixbyite materials have a broader fluorescence linewidth. Therefore, they are useful for the development of tunable lasers and ultra-short pulse lasers. The lasing wavelength is shifted slightly by changing the host material, as seen in Figure 4.6. Therefore, by combining these materials, the effective bandwidth can be broadened. In Yb-doped materials, concentration quenching is negligible. Therefore, Yb-doped Bixbyite materials are considered very promising for use as new types of solid-state laser materials in the future [20].

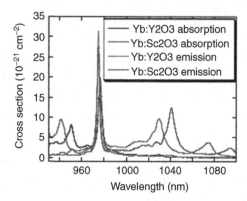

Figure 4.6 Absorption and fluorescence spectra for Yb-doped Y_2O_3 and Sc_2O_3 ceramics [20].

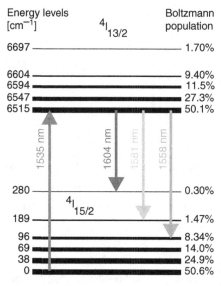

Figure 4.7 Energy level and Boltzmann population of Er-doped Sc_2O_3 ceramics [23].

In the case of Er-doped laser materials, the most common transition is $^4I_{11/2} \rightarrow {}^4I_{13/2}$, which emits a wavelength of 2.8 μm [21]. However, this transition mechanism is taking advantage of an energy transfer ($^4I_{13/2}, {}^4I_{13/2} \rightarrow {}^4I_{9/2}, {}^4I_{15/2}$) due to dipole interaction, and therefore high efficiency cannot be expected from this kind of laser system.

Recently, together with progress in semiconductor laser technology, highly efficient lasers have been studied using a direct pumping method with an LD (laser diode) excitation source. In the case of direct pumping of an Er-doped laser system, lasing at 1.6 μm with high efficiency has been reported [22].

Figure 4.7 shows the relationship between energy levels and Boltzmann population in Er-doped Sc_2O_3 ceramics. For instance, by direct pumping into the emitting level with

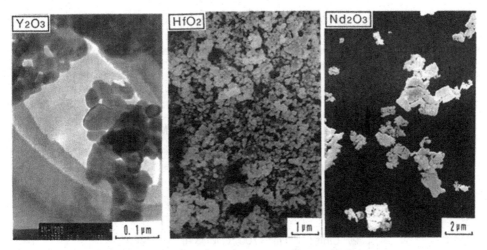

Figure 4.8 SEM photographs of Y_2O_3, HfO_2, and Nd_2O_3 powders used as raw materials.

a 1535 nm LD, it is possible to generate a 1558 nm laser [23]. In this case, the atomic quantum efficiency is simply $1535/1558 = 98.5\%$, and is therefore very important from the viewpoint of high efficiency technology.

4.3 Synthesis of optical grade sesquioxide ceramics [24]

Typical sesquioxide laser materials are Sc_2O_3, Y_2O_3, and Lu_2O_3. In this section, we describe the fabrication of Y_2O_3 ceramics as a representative example. SEM images of starting raw materials (Y_2O_3 and Nd_2O_3) and sintering aids (HfO_2) for $Nd:Y_2O_3$ ceramics are shown in Figure 4.8. The primary particle size of the main Y_2O_3 powders is very fine, around 50–60 nm, and the Nd_2O_3 powders are submicrometer particles.

A flow sheet of the fabrication process for $Nd:Y_2O_3$ ceramics is shown in Figure 4.9. First, in addition to the Y_2O_3 and Nd_2O_3 raw powder materials, a small amount of HfO_2 powder is added as a sintering aid. Then, a dispersing agent, binder and a suitable amount of ethanol are added, and the mixture is ball milled for 12 hours. The ball milled slurry is dried and granulated using a spray dryer. Granulated spherical powders of diameter 30–40 μm are obtained. They are pressed temporarily into a slab or cylinder shape, and then isostatically pressed with a pressure of 98–198 MPa using a CIP (cold isostatic pressing) machine. Then, the pressed powder compacts are calcined in air or oxygen atmosphere in order to remove organic components. Next, they are sintered in oxygen or hydrogen or under vacuum until the relative density reaches over 95%. In this step, the optical quality of the $Nd:Y_2O_3$ ceramics is opaque or mostly translucent. Additionally, these sintered samples are heat treated in a HIP (hot isostatic pressing) machine with a pressure of 198 MPa (in argon atmosphere) at 1700 °C for 3 hours. Finally, the optical grade $Nd:Y_2O_3$ ceramic samples are polished for use as laser gain media.

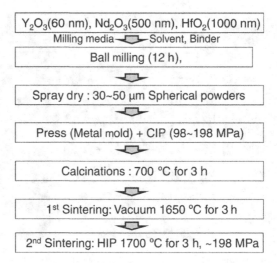

Figure 4.9 Flow sheet for the fabrication of optical grade Nd:Y$_2$O$_3$ ceramics.

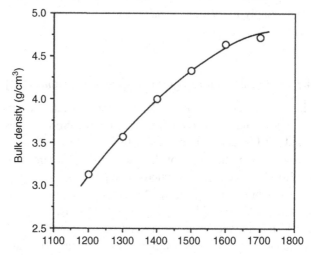

Figure 4.10 Relationship between the sintering temperature and relative density of Nd:Y$_2$O$_3$ ceramics.

The sintering behavior of the Nd:Y$_2$O$_3$ ceramics is shown in Figure 4.10. Relative densities at each vacuum sintering temperature were measured between 1200 and 1700 °C. The relative density was only 3.15 g cm^{-3} at 1200 °C, but increased with an increase in sintering temperature. At higher than 1600 °C, the relative density reached over 90%, which is suitable for HIP treatment.

Fracture surfaces of the starting powder compacts (green body) and the powder compacts sintered at 1100–1650 °C (soaking time 1 hour) are shown in Figure 4.11. Densification

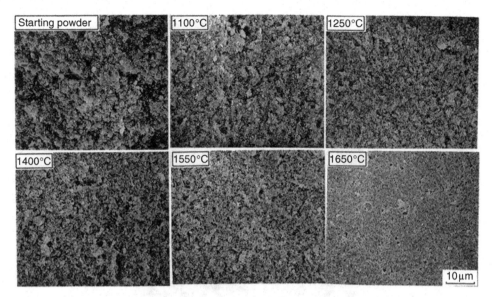

Figure 4.11 Fracture surfaces of Nd:Y_2O_3 ceramics heated to 1100 to 1700 °C and the starting powder compact (green body).

occurred up to 1550 °C but no rapid grain growth was observed in the prepared Nd:Y_2O_3 ceramics. For comparison, in Y_2O_3 ceramics without sintering aids, rapid grain growth occurred at temperatures higher than 1600 °C, which results in poor optical quality because many pores are trapped in the sintered body. However, in the Y_2O_3 ceramics prepared using a sintering aid, the grain size was smaller and almost no residual pores were observed.

Fracture surfaces of Nd:Y_2O_3 ceramics prepared with sintering aids (HfO_2) were investigated by SEM. Figure 4.12(a) and (b) shows the fracture surface of samples sintered at 1650 °C for 1 hour under vacuum and HIPed at 1700 °C for 3 hours at 196 MPa, respectively. In sample (a), small residual pores are observed in the scanning electron image, and the backscattering electron image confirms that the composition is not homogeneous. In sample (b), the grain size is around 20 μm, and there are almost no residual pores inside grains or at grain boundaries.

Fracture surfaces of Nd:Y_2O_3 ceramic prepared without sintering aids were also investigated using SEM and are shown in Figure 4.13. In the case of vacuum sintered sample (left side), it was confirmed that rapid grain growth occurred and the grain size was larger compared with the sample shown in Figure 4.12(a). Many pores were trapped in the sintered body, and even HIP treatment could not eliminate these residual pores (right figure). Therefore, scattering due to residual pores was very significant in Y_2O_3 ceramics prepared without sintering aids, lowering the optical quality.

The appearance of the Nd:Y_2O_3 ceramics prepared with sintering aids is shown in Figure 4.14. Their transparency is very high, comparable to the optical quality of the previously prepared Nd:YAG ceramics.

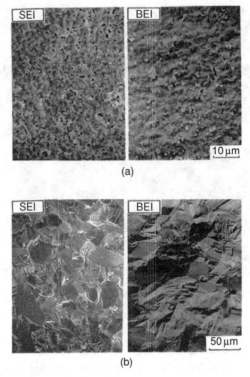

(a)

(b)

Figure 4.12 Fracture surfaces of Nd:Y_2O_3 ceramics (a) sintered at 1650 °C for 1 hour under vacuum and (b) HIPed at 1700 °C for 3 hours at 196 MPa. SEI, scanning electron image; BEI, backscattering electron image.

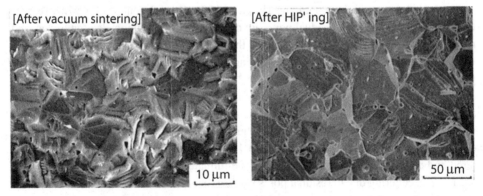

Figure 4.13 SEM photographs of Nd:Y_2O_3 ceramics prepared without HfO_2 synthesized by the same process as in Figure 4.12.

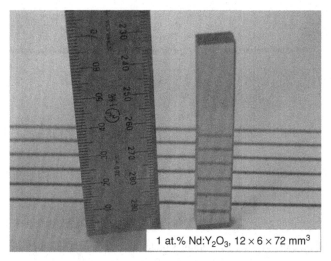

Figure 4.14 Appearance of Nd:Y$_2$O$_3$ ceramics.

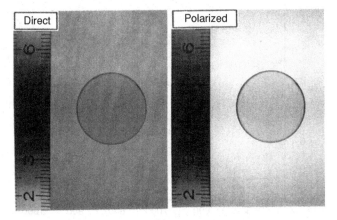

Figure 4.15 Photographs of Nd:Y$_2$O$_3$ ceramics taken under natural light and a polarizing plate.

The optical quality of the Nd:Y$_2$O$_3$ ceramics was observed on a macro scale under a polarizer and an interferometer. Figure 4.15 shows images taken under direct light and a polarizing plate. The material is free from double refractions (optically anisotropic phases), and the straight fringes at a macro level confirm the high optical homogeneity.

In addition, the optical quality of the Nd:Y$_2$O$_3$ ceramics was observed at a micro level under a polarized optical microscope. Figure 4.16 shows the microstructure under direct light and transmitted polarized light. No residual pores were observed in the sintered body, and the polarized image, which was black, showed that the material is optically isotropic.

Both surfaces of the Nd:Y$_2$O$_3$ ceramics were mirror polished and the in-line transmittance was measured. As shown in Figure 4.17, a small amount of scattering or absorption

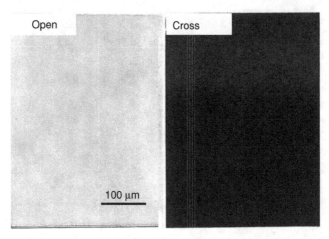

Figure 4.16 Microstructure of Nd:Y$_2$O$_3$ ceramics under direct and polarized light.

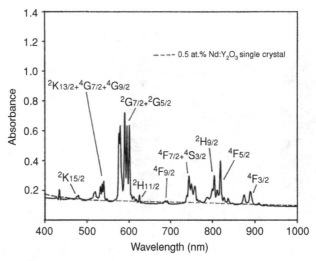

Figure 4.17 Absorption spectrum of Nd:Y$_2$O$_3$ ceramics and background level of 0.5 at.% Nd:Y$_2$O$_3$ single crystal for wavelengths in the region from the visible to the near infrared.

was observed in the visible wavelengths but the in-line transmittance was almost equivalent to that of a single crystal counterpart grown by the Verneuil process in the wavelength region around 1 μm (lasing wavelength region).

The infrared transmission spectrum of the Nd:Y$_2$O$_3$ ceramic at 4000–400 cm^{-1} (2.5–25 μm) is shown in Figure 4.18. It can transmit wavelengths up to 7 μm, and the base line was about 82%, which is very close to the theoretical transmittance.

Photographs of prepared Y$_2$O$_3$ ceramics doped with various types of laser active ions such as Nd, Yb, and Er are shown in Figure 4.19. All of the Y$_2$O$_3$ ceramics showed very high transmittance and extremely low optical scattering at lasing wavelengths.

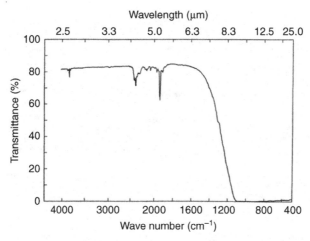

Figure 4.18 Optical transmission spectrum of Nd:Y_2O_3 ceramic in the infrared region.

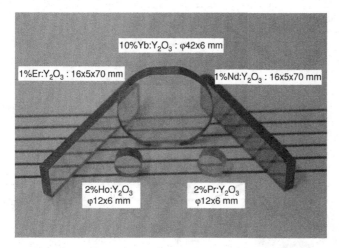

Figure 4.19 Appearance of transparent Y_2O_3 ceramics doped with various laser active ions.

Photographs of 1 at.% Er:Sc_2O_3 and 10 at.% Yb:Lu_2O_3 ceramics are shown in Figure 4.20. Both samples showed high optical quality. The former ceramic was tested for use as an eye-safe laser (1.5 μm lasing), and extremely high efficiency was achieved.

In order to investigate the quality of the materials at the atomic level, the 1 at.% Er:Sc_2O_3 ceramic was investigated by TEM and EDS (energy dispersive X-ray spectroscopy) analysis. The results are shown in Figure 4.21. Since the concentration of Er was low, it was not detectable. But it was confirmed that there was no segregation of Er or sintering aids at the grain boundaries, and the microstructure was a uniform structure.

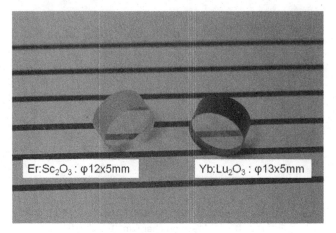

Figure 4.20 Appearance of 1 at.% $Er:Sc_2O_3$ and 10 at.% $Yb:Lu_2O_3$ ceramics.

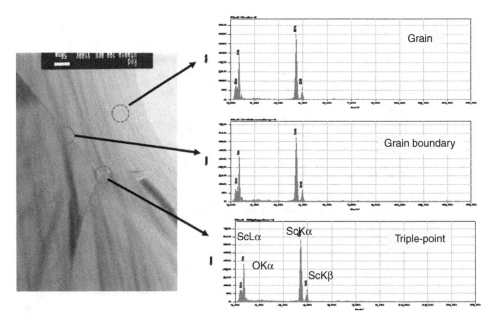

Figure 4.21 TEM image and EDS analysis of an inner grain and grain boundary of 1 at.% $Er:Sc_2O_3$ ceramic.

4.4 Optical quality and laser performance

It is extremely difficult to produce rare earth oxide single crystals such as Sc_2O_3, Y_2O_3 and Lu_2O_3 by the conventional melt-growth technology. There was a report of successful laser oscillation using an $Nd:Y_2O_3$ single crystal grown by the Verneuil process, but there were no details of the optical quality of such single crystal materials. Here, we describe the

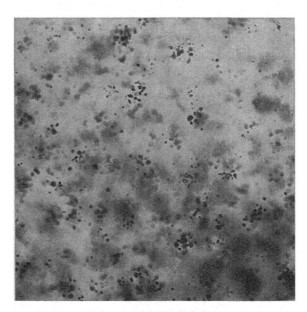

Figure 4.22 Optical transmission microstructure of $Nd:ThO_2-Y_2O_3$ ceramic prepared by Dr. Greskovich in 1974.

first laser oscillation using $Nd:ThO_2-Y_2O_3$ ceramics developed by Dr. Greskovich in 1974, and compare this with the characteristics of current Y_2O_3 ceramic laser materials. After 20~30 years since its invention, the fabrication technology of ceramic laser materials has progressed remarkably.

A transmitted microscopy image of the first $Nd:ThO_2-Y_2O_3$ ceramic, which produced successful pulse laser generation in 1974, is shown in Figure 4.22. Many residual pores and inclusions, called "orange peel," are observed in the material. The optical loss of this material and its advanced type was around 10–20% cm^{-1}. The optical loss of commercial Nd:YAG single crystal laser materials is below 0.2% cm^{-1}. As seen from this difference, it is clear that the optical quality of the Y_2O_3 ceramics developed in 1974 was not sufficient for use as a laser material.

The pulse laser characteristics of the $Nd:ThO_2-Y_2O_3$ ceramic pumped with a flashlamp is shown in Figure 4.23. While laser oscillation was successful, the laser oscillation efficiency was only about 0.1%. After 1974, the optical quality of the materials was improved and laser oscillations by flashlamp pumping were reported. However, the laser oscillation efficiencies were not sufficient to use these materials as laser gain media.

In Figure 4.24(a), (b), and (c), the appearance of large $Er:Y_2O_3$ ceramics (diameter 40 mm × thickness 8 mm), the birefringence distribution observed under a polarizer, and a wavefront image (refractive index distribution) measured by a Twiman–Green interferometer are shown, respectively. The optical transparency of the prepared $Er:Y_2O_3$ ceramics was very high, comparable to that of the single crystal counterpart. No double refractions were observed under a polarizer. Almost straight fringes were observed by interferometry,

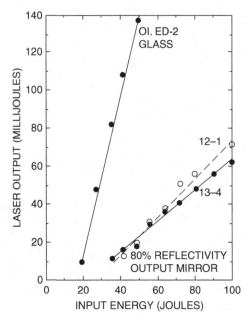

Figure 4.23 Pulse laser characteristics of Nd:ThO$_2$–Y$_2$O$_3$ ceramic pumped by a flashlamp in 1974. Reprinted with permission [1], copyright 1973, American Institute of Physics.

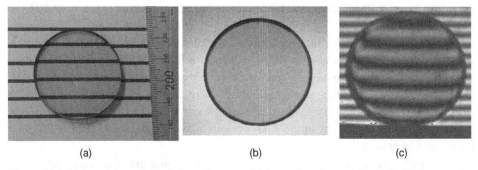

Figure 4.24 (a) Appearance, (b) polarizing image and (c) wavefront image of large Er:Y$_2$O$_3$ ceramic.

suggesting that the uniformity of the refractive index was very high throughout the materials. In addition, it was confirmed that the materials did not include any residual pores, only several inclusions were observed under an optical microscope.

The grain boundary region of the Er:Y$_2$O$_3$ ceramics was observed by SEM. As shown in Figure 4.25, the average grain size was within a few micrometers. Segregation of Er and sintering aids was not observed inside grains or at grain boundary regions.

Figure 4.26(a) shows a TEM image of a grain boundary of the Er:Y$_2$O$_3$ ceramic. A clear grain boundary, where two crystal orientations meet, with regular arrangement of lattice

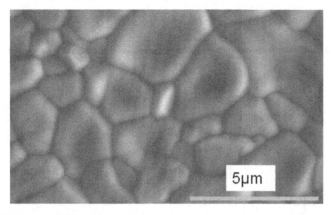

Figure 4.25 SEM image of Er:Y_2O_3 ceramic.

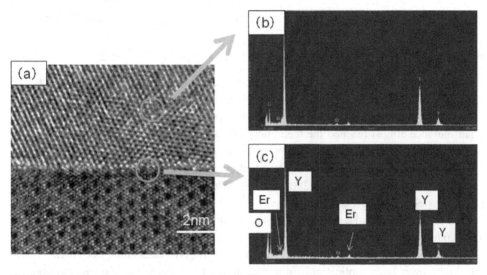

Figure 4.26 (a) Lattice structure near a grain boundary, and EDS analysis for (b) an inner grain and (c) a grain boundary of Er:Y_2O_3 ceramic.

structure, was observed without any secondary phases. EDS analysis results for an inner grain and a grain boundary of the Er:Y_2O_3 ceramic are shown in Figure 4.26(b) and (c), respectively. An electron beam probe 1 nm in diameter was irradiated onto an inner grain and grain boundary area for the EDS analysis. Only Er, Y and O elements were detected in the inner grain and grain boundary areas, and no sintering aids were observed in any regions of the material. Accordingly, there was no segregation due to the addition of a small amount of sintering aid, suggesting that the sintering elements are dissolved uniformly in the host materials.

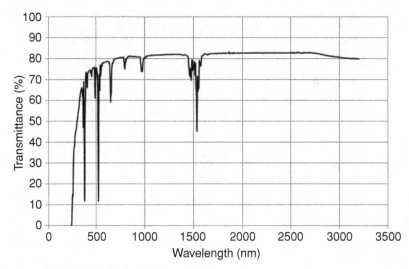

Figure 4.27 Transmission spectrum of 0.25 at.% Er:Y_2O_3 ceramic (9 mm thickness) between 200 and 3200 nm.

In-line transmittance of the Er:Y_2O_3 ceramics for wavelengths ranging from 200 to 3200 nm is shown in Figure 4.27. In the short wavelength region (i.e., visible region), the in-line transmittance was lowered due to absorption or scattering. But the in-line transmittance almost reached the theoretical value (about 82%) in the longer wavelength regions from 800–1000 nm, and the optical loss was very low.

A schematic diagram of the laser setup and the laser performance of 1.5 at.% Nd:Y_2O_3 ceramics are shown in Figure 4.28(a) and (b). When it was pumped with an 808 nm LD (laser diode) at 742 mW, an output power of 160 mW was achieved. In this case, the slope efficiency was 32%, and simultaneous laser oscillation of two wavelengths (1074.6 nm and 1078.6 nm) was observed [19].

In the case of Yb system laser materials, mainly disk type lasers have been investigated for high power laser applications. In particular, Yb:Lu_2O_3 ceramic is being investigated because its thermal conductivity does not decrease significantly with the addition of Yb ions. When it was pumped at 45.3 W, an output power of 32.6 W was achieved (slope efficiency 80%). Wavelength tunability was also suggested in this laser because it has a broad bandwidth (90 nm) at lasing wavelength [25]. Using this high output laser head, an ultra-short pulse femtosecond laser (pulse width 329 fs) with an average output of 40 W has also been realized [26].

However, high power laser generation using Yb:RE_2O_3 ceramics has not yet been achieved. Generally, the fracture strength of ceramic materials is higher than that of single crystal counterparts if they have the same crystal structure. Therefore, sesquioxide ceramic materials are expected to feature in the development of high power lasers in the future. Currently, ultra-short pulse laser generation based on sesquioxide materials has been

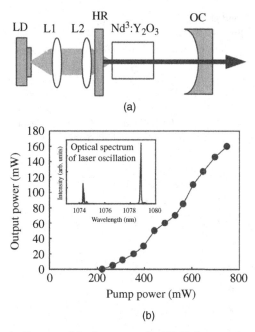

(a)

(b)

Figure 4.28 (a) Schematic diagram of the laser setup for Nd:Y$_2$O$_3$ ceramics. (b) Input–output curve for 1.5 at.% Nd:Y$_2$O$_3$ ceramic generated at two wavelengths of the $^4F_{3/2}$ to $^4I_{11/2}$ channel.

achieved. A pulse output power of 352 mW with 357 fs pulse width has been reported [27].

One of the advantages of ceramic lasers is that composite laser gain media can be easily fabricated compared to single crystal lasers. A schematic diagram of a setup for an ultra-short pulse laser using a sesquioxide composite made of Yb:Y$_2$O$_3$ and Yb:Sc$_2$O$_3$ ceramics is shown in Figure 4.29. Since the refractive indexes of Yb:Y$_2$O$_3$ and Yb:Sc$_2$O$_3$ are almost equal to each other, the Fresnel reflectance loss at the bonding interface can be reduced. Yb:Y$_2$O$_3$ and Yb:Sc$_2$O$_3$ have different center wavelengths and broad bandwidths so that the laser spectrum of such a composite can be significantly broader compared to conventional single structure laser gain media. As a result, a surprising ultra-short pulse laser generation of 53 fs was successfully demonstrated [28].

Absorption and emission cross-sections of Er:Sc$_2$O$_3$ ceramic at liquid nitrogen temperature (77 K) are shown in Figure 4.30 (see the energy diagram of Er:Sc$_2$O$_3$ ceramic shown in Figure 4.7). When it was excited with a fiber laser of 1535 nm wavelength, three fluorescence lines of 1604, 1581, and 1558 nm were generated.

Figure 4.31(a) and (b) show the laser oscillator setup and laser performance of Er:Sc$_2$O$_3$ ceramic at 77 K, respectively. The laser gain medium used was 0.25 at.% Er:Sc$_2$O$_3$ ceramic with a dimension of 9 × 2 × 5 mm^3. Both faces of the laser output surface were AR (anti-reflection) coated, and the laser oscillation experiment was performed under cryostatic conditions (cooled with liquid nitrogen). A fiber laser of wavelength 1535 nm was used

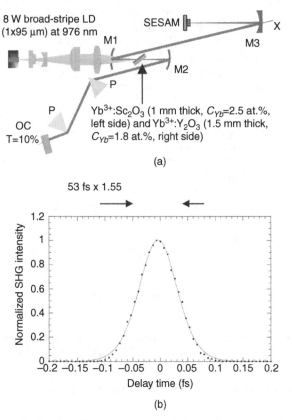

(a)

(b)

Figure 4.29 (a) Schematic diagram of the laser setup for Yb-doped Lu_2O_3. (b) Autocorrelation trace of 53 fs pulses with an average power of 1 W [28].

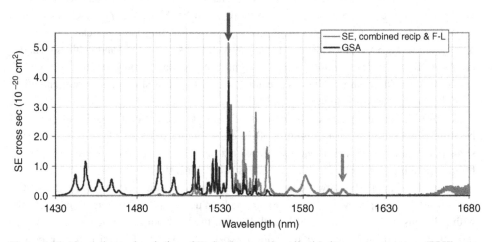

Figure 4.30 Absorption and emission of $Er:Sc_2O_3$ ceramic at liquid nitrogen temperature (77 K).

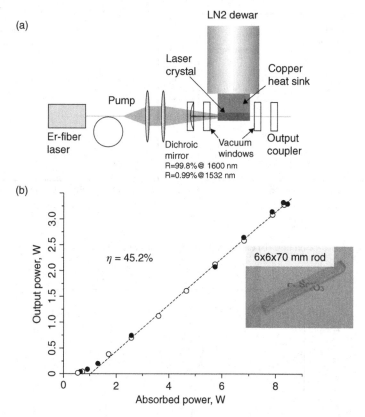

Figure 4.31 (a) Laser oscillator setup and (b) laser performance at 77 K for Er:Sc$_2$O$_3$ ceramic.

as an excitation source, and the output coupler used was $R = 96.6\%$ at 1558 nm (lasing wavelength). In this Er:Sc$_2$O$_3$ ceramic, the quantum defect was as low as 1.5%. Taking advantage of this low quantum defect, laser oscillation at 1558 nm (maximum output 3.3 W) with very high slope efficiency (45.2%) was realized [29].

Recently it has been found that the spectral bandwidth of the absorption and emission spectra of Nd-doped Y$_2$O$_3$ ceramics can be broadened by controlling the crystallite symmetry of the laser host material (i.e., Y$_2$O$_3$). Normally, the spectral bandwidth of Nd:Y$_2$O$_3$ is very narrow. Control of the spectral bandwidth enables many improvements in laser technology: (1) solving the technical problem of wavelength shifting and increasing spectral bandwidth of LDs during high power laser operation (due to strong excitation), (2) providing a high tunability laser, and (3) easy generation of short pulse lasers etc.

Figure 4.32(a) and (b) show absorption and emission spectra of normal and broad band type Nd:Y$_2$O$_3$ ceramics, respectively [30]. In the normal type ceramic, only narrow spectra were observed, but in the advanced type, broad spectra were confirmed, analogous to amorphous (glass) materials.

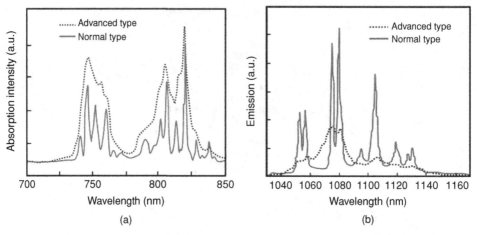

Figure 4.32 (a) Absorption and (b) emission spectra of normal and advanced Nd:Y$_2$O$_3$ ceramics.

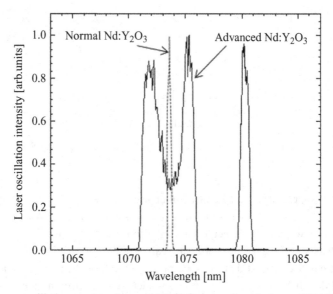

Figure 4.33 Laser oscillation spectra of normal Nd:Y$_2$O$_3$ ceramic and advanced Nd:Y$_2$O$_3$ ceramic.

Laser oscillation spectra of the above described normal and advanced Nd:Y$_2$O$_3$ ceramics are shown in Figure 4.33. Normal Nd:Y$_2$O$_3$ ceramic shows a very narrow spectral line (about 0.5 nm wide) and its laser oscillation wavelength is in the range 1073–1074 nm. In the advanced type, the spectral line width was about ten times broader than in the normal type [31].

Once the technology to control the spectral line width is established, the development of short pulse lasers (picosecond femtosecond) using Nd laser materials will accelerate.

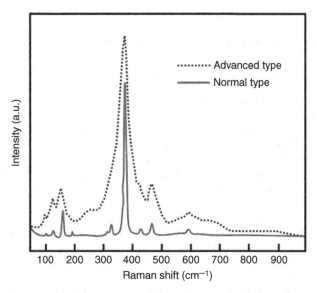

Figure 4.34 Raman spectra of normal and advanced Nd:Y$_2$O$_3$ ceramics.

Raman spectra for the above normal and advanced ceramics are shown in Figure 4.34. Although the XRD (X-ray diffraction) patterns of these two materials are similar – that is there is no difference in the crystal structure – the Raman spectra (phonon scattering) show that the peak positions were the same as each other but the peak widths (condition of phonon scattering) were different. Scientific understanding of this result is not clear yet but it can be considered that this difference will depend on the crystallite symmetry of the materials. Accordingly, advanced ceramic technology permits the control of ceramic materials at the atomic lattice level, and it is anticipated that this technology can enhance the spectroscopic properties of laser materials as well as the features of lasers.

In the above, methods for the fabrication of sesquioxide ceramics and their basic optical properties were described. Some sesquioxide ceramics have shown good laser performance. The deformation property (creep) of sesquioxide materials is larger than that of YAG materials at high temperature. Therefore, it is easier to fabricate large scale laser gain media by high pressure sintering, and sesquioxide ceramic laser materials are expected to feature in the development of high power lasers in the future.

References

[1] C. Greskovich and J. P. Chernoch, Polycrystalline ceramic lasers, *J. Appl. Phys.*, **44** (10) (1973) 4599–4605.
[2] T. Sekino and Y. Sogabe, *Rev. Laser Eng.* **21** (8) (1993) 827–831.
[3] F. Takei, S. Takasu, J. Ushizawa, and M. Sakurai, *Toshiba Rev.* **24** (12) (1969) 1–9.
[4] N. C. Chang, *J. Appl. Phys.* **34** (12) (1963) 3500–3504.
[5] R. A. Lefever, *Rev. Sci. Instrum.* **33** (1962) 1470.

[6] J. E. Geusic, H. M. Marcos, and L. G. Van Uitert, *Appl. Phys. Lett.* **4** (10) (1964) 182–184.

[7] G. Adachi, Physics and chemistry of yttrium compounds, *Ceramics* **23** (5) (1988) 430–436.

[8] R. Jagannathan, T. R. N. Kutty, M. Kottaisamy, and P. Jeyagopal, *Jpn. J. Appl. Phys.* **33** (1994) 6207.

[9] B. M. Walsh, J. M. McMahon, W. C. Edwards, N. P. Barnes, R. W. Equall, and R. L. Hutcheson, *J. Opt. Soc. Am. B* **19** (2002) 2893.

[10] P. H. Klein and W. J. Croft, *J. Appl. Phys.* **38** (1967) 603.

[11] M. Larionov, J. Gao, S. Erhard, A. Giesen, K. Kontag, V. Peters, E. Mix, L. Fornasiero, K. Petermann, G. Huber, J. A. der Au, G. J. Spuhler, F. Brunner, R. Paschotta, U. Keller, A. A. Lagatsky, A. Abdolvand, and N. V. Kuleshov, *OSA TOPS* **50** (2001) 625.

[12] Y. Sato, J. Akiyama, and T. Taira, *Opt. Mater.* **31** (2009) 720.

[13] K. Beil, S. T. Fredrich-Thornton, R. Peters, K. Petermann and G. Huber, *ASSP2009, WB28*, Denver, Co, USA.

[14] Y. Sato, I. Shoji, K. Kurimura, T. Taira, and A. Ikesue, *OSA TOPS* **50** (2001) 417.

[15] T. Förster, *Ann. Phys.* **2** (1948) 55.

[16] V. Lupei, A. Lupei, S. Georgescu, T. Taira, Y. Sato, and A. Ikesue, *Phys. Rev. B* **64** (2001) 092102.

[17] A. Lupei, V. Lupei, T. Taira, Y. Sato, A. Ikesue, and C. Gheorghe, *J. Lumin.* **72** (2003) 102–103.

[18] J. A. Williams-Byrd, L. B. Petway, W. C. Edwards, and M. Turner, *CLEO2003*, Baltimore, MD, USA (2003), paper CWG5.

[19] J. Lu, J. Lu, T. Murai, K. Takaichi, T. Uematsu, K. Ueda, H. Yagi, T. Yanagitani, and A. A. Kaminskii, *Jpn. J. Appl. Phys.* **40** (2001) L1277.

[20] M. Tokurakawa, A. Shirakawa, K. Ueda, H. Yagi, M. Noriyuki, T. Yanagitani, and A. A. Kaminskii, *Opt. Express* **17** (5) (2009) 3353–3361.

[21] G. Huber, W. Duczynski, and K. Petermann, *IEEE J. Quantum Electron.* **24** (1988) 920.

[22] Y. Sato, T. Taira, N. Pavel, and V. Lupei, *Appl. Phys. Lett.* **82** (2003) 844.

[23] N. Ter-Gabrielyan, L. D. Merkle, A. Ikesue, and M. Dubinskii, *Opt. Lett.* **33** (2008) 1524–1526.

[24] A. Ikesue, K. Kamata, and K. Yoshida, Synthesis of transparent Nd-doped HfO_2–Y_2O_3 ceramics using HIP, *J. Am. Ceram. Soc.* **79** (2) (1996) 359–364.

[25] R. Peters, C. Kraenkel, K. Petermann, and G. Huber, *Opt. Express* **15** (2007) 7075.

[26] C. R. E. Baer, C. Kraenkel, C. J. Saraceno, O. H. Heckl, M. Golling, T. Suedmeyer, R. Peters, K. Petermann, G. Huber, and U. Keller, *Opt. Lett.* **34** (2009) 2823.

[27] M. Tokurakawa, K. Takaichi, A. Shirakawa, K. Ueda, H. Yagi, S. Hosokawa, T. Yanagitani, and A. A. Kaminskii, *Opt. Express* **14** (2006) 12832.

[28] M. Tokurakawa, A. Shirakawa, K. Ueda, H. Yagi, M. Noriyuki, T. Yanagitani, and A. A. Kaminskii, *Opt. Express* **17** (2009) 3353.

[29] N. Ter-Gabrielyan, L. D. Merkle, A. Ikesue, and M. Dubinskii, Ultralow quantum-defect eye-safe $Er:Sc_2O_3$ Lasers, *Opt. Lett.* **33** (13) (2008) 1524–1526.

[30] A. Ikesue, *Rev. Laser Eng.* **37** (4) (2009) 248–253.

[31] T. Yoda, S. Miyamoto, H. Tsuboya, A. Ikesue, and K. Yoshida, *CLEO 2004*, San Francisco, Poster Session (2004), paper CThT59.

5

Synthesis of RE (Nd) heavily doped YAG ceramics

The Nd element is a four-level laser, which is incorporated into garnet, vanadate, and so on as the most general laser active ion. With regard to YAG crystal, the main laser host material, the Nd ion is substituted for the Y ion in the 8 coordination site of the host crystal. However, because Nd has a large ionic radius, it is not easy to produce Nd-doped homogeneous YAG single crystals using traditional melt-growth technology, the CZ method. Figure 5.1 shows the relationship between the ionic radius of rare earth elements and the segregation coefficient in the production of YAG single crystals by the melt-growth method. The value of segregation coefficient indicates the level of solubility of rare earth ions (the doping ions) in YAG single crystal as solid solution. The dopant can be dissolved and at a high concentration homogeneously when the value is 1. When the value is greater than 1, the concentration of doping ions is higher than the designed concentration at the beginning, then it becomes lower gradually as the crystal grows. On the contrary, when the segregation coefficient is smaller than 1, the solubility of doped ions in the YAG crystal is limited, it is lower than the designed concentration at the beginning and then it increases as the crystal grows. Therefore in most cases, it is very difficult to dissolve rare earth ions into the host material homogeneously. In the case of YAG crystal, Nd is used widely as a laser active ion. Since the segregation coefficient of Nd in YAG crystal is around 0.2, the maximum amount of Nd ions that can dissolve homogeneously in the YAG host crystal is about 1 at.%. When an amount greater than 1 at.% is doped, inclusions (secondary phases) are generated, and it is difficult to obtain a single crystal of sufficient optical grade for laser oscillation.

If Nd ions can be heavily and homogeneously doped into YAG materials, then this can contribute to the development of new laser functions such as enhancement of beam quality by improvement of the optical quality, production of microchips by densification, or the generation of single-mode lasers and so on. However, conventional single crystal growth technology has not succeeded in overcoming this problem. On the other hand, a polycrystalline laser medium produced by a ceramics process does not have these intrinsic problems derived from the synthesis method, allowing laser active ions to be dissolved homogeneously and at a high concentration. Since the melt-growth method introduces an interfacial surface (between solid and liquid phases), segregation of the doped ions may

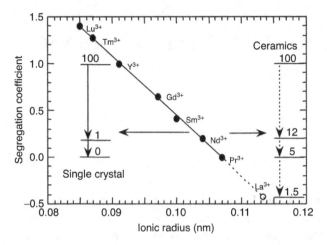

Figure 5.1 Relationship between the ionic radius of rare earth elements and the segregation coefficient.

occur. In contrast, the ceramic process does not lead to segregation of doping ions at the phase boundary between solid and liquid phases because sintering progresses by a solid-state reaction, diffusion and grain growth. Therefore, if polycrystalline laser gain media of optical grade can be produced by sintering, it will become possible to solve one of the problems of traditional laser technology.

This chapter describes how a high concentration Nd:YAG ceramic was easily produced using sintering technology. Section 5.1 introduces a production method for heavily doped Nd:YAG ceramic, its basic characteristics, and typical lasing characteristics. Additionally, it was revealed that SiO_2 as a sintering aid plays an important role in producing transparent and high concentration YAG ceramics. Section 5.2 describes the formation of high concentration Nd ceramic with optical homogeneity.

5.1 Production of heavily doped Nd:YAG and lasing characteristics

5.1.1 Abstract

Since a solid laser using Nd:YAG single crystal was first discovered in 1964 by the group of Guesic, many laser materials have been developed and some of them have been put to practical use. However, YAG is still the main solid-state laser material. Although the production of YAG single crystals started in 1964 using the Verneuil process as the easiest method, the CZ process is currently being utilized because of the very strict requirements for laser material. However, even the CZ process often does not produce single crystals of sufficient quality to be used as laser material. Recently, the growth process has become computer controlled [1] but a production technology which is able to solve the intrinsic problem of Nd:YAG single crystals (the formation of optically inhomogeneous parts such as a core and facets) has still not been developed.

Although it is relatively easy to produce homogeneous YAG single crystals without Nd, even under irregular growth conditions, the problems common to all single crystal growth methods, including the Verneuil and Bridgman processes in the case of Nd doped YAG single crystals, still remain. The problems [2–4] are principally classified as follows. (1) YAG single crystal doped with less than approximately 1 at.% Nd easily forms facets toward the rim area from the rotational axis and thus the optically homogeneous part is limited. (2) Only a small part of YAG single crystal doped with greater than approximately 1 at.% Nd can be utilized as laser material because segregations such as facets or inclusions become obvious owing to unstable dissolution of Nd. (Of course, although the Flax process can produce single crystals doped with a large amount of Nd, the optical quality is not suitable for laser application because it is impossible to produce a specific crystallite orientation and large scale crystal.) For the former problem, it is possible to remove the optically homogeneous parts from the grown crystal for laser application, but it is difficult to produce a large volume and the productivity is low. For the latter problem, although the synthesis of heavily doped Nd:YAG single crystal has not been achieved yet, it would be remarkably useful for applications demanding high beam quality such as compact laser systems for microchip development or single-mode oscillation by a substantial increase in the concentration of laser active ions (increase in absorbing coefficient for gain media and increase in laser gain). However, a disadvantage is the decrease in fluorescence lifetime [5] by concentration quenching. We have demonstrated direct oscillation of a single-mode laser without applying the etalon system of a conventional laser system. Although it is extremely important for laser applications to obtain a high energy conversion efficiency and to satisfy the demand for compact size and high output power, no breakthrough has been reported using YAG single crystal. Recently, vanadate single crystals with high absorption coefficient such as $Nd:YVO_4$ and $Nd:GdVO_4$ have been developed. It was reported that the technical problems of Nd:YAG single crystals mentioned above can be overcome with these vanadate crystals. However, the thermal conductivity of vanadate single crystal is about only half that of YAG, and its mechanical strength is insufficient.

Using the segregation coefficient, it is possible to determine whether laser active ions (Nd) can dissolve in YAG single crystal or not. Shiraki and Kuwano [6] and Monchamp [7] reported that the segregation coefficient of Nd in YAG is around 0.2, which means only approximately 1 at.% Nd can dissolve in YAG single crystal. When the segregation coefficient is less than 1, it is difficult to form a solid solution with the rare earth ion in YAG host crystal. This is mainly because of the large difference in the ionic radii of Y^{3+} and Nd^{3+}. (The Nd^{3+} ion is larger than the Y^{3+} ion [8] when Nd is substituted for Y in the YAG crystallite lattice.) Since YAG has a highly symmetrical crystallite lattice, only rare earth elements which have an ionic radius equal to or slightly smaller than Y^{3+} can be dissolved stably.

Theoretical analysis of Nd solid melt obtained from single crystal technology in the past has concluded that it is difficult to solve the problem caused by the limitation in crystallite lattice size in YAG. It is expected that the same problem will also occur in

YAG ceramics which are different in microstructure from single crystal (presence of grain boundaries) when a specimen containing a high concentration of Nd is prepared. This chapter introduces an unknown potential laser material which was impossible to produce using single crystal growth technology, namely the synthesis of transparent YAG ceramics and the optical and lasing characteristics of the specimens obtained.

5.1.2 Experimental

In this experiment, 99.99 mass% high purity Al_2O_3, Y_2O_3, and Nd_2O_3 powders were used, as described in previous chapters. The amount of each powder was adjusted to the YAG composition, the Nd concentration was weighed in the range 0.3–4.8 at.%, and 0.5 mass% TEOS was added. A suitable amount of ethanol was added to the powders, and they were mixed using a ball mill with high purity alumina balls for 12 hours. The prepared slurry was dried by atomizing using a spray drier; 50 μm diameter spherical granulated powder was obtained. Details such as the molding method were the same procedures as described in previous chapters. Green powder compact after pressing with CIP was sintered at 1750 °C for 20 hours under vacuum (1.3×10^{-3} Pa). The microstructure of sintered specimens was observed using reflection and transmission polarized optical microscopy, and SEM (EPMA).

The amount of Nd in the specimens was analyzed qualitatively and quantitatively by spectrum analysis with EDX or spectral photometry. The absorption behavior of the specimens in the range 400–1000 nm was also measured by spectral photometry with 0.5 nm slit width at 120 nm min^{-1} scanning rate. Fluorescence spectra were detected using an Si photodiode at 1064 nm with 800 nm beam split from a 150 W xenon lamp. Moreover, the duration of fluorescence for Nd ions was determined by a high speed oscilloscope (detector PbS) at 1064 nm after light flashing irradiation generated by a 500 W xenon flashlamp as light source. CW laser emitted from an output mirror as laser oscillation experiment was measured by a light power meter using an optical resonator as described in the previous chapter. Commercial 0.9 at.% Nd:YAG single crystal was applied as a standard specimen for general optical and lasing characteristics.

5.1.3 Results

Figure 5.2 shows micrographs taken by a reflection microscope of Nd:YAG ceramics (after thermal etching) with various Nd concentrations (0.3, 1.1, 2.4, and 4.8 at.%) sintered at 1750 °C for 20 hours under vacuum. Although all of the specimens consist of relatively homogeneous grains, several tens of micrometers in size, the average grain size has a tendency to decrease slightly as the Nd concentration increases. No pores and also no precipitates (secondary phases) as a result of excess addition of Nd (greater than 1 at.%) were observed under an optical microscope.

Figure 5.3(a) is a micrograph taken by a transmitting polarizing microscope of 4.8 at.% Nd:YAG ceramic. There are almost no remaining pores along the depth of the specimen under open-nicol and the specimen gives a complete dark-field under cross-nicol

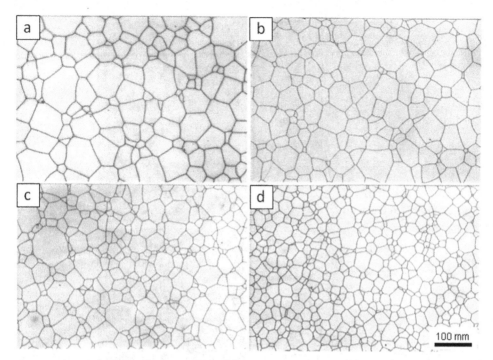

Figure 5.2 Surface structure of transparent Nd:YAG ceramics after thermal etching. The Nd concentrations are (a) 0.3, (b) 1.1, (c) 2.4, (d) 4.8 at.%, respectively.

observations. This reveals that optically homogeneous Nd:YAG ceramics can be synthesized without pores or precipitates even with high Nd concentration. Figure 5.3(b) is a micrograph taken with a transmission polarization microscope of Nd:YAG single crystal with the same composition prepared by the FZ (float zone) method. A lot of elongated inclusions several tens of micrometers in diameter are observed inside the single crystal. EDX (energy dispersive X-ray) and XRD (X-ray diffractometry) revealed that these inclusions were perovskite dissolving a lot of Nd as solid melt ($NdYAlO_3$). It is found that synthesizing optically homogeneous Nd:YAG containing a high concentration of Nd is not easy, even using the FZ process.

Figure 5.4 indicates the line profile result analyzed by EPMA of Nd:YAG containing 4.8 at.% Nd after thermal etching of the polished surface of the specimen. Although added Si as sintering aid is not detected because it was below the detection limit (320 mass ppm by ICP measurement), there is almost no difference between the inside grain and grain boundary in element distribution for Al, Y, and Nd. (The Nd distribution was also investigated using TEM-EDS; Nd is considered to dissolve into inner crystallite grains homogeneously because no Nd segregation was observed along grain boundaries.)

Table 5.1 indicates the results of spot analysis using EDX of particles and grain boundaries chosen randomly in 2.4 at.% Nd:YAG ceramics. Five spots are measured inside and at the boundary seen. It is seen that a homogeneous sintered body was formed

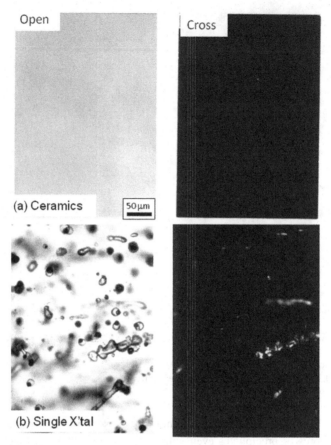

Figure 5.3 Open and cross-nicol photographs of 4.8 at.% Nd:YAG ceramics observed by a polarization microscope.

microstructurally and compositionally because there is almost no difference between the analysis data inside and at the boundary of particles.

Figure 5.5 shows the relationship between Nd concentration added to YAG ceramics and the amount of Nd detected from constituent particles of Nd:YAG ceramics using EDX analysis. Since the Nd concentration in constituent particles of Nd:YAG ceramics increases linearly with increasing Nd addition, it is found that Nd is distributed homogeneously in particles.

Figure 5.6 shows X-ray diffraction patterns of powder of 4.8 at.% Nd:YAG ceramic. No secondary phases except YAG (garnet) were detected, meaning that a homogeneous sintered body formed compositionally as determined by X-ray diffraction (scale larger than the X-ray wavelength).

Figure 5.7 shows absorption spectra (specimens 1 mm thick) of 0.9 and 4.8 at.% Nd:YAG. Background levels of absorption are lower than 0.1 for all the specimens. The host materials

Table 5.1 *Spot analysis inside particles and at the grain boundary of 2.4 at.% Nd:YAG ceramics using EDX*

	Grain				Grain boundary		
	Al_2O_3 (mass%)	Y_2O_3 (mass%)	Nd_2O_3 (mass%)		Al_2O_3 (mass%)	Y_2O_3 (mass%)	Nd_2O_3 (mass%)
Spot 1	42.49	55.21	2.30	Spot 6	42.57	54.98	2.45
Spot 2	42.50	55.19	2.31	Spot 7	42.55	55.01	2.44
Spot 3	42.53	55.08	2.39	Spot 8	42.49	55.21	2.30
Spot 4	42.51	55.25	2.24	Spot 9	42.55	55.04	2.41
Spot 5	42.50	55.19	2.31	Spot 10	42.52	55.15	2.33

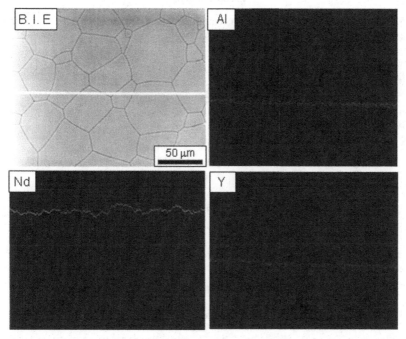

Figure 5.4 Backscattering photographs of 2.4 at.% Nd:YAG ceramics with line profiles for Y, Al, and Nd elements by EPMA (full scale Y, Al 5×10^3 pcs, Nd 1×10^2 pcs).

have quite good transparency after subtracting Nd absorption. Absorption by Nd ions in the 4.8 at.% Nd:YAG ceramic is much larger than that in the 0.9 at.% Nd:YAG ceramic.

Figure 5.8 shows the $^4F_{5/2}$ band of the absorption spectrum for 1 and 4.8 at.% Nd:YAG. The absorption coefficient of 1 at.% Nd:YAG is 8.5 cm^{-1} which is equal to the value for the same composition of Nd:YAG single crystal; that of 4.8 at.% Nd:YAG was 40.8 cm^{-1}. Since the absorption coefficient of the $^4F_{5/2}$ band increases in proportion to the increase

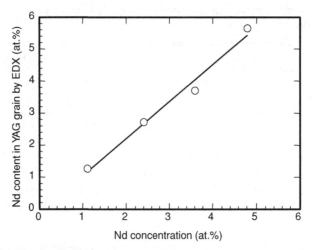

Figure 5.5 The relationship between Nd concentration (calculated content) and the amount of Nd detected in Nd:YAG ceramics using EDX analysis.

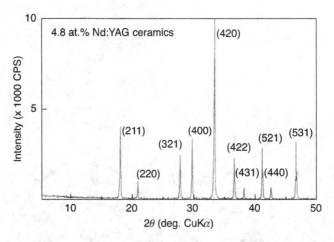

Figure 5.6 X-ray diffraction pattern of 4.8 at.% Nd:YAG ceramic.

in Nd concentration, spectrometry also gives proof that Nd is dissolving homogeneously in the YAG ceramics. In the case of Nd:YAG single crystal, the absorption coefficient is only around 8.5 cm^{-1} but Nd:YAG ceramic is improved with respect to the single crystal technology. Nd:YAG ceramics with a large absorption coefficient are contributing to laser technology and can achieve single longitudinal mode oscillation by a microchip laser or a compact resonance configuration as discussed below.

Figure 5.9 shows the amount of fluorescence at 1064 nm when Nd:YAG ceramics with varied Nd concentration are excited by a wavelength of 800 nm of spectrum split from a 150 W xenon lamp. The exponent of the vertical axis indicates the fluorescence intensity,

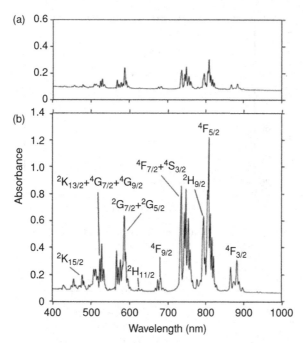

Figure 5.7 Absorption spectra for (a) 0.9 at.% Nd:YAG ceramic and (b) 4.8 at.% Nd:YAG ceramic from the visible to the near infrared wavelength region.

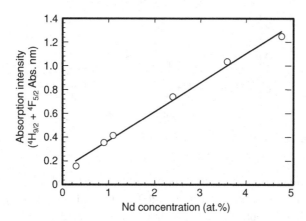

Figure 5.8 Relationship between Nd concentration and integrated absorption intensities of $^4H_{9/2} \rightarrow {}^4F_{5/2}$ bands.

with the intensity of a specimen of 0.9 at.% Nd:YAG single crystal defined as 100. The fluorescence intensity of specimens increases rapidly up to 1.1 at.% Nd concentration and reaches a maximum at 2.4 at.%. Then, it starts to decrease gradually at high Nd concentration above 2.4 at.%. This reason is the occurrence of concentration quenching for Nd ions as

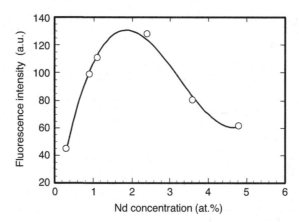

Figure 5.9 Fluorescence intensities of YAG ceramics with from 0.3 to 4.8 at.% Nd content excited by a xenon lamp at room temperature.

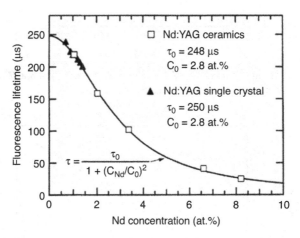

Figure 5.10 Dependence of fluorescence lifetime at 1064 nm on the Nd concentration in YAG ceramics.

the Nd concentration increases, resulting in shorter lifetime for each Nd ion. However, the maximum level of fluorescence occurs around 2–3 at.% Nd concentration, so producing YAG ceramics with this optimal Nd concentration would make it possible to obtain laser oscillation which is more powerful than that of Nd:YAG single crystals.

Figure 5.10 shows the fluorescence lifetime at 1064 nm using six types of YAG ceramic with varying Nd concentration. The fluorescence lifetime in the non-concentration quenching region (Nd concentration less than 1 at.% approximately) of Nd:YAG ceramics is in the range 230–205 μs. It can be concluded that the lifetime is comparable to 230–260 μs generated from 1 at.% Nd:YAG single crystal which has been reported. The Nd:YAG ceramics containing greater than 1 at.% Nd have a similar pattern compared with Nd:YAG single crystal because of the occurrence of concentration quenching (interaction among

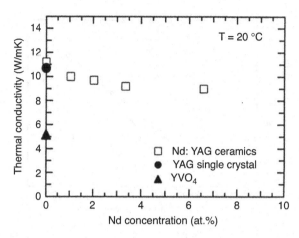

Figure 5.11 Thermal conductivity of YAG ceramics as a function of Nd concentration.

spins of Nd ions closer together). In contrast to single crystal material, the ceramics can work as lasing devices with high concentrations of Nd ions because there are no inclusions or optically inhomogeneous parts.

Figure 5.11 shows the thermal conductivity measured by a laser flash method for single crystal and ceramic specimens of YAG fabricated into samples of diameter 10 mm and thickness 2 mm. Although the thermal conductivity of ceramics deteriorates as the Nd concentration increases, even for 9.1 at.% Nd it reaches 9 W m^{-1} k^{-1}. Recently, vanadate single crystals (Nd:YVO$_4$, Nd:GdVO$_4$, etc.) with high absorption coefficient and good lasing efficiency have been attracting attention as microchip lasers, but the absorption coefficient of high concentration Nd:YAG is comparable or superior. Additionally, the high thermal conductivity (for Nd:YVO$_4$ single crystal this is $\sim$5 W m^{-1} k^{-1}) which is greater than 9 W m^{-1} k^{-1} as shown in this figure for high concentration Nd:YAG gives it the advantages of good chemical stability and mechanical characteristics compared with vanadate crystals.

Figure 5.12 shows the lasing characteristics for 1.1, 2.4, 4.8 at.% Nd:YAG ceramics, and 0.9 at.% Nd:YAG single crystal. None of the measured specimens were covered with an anti-reflective coating (AR coating). These data were measured in the 1990s. Although effective laser oscillation was not achieved due to insufficient LD excitation technology available to the author and his co-researchers in those days, it is possible to compare the relative material characteristics. The threshold value of laser oscillation for commercial 0.9 at.% Nd:YAG single crystal is approximately 280 mW (slope efficiency 24%) which is lower than the value for all the Nd:YAG ceramics. Although the threshold vales of 1.1 and 2.4 at.% Nd:YAG ceramics are much higher than that of 0.9 at.% Nd:YAG single crystal, the output value becomes higher than that of 0.9 at.% Nd:YAG single crystal if the input energy increases. The threshold value for laser oscillation and the slope efficiency for 1.1 and 2.4 at.% Nd:YAG ceramics are 309 mW and 28%, 366 mW and 40%, respectively. In CW oscillation, the values for 4.8 at.% Nd:YAG ceramics containing a large amount

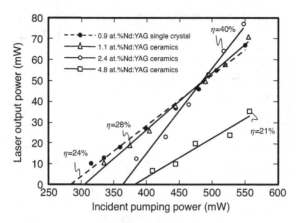

Figure 5.12 Laser output versus input energy for 1.1, 2.4, 4.8 at.% Nd:YAG ceramics, and 0.9 at.% Nd:YAG single crystal excited by an 808 nm diode laser.

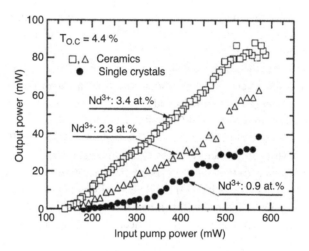

Figure 5.13 Performance of microchip lasers of 2.4, 3.6 at.% Nd:YAG ceramics, and commercial high quality 1 at.% Nd:YAG single crystal.

of Nd are 384 mW and 21%, inferior to the other specimens. The trend shown in laser oscillation is similar to the amount of fluorescence (due to the same CW excitation) at 1064 nm. Consequently, it is found that a higher fluorescence intensity gives good lasing characteristics. It can be confirmed that 1.1 and 2.4 at.% Nd:YAG ceramics are comparable or superior to 0.9 at.% Nd:YAG single crystal with respect to lasing characteristics under CW oscillation.

Figure 5.13 shows the lasing characteristics for specimens of 2.4, 3.6 at.% Nd:YAG ceramics, and commercial high quality 1 at.% Nd:YAG single crystal fabricated into less than 1 mm thick chip form. With increased excitation light at higher Nd concentrations,

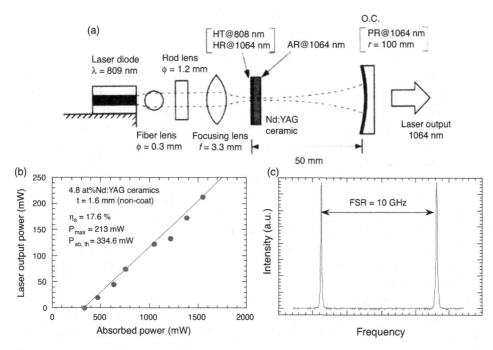

Figure 5.14 (a) Schematic of the diode pumped Nd:YAG ceramics laser. The surface of Nd:YAG ceramic was not covered with an AR coating (only optical polishing). (b) Single longitudinal mode output as a function of absorbed power. (c) Optical spectrum of single mode operation at 1064 nm measured by a confocal interferometer with 1 GHz free spectral range.

the oscillation characteristics of ceramic materials become better than those of single crystal.

Figure 5.14 shows a single longitudinal mode oscillation test using 4.8 at.% Nd:YAG. Relatively thick lasing media are used with ordinary Nd:YAG single crystal because of the low absorption coefficient of excitation, and an etalon is also generally applied as a mode selector. Although Nd:YAG single crystal can be used for single longitudinal mode oscillation, the lasing efficiency is low and the size of oscillator is big. Although Figure 5.14(a) shows a schematic of a resonator for single mode oscillation using high concentration Nd:YAG ceramic, thin media (1.6 mm thickness) can be used since this ceramic has a high absorption coefficient for exciting light (808 nm) which is 4.8 times greater than that of single crystal material, and it is able to generate single-mode oscillation without an etalon. Figure 5.14(b) shows the lasing characteristics of single-mode oscillation using 4.8 at.% Nd:YAG ceramic. The threshold value for oscillation is 334.6 mW and the slope efficiency is 17.6%. Additionally, Figure 5.14(c) shows the laser spectrum measured using confocal interferometry with 1 GHz free spectral range. The laser oscillated by polycrystalline ceramics is single spectrum. It was confirmed that ceramic material with grain boundaries can generate the highest coherency of lasing oscillation and the efficiency is also very high.

5.1.4 Discussion

5.1.4.1 Why Nd dissolves highly and homogeneously

As described before, it was easy to produce Nd:YAG ceramics having good optical characteristics derived from Al_2O_3, Y_2O_3, and Nd_2O_3 powders by a solid-state process. Although the added Nd concentration varied from 0.3 to 4.8 at.% in this experiment, all of the specimens were pore free and contained Nd homogeneously as solid-melt.

Since the segregation coefficient of Nd in YAG single crystal is very low (approximately 0.2) as mentioned in Secion 5.1.1, it is quite difficult to produce optically homogeneous YAG single crystal containing greater than around 1 at.% Nd by the CZ method. The optical characteristics of YAG single crystal containing greater than 1 at.% (maximum approximately 1.5 at.%) Nd deteriorate drastically, so it is difficult to use these materials in laser devices. If more than 1 at.% Nd is dissolved in YAG crystal, it is known that a precipitate is formed [1–3] inside the crystal of an Nd compound (primarily perovskite structure $(YNd)AlO_3$) even if the growth rate of the Nd:YAG single crystal is adjusted to lower than 0.5 mm^{-1} h (lower than 0.2 mm^{-1} h in some cases). According to Schneider's group, the ionic radius of lanthanoid series rare earth elements (R) forming stable aluminum garnet $(R_3Al_5O_{12})$ is 0.106 nm. The ionic radius of Nd^{3+} is 0.112 nm and the ionic radius of Y^{3+} is 0.102 nm, so comparison of the ionic radius shows that producing structurally homogeneous single crystals containing high concentrations of Nd is difficult. However, the results obtained for high concentration Nd:YAG ceramics in this experiment show that the existing problems of traditional single crystal growth can be solved to an extent.

It was confirmed that Nd can dissolve in ceramic material homogeneously in a concentration range that is impossible to produce using single crystal growth technology.

(1) The results from EPMA line analysis and EDX spot analysis reveal that the distribution of Al, Y, and Nd elements in the obtained Nd:YAG ceramic was homogeneous inside grains and at grain boundaries.

(2) Backscattering images also revealed that the Nd:YAG ceramic is uniform compositionally.

(3) EDX measurement of the Nd concentration in particles comprising the specimens revealed that the Nd concentration inside grains increased in proportion with the amount of Nd added to the Nd:YAG ceramics.

(4) Although the spectral linewidth is narrow and the optical absorption spectrum is sharp when Nd is dissolved into YAG crystal as solid melt (Nd substituted in the crystallite lattice of YAG), ceramics containing high Nd concentration also had sharp absorption lines and a large absorption coefficient.

(5) The excitation band of the Nd:YAG laser, namely the $^4F_{5/2}$ band of the obtained Nd:YAG ceramics showed increasing absorption coefficient in proportion to the quantity of Nd added and deterioration of the transparency of the specimens (elevation of background levels in the absorption spectrum) was not observed.

Judging from these facts, there is no reason why Nd does not dissolve in YAG ceramics homogeneously as solid-melt. Consequently, the added Nd is able to dissolve homogeneously in YAG ceramics up to 4.8 at.%, and it is concluded that the added Nd^{3+} ions substitute for Y^{3+} ions. As a result, YAG ceramics can contain five times more homogeneous dissolution of Nd than YAG single crystals.

This fact contradicts knowledge obtained from the basic research on YAG single crystal growth by the CZ method and from crystallographical investigation reported by Schneider's group. Nd:YAG single crystal obtained by the CZ method is produced at a high temperature which is higher than the melting point. Since Nd is considered as a kind of impurity in YAG crystal, the single crystal produced at high temperature can include more impurity inside the crystal, if focusing only on the amount of Nd dissolution. More than 1 at.% Nd is dissolved in YAG single crystal during crystal growth; excessive Nd cannot avoid forming inclusions or facets since the Nd:YAG single crystal becomes very unstable. It is considered that this effect (formation of impurity phase) can keep the YAG single crystal in a stable condition as host material. On the other hand, the synthesis temperature for ceramics by the solid phase method using Al_2O_3, Y_2O_3, and Nd_2O_3 powders is approximately 250–300 °C lower than the temperature for single crystal growth. This ought to be very disadvantageous for Nd dissolution in the solid-melt compared with single crystal. Therefore, the fact that YAG ceramic is able to dissolve a lot of Nd is not due to the temperature of the process.

The dissolution phenomenon of Nd as solid-melt originates in the difference in microstructure between single crystals and ceramics. Ceramics can be considered as an aggregation of small size grains (single crystals) having random crystallite orientation. There are grain boundaries at the crystallite interfacial surfaces; the adjacent grain boundary of each crystal has crystallite defects such as rearrangements compared with the center of the crystal. This produces a field at the adjacent grain boundary which is able to absorb the large stress caused by dissolution of Nd in YAG crystal. Although the solid-melt of Nd might also be explained by the effect of stress absorption at grain boundaries the following points contradict this idea.

(1) YAG ceramics can dissolve Nd at concentrations five times larger than is possible in single crystals. If the effect of grain boundaries makes it possible to dissolve large amounts of Nd as solid-melt, then the Nd concentration close to the crystallite grain boundaries should be extremely high. However, spot analysis by EDX and line analysis by EPMA revealed that there is no obvious difference between adjacent grain boundaries and inside grains in Nd:YAG ceramics.

(2) Measurement of the fluorescence lifetime revealed that the fluorescence lifetime deteriorates dramatically owing to the effect of concentration quenching (change in crystalline field) as the Nd concentration increases. If the Nd concentration inside crystallites is much lower than that at grain boundaries, it would be split into two components with a long fluorescence lifetime coming from the low Nd concentration part and a short fluorescence lifetime coming from the high Nd concentration part. However, this signal was not detected.

According to the above considerations, it is very unlikely that the dissolution of Nd as solid-melt would be caused by the difference in microstructure between single crystals and ceramics.

A third reason, can be given as the difference in composition between single crystals and ceramics. Nd:YAG single crystal is crystallized (single crystallization) from melted high purity Al_2O_3, Y_2O_3, and Nd_2O_3 powders. Schneider's experiment also investigated the relationship between the amount of Nd dissolved in YAG as solid-melt and the ionic radius using a solid phase method, as in the present experiment. The same composition was used for single crystal growth and no sintering aid was added. However, the Nd:YAG ceramics in the present experiment had a small amount of SiO_2 (pyrolysis of TEOS) added as sintering aid. The effect of SiO_2 on the dissolution of Nd in YAG single crystal and the problem of whether Si can be substituted in the YAG crystallite lattice have not been reported. SiO_2 is generally considered as an impurity in Nd:YAG single crystals. It can be expected to affect the crystalline field substantially. In particular, it is known that the optical absorption, fluorescence characteristics, and lasing characteristics of the Nd^{3+} ion are considerably altered by the effect of the crystalline field. It is very unlikely that an impurity material with the possibility of altering the characteristics of original YAG is added intentionally. (From a historical point of view, the lasing characteristics of Nd:YAG single crystal derived from raw materials of high purity are important, and raw material containing impurities has not been used.) Additionally, the addition of Si is not considered to be beneficial for crystallite defects because a positive ionic empty orbital is formed due to charge compensation of the crystallite lattice if Si^{4+} is substituted for Al^{3+} in the YAG lattice. However, the present experiment verified that it cannot be determined how an infinitesimal amount of added Si (approximately 300 mass ppm) affects the general optical and lasing characteristics of Nd:YAG ceramics. It is most probable that Nd is able to dissolve easily in YAG crystal because of the compositional influence of added Si. Experiments and theoretical analysis to determine whether Si affects Nd dissolution as solid-melt will be described in Section 5.2.

5.1.4.2 Significance of high concentration Nd:YAG ceramic lasers

Nd:YAG has enhanced physical properties such as chemical stability, thermal conductivity, and low threshold value (minimum energy input to obtain laser oscillation), it is able to withstand high power stably, and has features which make it possible to oscillate both modes. These properties are important features for solid lasers. The biggest drawback of YAG is the limited amount of Nd dissolved as solid-melt. As mentioned in Section 5.1.1, the limit of Nd dissolution in YAG single crystal produced using the CZ method is approximately 1 at.%. Laser devices using YAG single crystals are available only up to this concentration range. Recently, new types of laser have been developed by adding Nd to crystals such as GGG ($Gd_3Ga_5O_{12}$), YVO_4, $GdVO_4$, etc., and some exceeded YAG lasers in output characteristics [9]. Such material can dissolve more Nd as solid-melt because of the large segregation coefficient for Nd compared with the host materials. For example, the segregation coefficient of Nd:YVO_4 is approximately 0.6 [9], and up to 5 at.% Nd can be

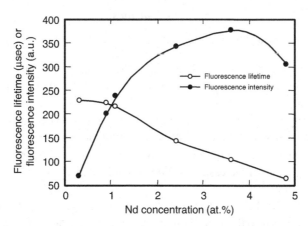

Figure 5.15 Laser power (fluorescence) intensities calculated from both the fluorescence lifetime and Nd concentration in YAG ceramics.

dissolved stably. If it is possible to introduce a large amount of Nd^{3+} ions into YAG crystal without deterioration of the optical characteristics, the potential drawback of YAG single crystal can be overcome.

Although in the present experiment up to 4.8 at.% Nd was dissolved homogeneously as solid-melt in YAG ceramics, the maximum slope efficiency (40%) occurred at 2.4 at.% under CW laser oscillation. Theoretically, it is considered that the amplifying efficiency of the laser increases as the amount of fluorescence increases if the optical level of the host material is constant. Figure 5.15 is a plot of the fluorescence intensity of Nd:YAG ceramics, calculated from the fluorescence lifetime and Nd concentration. The maximal value of fluorescence for Nd:YAG ceramics occurs for Nd in the range from 2.4 to 3.6 at.%. It is approximately twice the value of commercial 0.9 at.% Nd:YAG single crystal.

Slope efficiencies of 0.9 at.% Nd:YAG single crystal and 2.4 at.% Nd:YAG ceramics in CW laser oscillation are actually 24% and 40%, respectively; there is a difference of almost twice in amplifying efficiency after laser oscillation. However, the threshold value of oscillation for the latter is 366 mW, 86 mW larger than for the former. The reason why a high Nd concentration gives a large threshold value is that (1) the amount of scattering in the specimen is increased slightly with Nd addition (i.e. the specimen becomes slightly inhomogeneous optically), and (2) a deterioration of fluorescence lifetime is caused by Nd addition.

Traditional single crystal growth technology could not produce high Nd concentrations in YAG single crystals, so the basic characteristics of Nd:YAG laser in the concentration quenching range could not be measured. From the results of the present experiment, YAG ceramic has a maximal value of fluorescence for between 2 and 3 at.% Nd concentration approximately, and a high slope efficiency laser can be realized.

This laser is similar to or better than $Nd:YVO_4$ or $Nd:GdVO_4$ which have better output characteristics than single crystals.

5.1.5 Conclusions

The experimental results are summarized as follows.

(1) Transparent YAG ceramics containing more than 3 at.% (maximum 4.8 at.%) Nd concentration which is never achieved in Nd:YAG single crystals can be produced by a solid phase method with vacuum sintering.
(2) The reason why it is possible to produce a transparent YAG ceramic with high Nd concentration is considered to be the effect of a small addition of SiO_2.
(3) Although the integrated absorption intensity by Nd^{3+} ions in YAG ceramics increases as the Nd concentration increases, the fluorescence intensity is greatest at around 2.4 at.% Nd concentration.
(4) Although the threshold values of laser oscillation for 1.1 and 2.4 at.% Nd:YAG ceramics are somewhat larger than those of commercial 0.9 at.% Nd:YAG single crystal, the slope efficiency is superior to that of single crystal. In particular, the slope efficiency of 2.4 at.% Nd:YAG ceramic is approximately 40%, a good result almost twice that of single crystal.

5.2 Effect of impurity (Si) on Nd solid-melt in YAG ceramics

5.2.1 Introduction

It was mentioned in Section 5.1 that the optical quality of Nd:YAG ceramics containing an equal or larger Nd concentration compared with ordinary YAG single crystal is excellent, and laser oscillation can be achieved using these ceramics. Most of the Nd:YAG single crystals applied in solid-state lasers are produced by the Czochralski (CZ) method, but in some cases such crystals are grown by the Bridgman method for research purposes. The segregation coefficient indicates whether a particular added element can be incorporated homogeneously in the host crystal or whether it will remain undissolved in the melt. When the segregation coefficient of Nd in CZ grown YAG crystal is less than 1, Nd does not dissolve uniformly in the host crystal, and when the value is close to zero, Nd cannot dissolve in the host crystal. In the literature, the segregation coefficient of Nd in YAG single crystal grown by the CZ method is approximately 0.2 [6, 7].

It is technically difficult to dissolve even a small amount of Nd, less than 1 at.%, in the host crystal homogeneously. Increasing the amount of Nd causes the formation of optically inhomogeneous phases such as facets and inclusions, which results in a quality inappropriate for optical devices. Plans to improve the optical characteristics by the addition of Nd include (1) making a mild temperature graduation at the phase boundary between the liquid and the solid (phase boundary between the growing crystal and melt), and (2) although the production technique has reduced the crystallite growth rate close to a critical rate (0.1–0.3 mm h^{-1}, there is still no method for dissolving Nd stably as a solid-melt. The present study succeeded in synthesizing Nd:YAG ceramics for laser use [10, 11] which have similar or better quality compared with single crystal by a solid-state reaction method.

Although this material can dissolve up to 4.8 at.% Nd homogeneously and laser oscillation is possible, the reason why only sintered bodies are able to dissolve a much higher amount of Nd is not clear theoretically. Whereas only high purity Al_2O_3, Y_2O_3, and Nd_2O_3 are used as starting raw material for the production of YAG single crystal by the CZ method, a small amount of TEOS is added as sintering aid in the solid-state reaction method. This section describes the influence of the small amount of SiO_2 formed by pyrolysis of TEOS on the dissolution of Nd as solid-melt in YAG crystal.

5.2.2 Experimental

This experiment also used 99.99 mass% purity Al_2O_3, Y_2O_3, and Nd_2O_3 that were used before. A mixture of only these three kinds of powder (TEOS free), or these three powders and 1.5 mass% TEOS (adjusting the amount of Al_2O_3 based on the SiO_2 remaining in the sintered body derived from TEOS by pyrolysis) were weighed in the YAG composition. Additionally, the Nd concentration in the prepared sample ranged from 1.2 to 7.2 at.% and the mixture was blended (pot milled) for 12 hours using high purity Al_2O_3 balls, adding an appropriate amount of ethyl alcohol. The ethyl alcohol was evaporated by heating the prepared slurry using a hot stirrer. The obtained powder blend was molded into a powder compact 20 mm in diameter by uniaxial pressing at 10 MPa, then it was pressed using CIP at 140 MPa. The compact was sintered at 1750 °C for 10 hours under 1.3×10^{-3} Pa.

The microstructure of the sintered specimen was observed by reflecting, transmission polar microscopes and SEM, and analyzed by EPMA qualitatively and quantitatively to determine the distribution of constituent elements. Moreover, the phases formed in the sintered specimens were determined by powder X-ray diffraction (XRD), and a change in lattice constant associated with Nd dissolution was measured (the sample was scanned at a step width of 0.004° applying Si as an internal standard, accurate measurement of the lattice constant versus $\cos 2\theta \pm 0.0001$ nm). The specimens for measurement of the transparent spectrum were 10 mm in diameter and 1 mm thick disks polished both sides to smoother than $R_a = 10$ nm. The spectrum was measured in the range from 400 to 1000 nm by a spectral photometer with 0.5 nm slit width, and 120 nm min^{-1} scan rate as before. The amount of Si in each specimen was measured using ICP additionally.

5.2.3 Results

Figure 5.16 shows BEI (backscattering) images of specimens (fracture cross-section) with added 1.2, 2.4, 4.8, and 7.2 at.% Nd without Si addition. Each specimen consists of particles of size smaller than approximately 20 μm, a precipitate phase appears inside grains and at grain boundaries if the Nd concentration is higher than 2.4 at.%, and the amount increases as the Nd concentration increases.

Figure 5.17 shows the results (no correction with a standard specimen) and position of spot analysis using EDX of inside YAG grains, at grain boundaries, and the precipitate phase of TEOS free specimens as previously, with added 7.2 at.% Nd. The Nd concentration

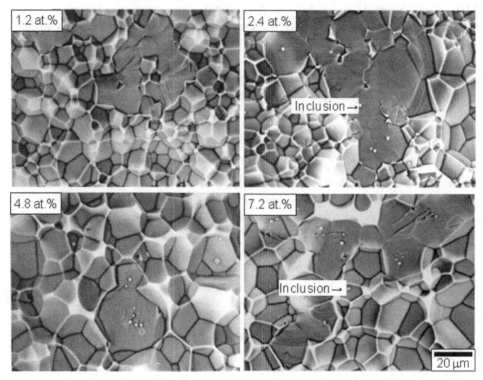

Figure 5.16 BEI images of YAG ceramics with varying Nd concentration without Si addition.

	Grain (a)	Boundary (b)	Inclusion (c)
Al_2O_3	39.0	39.4	25.1
Y_2O_3	55.6	54.7	34.6
Nd_2O_3	5.4	5.9	40.3

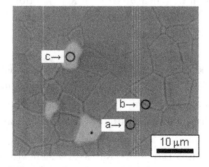

Figure 5.17 Spot analysis of Si undoped 7.2 at.% Nd:YAG ceramics by EDX.

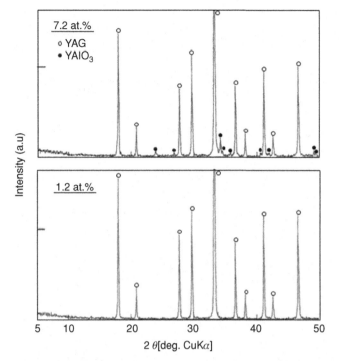

Figure 5.18 X-ray powder diffraction patterns of Si undoped 1.2 and 7.2 at.% Nd:YAG ceramics.

in the YAG particles forming the specimen is slightly lower than in the grain boundaries although it is within the error range of EDX analysis. On the other hand, surprisingly, the amount of Nd_2O_3 in the precipitate almost reached 40 mass%.

Figure 5.18 shows XRD patterns of specimens with Nd concentrations of 1.2 and 7.2 at.% as indicated in Figure 5.16. For the specimen with 1.2 at.% Nd, no precipitate is observed in the BEI image and only the YAG phase is present. The specimen with concentration of 7.2 at.% Nd has a peak belonging to the crystallite structure of the $YAlO_3$ phase excluding YAG. It was concluded by combining Figure 5.17 and this result that the precipitate observed in the specimen of high Nd concentration is $(YNd)AlO_3$ containing 40 mass% Nd_2O_3.

Figure 5.19 shows micrographs taken by reflection microscopes of specimens with added 0.2, 0.4, 0.8, and 1.5 mass% TEOS and 7.2 at.% Nd concentration. Although grain growth progresses as the amount of added TEOS increases in this series, the precipitated phase of $(YNd)AlO_3$ has a tendency to decrease. The precipitate disappears completely in the specimen with 1.5 mass% TEOS and it becomes a complete YAG uniform phase compositionally.

Figure 5.20 shows BEI images and line profile analytical results for Al, Y, and Nd elements in the specimen with added 1.5 at.% TEOS shown in Figure 5.19. Si was not analyzed because it was below the detection limit (930 mass ppm). This specimen is

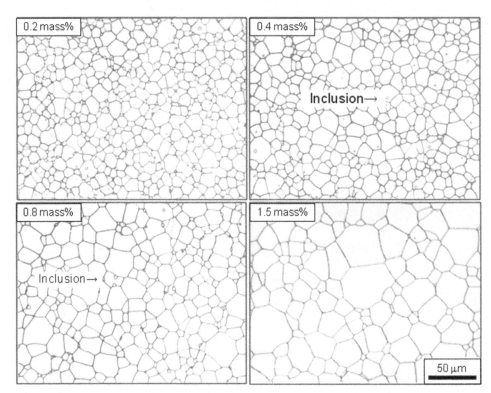

Figure 5.19 Surface of 7.2 at.% Nd:YAG ceramics prepared from the starting material with varying TEOS contents.

homogeneous and no precipitate was observed by BEI imaging; the concentrations of the elements from line profile analysis show almost no difference between inside grains and at the grain boundaries in the sintered body.

Figure 5.21 shows the interactive relationship between the amount of TEOS added and Si detected in the specimens by ICP. The amount of Si in the Si free specimen is 10 mass ppm. The amount of Si in the specimen tends to increase depending on the amount of TEOS added but it is very little, less than 930 mass ppm maximum.

Figure 5.22 shows the lattice constant for specimens with added 0 (Si: approximately 10 mass ppm), 0.8 (Si 570 mass ppm), and 1.5 (Si 930 mass ppm) mass% TEOS at 7.2 at.% Nd concentration. The lattice constant of the specimen containing 1.5 mass% TEOS, which can almost completely dissolve Nd with its large ionic radius, is supposed to be the maximum value. However, the theoretical relationship between the dissolution condition and the lattice constant was not observed. The crystallite lattice of YAG has a tendency to shrink slightly depending on the amount of TEOS added.

Figure 5.23 shows XRD patterns of specimens with reduced amounts of Y_2O_3 varying by 1.5 and 3.0 mass% of the stoichiometric composition (i.e., Al_2O_3 rich specimens) with 0.2 mass% TEOS and 7.2 at.% Nd concentration. The purpose of this experiment was to figure

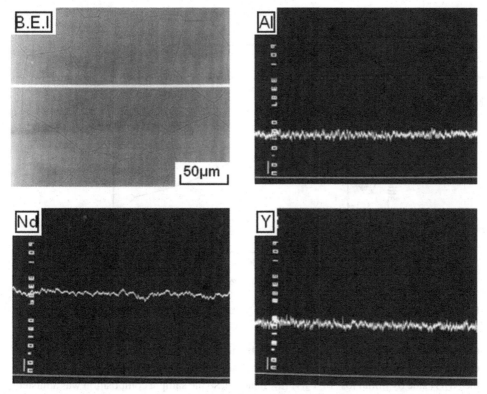

Figure 5.20 BEI images of Si doped 7.2 at.% Nd:YAG ceramics with line profiles for Y, Al, and Nd elements by EPMA.

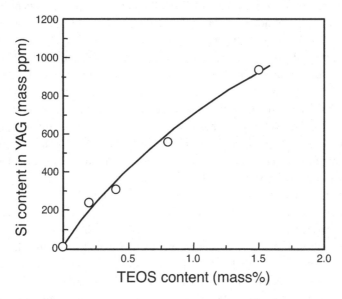

Figure 5.21 Relationship between the amount of TEOS added and Si detected in the specimens by ICP.

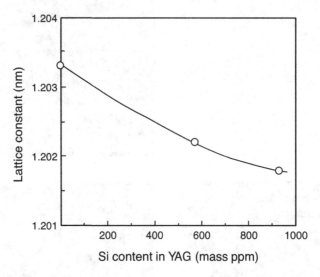

Figure 5.22 Lattice constant of 7.2 at.% Nd:YAG ceramics doped with varying amounts of TEOS.

out whether the precipitation of (YNd)AlO$_3$ mentioned before was caused by weighing error of the raw material or variation in chemical composition of the specimen shifting to the Al$_2$O$_3$ side by initial loss of Y$_2$O$_3$ raw material, so the sample composition was adjusted to the Al$_2$O$_3$ rich side intentionally. The specimen adjusted to chemically stoichiometric composition was YAG and minimal (YNd)AlO$_3$ phase; the other specimens adjusted to 1.5 and 3.0 mass% lack of Y$_2$O$_3$ were both (YNd)AlO$_3$ and Al$_2$O$_3$ phases except YAG. YAG, Al$_2$O$_3$, and (YNd)AlO$_3$ do not appear at the same time in the Y$_2$O$_3$–Al$_2$O$_3$ phase diagram; the reason for this will be described later.

Figure 5.24 shows the transmission spectra for specimens with no and 1.5 mass% TEOS and 7.2 at.% Nd concentration. The transparency of the specimen without TEOS addition is extremely poor (background level of transmission is 3–4%), and absorption at the specified wavelength by Nd is also slight. On the other hand, the specimen forming uniform YAG on the addition of TEOS in the structure has higher than 80% background level transparency, and a sharp light absorption pattern by Nd ion is also observed. This fact shows that transparent material was achieved as the result of a large amount of Nd dissolved in YAG ceramics as solid-melt by the addition of TEOS (SiO$_2$).

Figure 5.25 shows the appearance of the specimens presented in Figure 5.24. Although the specimen without added TEOS is opaque, TEOS addition makes the specimen transparent so it is possible to read the underlying letters clearly. The addition of TEOS makes it possible to synthesize more transparent YAG ceramics, and the author has succeeded in synthesizing a specimen with 10 at.% Nd concentration. However, increasing the Nd concentration above 6 at.% does not enhance the lasing characteristics.

Figure 5.26 shows micrographs taken by transmission polarized microscopy of specimens with 0.6 and 1.5 mass% TEOS and 1.2 and 7.2 at.% Nd concentration. The materials

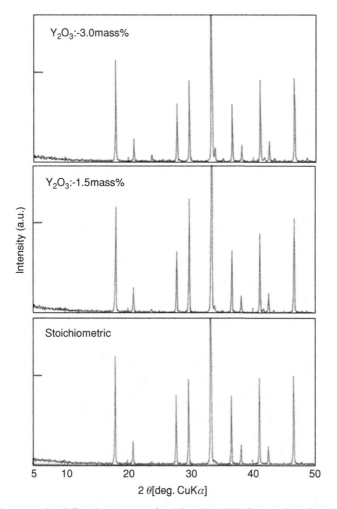

Figure 5.23 X-ray powder diffraction patterns for 7.2 at.% Nd:YAG ceramics of stoichiometric and Y_2O_3-poor compositions.

are perfectly transparent and optically isotropic bodies (cubic) since even the specimen containing 7.2 at.% Nd concentration is completely dark-field under cross-nicol.

Figure 5.27 shows YAG ceramics (Nd concentration 4.8 at.%, TEOS 1.0 mass%) fabricated into disks (14 mm diameter and 6 mm thick) irradiated with a He–Ne laser (wavelength 633 nm, output 5 mW) and the laser beam patterns projected onto a board after passing through the specimens. Scattering light inside the specimens can hardly be detected by visual observation (dark-field in the whole of the specimen) under laser beam irradiation. Not only is there almost no difference in light intensity between the laser beam projected onto the board after passing through the specimen and through air, but also the scattering light on the board can hardly be detected by visual observation. Although this means there

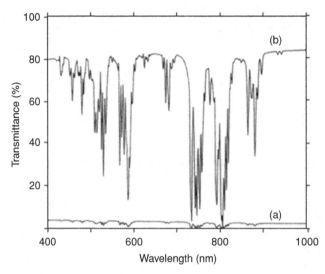

Figure 5.24 Optical transmission spectra for (a) Si undoped and (b) Si (930 mass ppm) doped 7.2 at.% Nd:YAG ceramics.

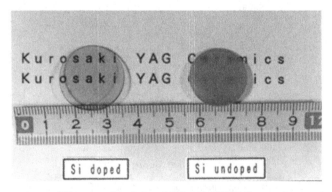

Figure 5.25 Appearance of Si undoped and doped 7.2 at.% Nd:YAG ceramics.

is almost no segregation (grain boundary phases) in the Nd:YAG ceramics containing very high concentrations of Nd, it suggests that the light scattering caused by grain boundaries or residual pores is also remarkably small.

5.2.4 Discussion

5.2.4.1 Why melt-growth technology cannot produce heavily doped Nd:YAG

It was described in Section 5.1 that it is possible to fabricate optically homogeneous Nd:YAG ceramics even when a large amount of Nd is added, and efficient laser oscillation can be produced. However, it is very difficult to dissolve Nd in YAG single crystal produced by ordinary single crystal growth technology (CZ method). It is difficult to grow the (211)

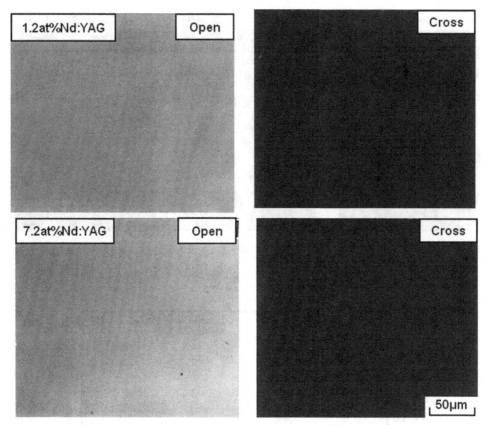

Figure 5.26 Micrograph taken by transmission polar microscopy of Si doped 1.2 and 7.2 at.% Nd:YAG ceramics under open and cross-nicol.

facet toward the periphery from the center of rotation in the crystal continuously, even in YAG single crystal containing around 1 at.% Nd. Additionally, it is generally known that a precipitation containing Nd as a major component is formed in the crystal, even if the crystal growth rate is minimized to around the limit, if more than 1 at.% Nd is added, and the quality of the single crystal becomes unsuitable for laser material [1–6]. This is attributed to the low segregation efficiency of around 0.2 of Nd dissolved in YAG single crystal produced via melted liquid in the CZ method. Figure 5.28 shows the relationship between the ionic volume (the value calculated by cubing the ionic radius, the lattice constant is doubtful) and the segregation coefficient in the lanthanoid rare earth series which can substitute for Y^{3+} ions of YAG single crystal, as reported by R. R. Monchanp [7]. The segregation coefficient of the ionic radius for Ho is similar to that of Y and is 1; the segregation coefficient becomes larger as the ionic radius of elements becomes smaller, i.e. moving towards Lu, meaning that these elements dissolve in YAG crystal as solid-melt more easily.

On the other hand, it is found that the segregation coefficient of each element decreases as the ionic radius becomes larger towards Dy^{3+}, Tb^{3+}, ..., Pr^{3+} from the turning point

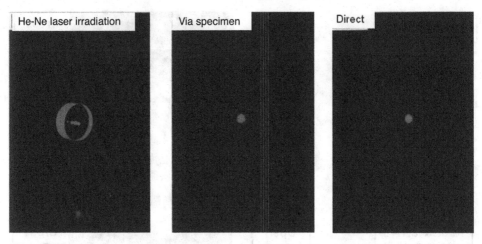

Figure 5.27 Appearance of 4.8 at.% Nd:YAG ceramics irradiated by a He–Ne laser (633 nm) beam and beam patterns after passing through the specimen or through air.

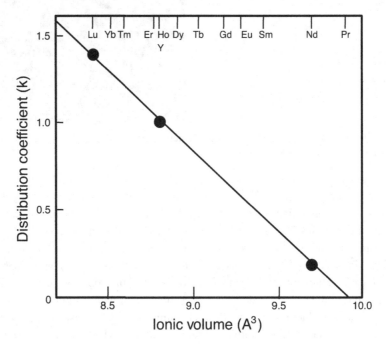

Figure 5.28 Plots of ionic volume versus distribution coefficients for RE^{3+} (lanthanoid rare earth) ions in $Y_3Al_5O_{12}$ single crystal reported by R. R. Monchamp.

of Y^{3+}. When the segregation coefficient is 1, all of the Y^{3+} ions can be substituted for a specified ion theoretically, the segregation coefficient can be expressed as follows:

$$C_x = kC_0 g \qquad (5.1)$$

where C_x, C_0, k, and g are the concentration of the added element, the initial concentration of melted liquid, the segregation coefficient, and the ratio of crystallizing melt. According to R. R. Monchamp, it was reported that the segregation coefficient of Nd in YAG crystal is 0.18, and the amount which can be practically dissolved in YAG single crystal is only 1 at.% approximately. The amount dissolved in the crystal homogeneously decreases dramatically from 100 to 1% when the segregation coefficient changes from 1 to 0.2. Pr which has a smaller segregation coefficient (a large ionic radius in other words) is hardly dissolved in YAG crystal.

The segregation coefficient has an essentially strong relationship with the ionic radius of the fluorescent element substituted into YAG crystal. Concerning this issue, Schneider's group reported that a lanthanoid series rare earth element R which can form aluminum garnet ($R_3Al_5O_{12}$) stably is limited to an ionic radius of 0.106 nm. The ionic radius of Nd^{3+} in 8 coordination is 0.112 nm; it is slightly bigger than the Y^{3+} ion, so it becomes very difficult to form YAG crystal stably. This is the reason why traditional single crystal growth technology cannot produce good optical quality crystal when relatively high concentrations of Nd are added into YAG single crystal.

5.2.4.2 Effect of Si addition for Nd dissolution

YAG single crystal formed by the CZ method described before can dissolve only a small amount of Nd, which is associated with the segregation coefficient. However, Nd:YAG ceramics with up to 7.2 at.% of Nd concentration, with good optical properties, can be produced. As a result, a specimen (2.4 at.% Nd:YAG) with an Nd concentration in the range that is generally difficult to produce in single crystals gave a slope efficiency of lasing characteristics (input energy versus output) almost twice that of 0.9 at.% Nd:YAG single crystal. Nd:YAG has Y^{3+} as the majority of R composing $R_3Al_5O_{12}$, and it becomes very difficult to stabilize the single crystal as the amount of Nd increases. The system without SiO_2 addition, namely Nd:YAG ceramics with composition very similar to the single crystal counterparts formed a precipitated phase analyzed as $(YNd)AlO_3$ after sintering. The amount of precipitation increased as the Nd concentration increased. This agrees with the result reported by Schneider that it was impossible to dissolve Nd ions stably when a garnet sintering body was produced by a solid phase process without Si addition. It was confirmed that the same phenomenon occurs even for a polycrystalline body produced using a different sintering step compared with single crystal.

However, in Nd:YAG ceramics with a suitable amount of SiO_2 it becomes easy to dissolve Nd as solid-melt. It was confirmed that an optically homogeneous sintered body can be produced with up to 7.2 at.% Nd without the formation of any precipitation such as $(YNd)AlO_3$. It is certain that the presence of SiO_2 causes the dissolution of Nd because the precipitated phase in the sintered body disappears as SiO_2 increases, as shown in Figure 5.19. The microstructure of the sintered body (average particle size and decreased precipitate) changed with an increasing amount of SiO_2, and the Nd concentration in YAG particles composing the sintering body increased associated with this change. Therefore, a suitable amount of addition of SiO_2 allows each YAG grain composing the sintering body to dissolve Nd largely and stably.

The lattice constants of the specimens with added 0.8 (Si quantity 570 mass ppm) and 1.5 (Si quantity 930 mass ppm) mass% of TEOS at 7.2 at.% Nd concentration shown in Figure 5.22 were smaller by 0.0011 and 0.0015 nm than those of the specimens without Si addition, respectively. The coordination numbers of the cations composing $R_3Al_5O_{12}$ are classified as R^{3+} is 8 and Al^{3+} is 4 and 6 in occupancy. Although it is well known that Si^{4+} substitutes for 4 coordination positions of garnet structure, the ionic radius of Si^{4+} is 0.026 nm when it is 4 coordination and that of Al^{3+} is 0.039 [12]. It is considered that the crystallite lattice of YAG shrinks because of the difference in the ionic radii and a cation vacancy also arises from the charge difference between both elements if Si^{4+} is substituted for Al^{3+}. Judging from the relationship between the measurement of the lattice constant (lattice constant of YAG shrinks with Si^{4+} addition) and the ionic radius, it can be concluded that at least a portion of added Si is dissolved in YAG crystal (the distribution of Si^{4+} is confirmed to be dissolved in the YAG crystallite lattice from TEM observation). Si^{4+} is substituted for Al^{3+} in the YAG lattice, and a decrease in the lattice constant occurs on the formation of the cation vacancy caused by the difference in the ionic radii and charges of these elements. It is considered that there is a strong relationship between this and the phenomenon that Nd with a large ionic radius is highly dissolved in YAG crystal. Si was not added into $R_3Al_5O_{12}$ in Schneider's experiment, and when Nd which has a large ionic radius is dissolved in YAG it causes the stability of the crystal to decrease. This study shows that it is reasonable to consider that the expansion of the lattice is canceled out when Nd^{3+} of 8 coordination substitutes for Y^{3+} if Al^{3+} is replaced with Si^{4+}. Consequently, the precipitation (secondary phase) observed in both Nd:YAG single crystal produced by the traditional single crystal growth technology and Nd:YAG ceramics without Si addition disappears with the effect of the addition of Si^{4+} (decrease in lattice constant) in Nd:YAG ceramics with added Si. Almost perfectly homogeneous ceramics can be synthesized with good optical properties up to high concentrations of Nd. Besides, although confirmation of whether the cation vacancy in Nd:YAG ceramics is formed by Si addition is possible by measuring the shift at an absorbing edge in the short wavelength range in the transmission spectrum (shifting to longer wavelengths), the electric resistance, or diffusion coefficient, and so on [13], have not been measured.

There are a lot of grain boundaries in ceramic materials, and it is known that many line defects (rearrangements) or point defects (cation, anion vacancies) reside near the grain boundaries. It might be thought that the crystalline defects increase with increasing dissolution of Nd. The following reasons show that this is not the case. (1) In YAG ceramics without Si addition the precipitate is formed. (2) There is almost no difference in Nd concentration between the grain boundary and inside the grain. (If the microstructure of the ceramics has high Nd concentration close to the grain boundaries causing obvious scattering, the material will be unsuitable for use in lasers.)

5.2.4.3 Inconsistency in the Y_2O_3–Al_2O_3 phase diagram

Although the inclusion ((YNd)AlO$_3$) in YAG ceramics disappears with the addition of Si and the amount of dissolution as solid-melt increases in YAG ceramics, it can also be considered that the precipitate of (YNd)AlO$_3$ phase is caused by compositional variability of the original Nd:YAG ceramics from the stoichiometric composition to the Y_2O_3 side (i.e., it becomes the chemical stoichiometric composition by a small amount of Si substituting for Al). It was found that three phases coexist with Al_2O_3 except YAG and (YNd)AlO$_3$ precipitate when the composition of Nd:YAG ceramics is shifted to the Al_2O_3 side as shown in Figure 5.23. Although these three phases do not coexist, judging from the Y_2O_3–Al_2O_3 series phase diagram [14], the specimen composition was shifted to the Al_2O_3 side from the chemical stoichiometric composition of YAG by the reduction of Y_2O_3. The Si composition system does not coexist, and the stoichiometry still remains (YNd)AlO$_3$, even for specimens in the YAG–Al_2O_3 system. It can be assumed that the precipitate is inevitable even though the composition (YNd)AlO$_3$ never appears in the phase diagram because it is very difficult to occur in the presence of YAG crystal containing a large amount of Nd (segregation of Nd makes it impossible to dissolve as (YNd)AlO$_3$ phase). There is no inconsistency from this point of view, and (YNd)AlO$_3$ is precipitated owing to the absence of Si as mentioned before, and not owing to compositional variability from the chemical stoichiometric composition. The occurrence of such a problem concerning dissolution as solid-melt is caused by the impossibility of drawing an equilibrium phase diagram between YAG and NdAG(Nd$_3$Al$_5$O$_{12}$) due to the absence of Nd$_3$Al$_5$O$_{12}$. It is not clear whether segregation is possible or not as for an ordinary solid-melt body which consists of two elements. Additionally, it is necessary to conduct experiments of single crystal growth in the preparation of high concentration Nd:YAG with added Si in order to figure out the possibility of improving the quality of the single crystals or the specific features of the solid-melt dissolution of Nd at high concentrations in YAG crystal by the addition of Si.

5.2.5 Conclusion

With regard to the effect of impurity (Si) on Nd dissolution as solid-melt in YAG ceramics, the dissolution of Nd at high concentrations in YAG ceramics can be summarized as follows.

(1) It becomes very easy for Nd to dissolve in YAG crystals as solid-melt with the addition of Si, and optically homogeneous YAG ceramics containing up to 7.2 at.% Nd can be synthesized.

(2) The reason why the addition of Si makes it easy for Nd to dissolve in YAG is considered to be that the expansion of the YAG crystallite lattice with Nd dissolution can be inhibited by Si.

References

[1] T. Sekino and Y. Sogabe, *Rev. Laser Eng.* **21** (8) (1993) 827–831.
[2] K. Shiraki, *Oyo Butsuri (Bull. Appl. Phys. Soc. Jpn.)* **38** (2) (1969) 177–182.
[3] F. Takei, S. Takasu, J. Ushizawa, and M. Sakurai, *Toshiba Rev.* **24** (12) (1969) 1–9.
[4] K. Mori, K. Sugibuchi, and K. Shiroki, *J. Jpn. Appl. Phys.* **11** (1972) 764.
[5] F. Inaba *et al.*, *Laser Handbook*, Asakura Pub., Tokyo (1987).
[6] K. Shiraki and Y. Kuwano, *J. Chem. Soc. Jpn.* **7** (1978) 940–944.
[7] R. R. Monchamp, *J. Cryst. Growth* **11** (1971) 310–312.
[8] S. J. Schneider, R. S. Roth, and J. L. Waring, *J. Res. Natl. Bur. Std.* **65A** (4) (1961) 345–374.
[9] Laser Society Japan (ed.), *Advanced Laser Technology*, Nikkei Technical Book (1992).
[10] A. Ikesue, T. Kinoshita, K. Kamata, and K. Yoshida, *J. Am. Ceram. Soc.* **78** (4) (1995) 1033–1040.
[11] A. Ikesue, K. Kamata, and K. Yoshida, *J. Am. Ceram. Soc* **79** (1996) 1921–1926.
[12] W. D. Kingrey *et al.*, *Introduction to Ceramics*, Uchida Rokakuho, Tokyo (1980) Japanese editon, pp. 55–56.
[13] S. Shirasaki, S. Matsuda, H. Yamamura, and H. Haneda, *Adv. Ceram.* **10** (1985) 474–489.
[14] F. M. Levin *et al.*, *Phase Diagrams for Ceramists*, Vol. II, *American Ceramics Society* (1969), Figure 23.44.

6

Optical scattering centers in polycrystalline ceramics

6.1 Introduction

It was demonstrated that 1 at.% doped Nd:YAG ceramics, which have the same Nd content as melt-growth Nd:YAG single crystals, have laser performance (CW oscillation) comparable to that of the single crystal lasers. However, when the amount of Nd doping is higher than 1 at.% in the Nd:YAG ceramic gain medium, it can be used for microchip lasers for high efficiency enhancement and single mode oscillation. It has been confirmed that Nd can dissolve 10 times more in YAG ceramics than in YAG single crystals [1, 2]. There are many advantages of ceramic materials: (1) the fabrication of large size gain media that cannot be realized using single crystal materials [3], (2) the fabrication of optically homogeneous gain media without the facets and core seen in melt-growth single crystals [3], (3) heavy doping of laser active ions uniformly in the gain media [4, 5], and (4) the fabrication of composite laser gain media that cannot be realized using single crystal materials [6], for example, round clad–core structures. Especially in the case of composite laser gain media, ceramic processing technology allows design flexibility of the gain medium [7], and it has become a very interesting technology for materials science. Also, it is expected to be developed as an important technology from the viewpoint of laser generation and functional improvement. If a ceramic processing technology which can guarantee stable performance is developed for the above mentioned new technologies, then ceramic optical materials will replace the single crystal materials which are expensive and require a long manufacturing time. The range of applications will be much broader, and possibly there is the potential to create a new market based on the unique properties of ceramic lasers.

As mentioned above, new functionalities and advantages are slowly being discovered for Nd:YAG ceramic laser materials compared to the traditional single crystal materials, but there are still no reports on how the microstructural factors affect the characteristics of the ceramic laser materials. Laser output characteristics and laser beam quality are greatly affected by optical scattering in the materials [8]. This chapter investigates what factors in the microstructure of ceramic materials might cause scattering, quantitatively and qualitatively.

In this chapter, it is pointed out that scattering occurs at the residual pores and microstructure near grain boundaries. Several types of 1.1 at.% Nd:YAG ceramic with different amounts of residual pores and grain boundary phases were prepared [9 –11]. By investigating the relationship between the amount of residual pores/grain boundary phases, the absorption coefficient (scattering coefficient) and laser oscillation performance in detail, it became clear that there is a correlation between the microstructure, scattering characteristics, and laser performance. Based on these relationships, the point of this chapter is to find out what are the main factors causing optical scattering, and to determine the microstructural refinement necessary for an improvement in laser oscillation efficiency, and modification of the fabrication process required to realize the microstructural refinement.

In addition, the most recent (2009) data, such as direct observations of micro-scattering in Nd:YAG ceramic gain medium by laser tomography, the relationship between the scattering (defect) density determined by laser tomography and the laser damage threshold of single crystals and ceramics induced by a high power pulse laser, and the shape of the damage, are discussed.

6.2 Experimental procedure

6.2.1 Sample preparation and conditions for laser oscillation experiments

The starting powder materials for these experiments were the same as those (Al_2O_3, Y_2O_3 and Nd_2O_3) described in the previous chapter. The raw powder materials were weighed to form the composition 1.1 at.% Nd:YAG, and two amounts of TEOS were selected as additive: 0.5 mass% and 1.0 mass%. The two types of mixtures were ball milled with high purity alumina balls for 12 hours. Then the ball-milled slurry was dried and granulated using a spray dryer. To adjust the diameter of the granulated powders, the rotating speed of the atomizer in the spray dryer was set to 8000 or 12 000 rpm, and two types of granule size were achieved: 50–150 μm or 50 μm.

In order to control the number of residual pores in the sintered ceramics, the two types of granulated powders with 0.5 mass% TEOS composition were mixed in a given ratio, and then pressed into tablets with CIP molding at 140 MPa, changing the packing density of each powder compact. Then the powder compacts were calcined at 600 °C for 3 hours, and sintered at 1750 °C for 1–20 hours in a vacuum furnace (1.3×10^{-3} Pa).

The granulated powders with 1.0 mass% TEOS composition were pressed into tablets and calcined under the some conditions as for the 0.5 mass% TEOS composition powder. In order to control the amount of grain boundary phase in the sintered ceramics, the calcined samples were sintered at 1750 °C for 20 hours in a vacuum furnace; here the cooling rate was set from 10 °C h^{-1} to 600 °C h^{-1} down to room temperature.

For in-line transmittance measurement, the surface roughness (R_a) of both faces of each tablet sample (diameter 10 mm) was machined to be within 2 nm. The surface roughness of the samples was measured by a non-contact surface roughness meter. Using a spectrophotometer, the in-line transmittance of each sample was measured in the wavelength region 400–1000 nm with a slit width of 0.5 nm, and a scanning speed of 120 nm min^{-1}.

A laser oscillator similar to that described in the previous chapter was used for CW laser oscillation experiments. An LD (808 nm, maximum output 600 mW) was used as an excitation source. The dimension of the ceramic gain medium sample was 10 mm diameter by 5 mm thickness. The surface finish quality of the sample was $R_a \leq 0.2$ nm, flatness $\lambda/10$, and parallelism ≤ 10 s. A 100% reflection mirror and 98% output mirror were put on both faces of the sample. The output power of CW oscillation was measured by an optical power meter. (The differences in the samples described here for laser oscillation testing compared to previous chapters are the dimensions of the samples and the configuration of the laser oscillator.) When measuring the scattering coefficient and laser oscillation performance, a commercial single crystal (0.9 at.% Nd:YAG) produced by the CZ method was used as a reference sample.

The data described in this chapter are from samples fabricated around 1995–1996, and the performance of recently (in 2010) fabricated YAG laser ceramics is greatly improved. The author started this study in the beginning of the 1990s, and an excitation system using LDs (laser diodes) was also introduced around that period. Thus, excitation by LD was not fully developed, and the absolute value of laser oscillation efficiency was not very high. However, it was possible to compare the results with the laser performance of commercial Nd:YAG single crystal as a reference.

In addition, scattering properties detected by laser tomography and laser damage of typical laser ceramic samples prepared from 1995 to the present date (by 2009) were investigated. Basically the same fabrication process was used for these ceramic samples, but the preparation of raw powder materials and fabrication of ceramic material are improving slightly from year to year. This has greatly improved the optical properties of laser ceramics.

6.2.2 Measurement of the absorption coefficient (scattering coefficient) and method of calculation

When light passes through a material, the intensity of the incident light is attenuated in the material by the characteristic absorption of the material and scattering centers in the material. The light absorption coefficient ($\alpha(\lambda)$) which includes both factors can be expressed by the following equation:

$$\alpha(\lambda) = -t^{-1} \ln[I(\lambda)/I_0(\lambda)T(\lambda)^2] \tag{6.1}$$

where $I_0(\lambda)$, $I(\lambda)$, $T(\lambda)$, and t are the intensity of the incident beam and of the beam after passing through the material at a given wavelength (λ), a correction factor for Fresnel loss, and the thickness of the material, respectively. $I(\lambda)/I_0(\lambda)$ can be calculated from a reciprocal of the transmittance of the measured wavelength (λ). However, in Nd:YAG, there is also the characteristic absorption of Nd ions. Therefore, only the transmittance of the background level (excluding the characteristic absorption) was used to calculate $I(\lambda)/I_0(\lambda)$.

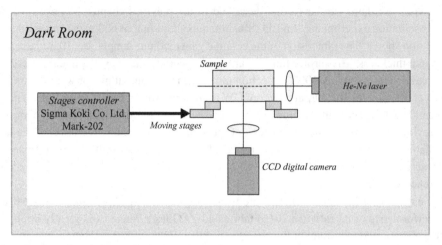

Figure 6.1 Experimental setup for laser tomography.

In the measurement of in-line transmittance, the Fresnel surface loss $\beta(\lambda)$ [12] can be estimated from the following equation:

$$\beta(\lambda) = (n(\lambda) - 1)^2 / (n(\lambda) + 1)^2 \tag{6.2}$$

where $n(\lambda)$ is the refractive index of the material at a given wavelength, and the value for YAG single crystal reported by W. L. Bond [13] was used for this calculation. The correction factor $T(\lambda)$ in Eq. (6.1) was set as $1-\beta$.

6.2.3 Measurement of scattering properties by laser tomography

Figure 6.1 shows the laser tomography setup used in this experiment. To detect the scattering centers, He–Ne laser (633 nm) and SHG green laser (532 nm) sources were used to irradiate the materials. To create a two-dimensional image of the scattering centers in the materials, a CCD camera was set perpendicular (90°) to the direction of the irradiating laser beam. The dimension of the sample was $8 \times 8 \times 20$ mm^3, and two opposite faces for laser irradiation and two other opposite faces for imaging with the CCD camera were polished to a mirror finish.

6.2.4 Laser damage testing by a high power pulse laser

The setup for laser damage testing is shown in Figure 6.2. Pulsed shot from an Nd:YAG laser ($\lambda = 1064$ nm, pulse width 8 ns, energy density 100 J cm^{-2}) was focused into the materials. In this experiment, only commercial high quality 1 at.% Nd:YAG single crystal and low loss (0.2% cm^{-1}) 1 at.% Nd:YAG laser ceramics were used. After damage testing, the samples were observed under an optical microscope.

Table 6.1 *Specification of the specimens used in this work*

	A1	A2	A3	A4	B1	B2	B3	B4
TEOS (mass%)	0.5				1.0			
Cooling rate (°C h^{-1})	40				600	150	40	10
Pore volume (vol. ppm)	1.5	3.2	6.8	9.3	1.4	1.6	1.3	1.5

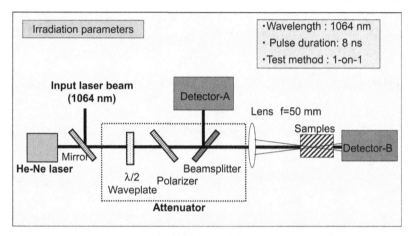

Figure 6.2 Schematic diagram for the laser damage experiment.

6.3 Results

Details of the samples prepared for this experiment and the measurement of optical properties are summarized in Table 6.1. Series "A" are the samples with 0.5 mass% TEOS. All these samples were sintered at 1750 °C for 20 h, and cooled down at 40 °C h^{-1}. The samples were prepared by mixing two types of granulated powders (spray dried powders, granule size 50–150 μm and <50 μm) to produce different packing densities, resulting in different residual pore densities (1.5–9.3 vol. ppm) in the final sintered body. Samples A1–A4 were prepared by mixing the 50–150 μm and <50 μm granules in given ratios of 100:0, 70:30, 30:70, and 0:100%, respectively. The pore density of each sample was estimated by counting the number of pores observed in an area of 350 × 500 μm^2 of 50 different views in total under an optical microscope.

Series "B" are the samples with 1.0 mass% TEOS, and all these samples were prepared using only the 50–150 μm granules, similar to sample A1. They were sintered at 1750 °C for 20 h but different cooling rates (600, 150, 40, and 10 °C^{-1} h) from the sintering temperature to room temperature were applied. The amount of Si in the samples of series A and B was 310–350 mass ppm and 640–720 mass ppm, respectively, measured by ICP-MS (inductively coupled plasma mass spectrometry). It was confirmed that the amount of Si was different in series A and B, relating to the amount of TEOS added. (The amount of

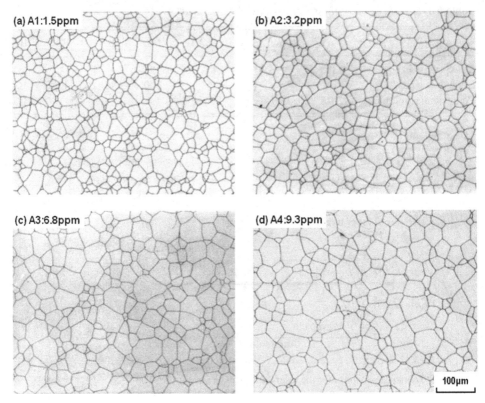

Figure 6.3 Surface of Nd:YAG ceramics with various pore volumes after thermal etching.

Si was similar in samples of each series.) Accordingly, series A samples have different amounts of residual pores and do not have secondary phases (grain boundary phase), and series B samples have the lowest amount of residual pores similar to sample A1 but include different amounts of grain boundary phases due to the different cooling rates of the samples.

Series A samples were mirror polished and thermally etched at 1600 °C for 30 min. The microstructures observed under a reflecting microscope are shown in Figure 6.3. There were no residual pores observed on the surface of the samples. Only the average grain size and grain size distributions were different from each other, and there was almost no difference in microstructure.

Figure 6.4 shows typical images of series A samples observed by transmission optical microscopy. Residual pores were observed in the depth direction of each sample, and the number of residual pores increased from sample A1 to A4. All samples were also observed under a cross-nicol polarizer (not shown here). Optically anisotropic phases were not observed. (That is, the only difference between the samples is the amount of residual pores.)

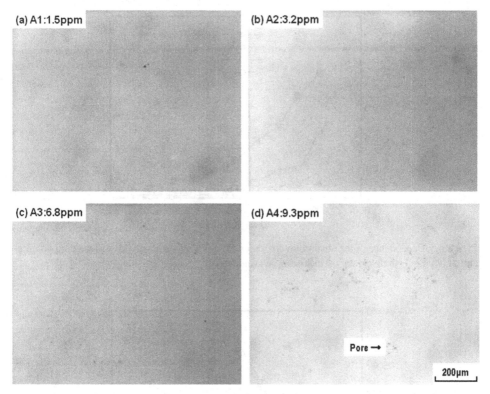

Figure 6.4 Transmission microscope photographs of Nd:YAG ceramics with various pore volumes.

Absorption coefficients of samples A1, A2, A4 and 0.9 at.% Nd:YAG single crystal calculated from Eq. (5.1) and Eq. (5.2) are summarized in Figure 6.5. The absorption coefficient became smaller with a decrease in the number of residual pores in the sample and also with an increase in measuring wavelength. The behavior of the absorption coefficient of sample A1 with the lowest amount of residual pores (ca. 1.5 vol. ppm) and the absolute values were very similar to those of the 0.9 at.% Nd:YAG single crystal.

CW laser oscillation of series A samples and the Nd:YAG single crystal are shown in Figure 6.6. In this laser oscillator, the laser threshold and slope efficiency of the Nd:YAG single crystal were 70 mW and 25.9%, respectively. The laser threshold and slope efficiency of samples A1, A2, A3, and A4 were 100 mW and 30.6%, 132 mW and 10.7%, 177 mW and 6.4%, and 226 mW and 4.5%, respectively. In sample A1, the laser output power was lower than that of the Nd:YAG single crystal at lower input power, but when the input power increased, the output power from sample A1 exceeded the output power of the single crystal sample. Among series A samples, only sample A1 which had the lowest residual pore volume (ca. 1.5 vol. ppm) exhibited laser performance which was equivalent to or superior to that of commercial Nd:YAG single crystals. From the results of series A

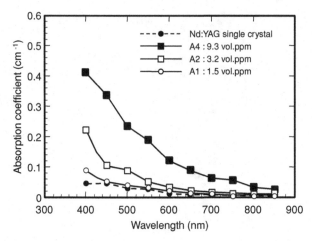

Figure 6.5 Dependence of the absorption coefficient of Nd:YAG ceramic with various pore volumes and of Nd:YAG single crystal on measuring wavelength.

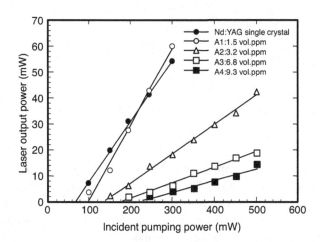

Figure 6.6 Laser output power versus input energy for 0.9 at.% Nd:YAG single crystal and 1.1 at.% Nd:YAG ceramic with various pore volumes excited with an 808 nm diode laser.

samples, it was clear that the number of pores in the sintered body is closely related to the laser threshold and slope efficiency.

Fracture surfaces of series B samples are shown in Figure 6.7 observed by SEM (scanning electron microscope). Grain boundary phases were not observed in sample B1 which was cooled at 600 °C h⁻¹ from the sintering temperature. The other samples (B2–B4) which were cooled down at rates of 150–10 °C h⁻¹ included many grain boundary phases between the Nd:YAG grains. It was confirmed that the amount of grain boundary phase increased when the cooling rate was slower.

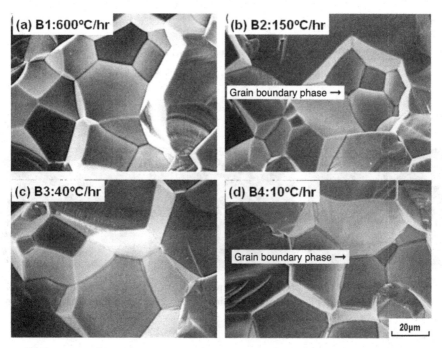

Figure 6.7 Fracture surfaces of 1 mass% TEOS doped Nd:YAG ceramics sintered at 1750 °C for 20 h and cooled down at rates of 10 to 600 °C h^{-1}.

Figure 6.8 shows open nicol and cross-nicol images of samples B1, B3 and B4 taken by a transmission polarized optical microscope. In these samples, residual pores were not observed in the depth direction of the sample, but, except for sample B1, needle-shaped grain boundary phases between Nd:YAG grains were confirmed in the depth direction under cross-nicol observation. Also, double refraction due to optically anisotropic phases was observed on the grain surfaces. Accordingly, it was confirmed that the grain boundary phases were segregated not only at the grain boundaries but also on the grain surfaces.

Figure 6.9 shows a spot analysis result of sample B4 near a grain boundary by EDX (energy dispersive X-ray spectroscopy). Spots X, Y, and Z indicate the inner grain, the surface of the grain, and the grain boundary, respectively, and the analysis results are summarized in the table below the micrograph.

There was no difference in basic composition (Al, Y, Nd) of the inner grain and surface of the grain. However, a large amount of SiO$_2$ (2.1 mass%) and Nd$_2$O$_3$ (1.7 mass%) was detected at the grain boundary. Thus, it can be assumed that the grain boundary phase is mainly composed of SiO$_2$ and Nd$_2$O$_3$. (Note that the size of the detected beam spot is about 1–2 μm and it diverges into the material. Therefore, the detected region may include both the grain boundary and part of the inner grain.)

Figure 6.10 shows a bright field image of samples B1 and B4 cooled down at 600 and 10 °C h^{-1}, respectively, observed by TEM (transmission electron microscopy). The

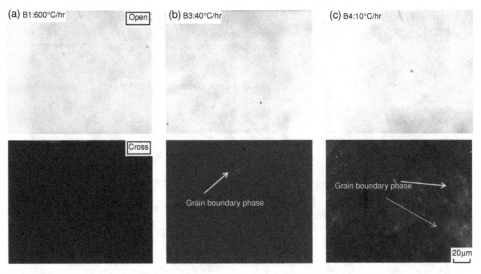

Figure 6.8 Open and cross-nicol photographs of specimens B1, B3 and B4 observed by a polarizing microscope. Optical anisotropic phases (extra grain boundary phases) are clearly seen in specimens B3 and B4.

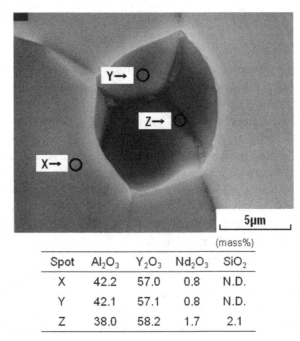

Spot	Al$_2$O$_3$	Y$_2$O$_3$	Nd$_2$O$_3$	SiO$_2$
				(mass%)
X	42.2	57.0	0.8	N.D.
Y	42.1	57.1	0.8	N.D.
Z	38.0	58.2	1.7	2.1

Figure 6.9 Spot analysis of specimen B4 around a grain boundary by EDX. Spots X, Y, and Z indicate the inner grain, surface of the grain and grain boundary, respectively.

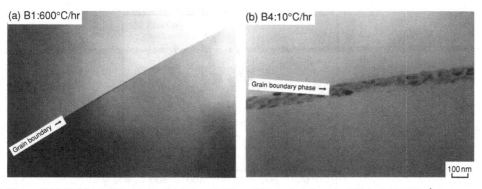

Figure 6.10 Bright-field images of specimens B1 and B4 cooled down at 600 and 10 °C h⁻¹, respectively, by transmission electron microscopy. The junction between faces of Nd:YAG grains is the observation position for both specimens.

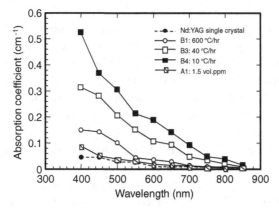

Figure 6.11 Dependence of the absorption coefficient of Nd:YAG single crystal and of Nd:YAG ceramic with various amounts of extra grain boundary phase on the measuring wavelength.

observation positions were the boundaries of Nd:YAG grains. The grain boundary of sample B1 was just a straight line, and no secondary phases other than YAG were seen at the observation position. Even if grain boundary phases are segregated in sample B1, the thickness of the grain boundary phase will probably only be a few nanometers. On the other hand, the segregation of elongated grain boundary phases (diameter ca. 10 nm × length ca. 50 nm) was observed at the grain boundary, of thickness approximately 80 nm. This result was compatible with the results of transmission optical microscopy and SEM.

Figure 6.11 shows the dependence of the absorption coefficients of Nd:YAG single crystal and Nd:YAG ceramic samples (B1, B3, B4, and A1) with various amounts of extra grain boundary phase on the measuring wavelength. (Sample A1 and the single crystal were used as reference.) The absorption coefficients of samples B3 and B4, which included the grain boundary phase, were much larger than that of sample B1, and it was clear that the

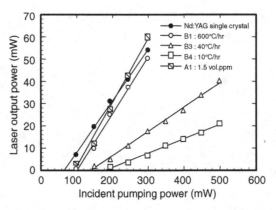

Figure 6.12 Laser output power versus input energy for 0.9 at.% Nd:YAG single crystal and 1.1 at.% Nd:YAG ceramics with various amounts of extra grain boundary phase excited with an 808 nm diode laser.

absorption coefficient becomes smaller with an increase in cooling rate (due to a decrease in the amount of secondary phase segregated at the grain boundary). However, the absorption coefficient of sample A1, which was prepared with less TEOS than the series B samples, was significantly smaller than that of sample B3 which was cooled down at the same rate, and also it was slightly smaller than that of sample B1 in which the grain boundary phase was scarcely observed. From this fact, it can be considered that sample B1 may include locally a very insignificant amount of grain boundary phase.

The laser oscillation performance of Nd:YAG ceramic samples (B1, B3, B4, and A1) and Nd:YAG single crystal are shown in Figure 6.12. The laser threshold of series B samples becomes smaller with an increase in cooling rate during the sintering process, and it was recognized that the slope efficiency also increased with an increase in cooling rate. This suggests that the laser output characteristic is also affected by the amount of grain boundary phase of the materials, since the amounts of residual pores in the materials were similar.

Among series B, sample B1 showed the best laser performance, but it was inferior to that of sample A1 and the Nd:YAG single crystal because its absorption coefficient was slightly larger. Therefore, it can be concluded that the slower cooling rate during the sintering process of Nd:YAG ceramics causes the segregation of a grain boundary phase and increases the scattering which is not favorable for laser materials.

The relationship between the average grain sizes of sample A1 (thickness = 10 mm) sintered at 1750 °C for 0.7, 1.5, 5, 10, 20, 40, 80, and 200 h under vacuum and the in-line transmittance at a wavelength of 1000 nm is shown in Figure 6.13. The grain size increased with an increase in holding time at the sintering temperature. When the grain size was larger than 20 μm (sintering time over 5 h), the residual pore volume of the sample was smaller than 3.0 vol. ppm, and the in-line transmittance reached a constant value of ca. 85% of the theoretical transmittance of YAG. This is because the number of residual pores in the material reached a constant level (pore volume between 3.0 and 1.2 vol. ppm), and

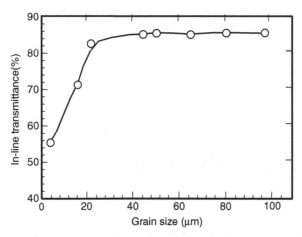

Figure 6.13 Relationship between the average grain size of specimen A1 sintered at 1750 °C for 1 to 200 h under vacuum and the in-line transmittance at a wavelength of 1000 nm.

the in-line transmittance became independent of grain size (20–95 μm). When the grain size was smaller than 20 μm (sintering time less than 1.5 h), the in-line transmittance was lower. This is not only influenced by the grain size. In these samples, densification and compositional homogenization have not yet occurred, and residual pores (>100 vol. ppm) and localized grain boundary phases ($YAlO_3$ and Al_2O_3) were observed. Accordingly, optical scattering occurred for these reasons.

Figure 6.14 shows transmission microscopy images of 0.9 at.% Nd:YAG single crystal grown by the Czochralski method. Inclusions tens of micrometers in size are observed, and the total amount of inclusions is extremely low. In the laser crystals produced in the 1990s, a small number of scattering objects were included even in commercial quality crystal rod which was cut from a good optical quality part of a single crystal boule grown by the CZ method. In addition, facets and inclusions were also observed in the single crystal laser rod, and they caused scattering in the crystal. Therefore, the absorption coefficient of single crystal is not always zero, as shown in Figures 6.5 and 6.11.

Figure 6.15 shows laser tomography images taken with a He–Ne laser (633 nm) of Nd:YAG ceramics with optical loss of 0.5% cm^{-1} and <0.1% cm^{-1} (the optical loss was measured using an Nd:YAG laser of 1064 nm), and a commercial Nd:YAG single crystal, showing the scattering condition in the materials. From the translucent aspect of the ceramics, the sample with optical loss of 0.5% cm^{-1} is very high grade, but its optical quality is not sufficient for use as a laser material. There were very few residual pores in the ceramic materials, and it was confirmed that there were no pores in the beam path of the He–Ne laser. The scattering object (white spot) is thought to be grain boundary scattering. But, in the sample with optical loss of <0.1% cm^{-1}, it was difficult to detect the scattering beam, and the level of scattering is equivalent to that of the single crystal which does not include grain boundaries.

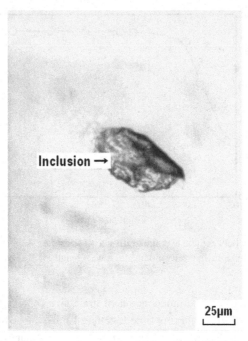

Figure 6.14 Transmission microscope photograph of 0.9 at.% Nd:YAG single crystal grown by the Czochralski method. A fairly small number of inclusions is observed in the commercial laser rod.

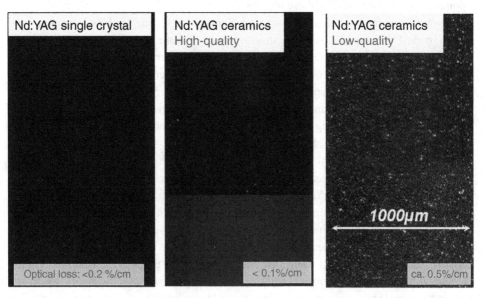

Figure 6.15 Laser tomography images taken with a He–Ne laser (633 nm) of Nd:YAG ceramics with optical loss of 0.5 and <0.1% cm^{-1} (the optical loss was measured using an Nd:YAG laser of 1064 nm), and a commercial Nd:YAG single crystal.

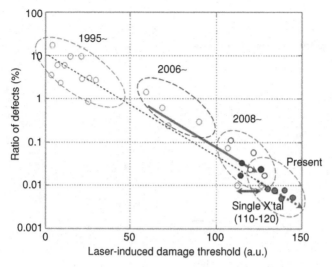

Figure 6.16 Laser-induced damage threshold of typical Nd:YAG ceramics with chronological data.

If we suppose that Rayleigh scattering by grain boundaries occurs in the ceramic materials, then scattering is directly proportional to the sixth power of the size of the scattering object. In this case, the scattering properties of ceramics will never be equivalent to those of high quality single crystals. Conversely, the optical loss of the high quality Nd:YAG ceramics prepared in this study is equivalent to or smaller than that of the single crystal counterpart. One of the reasons for this is that the size of structural defects near the grain boundaries which produce scattering is extremely small (in the case of a dislocation, the size is ca. 0.1 nm), and is sufficiently small compared with the laser oscillation wavelength (near 1000 nm). Therefore, Rayleigh scattering can be assumed to be extremely small in high quality Nd:YAG ceramics. Another reason is the dependence of the defect size on the measuring wavelength. The wavelength of the He–Ne laser used for laser tomography is 633 nm, and the optical scattering loss was measured with an Nd:YAG laser of 1064 nm. In the Nd:YAG sample with optical loss of 0.5% cm^{-1} (1064 nm), a lot of scattering was observed by laser tomography (633 nm) compared with the high quality ceramics. However, the scattering properties of these two samples may be similar to each other at the lasing wavelength (near 1000 nm).

Figure 6.16 summarizes the laser-induced damage threshold of typical Nd:YAG ceramics with chronological data prepared by the authors. In 1995, a highly efficient Nd:YAG ceramic laser was fabricated, but the ratio of defects (number of defects in the bulk sample) was very large in those ceramic samples measured at 633 nm (He–Ne laser) which is shorter than the lasing wavelength (1064 nm for Nd), and the LIDT (laser-induced damage threshold) value was small. However, this LIDT value has improved from year to year, and now it has reached the level of commercial high quality single crystals. The defect ratio was calculated

Figure 6.17 Internal microstructure of high quality Nd:YAG single crystal and ceramic irradiated with a high energy density YAG laser (pulse width: 8 ns, energy density 100 J cm^{-2}).

using the equation

$$\text{Ratio of defects} = \text{defect area/measured area.}$$

Defects include scattering due to macro defects such as facets in single crystals, and scattering due to nano-to-atomic level defects in ceramic grains. These experimental data were from measurements of scattering at 633 nm. Therefore, the ceramic with a small number of defects has a higher scattering result at 633 nm compared to scattering at the lasing wavelength (ca. 1000 nm).

A laser damage test was performed for high quality Nd:YAG single crystal and ceramic. A high energy density YAG laser (wavelength 1064 nm, pulse width 8 ns, energy density 100 J cm^{-2}) was irradiated into the material, and cracks due to laser damage were observed under a transmitted optical microscope. As shown in Figure 6.17, cracks were observed in the path of the giant-pulsed laser irradiation in both the single crystal and ceramic. It was noticed that the cracks in the single crystal were larger than those in the ceramic. This is because the ceramic has a large number of grain boundaries, and the damage resistance of the ceramic is larger than that of the single crystal [14]. Although we have not yet studied the details of parameters relevant for laser damage in single crystals and ceramics, it can be considered that the grain boundaries (difference in fracture resistance perhaps) suppress the spread of cracks in ceramics.

6.4 Discussion

Many kinds of laser materials have been developed conventionally, and it is obvious that the properties of the host materials greatly influence the laser characteristics. In addition, the quality of a given laser material also influences the laser output and laser beam quality. Therefore, it is necessary to control the manufacturing process strictly. Otherwise, it will be very difficult to produce high quality laser materials consistently. In the case of single crystal materials, Bernoulli's method was initially applied to the production of single crystals in the 1960s around the beginning of the development of solid-state lasers. However, the CZ method has become an important process for the production of commercial single crystals because of the need for strict quality control. John A. Carid *et al.* [15] investigated the relationship between optical loss (scattering) and laser oscillation efficiency using Nd, Cr co-doped GSGG (gadolinium scandium gallium garnet) and Nd-doped YAG single crystals. They reported that the laser oscillation efficiency is greatly affected by the losses of the optical resonator and laser host materials. Accordingly, we can evaluate the quality of laser host materials from the results of laser characteristics achieved by optimizing the optical resonator (using the same equipment) in a laser experiment.

The main difference in microstructure between ceramics and single crystals is considered to be the presence of grain boundaries and the absolute amount of residual pores. If we assume that optical scattering occurs at pores and grain boundaries in ceramics, it can be estimated that the optical loss in the optical resonator will be significantly increased with an increase in the absolute amount of such defects. If grain boundaries are the main cause of scattering, then the future of ceramic lasers is hopeless because basically ceramics consist of numerous grain boundaries. However, if the scattering is due to other defects (e.g., residual pores, secondary phases etc.), it is possible to control the number of such defects (scattering sources) using advanced ceramic technology. The scattering level of recently developed YAG ceramics is comparable to that of high quality Nd:YAG single crystals in terms of laser tomography measurement and laser damage resistance. Once clean grain boundaries are formed in ceramics, it is possible to obtain low optical loss material of 0.1% cm^{-1} or less, and it is possible to achieve high laser oscillation efficiency which is comparable or superior to that of single crystals. Conventionally, almost all materials scientists and laser scientists have considered that the grain boundaries in ceramics inhibit the generation of coherence beams such as lasers. However, the results of current ceramic laser experiments have reversed this conventional scientific or technological idea.

As described in Chapter 5, from the fact that the laser output characteristics of Nd:YAG ceramics are comparable to those of the single crystal counterparts, it can be assumed that the main scattering is not due to the presence of grain boundaries. However, it was difficult to justify this from the results achieved in the 1990s. In the research described here, different types of Nd:YAG ceramics with controlled amounts of residual pores and grain boundary phases were prepared, and their microstructures, scattering coefficients and laser output characteristics were investigated in order to distinguish the nature of scattering in ceramic laser materials.

Normally 0.5 mass% TEOS (finally it is decomposed into SiO_2) is added as a sintering aid to produce transparent Nd:YAG ceramics. But, in this study, an excess amount of TEOS (1 mass%) was added to prepare Nd:YAG ceramics with extra grain boundary phases. Here, it was confirmed that ca. 700 mass ppm of Si is included in this material (this is about

350 mass ppm higher than in normal Nd:YAG ceramics) by ICP analysis. By controlling the cooling rate from the sintering temperature to room temperature during the sintering process of these materials with excess Si included, samples with different amounts of grain boundary phases were produced. The sample B series included the same amount of Si, but there were two different Si positions depending on the cooling rate: Si remaining in the grains (solid solution), and Si segregated at the grain boundaries. In the case of normal material (350 mass ppm Si added to Nd:YAG ceramic), regardless of the very slow cooling rate (40 °C h^{-1}), no extra grain boundary phases other than YAG were observed under a transmitting optical microscope and SEM observation. However, the microstructure of Nd:YAG ceramics is changed (segregation of grain boundary phases occurred) by adding extra Si. As described in Chapter 3, it was confirmed that Si plays an important role in the homogeneous dissolution of large amounts of fluorescent Nd ions in the YAG host crystal. It can be estimated from the above experimental results that the amount of Si soluble in YAG crystal is very small. In the case of 1 mass% added Si in Nd:YAG ceramic, extra Si is oversaturated in the YAG crystal at the sintering temperature (1750 °C). It is considered that excess Si segregates at grain boundaries when the temperature decreases during the cooling process. Therefore, when the cooling rate is very slow, only the saturable amount of Si remains in the Nd:YAG grains and the excess Si is segregated at grain boundaries. When the cooling rate is faster, some excess Si which dissolved under non-equilibrium conditions remains in the Nd:YAG grains. Accordingly, samples with different amounts of grain boundary phases can be prepared with different cooling conditions.

As seen in Figure 6.11, it is clear that the scattering level is very different in the samples with different amounts of grain boundary phase, even though the samples have the same chemical composition. In samples B1 (cooled at 600 °C h^{-1}) and sample B4 (cooled at 10 °C h^{-1}), the absolute value of absorption coefficient differed by about three to eight times at each measuring wavelength. When samples with different amounts of grain boundary phases were pumped with an LD, the microstructure directly affected the absorption characteristics of the materials, and it also affected the laser characteristics. This means that a sample with low absorption coefficient (B1, grain boundary phases were not detected) has a low laser threshold and high slope efficiency. Likewise, a sample with a high absorption coefficient (B4, a large amount of grain boundary phases was detected) gives results unfavorable for use as a laser gain medium. Accordingly, it is clear that the grain boundary phase is an important factor that influences the laser characteristics (scattering characteristics as well). However, in Figure 6.9, an elongated grain boundary phase (length ca. 50 nm × diameter 10 nm) was confirmed in sample B4 by TEM observation. Since the size of scattering objects is much smaller than the measuring wavelength, the scattering intensity can be determined by applying Rayleigh's equation [16]:

$$S = (128\pi^5 d^6/3\lambda^4)[\{(n_2(\lambda)/n_1(\lambda))^2 - 1\}/\{(n_2(\lambda)/n_1(\lambda))^2 + 2\}]^2 \qquad (6.3)$$

where S, d, λ, and n_2 and n_1 are the cross-section for scattering of a particle, the radius of the scattering body, the measuring wavelength, and the refractive indexes of the host material and scattering body, respectively. From Eq. (6.3), the scattering intensity increases proportionally with the change in λ^{-4} and $[\{n_2(\lambda)/n_1(\lambda))^2 - 1\}/\{(n_2(\lambda)/n_1(\lambda))^2 + 2\}]^2$ and the number (or volume) of scattering centers. As shown in Figure 6.11, the scattering coefficient increased significantly with the decrease in measuring wavelength, especially at 400 nm in sample B4 with segregated grain boundary phase. The scattering coefficient ratio of sample B4 at both 400 and 800 nm is 13.9 (0.528 at 400 nm and 0.038 at 800 nm). According to Rayleigh's equation, the scattering loss coefficient at 400 nm becomes 16 times the value at 800 nm. The results obtained from the present study (B series) thus agree with Rayleigh's scattering theory. As seen in the above equation, the scattering intensity is also affected by the ratio of the refractive indexes of the host material and scattering body at a certain measuring wavelength. It is difficult to know the refractive index of the scattering body as its composition is unknown, but generally the variation in refractive index of the host material and scattering body is considered to be very small when the wavelength difference is about 400 nm (from 800 nm to 400 nm). Therefore, it can be estimated that the difference in refractive index ratio of the host material and scattering body at 400 and 800 nm is small, and hence the influence of the change in refractive index of the scattering body on the wavelength dependence of scattering is very small [13]. The absorption coefficient of sample B series varied in accordance with Rayleigh's scattering theory, and it was clear that the amount of scattering is determined by the microstructures of the samples. From these results, it can be concluded that the significant variation in the laser oscillation characteristics of sample B series, which have the same chemical composition, is mainly due to the difference in the scattering characteristics of the samples, and the amount of scattering is dependent on the difference in microstructure (increase or decrease in grain boundary phase).

Next, the results of sample A series, which have scattering caused by residual pores, will be discussed. In this case, the size of the scattering body (pores) is sufficiently larger than the measuring wavelength and thus Fresnel's scattering theory is applicable [17, 18]:

$$S = 3kV_f/8d \tag{6.4}$$

where k and V_f are a constant and the density of the scattering body, respectively. The absorption coefficient of sample A series increased proportionally with an increase in pore volume, in agreement with Fresnel's equation. However, it cannot explain the absorption coefficient of sample A series varying with measuring wavelength. In this experiment, the pore volume was calculated using a transmission optical microscope, and the minimum dimension of pores that can be observed under the microscope is a few micrometers. If there are pores smaller than the detection limit, then the scattering of sample A series will behave in accordance with both Fresnel's and Rayleigh's scattering theories. In this case, the change in behavior of the scattering coefficient of sample A series may resemble that of sample B series. The Si content in sample A series was as low as ca. 350 mass ppm, and it was confirmed that samples are free of grain boundary phases by transmission polarized optical microscope and scanning electron microscope (SEM) observation. Therefore, it is considered that the difference in microstructure of sample A series is simply the number of residual pores. In materials with grain boundaries, it is well known that the dislocation density is very high near the grain boundaries, and this defect affects the optical properties to a small extent. That is, the difference in crystal orientation is very large at the junctions (grain boundary) between grains so that high angle grain boundaries (HAGBs) are mostly formed. Therefore, disorientation (density variation) of the atomic order occurs, and regions of inhomogeneous refractive index are generated locally as a result. Theoretically, it can be considered that the intrinsic defects (grain boundaries) in ceramics cause light scattering. However, from the experimental results, it was recognized that the scattering level of Nd:YAG ceramics with very low amounts of residual pores is almost comparable to that of single crystals. From this fact, it was shown that scattering at the grain boundaries *(clean grain boundaries without secondary phase)* is negligibly small compared with the scattering at pores. Recently, the fabrication of high quality Nd:YAG ceramics has become possible, and the optical quality is superior to that of commercial Nd:YAG single crystals in terms of low scattering and high homogeneity etc. This means that the scattering due to grain boundaries is almost negligible at the lasing wavelength (around 1 μm). Accordingly, the presence of residual pores *(not grain boundaries)* should be addressed as one of the factors causing scattering in ceramics.

Samples with different grain size were prepared by changing the sintering time, and then their in-line transmittance was measured. As shown in Figure 6.13, the in-line transmittance became almost constant (85% at 1 μm) when the grain size ranged from 20 to 95 μm. In this case, the amount of residual pores in the samples was within 3.0–12.0 vol. ppm, and the scattering must depend only on the number of grain boundaries. In the samples with average grain size of 20 μm and 95 μm *(samples of the same dimension)*, the number of grain boundaries is larger in the sample with smaller grain size. When an incident beam passes through the samples, the number of encounters of the incident beam with grain boundaries will be about five times different. However, the results of in-line transmittance measurement showed that the scattering levels were almost equal. Basically, this suggests that the scattering is not mainly due to grain boundaries.

In addition to the basic measurement of transmittance, internal scattering of Nd:YAG ceramic samples with different levels of scattering and of high quality single crystals was observed directly by laser tomography. The measuring wavelength was 633 nm, which is shorter than the lasing wavelength. Therefore, it is considered that the scattering observed at 633 nm will be stronger than scattering during lasing. It was confirmed that when the internal scattering of Nd:YAG ceramics at 1064 nm is <0.1% cm^{-1}, the optical quality is comparable to that of the commercial high quality single crystal counterpart. When a He–Ne laser beam was irradiated into the samples with scattering level of 0.5% cm^{-1} (without residual pores), a beam line was detected along the beam. It is considered that such scattering occurred near grain boundary regions. In the samples without residual pores, the main scattering sites must be grain boundaries only. A perfect grain boundary, which does not cause scattering, must be clean and free of segregation of grain boundary phases that can be detected by transmission polarized optical microscopy, high resolution SEM (scanning electron microscopy) and HRTEM (transmission electron microscopy). Ceramics having perfect grain boundaries, as confirmed by these detection methods, have extremely low scattering characteristics and excellent optical quality that are comparable or superior to those of the single crystal counterpart.

Once the scattering characteristics of ceramics are comparable to those of single crystals, the next step is to produce ceramic materials with high resistance to high power pulse laser damage which is comparable to that of single crystal material. Regarding the optical damage in ceramics, we have not yet confirmed whether the fracture originated from the bulk (inside the grain) or the grain boundary. Generally, the dislocation density is very high around the grain boundaries, and it is likely that the fracture originates from the grain boundary region. Likewise, single crystal materials also include optically inhomogeneous parts (e.g., facets), and it is considered that laser damage originates from these inhomogeneous parts.

It is interesting to know which material has superior performance in terms of scattering characteristics and optical damage, whether it is single crystal or polycrystalline ceramic material. The single crystal material has been developed and improved for over 40 years, and the possibility of improving the optical quality further is now small. In the case of optical ceramics, the removal of residual pores and the formation of clean grain boundaries are essential for improving the optical quality. But the optical characteristics of ceramics can be further enhanced by a new idea. In Chapter 7, the fabrication of single crystal ceramics by sintering will be described. In these new ceramics, grain boundaries do not exist. Therefore, such new processes will probably be essential to the fabrication of optical ceramics with extremely low scattering loss.

6.5 Summary

In order to investigate the nature of scattering in Nd:YAG ceramics, many kinds of ceramic samples with different amounts of residual pores and grain boundary phases were prepared. Correlations between their microstructures, scattering characteristics, and laser oscillation performance were examined. As a result, the following conclusions were drawn.

(1) Light scattering in Nd:YAG ceramics was mainly caused by residual pores and grain boundary phases, and the amount of scattering depended on the absolute number of such defects. Therefore, it is necessary to control the number of pores and grain boundary phases to as low a level as possible in optical ceramics in order to achieve higher laser oscillation performance.
(2) The laser oscillation performance of Nd:YAG ceramics is comparable to that of the single crystal counterpart only when the amount of residual pores reaches approximately 1.5 vol. ppm and the material is free of grain boundary phases.
(3) Basically in the Nd:YAG ceramics, (clean) grain boundaries did not adversely affect the scattering characteristics and the laser oscillation performance.

References

[1] A. Ikesue, K. Kamata, and K. Yoshida, Effects of neodymium concentration on optical characteristics of polycrystalline Nd:YAG laser materials, *J. Am. Ceram. Soc.* **79** (1996) 1921–1926.

[2] A. Ikesue and K. Kamata, Role of Si on Nd solid-solution of YAG ceramics, *J. Jpn. Ceram. Soc.* **103** (5) (1995) 489–493.

[3] A. Ikesue, T. Kinoshita, K. Kamata, and K. Yoshida, Fabrication and optical properties of high-performance polycrystalline Nd:YAG ceramics for solid-state lasers, *J. Am. Ceram. Soc.* **78** (4) (1995) 1033–1040.

[4] A. Ikesue, T. Taira, Y. Sato, and K. Yoshida, High-performance microchip lasers using polycrystalline Nd:YAG ceramics, *J. Jpn. Ceram. Soc.* **108** (4) (2000) 428–430.

[5] I. Shoji, T. Taira, and A. Ikesue, Thermally-induced-birefringence effect of highly Nd^{3+}-doped $Y_3Al_5O_{12}$ ceramic lasers, *Opt. Mater.* **29** (2007) 1271–1276.

[6] A. Ikesue and Yan Lin Aung, Synthesis and performance of advanced ceramic lasers, *J. Am. Ceram. Soc.* **89** (6) (2006) 1936–1944.

[7] A. Ikesue and Y. L. Ang, Ceramic laser materials, *Nature Photon.* **2** (2008) 721.

[8] G. Adachi, K. Shibayama, and T. Minami, *New Technology for Advanced Materials*, Kagaku-Dojin, Kyoto (1987), p. 183.

[9] A. Ikesue and K. Yoshida, Scattering in polycrystalline Nd:YAG lasers, *J. Am. Ceram. Soc.* **81** (8) (1998) 2194–2196.

[10] A. Ikesue, K. Yoshida, T. Yamamoto, and I. Yamaga, Optical scattering centers in polycrystalline Nd:YAG laser, *J. Am. Ceram. Soc.* **80** (6) (1997) 1517–1522.

[11] A. Ikesue and K. Yoshida, Influence of pore volume on laser performance of Nd:YAG ceramics, *J. Mater. Sci.* **34** (1999) 1189–1195.

[12] L. D. Landau and E. M. Lifshitz, *Electromagnetism of Continuous Media*, Pergamon, Oxford (1975), p. 272.

[13] W. L. Bond, *J. Appl. Phys.* **36** (5) (1965) 1674–1677.

[14] A. A. Kaminskii *et al.*, *Crystallogr. Rep.* **50** (2005) 869; H. Yagi, T. Yanagitani, T. Numazawa, and K. Ueda, *Ceram. Int.* **33** (2007) 711–714.

[15] J. A. Caird, N. D. Shinn, T. A. Kircoff, L. K. Smith, and R. E. Widder, *Appl. Opt.* **25** (23) (1986) 4305–4520.

[16] H. C. van de Hulst, *Light Scattering by Small Particles*, John Willey and Sons, New York (1957).

[17] K. Hamano, *Fine Ceramics Handbook*, Asakura, Tokyo (1987), p. 822.

[18] J. G. J. Peelen, Microstructure of ceramics, *Br. Ceram. Soc.* (1975).

7

Advanced technologies in ceramics (composite, fiber, single crystal by sintering method, etc.)

Generally a laser gain medium made of crystal or glass has a uniform design and simple composition, but they have played an important role in solid-state lasers in the twentieth century. Conventionally, laser functionalities such as high power output, high beam quality and short pulse generation have been improved only by amplifying the emission generated from a high optical quality laser gain medium. The research and development of laser gain media principally involved improving the optical quality, growing large scale media, and synthesis of materials with new composition. In recent years, however, conventional technologies have become no longer able to contribute to the improvement of laser performance, and new functional laser elements have been required. For example, a composite laser element was demonstrated in 1995 in America which achieved a new level of performance (functionality) that could not be obtained using a simple laser element [1]. A composite laser element is formed by bonding different crystals with two or more different compositions. First, the bonding surfaces of the crystals are optically polished, then brought in contact with each other and heated at a high temperature to obtain a composite laser element. This conventional diffusion bonding technology is limited to flat surface bonding only, and the bonding strength is not high. On the other hand, composite laser elements made of polycrystalline ceramic materials can have any desired design because the ceramic technology is based on a powder process. Therefore, it is expected that ceramic technology can solve the problems and limitations seen in conventional crystal bonding. In this chapter, the difference between conventional single crystal composites and advanced ceramic composites will be described, together with the advantages of ceramic technology [2].

Glass laser gain media can be produced on a large scale with high optical quality, and they have been used as high energy short pulse laser sources for nuclear fusion experiments. Although a high energy laser is generated, poor thermal conductivity limits the practical applications in CW operation and pulse operation with a high repetition rate. Glass optical fibers have been developed and are now widely used in optical communication systems. They are replacing the conventional copper cables because, for example, they can transmit over long distances at high speed and have a large carrying capacity with low signal loss etc. Generally, optical fiber used for communication is made up of a core (carries the light pulses) and cladding (reflects the light pulses back into the core). Two different compositions of fused silica glass are placed in a double crucible and melted. This is called the double-crucible method. Then the melted glass is pulled into a wire to produce a clad–core optical fiber. This fiber production technology was used to develop a glass fiber laser gain medium. Laser active elements are doped only in the core; the cladding layer does not include any laser active elements and has a lower refractive index than the core. By using a long and narrow glass fiber gain medium (diameter several tens of micrometers), it was possible to generate a high quality beam because the heat generated can be removed along the long and narrow fiber. In addition, light propagation occurs smoothly inside the core, and a highly concentrated laser beam can be generated without the use of any focusing lens. If it is possible to adapt this glass fiber laser technology to crystalline materials, it will become an innovative technology for improving laser functionalities in the future. In this chapter, the fabrication of narrow ceramic fibers using advanced technology and an exploration of their laser performance are discussed.

In general, materials produced by the melt-growth process are single crystal, and materials produced by sintering are polycrystalline ceramics. In this century, however, this kind of general idea is changing. The word "ceramics" is derived from a Greek word "Keramos," and generally it refers to a polycrystalline material produced by sintering (without melting). Initially, Matsuzawa *et al.* found abnormal grain growth in a sintered body, and they used this phenomenon to grow a single crystal directly from a sintered body (polycrystalline ceramic) without melting. They succeeded in producing an Mn–Zn ferrite single crystal by sintering, and in the 1980s this material was used in magnetic tape heads [3–5]. Normally, ceramic materials have many grain boundaries and the grains drop out easily. Therefore, a new material, which had good wear properties, without grain boundaries, was required for magnetic tape heads. Similarly, the author considers that ceramic laser materials without grain boundaries produced by sintering will be very important for future solid-state laser technology.

Currently, the crystal structure of polycrystalline laser materials is limited to the "cubic" system. Current technology cannot produce non-cubic ceramic laser materials. Ceramic laser materials, typically Nd:YAG ceramics, include many grains and grain boundaries. Although the optical quality of each crystal grain is very high, there are many dislocations near the grain boundary regions where crystal grains with different crystal orientations contact each other. These dislocation defects at grain boundaries may cause scattering due to the change in refractive index, and a decrease in thermal conductivity due to the localized defection of lattice vibration, and may be sites of optical damage during high power laser operation. Therefore, grain boundaries are considered to be a possible major problem in the development of high performance laser elements (ideal laser crystals) in the future. Although the development of single crystal ceramics by sintering (SSCG, solid state crystal growth) is still in progress, the advantages of single crystals produced by sintering will be discussed in this chapter.

7.1 Composite technology

7.1.1 Problems of current single crystal composites

In order to improve the laser performance, two different crystals with different compositions but the same crystal structure are bonded to make a composite laser element. Such composite laser elements are used in the research and development field. First, the bonding surfaces of single crystals are polished to an optical grade, and then they are brought in contact with each other and heated at a high temperature (lower than the melting point) for diffusion bonding [6].

Using this composite fabrication technology, it has become possible to produce advanced laser elements with complicated designs, such as end-cap and clad–core designs, which cannot be produced directly by melt-growth (single crystal) technology. Very interesting results were achieved with advanced crystal composite laser elements. However, there are still some issues in these crystal composites from the viewpoint of materials science.

(1) The directions of the bonding crystals are different from each other, and it is difficult to produce one component.
(2) Although the bonding surfaces are polished to an optical grade before bonding, surface roughness at the nano-level can be detected. Thus, when these two bonding surfaces are contacted with each other, interstices are formed at the interface. Therefore, a perfectly bonded interface cannot be produced, resulting in poor thermomechanical properties in the composite element, so it is not resistant to (high power) laser damage.
(3) Conventional diffusion bonding technology is limited to the bonding of flat surfaces, and only simple design composite elements can be produced.
(4) The diffusion distance of laser active ions is limited, and it is not possible to control the microstructure of the bonding interface in the crystal composite.

Among these issues, (1) and (2) especially are intrinsic problems of crystal composites. Therefore, practical composite elements cannot be produced for industrial use unless these technical issues can be solved.

The technical problems of current single crystal composites in laser applications are summarized in Figure 7.1(a). In single crystal composites, generally the crystal system is the same but the compositions are different. For example, in a crystal composite made of Nd:YAG and YAG, the host crystals have the same garnet structure, but their compositions are different: Nd-doped YAG and undoped YAG. First, their bonding faces are polished to an optical grade. Then the polished faces are made to stick together (1) with or (2) without the use of adhesive materials (e.g., inorganic adhesive). Then the sample is heated at a high temperature (lower than the melting point) to obtain a diffusion bonded interface. Although the finishing of the bonding surfaces is very fine, between $\lambda/10$ and $\lambda/20$ (at $\lambda = 633$ nm), when one face is placed neatly on the other face, partially contacted points are generated. Since it is not possible to obtain a perfect flat surface even by high precision polishing, when the bonding faces are contacted many interstices of size $\lambda/10 \sim \lambda/20$ are generated. Therefore, the thermomechanical properties of the bonding interface are not sufficient.

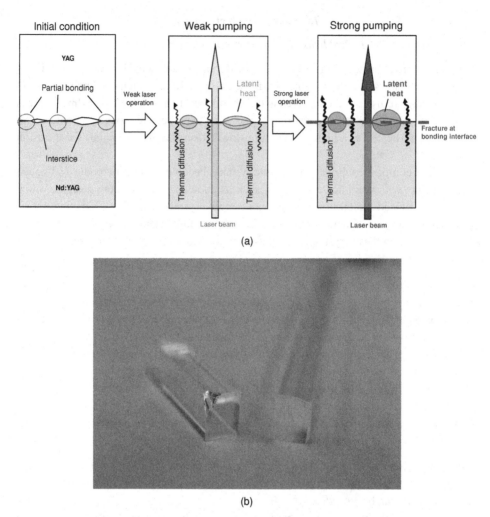

Figure 7.1 (a) Problems of current composite elements for laser application. (b) Fracture surface of a commercial composite (YAG–Yb:YAG single crystal) after irradiation with 940 nm LD source (input power 500 W). (The flat surface is a fracture surface at the bonding interface.)

When these crystal composites are tested for laser operation, it is possible to generate laser radiation at low power excitation. Because there is no chemical bond between the Nd:YAG and YAG crystals at the interface (covalent bond between Nd:YAG and YAG at the atomic level), the interface of this crystal composite has very low mechanical strength. In addition, it is expected that thermomechanical properties such as the thermal conductivity and thermal diffusivity will be very poor. Accordingly, this crystal composite is suitable for low power laser operation but when the input power is increased, the defects at the interface absorb thermal energy and the beam quality and lasing efficiency decrease. When

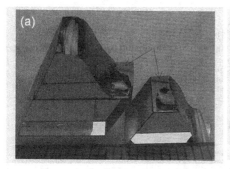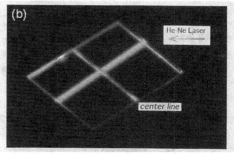

Figure 7.2 (a) Surface of bonded Ti:sapphire laser material after a fracture experiment. (b) Scattering condition when a He–Ne laser is irradiated into the bonding interface [7].

this composite is strongly excited for high power laser operation, the sample may break down due to thermal and mechanical damage in the defect regions.

Figure 7.1(b) shows a crystal composite made of YAG–Yb:YAG after a high power laser operation test. When the crystal composite sample was pumped with an LD of 500 W ($\lambda = 940$ nm), a fracture was generated from the bonding interface; this fracture had a very flat surface. This suggests that the bonding strength of the crystal composite produced by conventional diffusion technique is not high.

Another example of a crystal composite reported by Sugiyama *et al.* is shown in Figure 7.2 [7]. A Ti:sapphire crystal composite was prepared by a diffusion bonding technique. Figure 7.2(a) shows a fracture surface of the crystal composite after a Q-switch laser operation test. A flat surface was observed at most of the fracture surface. When breakage occurs in a material, normally an irregular fracture surface is observed, except for cleavage materials. Therefore, the flat area of the fracture surface in the figure suggests a non-bonding area. The bonding interface of the crystal composite was detected with a He–Ne laser. When a laser beam was irradiated perpendicular to the bonding interface, a scattering beam line along the bonding interface was observed as shown in Figure 7.2(b). This scattering beam line suggests that interstices are formed at the bonding interface. This result is consistent with the results shown in Figure 7.2(a), indicating that the bonding condition is not perfect.

Composite laser elements are very promising for improving laser performance. However, the conventional bonding technology used in making crystal composites limits the potential for high power and high functional lasers. For that reason, advanced ceramic bonding technology is required to overcome the technical issues of conventional crystal bonding technology.

7.1.2 Meaning of advanced ceramic composite

In this section, the difference between composites fabricated by conventional bonding technology and those fabricated using an advanced ceramic technology will be described

	YAG YAG Nd:YAG (Layer)	(Cylindrical)	(Waveguide)	Fiber (Clad-ocro)	Gradient
High power & efficiency (heavy obstruct of Nd^{3+} too)	◎	◎	◎	◎	◎
High beam quality low thermal loss low thermal double reflection	◎	◎	◎	◎	◎
Beam mode control	-----	○	○	◎	---
Beam pattern control	-----	○	○	---	◎
High function	◎	○	---	○	---
	○	○	○	◎	○

◎ :Very Good   ○ :Good   -- : Same as conventional

Figure 7.3 Typical examples of ceramic composite laser materials.

[8]. Some typical examples of advanced ceramic composites are summarized in Figure 7.3. Advanced ceramic technology allows the fabrication of composite laser elements with flat bonding interfaces such as (1) layer-by-layer, (2) square clad–core and (3) waveguide etc. which can be fabricated using conventional crystal bonding technology. It also allows the fabrication of composite laser elements of complicated design such as cylindrical clad–core elements which have a curved bonding interface that cannot be achieved using conventional crystal bonding technology. In the fabrication of ceramic fiber laser elements, it is possible to produce a ceramic fiber of 100 μm diameter end-capped with undoped YAG at both ends.

The doping profile of Nd ions in the composite rod shown in Figure 7.4 is not a step-by-step structure as in the conventional composite. This rod has a smooth Nd doping profile gradient along the length direction. It was not possible to produce such a smooth gradient composite using conventional crystal bonding technology, but advanced ceramic technology has made it possible. In the case of the ceramic process, for example, granulated powders of different concentrations of Nd-doped YAG are stacked in layers along the length direction, and then they are pressed into a powder compact. A composite laser element with smooth Nd gradient doping profile can be obtained after sintering this powder compact.

Two types of laser elements, an Nd:YAG single crystal with uniform doping profile of Nd ions, and an Nd:YAG composite with gradient doping profile of Nd ions (Figure 7.5), were prepared with the same dimensions. An LD of 808 nm irradiated from an end position, and the thermal distribution in the laser elements was inspected by a thermography camera. In the case of a uniform doping profile it was observed that heat generation is very high at the end position where LD light is pumped directly. In the case of a gradient doping profile, heat generation is very gradual throughout the sample. Therefore, by controlling the doping profile of laser active ions throughout the composite laser element, it has become possible to manage the thermal distribution in the laser element. Thus, advanced ceramic composite

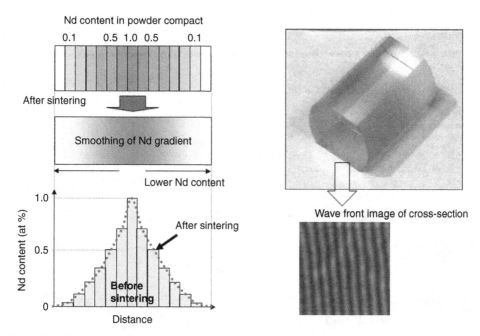

Figure 7.4 Appearance of YAG ceramics with smooth gradation of Nd ions.

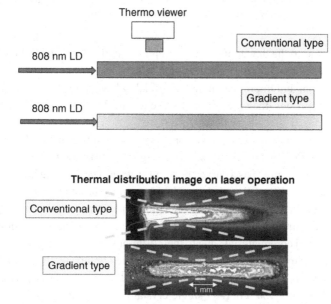

Figure 7.5 Comparative data of heat generation for YAG ceramics with homogeneous Nd distribution and smooth Nd gradation.

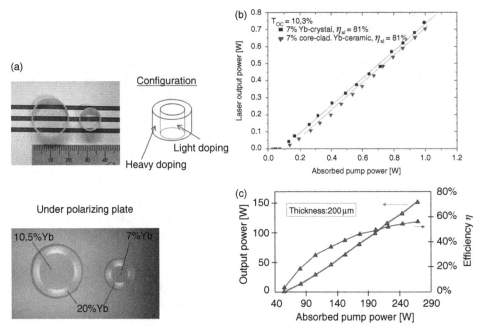

Figure 7.6 (a) Appearance of a cylindrical clad–core structure Yb:YAG ceramic. (b) Laser performance of high quality commercial Yb:YAG single crystal and the clad–core composite Yb:YAG ceramic pumped with a Ti:sapphire laser. (c) Laser performance of a thin disk of the clad–core 7%–20% Yb:YAG microchip ceramic composite pumped with a 940 nm LD.

technology can lead to the development of high power, high energy and high beam quality lasers that were not attainable using conventional technology. Although investigations of the thermal management and improvement in beam quality have not yet been performed using these ceramic composites, they are very attractive for the development of laser technology in the future.

An advanced clad–core design ceramic composite, the cladding material is 20% Yb:YAG and the core material is 7% Yb:YAG, is shown in Figure 7.6(a), together with a transmission polarized image. This sample has a curved bonding surface. The bonding interface cannot be seen by the naked eye. When it was inspected under a transmission polarizer, birefringence due to stress was observed in the cladding area only, and the core area was stress free. It is considered that such stresses were caused by the difference in thermal expansion coefficient between cladding and core materials due to their different chemical compositions. Only the core area contributes to laser generation in this clad–core composite, thus the birefringence in the cladding area is not a major issue for laser oscillation. The laser performance of a commercial high quality 7% Yb:YAG single crystal (not a composite) and the advanced clad–core composite (core 7% Yb:YAG) shown in Figure 7.6(a) is shown in Figure 7.6(b). Ti:sapphire laser was used as an excitation source. Both laser elements showed low lasing thresholds, and high slope efficiencies around 84%; the quantum efficiency of the Yb:YAG

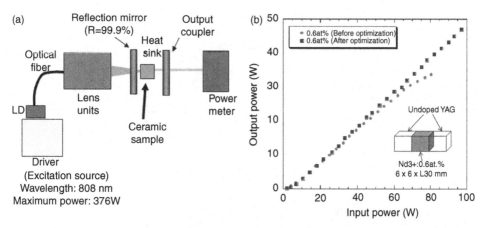

Figure 7.7 (a) Laser oscillation setup for the end-capped Nd:YAG ceramic composite pumped with an 808 nm LD, and (b) its laser performance after optimizing the laser oscillator.

laser was 91%. The above clad–core composite was processed into a microchip laser element with a thickness of 200 μm, and pumped with a 940 nm LD. The laser performance is shown in Figure 7.6(c). Maximum output power of 150 W and slope efficiency of 60% were achieved [9].

The laser oscillation setup and laser performance of an end-cap composite laser element made of YAG–0.6%Nd:YAG–YAG (dimension 6 × 6 × 30 mm^3) ceramics are shown in Figure 7.7(a) and (b), respectively. The left figure shows a schematic diagram of laser oscillation. The lasing threshold was very low, and stable CW laser oscillation up to 50 W with 50% slope efficiency was achieved. For reference, a single crystal composite, with the same configuration as the end-cap ceramic composite, which was prepared by conventional diffusion bonding technology, was also investigated. Lasing was not stable up to 50 W output. A reduction in output power and beam quality was also confirmed, suggesting imperfect bonding interfaces of the end-cap crystal composite.

The preparation of a waveguide laser element and its appearance (external dimension 10 × 64 × 1 mm^3) prepared using advanced ceramic composite technology are shown in Figure 7.8(a) and (b), respectively. Undoped YAG was used as the cladding layer, and 0.6% Nd:YAG was used as the core layer with a thickness of 200 μm. Under an optical microscope, no interstices and diffusive phases were observed at the bonding interfaces of YAG–Nd:YAG–YAG. A laser oscillation setup for this waveguide sample is shown in Figure 7.8(c). The incident pump light was propagated and amplified in a zig-zag path in the core layer of the waveguide element, and finally stable laser output up to 100 W was achieved. Although details of the laser performance and beam quality are being investigated, a high power output with very high beam quality has been achieved in recent experiments. A future plan is to adopt a side-pumping method using an 800 W LD in order to achieve around 400 W output power. In a current project, a large waveguide element

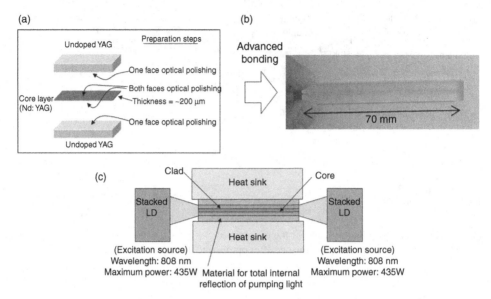

Figure 7.8 (a) Preparation steps for a waveguide structure, (b) photograph of a ceramic waveguide element, and (c) laser oscillation setup for a waveguide laser.

(100 mm × 20 mm core layer thickness 100 μm) was prepared and it will be tested for several kilowatt laser generation.

In order to investigate the bonding strength of advanced ceramic composites, a composite made of Nd:YAG and YAG was prepared, and four-point bending tests and fracture tests were conducted. As shown in Figure 7.9, the mechanical strength of the bonding interface is very good and fracture only occurred at positions away from the bonding interface. The fracture surface was not flat and was irregular in shape. The four-point bending strengths of the host materials (Nd:YAG and YAG ceramics) and of the advanced ceramic composite materials (Nd:YAG-YAG) were 220–290 MPa and 220–280 MPa, respectively. It is clear that the bonding strength of the ceramic composite is almost equivalent to the mechanical strength of the host material. This result suggests that the bonding condition in the advanced ceramic composite is perfect (ideal condition). Accordingly, the thermomechanical issues of the bonding interface seen in crystal composites will not be a problem in the case of advanced ceramic composites.

In addition, the microstructure around the bonding interface of the above ceramic composite was observed by a transmission optical microscope, SEM (scanning electron microscope), and TEM (transmission electron microscope), as shown in Figure 7.10. At the bonding interface, no interstices were observed by optical microscopy. No double refractions (stress caused by bonding) were observed under a polarizer. SEM observations also revealed that there are no secondary phases (abnormal microstructure) and no micro-interstices at the bonding interface at a micrometer level. Only a slight difference in grain size of Nd:YAG and YAG was recognized. In addition, using HR-TEM (high resolution

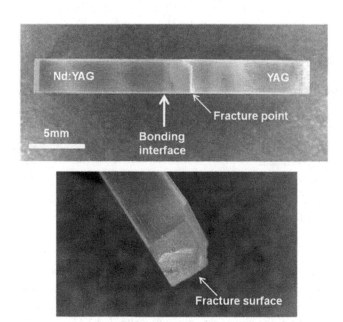

Figure 7.9 Fracture surface of Nd:YAG–YAG ceramic composite material. We can confirm that fracture did not occur at the bonding interface, and that the fracture surface is irregular (not a flat surface).

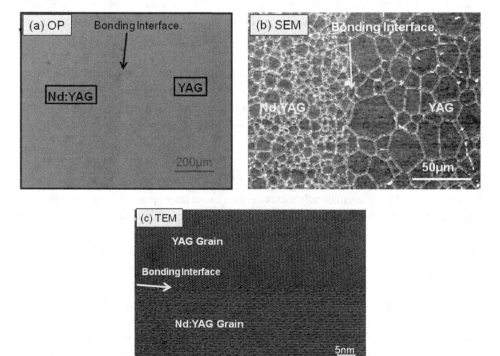

Figure 7.10 Microscopic observation near the bonding interface by (a) optical transmission microscopy (OP), (b) SEM and (c) TEM.

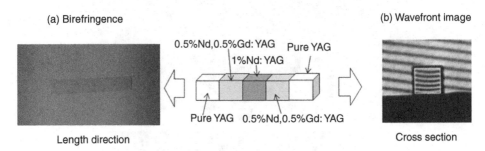

Figure 7.11 Inspection of Nd:YAG–YAG ceramic composite: (a) birefringence observed by polarizing plate and (b) wavefront image produced by interferometry.

TEM) observation, it was confirmed that the Nd:YAG and YAG ceramics are bonded at an atomic level and no interstices exist below nanometer level. Accordingly, the bonding condition of the advanced ceramic composite is significantly better than that of commercial crystal composites.

The optical quality of the above ceramic composite (Nd:YAG–YAG) was investigated on a macro scale. Figure 7.11(a) and (b) show a transmitted polarized image of the whole area of the composite, and a wavefront image taken using interferometry, respectively. There were no stresses along the composite caused by the different physical properties of Nd:YAG and YAG after bonding. In addition, straight fringes were confirmed in the wavefront image, suggesting that the optical homogeneity throughout the sample is very high.

The above results have confirmed that the advanced ceramic composite provides ideal bonding conditions, which can make perfect the inherent characteristics of composite laser elements. Therefore, it is certain that the advanced ceramic composite technology will contribute significantly to the advancement of solid-state laser technology in the future.

7.2 Ceramic fiber laser

The optical quality of glass materials is excellent for use in laser optics. However, the thermal conductivity of glass is very poor, and it is limited to use as a high energy driver laser for nuclear fusion experiments with very low repetition rate (for example, a few shots per day). One of the advantages of glass is that it can be processed into very thin and narrow wires, called fibers, above the glass transition temperature. Therefore, because of its geometry (several tens of micrometers diameter), although glass has poor thermal conductivity, it is possible to remove heat effectively from a glass fiber doped with laser active ions. This means that glass fiber laser technology is very attractive for the development of high power and high beam quality lasers. Since the core size can be reduced to the level of the lasing wavelength, thermal dissipation can be improved. Additionally, in the case of clad–core glass fibers, the refractive index of the cladding materials is lower than that of the core layer doped with laser active ions, so the generated laser beam can be reflected inside the

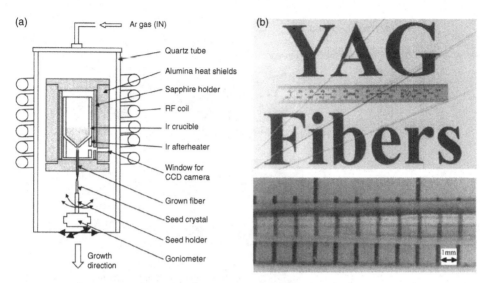

Figure 7.12 (a) Schematic diagram of the μ-PD method of producing crystal fiber optics. (b) Appearance of crystal fiber produced by the μ-PD method. Reprinted with permission from [10], copyright 2003, American Institute of Physics.

cladding area, and is confined in the core area. Therefore, it is possible to generate a high beam quality laser without using any focusing lens units.

Although the glass fiber laser has many advantages, it also has a weak point. That is low laser gain due to the limited doping concentration of laser active ions (e.g., Nd ions) in glass media. Therefore, the power generated is limited to ca. 1 W per unit length (i.e., 1 W cm^{-1}) of the glass fiber. It is clear that a glass fiber several tens of meters in length is necessary to generate several kilowatts laser emission. For that reason, if it is possible to produce crystalline fibers, the performance of fiber lasers is expected to be enhanced, and even a short fiber will be able to generate a high power output. However, there are still many difficulties in making crystal fiber lasers fit for practical use with current technology. Some studies on crystalline fibers are described below.

Normally crystal fibers are produced by a μ-PD (micro pulling down) method based on the conventional melt-growth method. Figure 7.12(a) and (b) show a diagram of the μ-PD method and the crystal fiber grown by this method, respectively [10]. As in the CZ method, high purity raw materials are melted in an iridium crucible. The melt is pulled down through a capillary tube at the crucible bottom and contacts with a small seed crystal. By pulling down the melt very slowly together with the seed crystal, a long and thin single crystal can be obtained. However, the diameter of the grown crystal is about 1 mm, and is relatively large on the scale of optical fibers. The optical quality of the grown crystal fiber is analogous to that of bulk crystal grown by the CZ method, and includes optically inhomogeneous regions such as facets and core. Therefore, it is very difficult to achieve laser oscillation using such crystal fiber materials. Recently, a research team from France

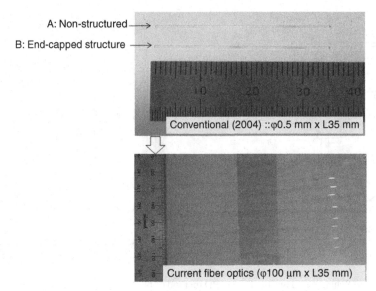

A: Non-structured ⟶

B: End-capped structure ⟶

Conventional (2004) ::φ0.5 mm x L35 mm

Current fiber optics (φ100 μm x L35 mm)

Figure 7.13 First demonstration of Nd:YAG ceramic fiber optics presented at the 2004 CLEO (Conference of Laser and Electro Optics) and a current fiber.

has reported laser oscillation using a μ-PD grown crystal fiber [11]. The diameter of the crystal fiber core was about 1 mm, and it also included defects seen in the bulk crystal. Therefore, laser oscillation from this crystal fiber was difficult to produce, and the laser oscillation efficiency was not very high.

The author has also attempted the fabrication of a ceramic fiber by advanced ceramic technology. The appearance of ceramic fibers reported at the 2004 CLEO Conference (Long Beach, CA, USA) [12], and produced in 2008 (unpublished data), is shown in Figure 7.13. In 2004, the diameter of the fiber core was around 500–1000 μm, its length was only about 50 mm and it was like a small rod. The dimension of the most recent ceramic fibers is around 100–200 μm with a length of 100–150 mm. In fact, the core diameter of the fiber should be nearly equal to the wavelength of the laser, but even the core diameter of current glass fibers is about several to several tens of micrometers. In the case of ceramic fibers, the core diameter should be reduced to less than 100 μm in order to achieve the special characteristics of an ideal fiber laser. There are still many technological problems in the production of ceramic fibers to be solved.

In 2004, the author reported two types of ceramic fibers. One had a simple design, and the other was an advanced fiber having an end-cap design. End-pumping was applied to compare their laser performance and its dependence on geometry. Figure 7.14(a) and (b) show their laser performance and the laser beam profile generated from an end-cap fiber, respectively. With the end-pumping technique, the simple design fiber easily reached output saturation conditions, but such saturation was not reached in the end-cap fiber. The slope efficiency was low (only about 23%) because the optical quality of the ceramic fiber was

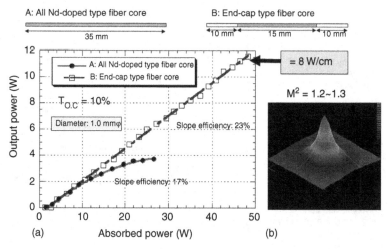

Figure 7.14 (a) Laser performance of non-structured and end-capped Nd:YAG fibers. (b) Beam profile emitted from end-capped fiber optics.

not very good. (The optical loss of this fiber was not measured but it is estimated to be 1–2% cm^{-1}.) However, it reached an output power of 8 W cm^{-1}, which is about eight times the output of a glass fiber laser. The beam profile pattern of the end-cap fiber shows that the generated laser beam is nearly TEM$_{00}$ mode (Gaussian mode), and its M^2 value (beam quality) was as good as 1.2–1.3, even though the development of ceramic fibers was at the early stage.

Using advanced ceramic technology, it was possible to produce not only a simple ceramic fiber, but also an advanced ceramic fiber (end-cap design: YAG–Nd:YAG–YAG) that cannot be produced using glass fiber technology. It was demonstrated that the end-cap design can control the thermal generation and power saturation during laser oscillation by the end-pumping technique. In addition, it is possible to produce a Q-switch function ceramic fiber. For example, by attaching a saturable absorber such as Cr^{4+}:YAG at one end of an Nd:YAG ceramic fiber, a functional composite fiber laser can be easily created. Accordingly, ceramic fiber laser technology allows the creation of innovative fiber lasers, but the key point of this technology depends on reducing the diameter of the ceramic fiber core as much as possible.

The optical quality of the most recent ceramic fibers was remarkably improved (optical loss of 0.5–0.2% cm^{-1}) and the fiber core was narrower than before. Although laser experiments on the most recent fibers are in progress, it is certain that their laser performance will also be significantly improved.

7.3 Single crystal ceramics produced by sintering

Generally, ceramics produced by sintering are polycrystalline materials, but this has changed in recent years. In the 1980s, Matsuzawa *et al.* succeeded in fabricating single crystal ceramic materials by sintering. They developed an advanced ferrite material

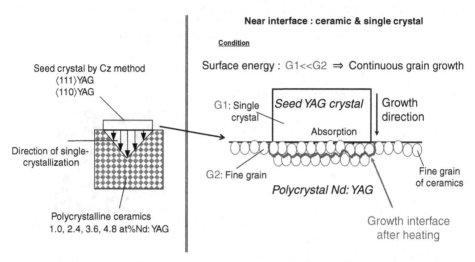

Figure 7.15 Illustration of the production of single crystal ceramic material by the sintering (non-melting) method.

which has high wear resistant properties for use as a magnetic tape head. By eliminating the grain boundaries of the ferrite material, the wear resistance of the magnetic tape head was remarkably improved, providing a longer product lifetime. However, their technology did not advance sufficiently for the development of optical materials.

In 2002, the author first succeeded in fabricating laser grade single crystal ceramic materials by sintering (SSCG, solid state crystal growth method) [13]. A schematic diagram of the single crystal sintering method is shown in Figure 7.15. First, a polycrystalline Nd:YAG ceramic material with a relative density of over 95% was prepared. Then, one face was polished and contacted with a seed crystal (single crystal YAG with any crystal orientation: $\langle 111 \rangle$, $\langle 100 \rangle$, $\langle 110 \rangle$). When the seeded ceramic was heated at a high temperature (below the melting point) of over 1700 °C, single crystallization occurred in the solid state due to continuous grain growth. When the surface energy of the fine grains of polycrystalline ceramic (E_p) is much greater than that of the seed crystal (E_s) (i.e., $E_p \gg E_s$), the fine grains tend to change to a thermodynamically stable condition at higher temperature. As a result, fine grains are absorbed into the seed crystal, and grain growth occurrs continuously. Finally the polycrystalline materials change to single crystal materials in a solid-state condition.

A trace amount of solid SiO_2 (in the form of colloidal silica) was added to prepare Nd:YAG sintered bodies. Then they were heat treated at 1780 °C for various soaking times: 0 h, 2 h and 5 h. Reflected microscopic images are shown in Figure 7.16. When the sintering time was 0 h, the microstructure had uniform grain size. When the sintering time was increased to 2 h, a few abnormal grains of 1 mm diameter were observed. And when the sintering was further increased to 5 h, it was found that the abnormal grain size increased. It is considered that the added colloidal silica served both as a seed crystal and as an accelerator for continuous grain growth. In solid-state crystal growth, the driving

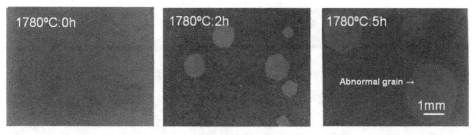

Figure 7.16 Change in microstructure of Nd:YAG ceramics doped with a small amount of SiO$_2$ additive sintered at 1780 °C for 0, 2, and 5 h soaking time.

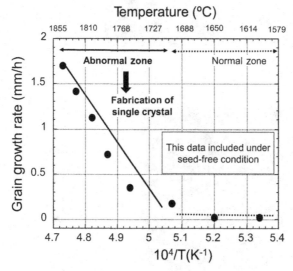

Figure 7.17 Relationship between the heat treatment temperature and rate of grain growth for Nd:YAG ceramic.

force for continuous crystallization is obtained from the formation and growth of abnormal grains.

The relationship between the heat treatment temperature and rate of crystal growth is shown in Figure 7.17. Crystallization started around 1700 °C, and the crystal growth rate increased with an increase in heat treatment temperature. In this experiment, a maximum growth rate of 1.7 mm h^{-1} was reached. Using the CZ method, the rate of crystal growth is generally about 0.2 mm h^{-1}. Therefore, greater or equivalent growth rate can be realized using the SSCG method.

The appearance of heavily doped Nd:YAG single crystals (2.4%, 3.6%, and 4.8% Nd:YAG) prepared by the SSCG method is shown in Figure 7.18(a). A typical reflected (polarized) microscopic image of a near growth interface of a 2.4% Nd:YAG sample is shown in Figure 7.18(b). It was confirmed that grain growth occurred from a seed crystal

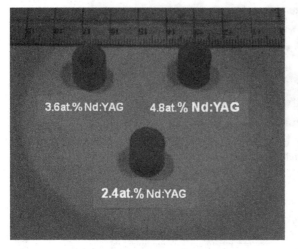
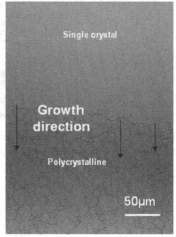
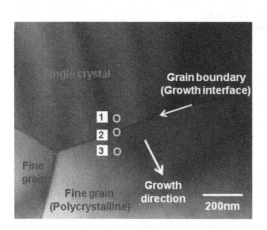
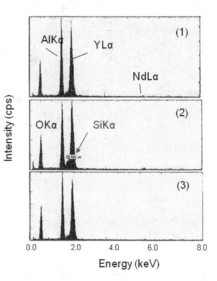

Figure 7.18 (a) Appearance of 2.4, 3.6, and 4.8 at.% Nd:YAG ceramic after heat treatment (after single crystallization). (b) Microstructure of 2.4 at.% Nd:YAG ceramic during single crystallization. (c) Bright field image of the growth interface between single crystal and polycrystalline materials observed by TEM-EDS analysis.

towards polycrystalline directions. It should be noted that heavily doped Nd:YAG single crystal can be produced by the SSCG method. Using the CZ (melt-growth) method, the segregation coefficient of Nd ions in the YAG host crystal is very small, and it is difficult to dope more than 1 at.% Nd ions homogeneously into the YAG host crystal. In the case of the SSCG method, since there is no solid–liquid interface, the concept of segregation is not relevant for this solid process. In this process, polycrystalline grains are gradually

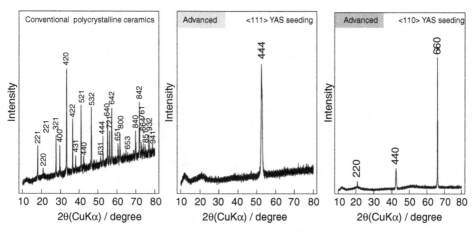

Figure 7.19 XRD patterns for regular polycrystalline, ⟨111⟩, and ⟨110⟩ seeded Nd:YAG material. We can confirm characteristic peaks for ⟨111⟩ and ⟨110⟩ seeded materials.

absorbed into the seed single crystal. Therefore, if the composition of each polycrystalline Nd:YAG grain is very homogeneous in the ceramic, then the homogeneity of the grown single crystal ceramics will also be very high. In Figure 7.18(c), a TEM (transmission electron microscopy) image and EDS (energy dispersive spectroscopy) analysis results of the growth interface between single crystal and polycrystalline material are shown. There was no boundary phase or secondary phase near the growth interface, and the growth interface was very similar to the grain boundary of the host polycrystalline ceramics. The EDS analysis results, however, showed a trace amount of Si (SiO_2) at the growth interface and at the grain boundaries, but not in the crystal grains. Studies on the crystal growth mechanisms are still under investigation, but it is certain that the Si component is closely involved in the growth mechanism.

Figure 7.19 shows XRD patterns of normal polycrystalline Nd:YAG ceramic, and the SSCG grown single crystal ceramics with ⟨111⟩ and ⟨110⟩ crystal orientations. In the case of normal sintered Nd:YAG ceramic without the seed crystal, random crystal orientations were observed. But in the case of the SSCG grown single crystal ceramics, the crystal orientations were the same as the seed crystal. Therefore, it was confirmed that single crystallization by sintering is technologically possible.

The appearance of SSCG grown Cr^{4+}:YAG single crystal ceramic bonded with a ⟨100⟩ oriented seed crystal is shown in Figure 7.20. Using this technology, it is possible to dope various types of laser active ions. This material was grown at 1700 °C, and the crystal growth rate was 4 mm h^{-1}. The crystal growth rate can be controlled mainly using (1) the seed crystal orientation, (2) the grain size of the polycrystalline ceramic, and (3) the heat treatment temperature. The maximum growth rate achieved was 5 mm h^{-1}.

Figure 7.21 shows a comparison of the laser performance of the SSCG grown Nd:YAG single crystal ceramic and that of pore-free normal Nd:YAG polycrystalline ceramic. In

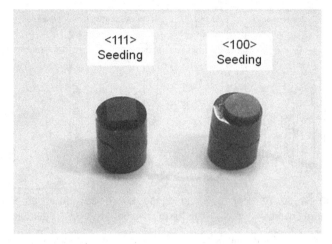

Figure 7.20 Appearance of a $Cr^{4+:}$YAG single crystal after heat treatment.

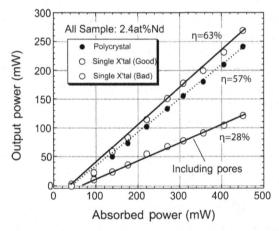

Figure 7.21 Comparative laser performance of polycrystalline and single crystal material pumped by a Ti:sapphire laser.

both cases, the Nd concentration was 2.4 at.%. A Ti:sapphire laser was used as an excitation source. A 99.9% reflection mirror and an output mirror with 5% transmission were used. The laser gain media samples were AR coated but the laser oscillation test was performed without optimizing the laser resonator. Both samples reached a slope efficiency of almost 60%, but the single crystal ceramic showed about 6% higher efficiency than the polycrystalline ceramic. This improvement was probably because scattering due to grain boundaries does not occur in the single crystal ceramic material.

The above process produced bulky single crystal ceramic materials. In recent years, research has been focused on quantum effect devices such as microsphere lasers. Glass spheres [14], semiconductor materials based on lithograph technology [15], high-polymer materials [16] and low melting point crystalline materials such as Sm:CaF$_2$ [17] have been studied as microsphere lasers. Because high melting point materials such as Nd:YAG cannot be produced directly from the melt as single crystals, it was believed that these materials cannot be produced as spherical single crystals. However, using advanced ceramic technology, spherical Nd:YAG single crystal ceramic materials can be produced by a simple process [18]. First, spherical powder compacts are formed. Then, spherical polycrystalline material is produced by sintering. Finally, single crystallization of the polycrystalline material is carried out in the solid state without using a seed crystal. Details of the fabrication technology for single crystal Nd:YAG ceramic microspheres are described below.

First, compacts of spherical powder (Nd:YAG composition), several tens of micrometers in diameter, were prepared. Then, the spherical powder compacts were heated at different temperatures between 1200 and 1750 °C (soaking time 1 minute). SEM images of the microstructures of the spherical powder compacts at each heating temperature are shown in Figure 7.22(a). With an increase in heating temperature, it was confirmed that densification and grain growth occurred as seen in the microstructure of normal sintered bodies – one sphere is composed of many fine crystal grains with randomized crystal orientation. Figure 7.22(b) shows the change in microstructure (observed by SEM) when the sphere was sintered at 1750 °C for 1, 5, 10, and 20 minutes. When the sphere was sintered at 1750 °C for 2 h, grain growth occurred continuously and only a few (two or three) pieces of crystal grain were observed in one sphere. When the sphere sample was sintered at 1750 °C for more than 2 h, the grain boundaries disappeared completely, resulting in a single crystal ceramic microsphere. However, necking between neighboring spheres occurred because of the reactive sintering process. In order to solve this necking problem, a dynamic sintering technique was invented. In this technique, sintering and crystal growth occur simultaneously in a rotating chamber. Real spherical shaped transparent Nd:YAG single crystal ceramic microspheres were fabricated successfully using the dynamic sintering technique.

A SEM image of a microsphere prepared by the dynamic sintering technique and a TEM image (lattice image) of the cross-section (thin section of an ion-milled surface) of the microsphere are shown in Figure 7.23(a) and (b), respectively. The SEM image shows that the Nd:YAG single crystal ceramic microsphere has a real spherical shape without necking. The TEM image confirms that the materials include no crystal defects such as microvoids and lattice defects. The optical quality of the microspheres was inspected using a transmission polarized optical microscope. As shown in Figure 7.23(c), transmitted light can be observed in the middle of the sample, and no double refraction was seen, suggesting that the microsphere is fully transparent. Accordingly, it is considered that the quality of the prepared microspheres is suitable for use in a microsphere laser, but laser oscillation tests have not yet been performed.

Interestingly, the single crystal ceramic microsphere can be produced without a seed crystal. The mechanism of single crystallization in microspheres is shown in Figure 7.24. First, spherical shaped powder compacts were prepared, and then they were sintered at high temperature to achieve fully dense Nd:YAG polycrystalline ceramic. Here, colloidal silica

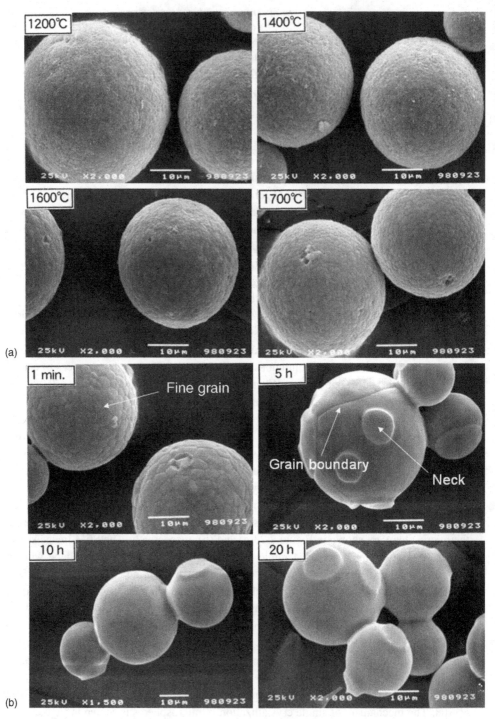

Figure 7.22 Change in microstructure of Nd:YAG microsphere ceramic sintered at (a) 1200 to 1700 °C for 1 min soaking time, and (b) 1780 °C for various soaking times. The microsphere Nd:YAG changes to single crystal form after 5 h soaking time at 1750 °C.

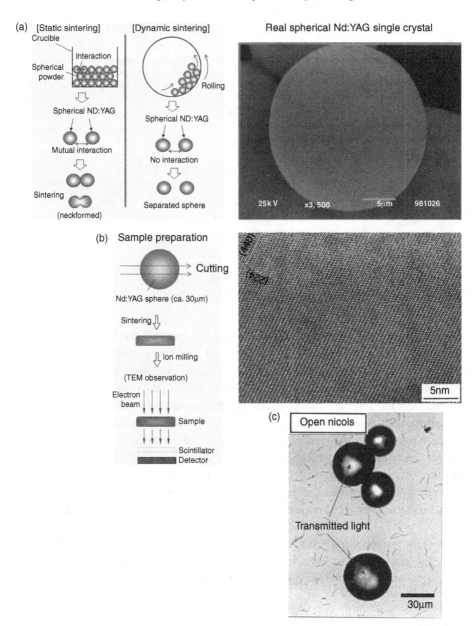

Figure 7.23 (a) Single crystal Nd:YAG microsphere without necking prepared by the dynamic sintering method. (b) Lattice structure of single crystal Nd:YAG microsphere by high resolution TEM. (c) Transmission polarized image of microspheres.

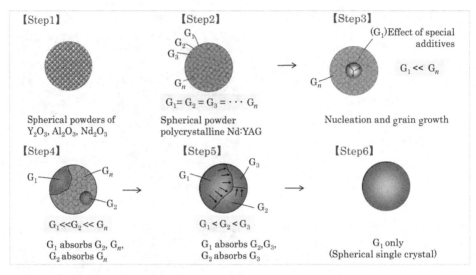

[Step1]

Spherical powders of
Y_2O_3, Al_2O_3, Nd_2O_3

[Step2]

$G_1 = G_2 = G_3 = \cdots G_n$

Spherical powder
polycrystalline Nd:YAG

[Step3]

(G_1)Effect of special
additives

$G_1 \ll G_n$

Nucleation and grain growth

[Step4]

$G_1 \ll G_2 \ll G_n$

G_1 absorbs G_2, G_n,
G_2 absorbs G_n

[Step5]

$G_1 < G_2 < G_3$

G_1 absorbs G_2, G_3,
G_2 absorbs G_3

[Step6]

G_1 only
(Spherical single crystal)

Figure 7.24 Illustration of the mechanism of single crystallization in microspheres.

(solid SiO_2) was initially added as a sintering aid, but finally it served both as a sintering aid during the sintering process, and as a crystal nucleating agent and crystal growth promoting agent during the single crystallization process. Abnormal grain growth occurred in the fully dense Nd:YAG polycrystalline ceramic sample, then continuous grain growth occurred from the abnormal grains to the other fine grains, and finally the polycrystalline form changed into a single crystal form. Although it is assumed that a number of abnormal grains can be created at different places in the sphere sample at the same time or at slightly different times, the surface energies of the crystal grains are different from each other. Therefore, based on the laws of physics, the surface energy of the abnormal grains is smaller than that of the other fine grains, and the fine grains are absorbed into the abnormal grains. Finally, the spherical polycrystalline material is transformed into the single crystal state.

One of the features of the SSCG method is that no solid–liquid interface is formed during the crystal growth process. Therefore, it is thought that it will be possible to produce the following single crystal ceramic materials by the SSCG method in the future:

(1) single crystals heavily doped with a small segregation coefficient dopant (e.g., Nd ions in YAG);
(2) bulky single crystals with homogeneous composition and high optical quality without the core and facets normally seen in conventional melt-growth crystals;
(3) single crystal materials (high melting point materials) that cannot be produced by the melt-growth process;
(4) single crystal composites with complicated design (e.g., layer-by-layer, clad–core etc.);
(5) single crystals with a special structure (e.g., microspheres).

Since the single crystal ceramic materials produced by the SSCG method have no grain boundaries, it is considered that this kind of material is the best quality for laser materials. This technology is summarized and discussed in Chapter 9.

7.4 Summary

The technologies used for composite lasers, fiber lasers, and single crystals by sintering described in this chapter are innovative and still developing technologies that could not be imagined in the laser technology of the twentieth century. These technologies are only those invented by the author, but it is also possible that many new innovative technologies will also be created in the future.

Regarding fiber lasers, it is still necessary to reduce the core diameter of the ceramic fiber. However, it was demonstrated that ceramic fibers have high laser gain, and high power operation can be realized from short lengths. The possibility of Q-switch function ceramic fiber lasers for pulse operation was also suggested.

The ceramic process offers greater design flexibility, and it was proved that this non-melting process can provide a totally different technological field from conventional melting processes such as glass making and single crystallization by the melt-growth process. It was recognized that the laser oscillation efficiency of the SSCG crystal was about 6% higher than that of normal Nd:YAG ceramic. This is not the only advantage of the SSCG crystals. Generally, even pore-free ceramics with clean grain boundaries include crystal defects because the grain boundaries are themselves defects in the crystal structure – generally speaking they are composed of an array of dislocations. Therefore, such defects may be easily attacked by laser damage. Of course, there have been reports of kilowatt laser generation from polycrystalline ceramic materials. But the elimination of grain boundaries by the SSCG method is very significant for the development of higher output lasers in the future, especially in the development of a laser gain medium that can generate large power from a single gain medium which has extremely high laser damage threshold. It has also been confirmed that the pore-free Nd:YAG single crystal ceramic produced by the SSCG method does not include the core and facets seen in conventional melt-growth crystals, and the dislocation density is also very low. It is also possible to produce a heavily doped Nd:YAG single crystal for laser operation. In that sense, the crystal grown by the SSCG method not only provides superior performance and allows new designs for functional devices, it may also be very important for future laser technology.

References

[1] US Patent No. 5441803.
[2] A. Ikesue and Y. L. Ang, Ceramic laser materials, *Nature Photon.*, **2** (2008) 721.
[3] S. Matsuzawa and S. Mase, US Patent No. 4339301 (NGK Inslators Ltd. (Nagoya JP) (1982).

[4] S. Matsuzawa and S. Mase, US Patent No. 4402787 (NGK Inslators Ltd. (Nagoya JP) (1983).

[5] S. Matsuzawa and S. Mase, US Patent No. 4519870 (NGK Inslators Ltd. (Nagoya JP) (1985).

[6] US Patent Nos. 5852622, 58546638.

[7] A. Sugiyama *et al.*, Direct bonding of Ti:sapphire laser crystals, *Appl. Opt.* **37** (12) (1998) 2407.

[8] A. Ikesue and Yan Lin Aung, Synthesis and performance of advanced ceramic lasers, *J. Am. Ceram. Soc.* **89** (6) (2006) 1936–1944.

[9] E. Pawlowski *et. al.*, *Proc. SPIE* **7578** (2010) 15–18.

[10] A. Yoshikawa, G. Boulon, L. Laversenne, H. Canibano, K. Lebbou, A. Collombet, Y. Guyot, and T. Fukuda, *J. Appl. Phys.* **94** (2003) 5479.

[11] D. Sanghera, N. Aubry, J. Didierjean, D. Perrodin, F. Balembois, K. Lebbou, A. Brenier, P. Georges, O. Tillement, and J. M. Fourmigue, *Appl. Phys. B* **94** (2009) 203–207.

[12] A. Ikesue and Yan Lin Aung, Progress in ceramic Nd:YAG laser, *Laser Source Technology for Defense and Security*, **6552** (2007) 655209–655215.

[13] A. Ikesue, T. Yoda, S. Nakayama, T. Kamimura, and K. Yoshida, 25th *Annual Meeting of Japanese Laser Society* (2005), Vol. 15, I 12.

[14] B. Lu *et al.*, High-order resonance structures in a Nd-doped glass microsphere, *Opt. Commun.* **108** (1994) 13–16.

[15] S. L. MacCall *et al.*, Whispering-gallery mode microdisk lasers, *Appl. Phys. Lett.* **60** (3) (1992) 289–291.

[16] M. Kuwata-Gonokami *et al.*, Lasing and intermode correlation of whispering gallery mode in dye-doped polystyrene microsphere, *Mol. Cryst. Liq. Cryst.* **216** (1992) 21–25.

[17] C. G. B. Garret *et al.*, Stimulated emission into optical whispering modes of spheres, *Phys. Rev.* **124** (1961) 1807–1809.

[18] A. Ikesue, T. Yamamoto, Y. Sato, T. Kamimura, K. and Yoshida, *25th Annual Meeting of Japanese Laser Society* (2005), Vol. 15, I 14.

8

Current R&D status of ceramic lasers worldwide

The first ceramic laser was developed by S. E. Hatch using Dy:CaF$_2$ material under cryogenic conditions, and then C. D. Greskovich reported pulse laser oscillation using Nd:ThO$_2$–Y$_2$O$_3$ ceramic in 1974. However, there were no reports of efficient ceramic laser oscillation (pulse and CW operation) at room temperature until the author developed a highly efficient polycrystalline Nd:YAG ceramic laser in 1995. It is historically true that the ceramic laser boom started a few years after this achievement. It was demonstrated that ceramic laser materials have many technological and performance characteristics that are comparable with or superior to those of commercial single crystal laser materials. Therefore, materials scientists and laser scientists around the world have shifted their attention from single crystals to polycrystalline ceramic materials. When the author first presented the Nd:YAG ceramic laser in 1994 at a conference in Japan, there were only a few people who listened to the presentation, and most comments and questions were negative. Also in 1995, the author published the paper in domestic and overseas journals but there were no researchers who showed interest in the paper. The author has collaborated with Dr. Taira of IMS (Institute of Molecular Science), and demonstrated the fabrication of heavily doped Nd:YAG ceramics, a characteristic of ceramic laser materials, and highly efficient microchip laser and single axial longitudinal mode oscillation using this laser element. In addition, green and blue lasers were also demonstrated although because ceramic laser gain media have grain boundaries it was assumed that short wavelengths could scarcely pass through. It seems that some national laboratories and private industries developed ceramic laser materials after the author's publication in 1995, however there have been no technological competitors until this century.

In 2002, the author was invited to an annual meeting of the American Ceramic Society held in St. Louis, USA. This was called the "Kingery–Coble–Bruck Symposium" in honor of the three great researchers who have modernized ceramic technology in the USA. The person who invited me to the symposium was Dr. Greskovich, a chairman of the symposium. At that time, the meeting room was very crowded with many attendees and several people were standing. It seemed like the beginning of the boom in ceramic lasers. In 2005, LCS-1 (First Laser Ceramic Symposium) was held during the annual meeting of the European Materials Research Society (E-MRS) in Warsaw, Poland. Many European and Japanese researchers presented papers on the results of ceramic laser studies. Among these

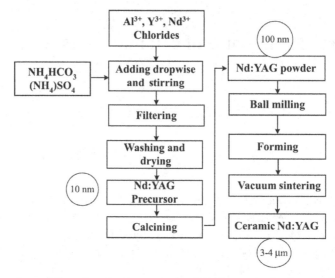

Figure 8.1 Experimental flow sheet for the synthesis of YAG ceramics by the wet-chemical precip-itation method of Ueda's group. Image reproduced with kind permission from Springer Science + Business Media [4].

presentations, however, successful laser oscillation was described by only two groups from Japan (the Ikesue group and the group of Professor K. Ueda). After this meeting, the LCS was held annually in Japan (Tokyo, 2006), France (Paris, 2007), China (Shanghai, 2008), Spain (Bilbao, 2009) and Germany (Munster, 2010). After 2005, successful laser oscilla-tion using ceramic gain media was reported from China and the USA, and international competition in the development of ceramic laser materials has now become increasingly important. But, as of April 2010, Japan is still leading in terms of the quality of materials and technological originality. This suggests that the level of ceramic technology is very high and also the industrial scale of the ceramics business is very big in Japan. Although YAG (yttrium aluminum garnet) remains a major practical laser host material, other laser materials are also being developed worldwide. In Sections 8.1 to 8.4, the achievements of various groups (except the author's group) are introduced. In Sections 8.5 to 8.7, the latest results on the development of devices for medial diagnosis, automobiles, and energy development using YAG ceramic laser gain media are described.

8.1 Garnet system materials

At LCS-1, the other technology for the fabrication of ceramic lasers was reported by Professor Ueda's group. They described the fabrication of YAG raw powder materials by a co-precipitation method (urea method), but did not give details of the wet process.

The fabrication process of YAG system ceramics devoloped by Ueda *et al.* is shown in Figure 8.1 [1, 4]. Fabrication conditions were not described, and the details of the process

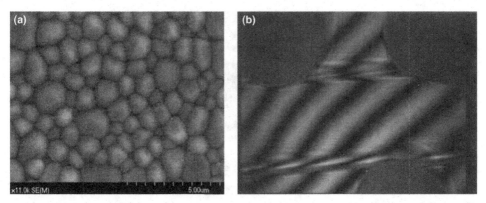

Figure 8.2 Microstructure revealed by SEM and a wavefront image of Nd:YAG ceramic produced by interferometry. (a) Reprinted with permission [1]. (b) Image reproduced with kind permission from Springer Science + Business Media [4].

are not clear. The key points of their process are (1) fabrication of homogeneous single phase YAG by a co-precipitation method, and (2) molding by a casting method.

In the dry process, Al_2O_3 and Y_2O_3 raw powders are blended, and then the powder mixture is heated at a high temperature for reactive sintering. Using this solid-state reaction method, the densification is generally not perfect, and the composition is not homogeneous. Therefore in the wet process, in order to overcome the drawbacks of the dry process, sub-micrometer size homogeneous YAG single phase powders are prepared by co-precipitation in advance of the sintering step. In addition, a slip casting method is applied in molding. First, raw powders are dispersed in a solvent by pot milling, and then the slurry is poured into a porous mold (generally a plaster mold). The slip casting method is good for mass production at low cost and can produce special shapes such as hollow pipes and crucibles that are hard to shape directly by other molding methods. However, most laser crystals are of bulk type, and it is not certain that slip casting is the best method for molding laser ceramic materials. The fundamental idea of the fabrication of high performance ceramics is that in the wet process homogenized raw powders are used to produce homogeneous sintered bodies. However, YAG is a stoichiometric material (i.e., YAG is a line compound in the phase diagram, and the limit of solid solubility is extremely small), and it is difficult to control the precise composition in the wet process. Therefore, from the technological and economical viewpoints, it is important to consider the yield percentage of YAG raw powder as well as the YAG sintered bodies.

Figure 8.2 shows a micrograph and a wavefront distortion image (measured by Twyman Green interferometry) of Nd:YAG ceramic prepared by the wet process [4]. The average grain size of the Nd:YAG ceramics is several micrometers, and a fully dense microstructure without residual pores can be confirmed. In the wavefront distortion image, the P_v value is up to $\lambda/9.6$, and this value is better than that of the commercial single crystal counterparts. (In this wavefront image, the observed direction is parallel with the direction of facet formation

Figure 8.3 Large size Nd:YAG ceramic ($100 \times 100 \times 20$ mm^3) with Sm:YAG ceramic layer cladding for a heat capacity laser. Reprinted with permission [1].

of the single crystal. Therefore, the observed direction is preferably perpendicular to the direction of facet formation.)

Figure 8.3 shows a large slab ($100 \times 100 \times 20$ mm^3) of Nd:YAG ceramic clad with Sm:YAG ceramic. Lawrence Livermore National Laboratory (LLNL) tested this material for use as a solid-state heat capacity laser (SSHCL), and they achieved a high average output power of 67 kW [5].

Penn State University fabricated a composite ceramic Er:YAG rod by tape casting for the first time. The length of the rod is 62 mm, and the composite consists of three segments: 0.25% Er:YAG, 0.5% Er:YAG and undoped YAG. N. Ter-Gabrielyan *et al.* of ARL (Army Research Laboratory) demonstrated this ceramic for laser operation under resonant pumping. They achieved a maximum output power of 6 W with a slope efficiency of 57% (see Figure 8.4). The ceramic composite was prepared by tape casting, but the details of the fabrication process are unknown [6].

The garnet system laser host material is usually YAG ($Y_3Al_5O_{12}$), but the author's group developed a disordered polycrystalline ceramic made of YSAG ($Y_3Sc_xAl_{5-x}O_{12}$), which is a solid solution of $Y_3Al_5O_{12}$ and $Y_3Sc_2Al_3O_{12}$. Nd:YSAG and Yb:YSAG have wider emission bands than YAG host material, and therefore ultra-short pulse laser oscillation could be demonstrated using these ceramic materials in the latter half of the twentieth century.

Figure 8.5 shows the difference in emission spectrum between Nd:YAG single crystal and Nd:YSAG ceramic [7]. The Nd:YAG single crystal had a narrow spectral width, while a broad spectral width was confirmed in the Nd:YSAG ceramics, especially when x varied from 0.3 to 2.0. The spectral bandwidth was the largest at $x = 1$ because in this case 50% of the six-coordinated Al sites in the garnet structure are replaced with Sc, the disorder in the crystal has reached the maximum level, and the characteristic narrow emission band of Nd was considerably broadened. Laser materials having a broad spectral width can be used for the generation of ultra-short pulse lasers and high tunability lasers. By taking advantage

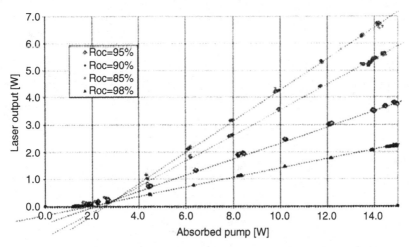

Figure 8.4 Laser performance of Er:YAG composite ceramic consisting of three segments [6].

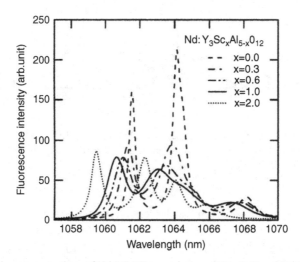

Figure 8.5 Emission cross-sections of Nd:YSAG ceramic ($x = 0.3$ to 2.0) and Nd:YAG single crystal ($x = 0$) at room temperature. Image courtesy of The Ceramic Society of Japan [2].

of this property, the author's group has succeeded in the generation of a 10 ps short pulse laser using Nd:YSAG ceramic in a mode-locked system [8, 9].

The slope efficiency of the Nd:YSAG ceramic laser was about 40% under CW operation. Since it does not require an excitation source of SHG from an Nd:YAG laser as in Ti:sapphire lasers (and it is possible to pump it directly with an 808 nm LD), the laser oscillator made of Nd:YSAG ceramic is very compact and has a high laser conversion efficiency.

In Figure 8.6, absorption and emission spectra for 25% Yb:YAG single crystal and 15% Yb:YSAG ceramic are shown [10]. Since the Yb system already has a broad spectrum, the

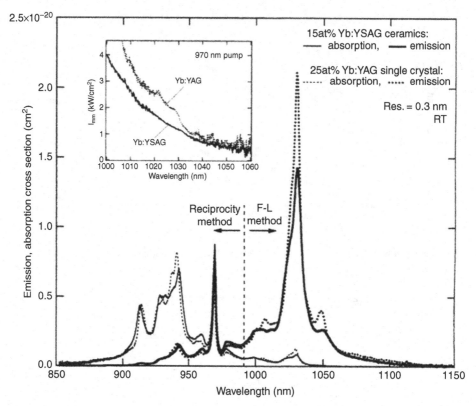

Figure 8.6 Absorption and emission cross-sections of $Yb:Y_3Sc_1Al_4O_{12}$ (Yb:YSAG) ceramic and Yb:YAG single crystal at room temperature. Reprinted with permission from [10], copyright 2004, American Institute of Physics.

broadening of the spectral width is not as large as in the Nd system, but it can be seen that the YSAG system has broader absorption and emission spectral width.

A setup for a mode-locked laser is shown in Figure 8.7(a) [11]. Since the principle of mode-locked lasers is commonly known, it is not described here. Ti:Sapphire laser was used as an excitation source and 15% Yb:YSAG ceramic (diameter 10 mm × thickness 1 mm) was used as a laser gain medium. Figure 8.7(b) shows the laser performance (input versus output characteristics) of the Yb:YSAG ceramics under CW operation. The slope efficiency varied significantly with the condition of the output coupler (i.e., transmittance of the output mirror). A maximum slope efficiency of 72% was achieved by optimizing the output coupler condition. Figure 8.7(c) shows a spectrum of short pulse laser emission achieved with 0.5% output coupler in the mode-locked system. Ultra-short pulse oscillation of around 280 fs was achieved. The average output power, peak power, and repetition frequency were 62 mW, 2.2 kW, and 101 MHz. This value was a world record for pulse laser generation using ceramic gain medium.

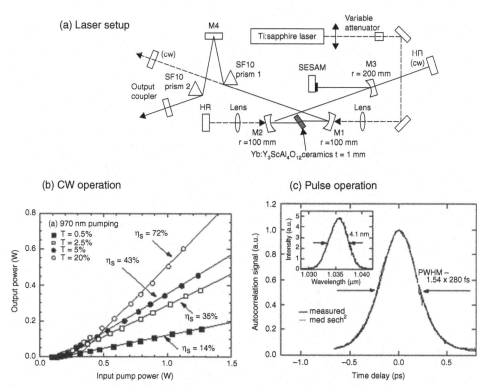

Figure 8.7 (a) Schematic diagram for a mode-locked system. Reprinted with permission from [11], copyright 2004, American Institute of Physics. (b) Laser performance under CW operation and (c) spectrum for the shortest pulse generation with 0.5% output coupler for 15% Yb:YSAG ceramic. The solid line indicates the ideal $sech^2$ shape. Reprinted with permission from [11], copyright 2004, American Institute of Physics.

8.2 Perovskite system materials

Tanaka *et al.* reported that the addition of Zr in a perovskite type BMT ($Ba(Mg,Ta)O_3$) material changed the crystal structure of the material from tetragonal to cubic. In addition, they achieved laser oscillation from this material doped with laser active ions (Nd). Although the details of the fabrication process, chemical composition, and optical quality of this material are unknown, they reported the following technical data [12, 13].

Figure 8.8 shows a reflected micrograph of the $Ba(MgZrTa)O_3$ ceramic. The average grain size of this material is approximately 20 μm. Residual pores and impurity phases are not detected on the observed surface. However, the transmittance is a few percent lower than the theoretical transmittance, and it can be assumed that a certain amount of scattering occurs in the material.

The emission spectrum of 1 at.% Nd doped $Ba(MgZrTa)O_3$ ceramic is shown in Figure 8.9. Typically Nd system materials have three types of emission, $^4F_{3/2}-^4I_{9/2}$, $^4F_{3/2}-^4I_{11/2}$, and $^4F_{3/2}-^4I_{13/2}$. Noticeably, none of the emission peaks of this material are as sharp as

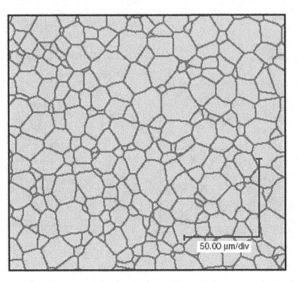

Figure 8.8 Reflection microscopy photograph of Ba(MgZrTa)O₃ ceramic [13].

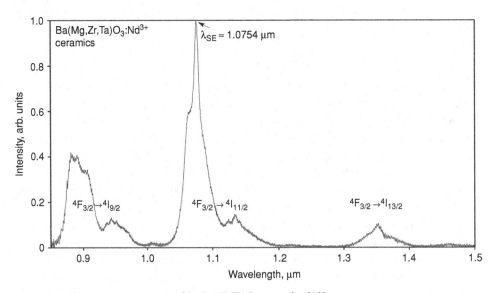

Figure 8.9 Fluorescence spectrum of Ba(MgZrTa)O₃ ceramics [13].

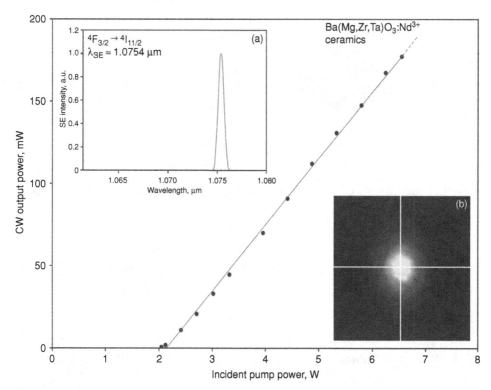

Figure 8.10 Laser performance of Ba(MgZrTa)O₃ ceramic under CW operation pumped by an 807 nm LD. The graph is courtesy of John Wiley and Sons [13].

seen in normal Nd:YAG materials, but they have a broad emission spectrum, characteristic of disordered materials.

This 1 at.% Nd doped Ba(MgZrTa)O₃ ceramic was pumped with an LD of 807 nm, and demonstrated CW laser operation. Figure 8.10 shows a lasing spectrum and the output beam profile under CW operation. The maximum output power of 177 mW was achieved at an incident power of 6.5 W, and the slope efficiency was about 5%. Single lasing spectrum (wavelength 1.075 μm) was achieved, and a relatively high beam quality with nearly TEM$_{00}$ mode was confirmed. The line width of the fluorescence emission spectrum of this material is about 30 times broader than that of normal Nd:YAG, and it is expected to be used in the development of tunable lasers and short pulse lasers in the future.

8.3 Non-oxide system (II–VI compound) materials

Chromium-doped II–VI materials such as Cr^{2+}:ZnSe have received a lot of attention in recent years for the development of tunable lasers in the mid-infrared wavelengths (1.7–3.0 μm). S. Mirov *et al.* reported a transparent polycrystalline Cr-doped ZnSe ceramic

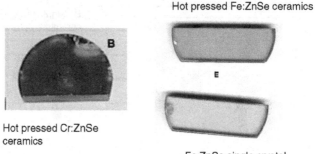

Hot pressed Fe:ZnSe ceramics

Hot pressed Cr:ZnSe
ceramics

Fe:ZnSe single crystal

Figure 8.11 Appearance of Cr- and Fe-doped ZnSe single crystal and ceramic formed by hot pressing [15].

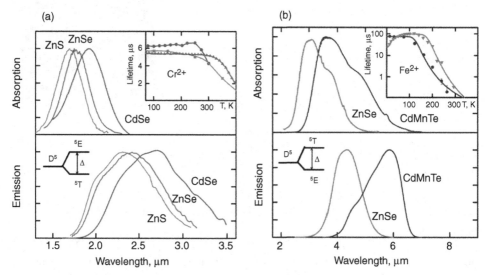

Figure 8.12 Absorption and fluorescence spectra of (a) Cr-doped ZnSe, ZnS, and CdSe and (b) Fe-doped ZnSe and CdMnTe [15].

fabricated by hot pressing 1 mol% CrSe doped ZnSe at 1400–1500 K at an axial compression of up to 350 MPa [14, 15].

Figure 8.11 shows the appearance of Cr:ZnSe and Fe:ZnSe single crystals and hot-pressed ceramics. The Cr-doped sample has a red color, and the Fe-doped sample has a yellow color. There is no difference in appearance between the single crystals and poly-crystalline ceramics.

The absorption and emission spectra of Cr-doped ZnSe, ZnS, and CdSe crystals are shown in Figure 8.12(a), and Fe-doped ZnSe and CdMnTe crystals are shown in

Figure 8.12(b). The Cr^{2+} and Fe^{2+} ions have very broad absorption and emission spectra in the II–VI host materials. Normally, such spectroscopic properties cannot be achieved in oxide host materials doped with rare earth elements. Therefore, these II–VI host materials are very important for the development of tunable lasers over a broad mid-infrared spectral range. They achieved 2 mJ output in pulse operation and 250 mW output in CW operation using the fabricated ceramics. In addition, an output of 150 mW was achieved in single mode oscillation. A high quality Cr:ZnSe crystal was tested for CW operation, using an Er:glass fiber (1.5 µm) laser as an excitation source. A high slope efficiency (43%) and maximum output power of 12.5 W (incident power of 29 W) was achieved. Therefore, the production of high quality ceramic laser materials and power scaling will be challenging problems in the future.

8.4 Fluoride system materials

T. T. Basiev *et al.* succeeded in fabricating a color center laser using $LiF:F_2^-$ ceramic (i.e., materials with an excess amount of F^- ions) [16]. They achieved 22% slope efficiency in quasi-CW oscillation mode by exciting the materials with an LD (semiconductor laser). This is an example of a recently reported polycrystalline laser. In this sense, ceramic technology is very important for the fabrication of new materials that are difficult to produce by the conventional melt-growth process.

P. Aubry *et al.* reported on the fabrication of Yb-doped CaF_2 ceramic [17]. The raw powder material was generated by the following chemical reaction, and they succeeded in producing very fine raw powders with an average grain size around 20 nm of $Yb:CaF_2$:

$$0.9Ca(NO_3)_2 + 0.1Yb(NO_3)_2 + 2.1HF = Ca_{0.9}Y_{0.1}F_{2.1} + 2.1HNO_3.$$

The raw powder was heated at 400 °C to remove impurities adsorbed at the particle surfaces, and then it was molded and sintered in vacuum followed by HIP post-treatment (hot isostatic pressing) in order to produce a transparent sintered body. The appearance of the fabricated 10% $Yb:CaF_2$ ceramic and its in-line transmittance (800–1200 nm) are shown in Figure 8.13.

Although the thickness of the material was not described in their paper, the transmittance at 1200 nm is as low as 55%. Thus, this material is not appropriate for use in optics. The fluorescence emission of the fabricated ceramic excited by an LD (980 nm) is shown in Figure 8.14. In addition to the upconversion bands of Er^{3+} ions around 540 nm ($^4S_{3/2}-^4I_{15/2}$) and 650 nm ($^4F_{9/2}-^4I_{15/2}$), emission due to Tm^{3+} ions at 470 nm can be observed. Therefore, in order to achieve good quality ceramic laser materials, it is necessary to eliminate trace impurities (ppm levels) and improve the transmittance.

Examples of typical ceramic lasers made of garnet, perovskite, II–VI compounds, and fluoride materials explored worldwide have been introduced in the above. It is clear that

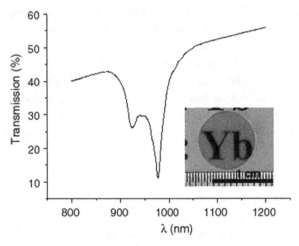

Figure 8.13 Appearance and transmission spectrum of 10% Yb:CaF₂ ceramic. Reprinted with permission from Elsevier [17].

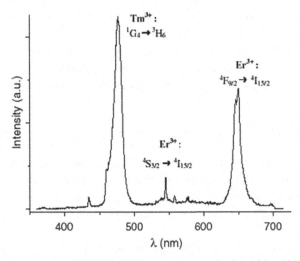

Figure 8.14 Emission spectrum of 10% Yb:CaF₂ ceramic on excitation with a 980 nm LD. Reprinted with permission from Elsevier [17].

ceramic laser gain media will become very important in the future, and currently there is still no research team able to meet entirely the demands for high power output, high beam quality and high conversion efficiency. The development of polycrystalline ceramic laser materials which are equivalent in quality to single crystal laser materials requires the development of ceramics with an ideal and perfect microstructure (i.e., formation of a defect-free microstructure except for clean grain boundaries). It is not surprising that no materials scientist has yet mastered this technology.

8.5 Applications in the fields of biotechnology and medical technology

Many excellent applications based on YAG ceramic lasers have been reported. Dr. Tanaka Koichi, a winner of the 2002 Nobel Prize in Chemistry, developed a simple and accurate method for determining the mass of proteins (a soft ionization method called *soft laser desorption*) in 1987 [18]. His method is helping to advance the life sciences and medical therapy in today's world. At that time, he adopted an N_2 (gas) laser as a light source for ionization because it has some advantages. (1) It generates ultraviolet (UV) wavelengths, and almost all of the polymers (inspection sample) can absorb ultraviolet laser light. (2) It has enough pulse laser energy to ionize the inspection sample, and a nanosecond level short pulse laser can irradiate the sample at several tens of hertz. Therefore, the N_2 laser was considered to be suitable in a TOF-MS (ionization time-of-flight mass spectrometry) analyzer.

Nowadays, ionization using a soft laser system has improved greatly, and the analyzer has been modified as a MALDI/TOF-MS (matrix-assisted laser desorption/ionization time-of-flight mass spectrometry) type for analyzing the mass of organic substances. This type of analyzer has become important for the analysis of human and animal proteins, and is also essential for the development of new medicines and tailor-made medical treatment.

However, the N_2 laser has some drawbacks that limit the performance of the MALDI system analyzer. These are (1) poor beam quality, (2) low efficiency, (3) short durability (about 50 shots in total), and (4) low repetition rate of the pulse laser (ca. 50 Hz at maximum). In addition, a high voltage is applied for excitation of the N_2 gas, and this generates electromagnetic noise (disturbance) due to a discharging phenomenon. This may cause a malfunction during operation of the analyzer. This problem has hindered the spread of MALDI/TOF-MS devices. Moreover, recently high throughput and two-dimensional mapping have been demanded for the development of high performance MALDI/TOF-MS, and a new laser source which can replace the existing N_2 laser is required.

Solid-state lasers such as Yb- or Nd-doped YAG lasers generate fundamental wavelengths around 1 µm, and their THG (third harmonic generation) wavelength is very close to the lasing wavelength (337 nm) of the N_2 gas laser. Therefore, the YAG laser is being investigated as a replacement for the N_2 laser. THG lasers based on diode pumped solid-state lasers (DPSSL) can overcome the drawbacks of N_2 gas lasers and can also be downsized to a table-top size. From the viewpoint of industrial products, either the Yb or Nd laser system could be suitable for MALDI/TOF-MS application. The Yb system has high quantum efficiency, and importantly it is possible to generate a short pulse laser because it has broad spectral width. However, the Yb system is a three-level laser (or quasi-four-level laser), and high technological skill (for strong excitation) is necessary for laser generation. In addition, the price of the LD (940 or 970 nm) excitation source is still very high.

Table 8.1 *Specification of the solid-state laser for the MALDI/TOF-MS analyzer*

Wavelength	Close to N₂ laser, 337 nm
Pulse energy	$>100\ \mu J$
Pulse width	<5 ns
Repetition rate	~ 10 kHz (variable)
Beam quality	$<1.3 M^2$
Required shot numbers	$\sim 2 \times 10^9$
Other requirements	low power consumption, low heat generation, low cost, portability, high efficiency etc.

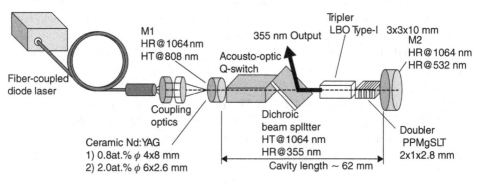

Figure 8.15 Experimental setup for a laser oscillator for the MALDI/TOF-MS analyzer [19].

On the other hand, the Nd system has slightly lower quantum efficiency compared to the Yb system but it does not require such technological skill for excitation. And, recently the 808 nm LD excitation source has come into commercial mass production, and the price is also reasonable. However, the limitation of Nd-doped YAG crystal is that only 1 at.% Nd ions can be doped homogeneously into the YAG host crystal. Therefore, it cannot be used as a microchip laser gain medium because the concentration of Nd ions is low and the thickness of the microchip is not sufficient to absorb the LD excitation source completely. However, ceramic technology allows the fabrication of heavily doped Nd:YAG ceramic laser gain media, which can be used to construct microchip lasers. Therefore, it is expected that heavily Nd-doped YAG ceramic microchip lasers will replace the conventional N₂ gas laser [19]. Table 8.1 summarizes the specifications of the solid-state laser required for the MALDI/TOF-MS analyzer. It is very difficult to meet all of these required specifications.

A schematic diagram of the laser oscillator for the MALDI/TOF-MS analyzer is shown in Figure 8.15. An 808 nm LD of 200 μm diameter fiber coupler was used as an excitation source (details of the oscillator system are omitted). 0.8 at.% and 2.0 at.% doped Nd:YAG ceramics were used as gain media, and for SHG (second harmonic generation) and THG (third harmonic generation), a QPM (PPMgSLT, periodically poled MgO doped

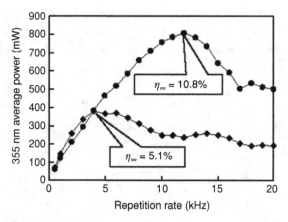

Figure 8.16 Relationship between the pulse energy and repetition rate for 0.8% Nd:YAG single crystal and 2% Nd:YAG ceramic [19].

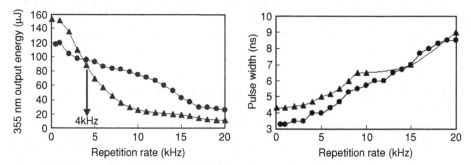

Figure 8.17 Comparative data with average output power of the TH wave for 0.8% Nd:YAG single crystal and 2.0% Nd:YAG ceramic [19].

stoichiometric LiTaO$_3$ crystal) element and LBO (lithium triborate, LiB$_3$O$_5$) crystal was used, respectively.

In Figure 8.16, the relationship between the pulse energy and repetition rate for 0.8% Nd:YAG single crystal and 2% Nd:YAG ceramic during 355 nm laser oscillation is shown. For all of the measurements, the incident energy was fixed at 7.4 W. Regarding the pulse energy (output), heavily doped 2.0 at.% Nd:YAG ceramic was superior to the 0.8 at.% Nd:YAG with an increase in repetition rate, but there was no difference in their pulse width.

The relationship between the average output power and laser repetition rate of the TH (third harmonic) wave is shown in Figure 8.17. When the 0.8 at.% doped Nd:YAG ceramic was oscillated at a repetition rate of 4 kHz, the maximum average output was 380 mW. On the other hand, an output power of 806 mW was achieved when the 2.0 at.% Nd:YAG was oscillated at a repetition rate of 12 kHz. The beam divergence was approximately 2.6 mrad and the M^2 value was 1.3 at a repetition rate of 1 kHz (pumped with 6 W). This laser beam quality meets the required specification for the MALDI/TOF-MS analyzer.

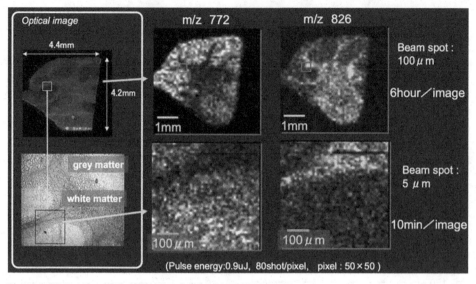

Figure 8.18 Results of application of the MALDI/TOF-MS analyzer using a ceramic laser. The ceramic laser is superior to the conventional N_2 laser system in terms of resolution and measuring speed [19].

Figure 8.18 shows the result of mass spectroscopy of rat brain tissue using the above developed microchip laser. A piece of rat brain tissue on a matrix (DHA) was scanned with the microchip laser, and mass spectroscopy was performed at each point of the brain tissue. The different mass peaks of targeted fat substances were extracted from the results of mass spectroscopy, and then the intensity of the ionization signal was converted to the image brightness for display as a two-dimensional image. The focusing diameters of the laser beam were set to 100 μm and 5 μm, and it was confirmed that the 5 μm laser gave a high resolution two-dimensional image. In the conventional N_2 gas laser system, it takes about 6 hours to scan one image, and the image is only low resolution. With the microchip ceramic laser system, the scanning time was reduced to about 10 minutes, and it was confirmed that the ceramic microchip laser is an excellent laser source for the MALDI/TOF-MS device.

8.6 High intensity lasers for engine ignition [20]

Conventionally, spark plugs have been used to ignite fuel in combustion engines. The spark plug fits into the cylinder head of the internal combustion engine. It includes a center electrode and a ground electrode. When it is connected to high voltage, discharge occurs between the electrodes. This discharge process generates a small ball of fire (about 6000 K) in the spark gap, and then the gases (gasoline) burn on their own.

A flame kernel ignited by a conventional spark plug is shown in Figure 8.19(a).

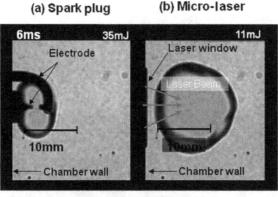

Figure 8.19 Ignition image for (a) a conventional spark plug and (b) a laser igniter [20].

For practical reasons spark-plug contacts are usually positioned relatively close to the wall of the combustion chamber, as opposed to at its center. Therefore, most of the combustion energy generated during ignition is lost due to a cooling effect from the electrode. Because of the design of the spark plug, it is very difficult to ignite in the center of the chamber. As a result, incomplete combustion occurs in the chamber and pollutants such as carbon monoxide and nitrogen oxides are released together with exhaust gas. In order to solve this problem, many ignition methods which might replace spark plugs have been investigated. For example, a plasma jet method was tested but the design of the ignition system was similar to that of the spark-plug system and the original problem still could not be resolved.

In recent years, people have become increasingly conscious of major environmental issues such as global warming and air pollution by harmful exhaust gases from engines. Therefore, new technologies for increasing the combustion efficiency of car engines and reducing the amount of harmful gas emissions into the environment are much sought after. Recently, several automobile companies from Japan, Europe and America have developed a new ignition system which can replace the conventional spark plug, and they have focused on the laser ignition system. Figure 8.19(b) shows an image of laser ignition. In this laser ignition system, since electrodes are not included, it is expected that the conventional technological issues can be solved. With a laser, it is possible to choose any position where fuel ignition is desired within the chamber. In fact there is great potential to create multiple-spark laser ignition systems, which could increase combustion efficiencies and shorten the combustion time even further by igniting the fuel at different points within the chamber. The other benefit of a direct injection engine is to reduce the harmful NO_x emissions.

On the other hand, the major advantage of the conventional spark plug is the dimension. Compared to the size of the spark plug (length ca. 10 cm), the conventional laser unit was very large (over 1 m length), and it was not possible to install it in a car engine. The laser intensity required at a focused point for ignition is 100 GW cm^{-2}. Ignition can be

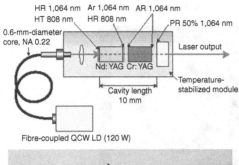

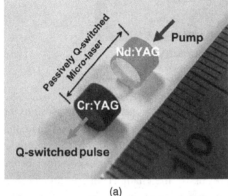

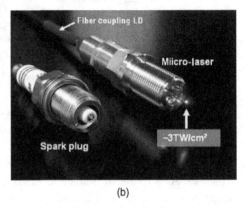

(a)   (b)

Figure 8.20 (a) Laser oscillator setup for a car igniter. Reprinted with permission [3]. (b) Appearance of a conventional spark plug and laser igniter [20].

achieved using a pulse laser but it requires a short pulse width (high peak value) and high beam quality (excellent beam focusing). In addition, a laser output power of over 10 mJ is necessary for stable combustion of fuel. The required repetition rate of the laser is about 100 Hz.

A schematic diagram of a micro-laser spark plug for ignition in a car engine is shown in Figure 8.20(a). A comparison of the micro-laser spark plug and a conventional spark plug is shown in Figure 8.20(b). Basically, the compact laser spark plug consists of Nd:YAG for laser generation and Cr^{4+}:YAG for passive Q-switching. This compact laser system can provide a highly efficient laser.

In Figure 8.21, Schlieren photographs (taken by a high speed camera at 25 000 frames per second) of the early stage (after 6 ms) of ignition and combustion in a cylinder are shown. The ignition of propane (C_3H_8) gas by (a) a conventional spark plug and (b) a micro-laser was compared. The energy applied to the spark plug was 35 mJ, and the energy applied to the micro-laser was 11 mJ. Immediately after ignition, the area of the flame produced by laser ignition was about three times larger than the area of the flame produced by the conventional spark plug. Also, energy applied to the micro-laser to achieve ignition was only 1/3 of that required by the spark-plug system. Regarding the ignition of gasoline, only a pulse energy of 4 mJ, peak power of 6–15 MW, and short pulse width of 300 ps

(a)   (b)
Spark plug   **Micro-laser**

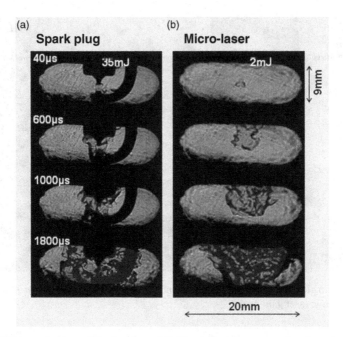

Figure 8.21 Schlieren images of the ignition of C_3H_8 gas by (a) a conventional spark plug and (b) a ceramic laser plug [20].

were required. The light-to-light conversion efficiency increased to 23%. The combustion efficiency for gasoline and the emission of harmful gases such as nitrogen oxides are currently under investigation. Once the technical differences of the micro-laser ignition system compared with the spark-plug system become clear, and if there is no problem of durability on installation in a car engine, then it will be possible to use this system in industry.

8.7 Investigation of solid-state lasers as solar pump lasers

[http://www.jaxa.jp/article/interview/vol53/index_e.html]

In Japan, for the effective usage of solar energy, JAXA (Japan Aerospace Exploration Agency) has explored the possibilities of (1) a space solar power system using photovoltaic solar cells and (2) a space solar power system using lasers (L-SSPS).

For (1), it is planned to launch a satellite equipped with solar panels into outer space (several hundreds of kilometers above the ground), and then transfer the generated power to Earth by microwave transmission technology. A space power plant is planned because a satellite orbiting the Earth can receive sunlight 24 hours a day, so that electrical power can be generated continuously from solar energy. However, there are still many issues in building this kind of system: (a) the possibility of efficient power generation in space where electromagnetic radiation is very strong, (b) construction technology for a large space

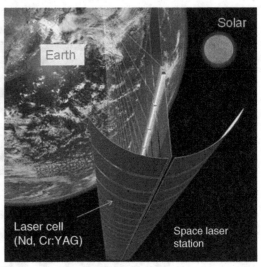

Figure 8.22 Schematic images for the development of a space solar laser (solar pumped laser in space) using ceramic laser gain media.

station (solar power plant) that can generate electricity and have rocket launch capability, (c) methods to transfer energy from a space station to the Earth, and (d) immeasurable costs for the completion of this system.

For (2), it is planned to transform solar energy into laser energy in the space station using solid-state laser technology, and then transfer the generated laser energy to the Earth. Figure 8.22 shows the basic concept developed by JAXA for a space solar power system using lasers (L-SSPS). This technological development has similar problems to the above system using solar cells. But, once we can see the possibility of this L-SSPS technology, it will become an innovative energy development for human beings.

Since solar rays are white light including electromagnetic waves and strong ultraviolet rays, it is not easy to achieve laser generation from sunlight using solid-state laser gain media. Strong ultraviolet rays cause solarization, and can easily damage the laser host materials. The size of commercial single crystal YAG laser materials produced by the conventional CZ method is very limited, and they cannot meet the required dimension for this solar power system. Glass laser gain media can be produced in a larger size, but the basic problem then is low thermomechanical properties and thus pulse operation with high repetition rate and CW operation with high output power are not possible. Therefore, there is only one way to generate high output power using ceramic laser gain media for this space solar system.

In the case of excitation by white light, it is not possible to achieve a satisfactory efficiency from normal Nd:YAG and Yb:YAG laser gain media. The Nd system can only absorb a few percent of the solar energy. The excitation level of Nd exists only between the $^4F_{5/2}$ (excitation 808 nm) and $^4F_{3/2}$ (885 nm) bands, and most solar radiation is shorter

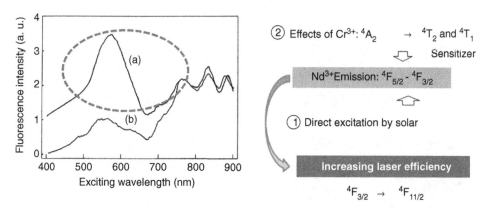

Figure 8.23 Absorption and emission characteristics as a function of excitation wavelength for (a) Nd-doped YAG and (b) Nd, Cr co-doped YAG ceramics.

than these wavelengths. Therefore, when a normal Nd:YAG gain medium is pumped with solar energy, only a few percent of solar energy is converted to laser energy and most of the solar energy is converted to thermal energy due to non-radiative transitions. In order to solve this problem, a sensitizer which can absorb the energy above the $^4F_{5/2}$ band (i.e., wavelengths shorter than 808 nm) and then transfer energy to the excitation bands of Nd ($^4F_{5/2}$ to $^4F_{3/2}$) is necessary. In 1995, the author succeeded in fabricating an Nd^{3+} and Cr^{3+} co-doped YAG ceramic gain medium for an efficient lamp pumping system [21]. Currently, this technology is expected to be used in the development of a space solar power system using lasers (L-SSPS) in Japan. In this project, laser oscillation experiments on the ground are underway, and finally it is planned to generate 100–400 MW in the space solar power system, and then transfer this energy to the Earth.

Figure 8.23(a) and (b) show the absorption characteristics of a normal Nd:YAG single crystal and an Nd, Cr co-doped YAG ceramic, and the dependence on excitation wavelength of the emission intensity for the $^4F_{3/2}-^4I_{11/2}$ transition in the Nd ion, respectively. By co-doping with Cr^{3+} ions, absorptions near 590 nm and 430 nm attributable to $^4A_2-^4T_2$ and $^4A_2-^4T_1$ transitions were confirmed, and red emission due to these absorptions was observed. Thus, efficient excitation for the $^4F_{5/2}$ band in the Nd^{3+} ion has become available. As shown in Figure 8.23(b), it was confirmed that the emission intensity of Nd, Cr co-doped YAG ceramic was 2–3 times higher than that of normal Nd:YAG ceramic because the Cr^{3+} ions play an important role in sensitizing.

The appearance of Nd, Cr co-doped YAG ceramic is shown in Figure 8.24(a). Due to the formation of Cr^{3+} ions, the sample was green in color. Although a large scale laser gain medium is required for the space solar system in the future, small scale gain media are necessary for ground experiments. Even in this case, about $300 \times 300 \times 30$ mm^{-3} gain media (at a maximum) are necessary for CW laser operation of 300 kW (at a maximum). Figure 8.24(b) shows the fabrication process for large scale Nd, Cr:YAG ceramic gain media. The demand characteristics for laser gain media are very strict, and it is impossible

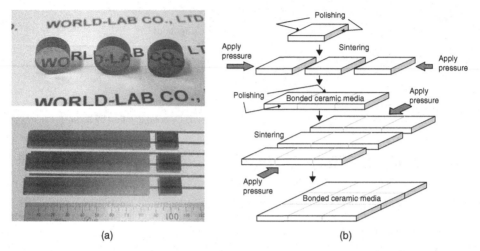

(a)   (b)

Figure 8.24 (a) Appearance of Nd, Cr co-doped YAG ceramics, and (b) image of the fabrication process of large scale ceramics by bonding for high power (300 kW) laser generation.

to fabricate such a large gain medium directly using current ceramic processes. As shown in the figure, smaller scale gain media will be prepared first, and then these gain media will be bonded to form a large scale gain medium.

The developmental status of ceramic laser materials worldwide, and some application technologies using the advantages of ceramic lasers in Japan have been introduced in this chapter. Since ceramic materials have advantages and characteristics that glass and crystals do not possess, it is expected that research and development on both materials technology and application technology will be performed more actively worldwide, and such technology will come into practical use in the near future.

References

[1] T. Yanagitani and H. Yagi, *Rev. Laser Eng.* **36** (9) (2008) 544–548.
[2] Y. Sato *et al.*, Spectroscopic properties and laser operation of Nd:$Y_3ScAl_4O_{12}$ poly-crystalline gain media, solid solutions of Nd:$Y_3Al_5O_{12}$ and Nd:$Y_3Sc_2Al_3O_{12}$ ceramics 2, *J. Ceram. Soc. Jpn. Suppl.* 112 (2004) S313–S316.
[3] M. Tsunekame, T. Taira, *et al.*, Technical digest on CD-ROM, *ASSP*/Nara 2008, paper MB4.
[4] J. Lu, H. Yagi, K. Takiuchi, T. Uematsu, J. F. Bission, Y. Femg, A. Shirakawa, K. Ueda, T. Yanagitani, and A. A. Kaminskii, *Appl. Phys. B* **79** (2004) 25.
[5] R. Yamamoto *et al.*, *Proc. Adv. Solid-State Photonics*, Nara (2008), paper WC5.
[6] N. Ter-Cabrielyan, L. D. Merkel, M. Dubinskii, E. R. Kupp, G. L. Messing, Tape cast composite Er:YAG Laser, OSA7s 93rd Annual Meeting, *Frontiers in Optics* 2009, Postdeadline paper, AWC6P-pdf, San Jose, CA, USA.
[7] Y. Sato, T. Taira, and A. Ikesue, Spectral parameters of Nd^{3+}-ion in the polycrystalline solid-solution composed of $Y_3Al_5O_{12}$ and $Y_3Sc_2Al_3O_12$, *Jpn. J. Appl. Phys.* **42** (1) (2003) 5071–5074.

[8] Y. Sato, I. Shoji, T. Taira, and A. Ikesue, The polycrystalline $Y_3Sc_xAl_{(5-x)}O_{12}$ as designable laser medium, presented at the *Conf. Laser Electro-Optics*, Baltimore, MD, June 6, 2003, paper CFG4.

[9] A. Ikesue, Yan Lin Aung, T. Taira, T. Kamimura, K. Yoshida, and G. L. Messing. Progress in ceramic lasers, *Mater. Res. Annu. Rev.* **36** (2006) 397–429.

[10] J. Saikawa, Y. Sato, T. Taira, and A. Ikesue, Absorption, emission spectrum properties, and efficient laser performances of $YbY_3Sc_1Al_4O_{12}$ ceramics, *Appl. Phys. Lett.* **85** (11) (2004) 1898–1900.

[11] J. Saikawa, Y. Sato, T. Taira, and A. Ikesue, Passive mode locking of a mixed garnet $Yb:Y_3Sc_1Al_4O_{12}$ ceramic laser, *Appl. Phys. Lett.* **85** (2004) 5845–5847.

[12] A. A. Kaminskii *et al.*, *Laser Phys. Lett.* **4** (11) (2007) 304–310.

[13] A. A. Kaminskii *et al.*, *Laser Phys. Lett.* **6** (4) (2009) 819–823.

[14] A. Gallian, V. V. Fedorov, S. B. Mirov, V. V. Badikov, S. N. Galkin, E. F. Voronkin, and A. I. Lalayans, *Opt. Express* **14** (2006) 11694.

[15] S. Mirov, V. Fedorov, I. Moskalev, D. Martyshkin, and C. Kim, Progress in Cr^{2+} and Fe^{2+} doped mid-IR laser materials, *Laser & Photonics Reviews*, **4** (2010) 21–41.

[16] T. T. Basiev, M. E. Doroshenko, P. P. Fedorov, V. A. Konuyshkin, S. V. Kouznetsov, and V. V. Osiko, *3rd Laser Ceramics Symp.*, Paris October 8–10, 2007, paper O-L-1.

[17] P. Aubry, A. Bensalah, P. Gredin, G. Patriarche, D. Vivien, and M. Mortier, *Opt. Mater.*, **31** (2009) 750–753.

[18] K. Tanaka, H. Waki, Y. Ido, S. Akita, Y. Yoshida, and T. Yoshida, *Rapid Commun. Mass Spectrom.* **2** (1988) 151.

[19] Y. Ido *et al.*, *Rev. Laser Eng.* **37** (4) (2009) 290–295.

[20] T. Taira *et al.*, *J. Combust. Soc. Jpn*, **51** (158) (2009) 288–294.

[21] A. Ikesue, K. Kamata, and K. Yoshida, 'Synthesis of Nd^{3+}, Cr^{3+}-codoped YAG ceramics for high-efficiency solid-state lasers, *J. Am. Ceram. Soc.* **78** (9) (1995) 2545–2547.

9

The future of ceramic technology

Chapters 1–7 described the basic technology of ceramic laser materials and laser generation, and typical applications of ceramic lasers. In 1995, it was reported that a coherence beam can be amplified efficiently in polycrystalline ceramic material. Since that report, research and development of ceramic lasers has been very active, and many innovative laser technologies that cannot be imagined using conventional glass and single crystal laser technologies have been developed [1–3].

Polycrystalline ceramic materials are normally composed of many microcrystallites of varying size (several micrometers to several tens of micrometers), with random orientations of these crystallites. The crystallites are often referred to as grains. The areas where these crystallite grains meet are known as grain boundaries. Until the second half of the twentieth century, polycrystalline ceramic materials were not included as laser materials because even a normal translucent or transparent ceramic included many scattering objects such as grain boundary phases and residual pores. Therefore, at that time, materials scientists and physical scientists (laser scientists) could not imagine that such ceramics would become new optical materials.

If one knows the principles of Rayleigh and Mie scattering theory, one can easily estimate the scattering level of polycrystalline ceramics, which have many scattering sources such as grain boundaries and residual pores. Therefore, it would be obvious to any scientist that the development of ceramic materials for laser generation would be pointless. Well then, the readers may question why the author developed ceramic lasers. The answer to that question is very simple. The author succeeded in obtaining laser oscillation in 1991 (before the public announcement in 1995) using polycrystalline ceramic materials. At that time, the author was neither a ceramics expert nor a laser scientist, but rather an ordinary refractory engineer. Although a non-expert engineer, the desire to know the true nature of the grain boundaries in ceramics and the challenge of achieving success in this area were the driving force for the author to develop ceramic lasers. In addition, the strongest driving force was the achievement of laser oscillation using a polycrystalline ceramic which was reported by Dr. Greskovich [4–6]. At that time, many materials scientists focused only on the exploration of new materials and nanotechnology etc., and there were only a few researchers who were exploring the intrinsic (latent) performance of the materials. In the 1990s, blue LEDs (light emitting diodes) using GaN (gallium nitride) were developed and

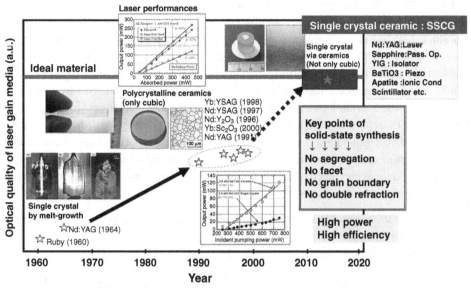

Figure 9.1 Chronological diagram showing progress in the development of solid-state laser material.

now blue lasers are very actively researched worldwide. In fact, the GaN material was noted 20 years ago, but many crystal defects were included in single crystal GaN layers. Therefore its performance for use as an LED was very poor, and the development was halted. However, Professor S. Nakamura discovered that the intrinsic (latent) performance of the material could be extracted by reducing the number of crystal defects [7]. This basic concept is common to the development of ceramic laser materials. It is well known that normal ceramic material includes many defect textures. But, there was almost no research exploring the original performance of the materials. Although the field of R&D of ceramic lasers is different from that of blue ray LEDs, the basic concept used by the author is very similar to that of Professor S. Nakamura. Accordingly, as a materials scientist, it can be said that if a researcher has much common knowledge, then he can only think and work within his knowledge limit and cannot think beyond his field so that new concepts can never emerge – that is to say, a great deal of knowledge often adversely affects the creation of innovative ideas. Where there is a technology which exceeds a general theory and a way to overcome difficulties using classical theory, such a new idea can only be reflected in the technology itself. If you have reasonable knowledge, and are interested in questioning technology and science, then this is the root of R&D.

Chronological data on solid laser materials are summarized in Figure 9.1. Dr. Maiman discovered the ruby laser in 1960 [8, 9], and then Nd:YAG, Ti:sapphire and Nd:YVO$_4$ lasers etc. were used at room temperature. All of these gain media were single crystals grown by the melt-growth method.

Before the 1990s, the possibility of ceramic lasers was suggested [10–12], but actually the birth of ceramic laser technology occurred in the 1990s when the author first reported an efficient ceramic laser oscillation at room temperature. Nowadays, polycrystalline ceramic laser materials are being investigated as novel materials that can overcome the technical problems of conventional melt-growth single crystal materials, for instance, (1) power scaling using a large scale medium [13], (2) creating new functionalities by advanced ceramic composite technology [14, 15], (3) creating new materials for new functionalities [16–18] etc. Therefore, it is expected that ceramic laser technology will become a core technology of next-generation lasers. However, the ultimate goal of solid-state lasers is not a "polycrystalline ceramic gain medium." It has been demonstrated that ceramic technology has many advantages compared to conventional single crystal technology, but there are still unresolved technical issues in ceramic laser technology. These are

(1) the crystal structure of the ceramic materials for use as a gain medium is limited to cubic systems, and
(2) many grain boundaries exist in polycrystalline ceramic materials.

Once a stable technology to produce high quality single crystal ceramic materials by sintering is established, the above two technical issues will be automatically resolved. Accordingly, in addition to cubic systems, it will be possible to produce single crystal ceramic materials with hexagonal or tetragonal structure. Basically, the dislocation (defect) density is very high near the grain boundaries in polycrystalline ceramic materials. Although ceramic laser materials can withstand existing high power lasers and high power density lasers, the grain boundaries may become an initial source of laser induced damage in the case of higher power lasers in the future. It is also possible to produce a grain-boundary-free single crystal by the conventional melt-growth method. But this method has the problems of size limitation and optical inhomogeneity. Therefore, single crystal ceramics produced by a sintering method possess the advantages of polycrystalline ceramic materials, and they do not have the weak points (i.e., grain boundaries) of polycrystalline ceramic materials. It can be assumed that single crystal ceramic laser materials will become ideal materials for solid-state laser technology in the future because they can offer a fusion of the performance of conventional single crystal lasers and the ceramic lasers of recent years [19–21]. Although polycrystalline ceramic materials have become a paradigm of next-generation solid-state lasers, the author considers that single crystal ceramic laser material will become an ultimate goal of solid-state laser materials because it does not include optically inhomogeneous parts such as the core and facets of melt-growth single crystals.

Since the author succeeded in the development of ceramic laser materials, the optical quality of the ceramic laser gain media has improved significantly. In solid-state lasers, laser active ions doped in the gain medium are continuously excited with a flashlamp, and the stimulated radiation is amplified in the gain medium by reflecting between two mirrors in order to achieve strong laser generation. In this process, the amplified laser beam is passed inside the gain medium a great number of times. Therefore, the gain medium must have extremely low scattering and very high optical quality. It is possible to apply this

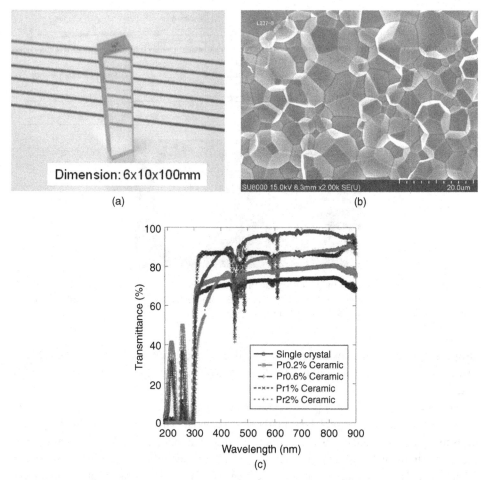

Figure 9.2 (a) Appearance of optical grade Pr:LuAG ceramic, (b) its microstructure under SEM observation, and (c) the emission spectrum of the optical grade Pr:LuAG ceramic compared with a single crystal counterpart for a PET scanner.

technology of material synthesis, which can meet the strict requirements of optical quality, to the fabrication of optical materials.

The appearance of an optical grade Pr:LuAG ($Pr_xLu_{3-x}Al_5O_{12}$) ceramic, its microstructure under SEM observation and the emission spectrum for the 5s–4f electron transition, compared with a single crystal counterpart, are shown in Figure 9.2 [22]. The transmittance of the ceramic materials was equivalent to that of the single crystal counterpart grown by the melt-growth process. However, the ceramic material can have a higher amount of Pr ion doping, and is almost free from the birefringence (double refraction) generally seen in melt-growth crystals. The density of this material is 6.7 g cm^{-3}, and because of this density, it can be used as a barrier to stop radioactive rays such as X-rays, r-rays etc. It has a

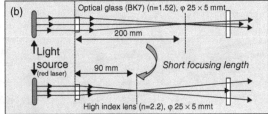
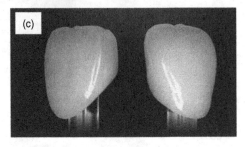

Figure 9.3 (a) High refractive index lens made of transparent cubic-zirconia ceramic; (b) schematic diagram of short focusing by a high index lens, and (c) artificial teeth made of translucent tetragonal-zirconia ceramic with higher mechanical strength.

strong fluorescence intensity and a very short decay time. Therefore, this material is being investigated for use as a scintillator for high-speed PET (positron emission tomography) machines.

Another application of optical ceramic materials is as special purpose lenses. Figure 9.3(a) shows a special lens made of transparent zirconia ceramic [23]. Typical glass materials such as BK-7 have a refractive index of around 1.5 to 1.7, which has limited the applications of lenses and prisms. This zirconia ceramic has a high index of around 2.2 to 2.3, and the raw materials are not so expensive. Thus, it has attracted attention for use in short focus lenses. To compare the focal length of this zirconia ceramic with commercial BK-7 glass, they were processed into lenses of the same dimension. The measured results are shown in Figure 9.3(b). It was found that the focal length of the ceramic was about half that of the BK-7 glass. Accordingly, this type of ceramic lens is expected to be used as a short focus lens in digital cameras (to develop a compact digital still camera) in the near future. Figure 9.3(c) shows an artificial tooth made of translucent zirconia ceramic [24]. Normally, high toughness zirconia materials have a tetragonal structure and are almost opaque. Using the conventional technology, translucent zirconia ceramic was produced by adding a large amount of Y_2O_3 (as a stabilizer) and sintering aids (TiO_2), because it was necessary to change the crystal structure from tetragonal to cubic in order to achieve translucency. As a result, the mechanical strength (especially toughness characteristics) was significantly lowered. However, for the non-translucent zirconia ceramics used as artificial teeth, optical scattering on the surface of the materials was very large and it looked a whitish color. Therefore, there was an aesthetic problem (it looked very different from

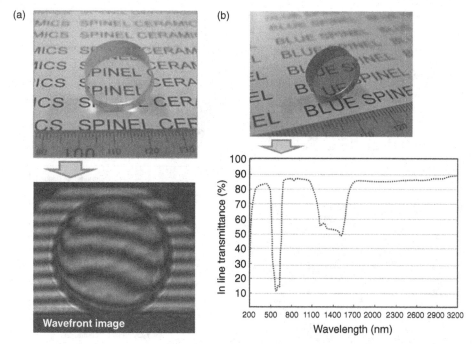

Figure 9.4 (a) Appearance of undoped spinel ceramic, and its wavefront image. (b) Appearance of cobalt-doped spinel ceramic and its in-line transmittance curve from 200 nm to 3200 nm.

natural teeth). However, with our currently developed technology, it is not necessary to change the crystal structure of the material. By reducing the number of structural defects (voids) in the material, it was possible to improve the translucency as well as the mechanical strength. The bending strength of the material at room temperature is around 1000–1200 MPa, and it is planned for commercial development because it is very promising for tooth implantation.

Undoped spinel ceramics and cobalt-doped spinel ceramics were produced by advanced ceramic processing technology, and their optical properties were also investigated (see Figure 9.4) [25]. Almost straight fringes were confirmed in the wavefront image of the undoped spinel ceramic, suggesting that the optical homogeneity throughout the sample is very high, and it is applicable as a passive lens. In-line transmittance of the Co:spinel (thickness 10 mm) almost reached the theoretical value. The optical loss of the Co:spinel was about 0.1–0.2% cm^{-1}, and this quality is also equivalent or superior to that of commercial Co:spinel single crystals. This material is promising for application as a Q-switch element in an eye-safe laser (1.5 μm region). We have confirmed that the laser damage threshold of this ceramic material is about three times higher than that of the commercial single crystal. Details of this investigation will be reported elsewhere.

The appearance of SSCG grown YIG ($Y_3Fe_5O_{12}$) single crystal ceramic and its microstructure near the growth interface are shown in Figure 9.5. Figure 9.5(c) shows

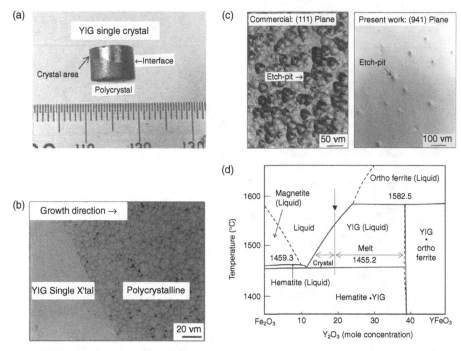

Figure 9.5 (a) Appearance of yttrium–iron–garnet single crystal produced by sintering, (b) microstructure near the growth interface between the single crystal and polycrystals, and (c) etch-pit images of YIG single crystals (left, commercial crystal produced by the float zone method; right, author's work produced by sintering). (d) Phase diagram for the Fe_2O_3–Y_2O_3 system.

etch-pit images [26] of this single crystal ceramic (right) and of a commercial YIG single crystal produced by the FZ (float zone) method. The growth rate of the SSCG grown YIG crystal was as fast as 1 mm h^{-1}, but the dislocation density (estimated from the etch-pit density) was about 10^2 cm^{-2}. This value was about two or three orders of magnitude smaller than that of commercial YIG single crystals. Bismuth-doped YIG single crystal is produced commercially by the LPE (liquid phase epitaxial) method, and it has been widely used as an isolator in optical communication [27]. As seen in the phase diagram of the Fe_2O_3–Y_2O_3 system [28, 29] shown in Figure 9.5(d), it is extremely difficult to produce this material by the melt-growth process because a peritectic point exists. Even if it can be produced by the melt-growth process, the size of the crystal will be very small. Another problem is the difficulty of doping the volatile Bi into YIG crystal, and the crystal growth rate is very slow (a few micrometers per hour). (It takes about one week to grow a crystal of 1 mm thickness.) In addition, a GGG ($Gd_3Ga_5O_{12}$) single crystal wafer has to be grown on it. Therefore, the current technology has both technical and cost problems. If it is possible to produce Bi-doped YIG single crystal ceramics by sintering in the future, it will provide a potential breakthrough for the current technology.

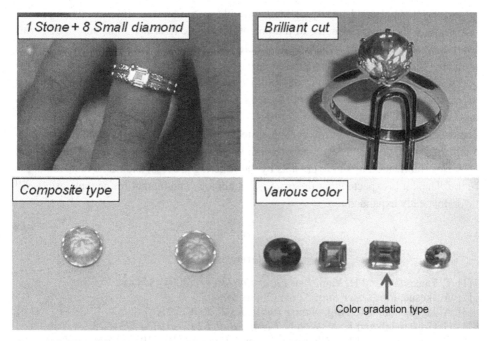

Figure 9.6 Ceramic jewelry composed of garnet, spinel, zirconia and so on.

Another application of optical ceramics is making synthetic jewelry [30]. Some ceramic jewelry is shown in Figure 9.6. Ceramic stones were cut and polished, the same as commercial gem stones. The quality of ceramic gem stones is very high, and it is very difficult to distinguish them from natural stones in appearance. Since ceramic material has design flexibility, uniquely designed gem stones can be produced, for example, (1) a colored central area in a non-colored transparent background, (2) color gradation across the stone, (3) color variation, i.e., a composite made of different color stones, and (4) a mosaic structure, i.e., stone composed of many different color dots etc. Therefore, ceramic jewels are very interesting for future jewel markets because these designs are so different from conventional jewelry.

The main concept for the development of ceramic laser materials is based on the fabrication of materials with defect-free structure, resulting in laser materials, and other optics. Therefore, this technology for producing ceramic laser materials can be extended not only to other optical materials but also to the synthesis of structural materials made of ceramics. For example, for engineering ceramics such as Al_2O_3 and SiC, the mechanical strength is very important. Generally, fracture occurs at structural defects such as residual pores. In fact, current ceramic materials include many defects, and their mechanical strength is much lower than their proper strength. Previously, Al_2O_3 and AlN ceramic materials normally used in IC (integrated circuit) and LSI (large scale integration) substrates included many structural defects (i.e., grain boundary phases and voids). After resolving these technical issues, the thermal conductivity of these ceramic materials was significantly improved. Accordingly, the key point in the development of ceramic laser materials described in this book is how to eliminate such structural defects in ceramics, but actually this technical problem is common to all other ceramic materials.

The author is very proud and honored that the ceramic laser has attracted a great deal of attention worldwide from materials scientists and solid-state laser scientists and physicists since the first report in 1995 [31, 32], and nowadays this success has increased the competition among countries involved in the technological development race. Younger generation researchers should develop this technology further, and the remaining technological problems (fabrication of single crystals by the sintering process) should be resolved in the future. Then, it is expected that the material technology and market for new materials will be dramatically expanded.

References

[1] T. Yanagitani and H. Yagi, *Rev. Laser Eng.* **36** (9) (2008) 544–548.

[2] A. Ikesue, *Rev. Laser Eng.* **37** (4) (2009) 248–253.

[3] A. Ikesue and Y. L. Ang, *Nature Photonics* **2** (2008) 721.

[4] C. Greskovich and K. N. Woods, *Ceram. Bull.* **52** (5) (1973) 473–478.

[5] C. Greskovich and J. P. Chernoch, *J. Appl. Phys.* **44** (10) (1973) 4599–4605.

[6] C. Greskovich and J. P. Chernoch, *J. Appl. Phys.* **45** (10) (1974) 4495–4502.

[7] S. Nakamura, T. Mukai, and M. Senoh, Candela-class high-brightness InGaN/AlGaN double-hetero structure blue-light-emitting diodes, *Appl. Phys. Lett.* **64** (13) (1994) 1687–1689.

[8] T. H. Maiman, *Nature (London)* **187** (1960) 493–494.

[9] T. H. Maiman, *Phys. Rev. Lett.* **4** (11) (1960) 564–566.

[10] S. E. Hatch, W. F. Parson, and R. I. Weagley, *Appl. Phys. Lett.* **5** (1964) 153.

[11] C. Greskovich and J. Chernoch, *J. Appl. Phys.* **44** (1973) 4599; M. Sekita, H. Haneda, T. Yanagitani, and S. Shirasaki, *J. Appl. Phys.* **67** (1) (1990) 453–458.

[12] G. de With and H. J. A. van Dijk, *Mater. Res. Bull.* **19** (1984) 1669–1674.

[13] R. M. Yamamoto *et al.*, *Adv. Solid State Photonics*, Nara (2008), paper WC3.

[14] A. Ikesue, Yan Lin Aung, Synthesis and performance of advanced ceramic lasers, *J. Am. Ceram. Soc.* **89** (6) (2006) 1936–1944.

[15] Y. Sato, A. Ikesue, and T. Taira, Tailored spectral designing of layer-by-layer type composite Nd:$Y_3ScAl_4O_{12}$/Nd:$Y_3Al_5O_{12}$ ceramics, *IEEE J. Select. Topics Quantum Electron.* **13** (3) (2007) 838–843.

[16] A. Ikesue, *Rev. Laser Eng.* **37** (4) (2009) 248–253.

[17] N. Ter-Gabrielyan, L. D. Merkle, A. Ikesue, and M. Dubinskii, Ultralow quantum-defect eye-safe Er:Sc_2O_3 lasers, *Opt. Lett.* **33** (13) (2008) 1524–1526.

[18] M. Tokurakawa, A. Shirakawa, K. Ueda, H. Yagi, M. Noriyuki, T. Yanagitani, and A.A. Kaminskii, *Opt. Express* **17** (2009) 3353.

[19] A. Ikesue, Yan Lin Aung, T. Yoda, S. Nakayama, and T. Kamimura, Fabrication and laser performance of polycrystal and single crystal Nd:YAG by advanced ceramic processing, *Opt. Mater.* **29** (2007) 1289–1294.

[20] A. Ikesue, T. Yoda, S. Nakayama, T. Kamimura, and K. Yoshida, *25th Annual Meeting for Laser Society Japan*, Keihanna (2005), paper I12.

[21] A. Ikesue, T. Yamamoto, Y. Sato, T. Kamimura, and K. Yoshida, *25th Annual Meeting for Laser Society Japan*, Keihanna (2005), paper I14.

[22] JP Patent: 2010–235388, Transparent ceramic, its producing method, and devices using the transparent ceramics.

[23] JP Patent: 2010–047460, Transparent ceramic, its producing method, and optical element using the transparent ceramics.

[24] JP Patent Filing no. 2009–225658, Zirconia ceramics and its producing method.

[25] PCT/JP2008/50858, Transparent spinel ceramics, its producing method, and optical element using the transparent spinel ceramics.

[26] A. Ikesue, unpublished work.

[27] K. Sato, *Bull. Ceram. Soc. Jpn.* **28** (2) (1993) 124–129.

[28] *Phase Diagram for Ceramics*, vol. 1, Figure 98, American, Ceramics Society (1964).

[29] T. Okada, *Bull. Ceram. Soc. Jpn.* **23** (5) (1988) 468–470.

[30] A. Ikesue, *Fall OSA Optics & Photonics Congress, Frontiers in Optics 2009, Advances in Optical Materials* (2009).

[31] A. Ikesue, T. Kinoshita, K. Kamata, and K. Yoshida, Fabrication and optical properties of high-performance polycrystalline Nd:YAG ceramics for solid-state lasers, *J. Am. Ceram. Soc.* **78** (4) (1995) 1033–1040.

[32] A. Ikesue, I. Furusato, and K. Yoshida, Fabrication of polycrystalline, transparent YAG ceramics by a solid-state reaction method, *J. Am. Ceram. Soc.* **78** (1) (1995) 225–228.

10

High resolution optical spectroscopy and emission decay of laser ceramics

The characterization of the static and dynamic spectroscopic properties of solid-state laser materials based on transparent materials doped with laser active ions is a crucial step in the development and optimization of laser solutions and tailoring them according to the needs of specific applications, along the deterministic chain technology–composition–structure–properties–functionality. High resolution optical spectroscopy of laser materials provides data on the variety, nature and structure of the doping centers, on the quantum states of the laser ions and on their interaction with electromagnetic radiation, with two major areas of relevance: (i) they can be used as very selective and sensitive microstructural methods, and (ii) they provide the relevant data (energy levels, transition cross-sections, dynamics of emission) for modeling of laser emission processes.

The crystalline host materials offer definite crystallographic sites for substitution, concerning the number of and distance to the anions of the first coordination sphere and the bond angles. Additionally, the crystal field can be influenced by the characteristics of other coordination spheres, especially the first cationic sphere, the main differences being caused by accidental or inherent variations in composition, such as lattice defects, doping ions in near sites, impurities, charge compensators, compositional disorder. These differences can determine different optical spectra (positions of lines, transition probabilities) that can be resolved by experiment, leading to a multicenter structure of the spectra for each type of site. Accordingly, not only will the structural center correspond to the first anionic coordination sphere, which defines the site, but it will also include the farther coordination spheres that can induce spectrally resolved crystal field differences. In the case of accidental variations of the cationic coordination spheres, the multicenter structure can be considered as composed of a normal center accompanied by perturbed centers originating from this center.

The high resolution optical spectra of many host materials with unique and well-defined crystallographic sites available for substitution indeed show satellite structures that reflect the existence of multicenter situations. The dependence of these structures of spectral satellites and of the emission dynamics on composition, doping concentration and conditions of fabrication enables unambiguous identification of the source of perturbation. In the case of RE-doped garnet materials, particularly for Nd:YAG, such satellite structures have been observed and examined both in single crystals [1–5] and in ceramics [6–10]. It was found that some of these structures are similar and depend only on the doping concentration, whereas the relative intensities of other structures are different in these types of materials, but independent of doping. The existence of grain boundaries and segregation [11, 12] and diffusion of the doping ions or of foreign ions from the sintering aid in ceramics could favor additional structural sites or new perturbed centers and modify the distribution of the doping ions. Such a situation would be manifest at the grain scale, and this recommends additional investigation by spatially resolved spectroscopic methods; nevertheless, the effect on the global properties will be determined by their relative extent.

This chapter concentrates on the spectroscopic properties of ceramic laser materials and on the possible differences from the corresponding single crystals, such as:

- the variety, nature and relative concentration of the structural centers, including the search for the existence and characterization of new types of centers such as defective centers at the grain boundaries or centers induced by the presence of foreign ions from the sintering aid;
- the quantum states (energy levels, transition cross-sections, line shape and width) of the doping ions in ceramics in comparison with the corresponding single crystals and the characterization of these states in the new materials;
- the distribution of the doping ions at the available sites and the effect of doping segregation and diffusion; characterization of the spatial variation of doping concentration using microscopic methods and correlation with the global measurements;
- the de-excitation processes, energy transfer, emission decay and quantum efficiency in ceramics.

10.1 Structural characterization of doped ceramics by optical spectroscopy

10.1.1 Multisite materials

10.1.1.1 Garnets

Although the doping RE^{3+} ions show a preference for the large dodecahedral c-sites of the garnets $A_3B_2C_3O_{12}$, the absorption spectra of various melt-grown garnet crystals show the presence of a small amount of RE^{3+} ions in octahedral a-sites. Due to the inversion symmetry C_{3i} of these sites, spectral lines can only be observed for transitions that allow magnetic dipole transitions, and such lines were observed in several cases, such as for Nd^{3+} and Er^{3+} in YAG [4, 13]. The presence of Er in the octahedral sites of the YAG crystals was also inferred from the specific perturbing effect on the spectra of Er in dodecahedral sites [5], as will be explained later. Investigation of the garnet ceramics did not reveal RE^{3+} ions in a-sites and this suggests differences in the doping mechanism. However, both single crystal and ceramic garnets can be considered as multisite systems for doping with smaller ions, such as the 3d-ions which, based on ionic radius arguments, can occupy both the octahedral a-sites and the tetrahedral d-sites, although in many cases a definite

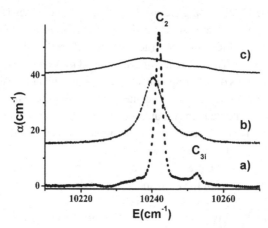

Figure 10.1 The presence of two sites in the $^2F_{7/2}$ (1) – $^2F_{5//2}(1)$ absorption spectra of Yb^{3+}:Y_2O_3 ceramics at different temperatures: (a) 10 K, (b) 80 K and (c) 300 K.

preference for one of these is manifested due to the particular electronic structure of the doping ion.

10.1.1.2 Cubic sesquioxides

Since the size and effective charge of the C_2 and C_{3i} sites in the cubic sequioxides R_2O_3 (R = Y, Lu, Sc) are similar, substitution of both sites by the doping RE^{3+} laser ions could be expected, although for several cases a preference for one of the sites was claimed. In principle, the empty oxygen polyhedra placed alternately with those occupied by cations along the $\langle 100 \rangle$ directions could accommodate interstitial doping ions or accidental ions or defects, although additional charge compensation would be necessary. However, no obvious presence of RE^{3+} ions in such interstitial sites was reported. Forced electric dipole transitions are allowed only for the RE^{3+} ions in the C_2 sites, whereas magnetic dipole transitions are allowed for both centers in transitions with $\Delta J = 0, \pm 1$. The optical spectra for a large variety of RE^{3+} ions in C_2 sites of the sequioxide crystals and transparent ceramics have been reported. However, because of the selection rules, only partial spectra for several ions in the C_{3i} sites have been clearly identified and in most cases these refer to the transitions between the ground manifold and the nearest excited manifold [14].

A high resolution low temperature spectroscopic investigation of Yb-doped Sc_2O_3 and Y_2O_3 ceramics revealed both the C_2 and C_{3i} centers, similar to previous reports [15, 16] on single crystals. These centers are clearly seen in the zero phonon $^2F_{7/2}(1) \rightarrow {}^2F_{5/2}(1)$ absorption, as shown in Figure 10.1 for Yb:Y_2O_3. The line of the C_{3i} center is shifted by ~ 11 cm^{-1} towards the higher energies compared with the C_2 line; a similar situation holds for the Yb-doped Sc_2O_3 ceramic [17]. Since the $^2F_{7/2}(1) \rightarrow {}^2F_{5/2}(1)$ absorption line of the C_2 center contains contributions from electric and magnetic dipoles, whereas in the case of the C_{3i} center only the latter contributes, an accurate evaluation of the relative occupancy of the two sites with Yb ions is not possible.

Table 10.1 *The nearest-neighbor (n.n., sphere 1) and next-nearest-neighbor (n.n.n., sphere 2) cationic coordination spheres in garnet lattices (R in units of a/8)*

Site	Surrounded by	Sphere 1		Sphere 2	
		m	R	m	R
A (c-site)	A (c-site)	4	$(6)^{1/2}$	8	$(14)^{1/2}$
	B (a-site)	4	$(5)^{1/2}$	4	$(13)^{1/2}$
	C (d-site)	2	2	4	$(6)^{1/2}$
B (a-site)	A (c-site)	6	$(5)^{1/2}$	6	$(13)^{1/2}$
	B (a-site)	8	$2(3)^{1/2}$	6	4
	C (d-site)	6	$(5)^{1/2}$	6	$(13)^{1/2}$
C (d-site)	A (c-site)	2	2	4	$(6)^{1/2}$
	B (a-site)	4	$(5)^{1/2}$	4	$(13)^{1/2}$
	C (d-site)	4	$(6)^{1/2}$	8	$(14)^{1/2}$

10.1.2 Multicenter structure and inhomogeneous broadening of spectra

The global picture of the perturbations determined by irregularities in the cationic coordination spheres around the doping RE^{3+} ions is determined by the variety of the perturbing centers, as well as by that of the available lattice sites and by their occupation by the perturbing centers, and reflects the structure of the host material. If the crystal field perturbation is isotropic, all perturbing centers from a coordination sphere around the RE^{3+} ion produce identical perturbations; however, in the case of anisotropic perturbations, each coordination sphere can be further divided into subspheres of identical perturbation.

The radius R and number of sites m for the nearest cationic coordination spheres around the cations in the garnet lattice are given in Table 10.1 in terms of $a/8$, a being the cubic lattice parameter (for undoped YAG, $a \approx 1.200$ nm). The unit cell contains eight units of $A_3B_2C_3O_{12}$ and thus the density of the c-sites, $24/a^3$, in YAG equals 1.389×10^{22} cm^{-3}.

The unit cell of the cubic sesquioxides contains 16 R_2O_3 compositional units, i.e. 32 R^{3+} cations. The lattice parameter a is equal to 0.9857 nm for Sc_2O_3, 1.016 nm for Lu_2O_3 and 1.0602 nm for Y_2O_3 and, despite the alternative empty sites along the 100 directions, the density of the R^{3+} ions is high, 3.338×10^{22} cm^{-3} for Sc_2O_3, 2.852×10^{22} cm^{-3} for Lu_2O_3 and 2.687×10^{22} cm^{-3} for Y_2O_3 [15, 16]. In contrast to the CaF_2 lattice, which has a unique cubic cationic site and all 12 n.n. cations are placed on a coordination sphere of radius $R = \sqrt{2}d$, where d is the fluorine cube edge, in the case of the cubic sesquioxides the existence of these two sites induces a slight variation in the cation–cation distance and the first coordination sphere is composed of several sub-spheres, as shown in Table 10.2.

Tables 10.1 and 10.2 enable calculation of the occurrence probabilities for perturbing centers in the various coordination spheres around the doping ion. For random distribution

Table 10.2 *The n.n. (sphere 1) and n.n.n. (sphere 2)*
cationic coordination spheres in the cubic
sesquioxides (R *in units of* a)

		Sphere 1		Sphere 2	
Site	Surrounded by	m	R	m	R
C_2	C_2	4	0.339	6	0.5
		4	0.378		
	C_{3i}	2	0.332	0	–
		2	0.376		
C_{3i}	C_2	6	0.332	0	–
		6	0.376		
	C_{3i}	0	–	6	0.5

of the perturbing centers, the probability of having n occupied sites out of m available sites in a sphere can be estimated using Eq. (2.24). Comparison of the calculated probabilities with the measured relative intensities of the spectral satellites in the absorption spectra could provide unique information on their origin. The various satellite structures showing simultaneously in a laser material can be further particularized by their emission dynamics.

10.1.2.1 Centers independent of doping

10.1.2.1.1 Centers connected with the nature of the fabrication process The absorption lines of RE-doped melt-grown YAG crystals show up to four satellites P_i of equal intensity, independent of RE concentration C_{RE} [5]. In the flux-grown crystals such satellites are practically absent, whereas in ceramics they are very weak, as illustrated in Figure 10.2 for the absorption transition $^4I_{9/2}(1) \rightarrow {}^4F_{9/2}(1)$ of Nd^{3+} in YAG. Figure 10.3 shows four satellites P_i in the absorption spectrum $^4I_{15/2}(1) \rightarrow {}^4I_{13/2}(1)$ of Er^{3+} in Czochralski grown YAG crystal and their absence in ceramics, and a similar situation was observed for other RE^{3+} ions (Pr^{3+}, Sm^{3+}, Ho^{3+} and so on) in YAG. In the case of the Er:YAG crystals it was observed that on increasing the Er concentration to 100%, the satellites P_i broaden and a new set of four satellites P_i' is gradually installed, with a slightly larger splitting from the main line N [5]. The relative intensities of the satellites P_i can thus be connected with the compositional differences between the host garnet materials produced by various fabrication techniques.

As mentioned in Chapter 2, the melt-grown garnet crystals show departures from the stoichiometric composition, manifested in excess A^{3+} ions that enter in the octahedral a-sites normally occupied by the ions B^{3+} (A(a) "antisites"), as shown schematically in Figure 10.4. The relative concentration of these antisites is small, several percent of the a-sites, depending on composition: in YAG, these antisites are $Y^{3+}(a)$ ions, with relative concentration $\sim 1.75\%$ of the a-sites in the Czochralski grown crystals; thus the melt-grown

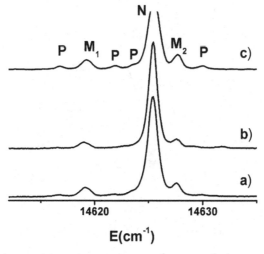

Figure 10.2 Comparative satellite structure in 10 K $^4I_{9/2}(1)$ – $^4F_{9/2}(1)$ absorption of Nd:YAG (a) ceramic, (b) flux crystals and (c) Czochralski crystals.

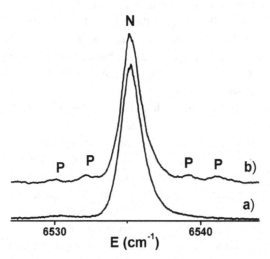

Figure 10.3 Satellite structure observed in 10 K $^4I_{15/2}$ (1) – $^4I_{13/2}$ (1) absorption of Er^{3+}(5 at.%):YAG: (a) ceramic and (b) Czochralski crystals.

garnet crystals are in fact A(*a*)-doped garnets. However, these defects have very reduced occurrence in the low temperature flux-grown crystals. By assuming that the excess A^{3+} ions occupy at random any *a*-site, the occurrence probability for such an A(*a*) ion ($n = 1$) in the first coordination sphere of *a*-sites around the RE^{3+} *c*-site can be calculated using Eq. (2.24) for *C* equal to the relative concentration of A(*a*) centers; as shown in Table 10.1, in this case $m = 4$. If the perturbations induced by the ions A(*a*) from any of these four n.n. sites at the *c*-site are equal, a unique satellite of relative intensity proportional

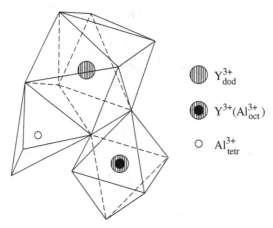

Figure 10.4 The non-stoichiometric defect in high temperature YAG: Y^{3+} in Al^{3+} octahedral sites.

to $4C(1-C)^3$ would be expected; however, if these differ (anisotropic perturbation), four satellites can be produced, each with relative intensity $C(1-C)^3$. By comparing the measured relative intensities of the satellites P_i in transitions with the matrix element $\langle 4f^n a J \parallel U^{(2)} \parallel 4f^n a' J' \rangle = 0$ to reduce the effect of the Ω_2 Judd–Ofelt parameter on the line strength S_{mn}, such as the absorption $^4I_{9/2} \rightarrow {}^4F_{3/2}$ in the case of Nd^{3+}, with the calculated occurrence probability, Eq. (2.24), it was found that they correspond to the concentration of the centers $A(a)$ determined by chemical analysis. This model enables the connection of the new P_i' satellites in the heavily doped Er:YAG crystals with the perturbing effect of the Er^{3+} ions entering in octahedral a-sites on the Er^{3+} ions from the dodecahedral sites, and thus they correspond to n.n. $Er^{3+}(c)$–$Er^{3+}(a)$ pairs. This confirms that in the melt-grown YAG crystals the doping Er^{3+} ions could occupy, besides the dodecahedral sites, a small fraction of octahedral a-sites. The probability of $Er^{3+}(a)$ centers in YAG is similar to that of $Y^{3+}(a)$ centers so their summed occurrence is practically constant along the whole doping range in Er:YAG. The four distinct perturbations can be explained by the strain induced effects caused by the $A(a)$ ions [18] or with the crystal field superposition model considering the additional low symmetry crystal field components determined by the distortion of the O^{2-} dodecahedron around the c-site [7] due to the n.n. $A(a)$ centers. Besides the fabrication process, the presence of the $A(a)$ centers in garnets depends on the particularities of the electronic structure of the A and B ions: in the scandium garnets such as GSGG the centers $A(a)$ are absent and the RE^{3+} spectra do not show P satellites. When doping the garnets with Cr^{3+}, which shows a preference for the a-sites, the concentration of the $A(a)$ centers is drastically reduced; nevertheless, Cr doping introduces new P-like centers.

The emission spectra to the ground manifold under non-selective excitation show P satellite structure similar to the absorption spectra and the practically unaltered relative intensities with respect to the unperturbed line N indicate similar emission quantum efficiency. The selectively excited luminescence reveals the individual emission spectra of these satellites, as shown in Figure 10.5 for the 10 K emission from the Stark level

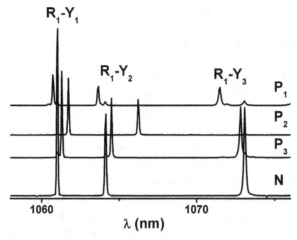

Figure 10.5 Part of the $^4F_{3/2} - {}^4I_{11/2}$ emission spectra of the isolated N center and of the P_i centers under selective excitation of Nd(1 at.%):YAG crystal at 10 K.

$^4F_{3/2}(1)$ to the manifold $^4I_{11/2}$ with selective excitation in the $^4I_{9/2}(1) \rightarrow {}^4F_{3/2}(1)$ transition in a Nd:YAG crystal. Only three P satellites are present in these spectra since the fourth is not resolved in the absorption transition used for excitation.

The spectral shifts of the P_i satellites with respect to the main line N depend on the composition of the garnet, on the ionic radius mismatch between the octahedral ionic radii of the ions A^{3+} and B^{3+}, on the nature of the doping ion and on the electronic transition. In the case of YAG, these shifts could be up to 15–20 cm^{-1} for the large ions Pr^{3+} and Nd^{3+}, but are less than 10 cm^{-1} for the small RE ions Er^{3+} and Tm^{3+}. The unresolved perturbations produced by the farther $A(a)$ sites contribute to asymmetric inhomogeneous broadening of the lines.

The antisite $A(a)$ centers in garnets can influence the properties of these materials:

- the shifted P_i satellites can prevent some of the doping ions absorbing the radiation from narrow-band excitation sources; however, they can still participate as acceptors in the parasitic self-quenching of the emission of the excited ions or act as traps of excitation;
- the P_i emission satellites can leave some of the excited structural centers outside the laser emission; such centers will contribute to enhanced heat generation and luminescence;
- the unresolved effects of $A(a)$ centers can determine asymmetric inhomogeneous broadening of the spectral lines in melt-growth crystals compared with ceramics;
- the $A(a)$ centers could contribute to reduction of heat conduction;
- the perturbation of the crystal field by the $A(a)$ centers can modify the positions of the Stark levels of the excited configuration $4f^{n-1}5d$ of the RE^{3+} ions;
- the charge-transfer bands of the perturbed centers can be different from the unperturbed centers N; these centers can introduce new bands below the conduction band of the crystal and thus they can influence the scintillator properties of the RE-doped garnets.

The extremely low intensity of the P_i satellites in ceramics compared with the single crystals indicates that their composition is much closer to the ideal garnet composition and this can explain some of their improved performance.

10.1.2.1.2 Search for perturbations induced by the inversion of sites in garnets The intensities of some EXAFS lines in YAG and an apparent global distortion of the symmetry of YAG to trigonal claimed in reference [19] were tentatively assigned to a large-scale ($\sim$10%) inversion of sites between the A and B ions in YAG. Normally, such inversion of sites would produce severe disordering of the structure of the YAG crystals or ceramics, leading to very intense and strongly shifted spectral satellites induced by the perturbing centers A(a) and B(c). However, no additional satellites confirming the existence of sizable concentrations of B(c) centers were observed, and the intensity of the observed satellites induced by the A(a) centers in crystals was completely explained, as described above, by the excess Y^{3+} ions entering in the a-sites. Thus the high resolution spectroscopy of RE^{3+}-doped YAG does not confirm the large-scale inversion discussed in reference [19].

10.1.2.1.3 Search for perturbed centers connected with the Nd^{3+} "dead sites" in garnets The existence of "dead sites" of Nd^{3+} that can reduce the laser efficiency and increase heat generation by reducing the pump level efficiency η_p due to transfer of excitation from levels above the emitting level $^4F_{3/2}$ to hydroxyl ions or to color centers induced by hydroxyl was postulated to explain the non-systematic variation of the quantum efficiency $\eta_q = \eta_p \eta_{qe}$ in Nd:YAG crystals [20–22]. Due to the high energy vibration quanta ($\sim$3200 cm^{-1}), the hydroxyl ions could contribute to efficient non-radiative de-excitation of nearby Nd ions. Such de-excitation should influence non-systematically the emission lifetime of the level $^4F_{3/2}$ too, as observed in several Nd-doped materials; however, no such effects were observed for Czochralski grown Nd:YAG crystals. On the other hand, since the absorption of the color centers in YAG is in the ultraviolet–visible range, at wavelengths shorter than 500–600 nm, their involvement as acceptors in energy transfer and in reduction of the efficiency η_p could only be invoked for short wavelengths and not for 809 nm diode laser pumping.

Taking into account the excellent properties of YAG as laser material, unambiguous identification and characterization of such "dead sites" would be of major relevance, and high resolution spectroscopic investigation could be of significance. The high density of energy levels above the emitting level $^4F_{3/2}$ of Nd^{3+}, determines very fast (10^8–10^9 s^{-1}) electron–phonon relaxation between these levels, so an extremely fast parasitic de-excitation of these levels would be necessary to divert the flow of excitation from reaching level $^4F_{3/2}$. This would require a very short distance between the parasitic de-excitation center and the excited Nd ion, leading to strong crystal field perturbations and to strongly shifted spectral with non-systematic sample dependent intensities. Indeed, the starting point for the model of dead sites was the observation of multicenter structure with four additional centers in the emission spectra of Nd:YAG crystals on changing the laser excitation wavelength slightly [20]. These centers were attributed to the perturbing effect of hydroxyl ions, although the infrared absorption spectra revealed only very low concentrations of hydroxyl. Moreover, it was found later that the relative intensities of these additional lines and the emission decay were fairly constant in various Nd:YAG crystals of similar Nd concentration, without a definite relation with the quite large variation of the emission quantum efficiency under visible laser excitation. Unfortunately, the multicenter structure reported in references [20, 21] was observed in emission spectra and no investigation of absorption spectra was reported, so a realistic assessment of the relative concentrations of the centers is not possible. Some of the emission lines in the transition $^4F_{3/2}(1) \rightarrow {}^4I_{9/2}(1)$ for these additional centers [20] are very close to the positions of the P_i satellites in the absorption spectra of Nd:YAG crystals [1–4, 8]: although these spectra show several non-identified very weak additional lines in the vicinity of $^4I_{9/2}(1) \rightarrow {}^4F_{3/2}(1)$ absorption, by at least an order of magnitude less intense than the satellites P_i and M_i, no Nd^{3+} spectral satellite that could be connected unambiguously with the presence of near hydroxyl ions or of hydroxyl-induced color centers was identified. Similarly, in the Nd:YAG ceramics only M_i satellites are obvious. Thus, the high resolution spectroscopy of doped YAG crystals and ceramics does not provide any evidence in favor of the model of high concentrations of "dead sites." Nevertheless, the color centers in Nd:YAG could contribute to reduction of the global emission intensity, compared with the incident power, by parasitic absorption that competes with the absorption in the laser active ion and thus in fact they would reduce η_a rather than η_p; this process could be active for lamp or visible laser pumping but quite inefficient for the near-infrared diode laser pumping.

Color centers based on oxygen vacancies with one or more trapped electrons are common defects in the YAG materials produced in inert or reducing atmosphere: the O^{2-} vacancies with two trapped electrons are called F color centers, and trapping of one or three electrons produces F^+ and respectively F^- centers. Such centers have been invoked to explain the increased lamp absorption of Nd:YAG ceramics and the reduction of this loss by suitable filtering of the lamp, and it was assumed that they are determined by utilization of a sintering aid containing SiO_2 [23]. Subsequent investigation of the as-sintered and air-annealed YAG ceramics showed correlated reduction of the absorption spectra and EPR signals of the F and F^+ centers and the appearance of the 256 nm O^{2-}–Fe^{3+} charge transfer band, without any apparent influence of the concentration of SiO_2 in the sintering aid; moreover, no such spectra are seen in very pure ceramics, with negligible Fe impurity [24]. It is then obvious that the formation of F centers is favored by the presence of doping ions, such as Yb, or accidental impurities (such as Fe) with variable ($+2$ – $+3$) valence: fabrication in vacuum or a reducing atmosphere would favor the lowest valence with simultaneous formation of O^{2-} vacancies and color centers, but these can be eliminated by subsequent annealing in air or oxygen. The contribution of impurities to the color centers in the oxide materials suggests the utilization of very pure raw materials; at the same time, in the case of active ions with variable valence such as Yb, careful final annealing in air may be necessary.

10.1.2.1.4 Defective centers at the grain boundaries in ceramics The crystal field acting on the doping RE^{3+} entering in the defective structural sites at the grain boundaries of

Table 10.3 *Shannon radius (nm) and atomic*
mass for ions from the sintering aids used for
garnet and sesquioxide ceramics

Ion	Ionic radius (nm)			Atomic mass
	r_8	r_6	r_4	
Th^{4+}	0.119	0.108		232.038
Hf^{4+}	0.097	0.085		179.49
Zr^{4+}	0.098	0.086		91.22
Si^{4+}		0.064	0.04	28.086
Mg^{2+}	0.103	0.086		24.312

ceramics may be different from the normal (internal) center; the global relative concentrations of these centers depend on the ratio (surface/volume) of the grain, as well as on the depth of the defective region. In the case of nanograins this ratio can be fairly large and thus quite high concentrations of defective sites would be possible. However, in laser quality transparent ceramics, with micrometer or tens of micrometer grain size and with very shallow grain boundaries (a few nanometers), this ratio could be too small to produce concentrations of defective doping centers observable by the conventional global spectroscopic methods. Indeed, high resolution investigation of RE-doped coarse-grain ceramics did not reveal any sizable new lines that could be linked with such grain boundary defective centers [10, 25].

10.1.2.1.5 New perturbed centers generated by the sintering process The sintering aids that control the densification of ceramics and the grain growth contain cations whose valence differs from that of the host cations, although the ionic radii are quite close. These ions can be incorporated in the host lattice and create vacancies that enhance the diffusion of the larger ions, contributing in this way to the sintering process. The doping ions and mixing of cations in solid solutions can play the role of sintering aids in many cases, but the effect is usually much smaller than that of the external aliovalent aids; nevertheless, the effect can be enhanced by the joint action of these factors. The sintering aids for garnet and cubic sesquioxide ceramics are usually oxides of four-valence ions. Incorporation of these ions in the host lattice is determined by the ionic radius mismatch with the substituted cation and by the bond angle preferences of the cations from the sintering aid. Table 10.3 gives the Shannon ionic radius and the atomic mass of several such cations.

In the case of Nd:YAG ceramics produced by solid-state synthesis it was found that 0.14–0.28 wt.% SiO_2 can act as a sintering aid to control the grain growth and to achieve transparency similar to that of single crystals [26–28]. Since then this aid has become standard and is supplied by mixing the raw material with TEOS (tetraethyl orthosilicate). Numerous recent papers discuss the role and effects of the SiO_2 sintering aid on the sintering of YAG or Nd:YAG ceramics [29–36]. It was thus found that the kinetics of grain growth in the system SiO_2–YAG at temperatures above $\sim$1550 °C follows a cubic rate law, in contrast to pure YAG where this law is quadratic. For YAG ceramics with no or little Nd, the grain sizes for sintering temperatures up to 1650 °C remain below 2 μm, but increase strongly to more than 20 μm at higher temperatures; for YAG ceramics with high Nd content (5–9 at.%) a strong increase in size, to $\sim$8–10 μm, occurs between 1520 and 1650 °C, then levels off. Whereas the kinetic growth law in pure YAG is specific to solid-state mass transport, the cubic law in the presence of SiO_2 indicates the presence of liquid-phase controlled transport or of a solid-state solute drag mechanism. The liquid-phase transport model is facilitated by the known liquid phase in the ternary system Y_2O_3–Al_2O_3–SiO_2 above 1400 °C. Generally, no obvious presence of foreign phases at the grain boundaries or triple points in high quality transparent ceramics has been reported for high temperature sintered ceramics, except for an amorphous layer a few nanometers thick containing SiO_2 from a high TEOS concentration in the case of slowly cooled Nd:YAG ceramics. Together with the evolution of the grain sizes above 1650 °C (practically inversely proportional to the Nd concentration), this favors the model of Nd solute drag size limiting effects in ceramics with less than 0.28 wt.% SiO_2 which are sintered at high temperatures [36]. Nevertheless, some Si can be lost by evaporation at high temperatures.

This mechanism of sintering assumes that at least some of the Si^{4+} ions from the sintering aid remain as substitutional impurities in the laser material lattice. The ionic radius r_4 of Si^{4+} is close to that of Al^{3+} (see Table 2.1), so some Si^{4+} could be expected to substitute for Al^{3+} ($Si^{4+}(d)$ centers) as discussed in the literature for Si-doped YAG [37]; this model was extended to all Si^{4+} ions in many subsequent works. Substitution of $Al^{3+}(d)$ ions would correspond to the preference of Si^{4+} for tetrahedral bonds. Generally, in the technology of garnet ceramics the sintering aid is mixed with the stoichiometric garnet material, i.e. with a material of $Y_3Al_5O_{12}$ composition and with neutral charge. If all the Si^{4+} from the 0.28 wt.% SiO_2 sintering aid enters the YAG, it will replace 0.66 at.% Al^{3+} ($\sim$1% of the Al^{3+} ions in tetrahedral d-sites) and will destroy the charge neutrality (+0.132) and stoichiometry (excess of Al^{3+}). The electric neutrality and the excess Al^{3+} ions will then impose Y^{3+} vacancies ($\sim$0.37% of the c-sites) in the garnet structure. These vacancies can be of major importance for the diffusion of the large ions (Y^{3+}, Nd^{3+}) that governs sintering. Taking into account the large negative extra charge introduced by these vacancies, a certain correlation of placement of the $Si^{4+}(d)$ centers around vacancies cannot be excluded. Obviously, such a complex situation will limit the amount of Si^{4+} ions that can be incorporated in the YAG ceramics. The solubility of Si^{4+} in YAG was dramatically increased (an order of magnitude) by co-doping with divalent ions such as Ca^{2+} or Mg^{2+} [38, 39] in materials where the composition of the raw material was adjusted from the start to obey the global stoichiometric garnet structure and charge neutrality. It was also shown that the mixed sintering aid B_2O_3–SiO_2 (0.34–1.35 mol%) determines efficient liquid phase sintering of Nd:YAG transparent ceramic at lower temperatures (1600 °C) and the amount of boron residing in the final ceramic is negligibly small (ppm range) [40]. The almost exclusive presence in the crystalline environment of Si^{4+} in YAG ceramics was confirmed by the narrow linewidth in the magic angle spinning NMR of $^{29}SiO_2$ doped material and the chemical shift was consistent with tetrahedral coordination [36].

It would then be expected that in the RE^{3+}-doped YAG ceramics the positive extra charge of the center $Si^{4+}(d)$, together with the charge compensator, will produce very strong crystal field perturbations at the RE^{3+} c-sites; the effect will be favored by the very short n.n. distance between the d- and c-sites (Table 10.1). However, although such perturbations should produce resolved spectral satellites, no obvious new satellite of sizable intensity that could be linked unambiguously with the perturbing effect of the substitutional Si^{4+} ions in the d-sites was observed in the global optical spectra of these ceramics [10, 25] and this could indicate that the amount of Si^{4+} ions residing in the Nd:YAG ceramics is quite low. It is thus apparent that the use of the TEOS sintering aid in the fabrication of YAG transparent ceramics has no marked influence on the global spectroscopic properties of the doping RE^{3+} ions; moreover, the close atomic mass of Si and Al prevents sizable alteration of heat conductivity. However, partial segregation of Si^{4+} at the grain boundaries could take place, leading to locally enhanced concentrations of perturbed RE^{3+} centers. Such a situation could, in principle, be revealed using spatially resolved methods of investigation.

Following the first success in the fabrication of transparent cubic sesquioxide ceramics [41–43] it became evident that the sintering aid is essential in the technological process. For these materials the usual sintering aids are dioxides of tetravalent elements SaO_2 (Sa = Zr, Hf, Th) and the amount is much larger than for garnets, several weight percent. Moreover, the reduced volatility of these compounds at the sintering temperature could determine high concentrations of Sa elements residing in the final product. Given the ionic radius (Table 10.3) and preference for cubic bonds, the Sa^{4+} ions could substitute the trivalent host cations Sc^{3+}, Lu^{3+} or Y^{3+} of these ceramics. The resulting positive extra charge would require charge compensation: besides cationic vacancies, this could be facilitated by the empty sites in the O^{2-} polyhedra surrounding the RE^{3+} ion, which could accommodate additional O^{2-} ions. Each additional O^{2-} ion will then introduce localized negative extra charges that can favor clustering of the Sa^{4+} ions. Although the substitutional model is generally accepted, the possibility of some of the Sa^{4+} ions residing inside the empty oxygen polyhedra of the sesquioxide lattice should not be completely excluded; charge compensation can be made by two additional O^{2-} ions occupying the empty corners to complete eight O^{2-} coordination. As expected from the ionic radius and bond-angle preference of Si^{4+}, no sintering aid effect was observed on sintering Y_2O_3 in the presence of TEOS [44].

The localized positive extra charge of Sa^{4+} and the negative charge introduced by the charge compensator coupled with the tight packing of the cationic sites and with the short ion–ion distances could produce strong crystal field perturbations at the RE^{3+} ion sites, leading to strongly shifted new spectral lines; moreover, these perturbations could alter the inversion symmetry of the near C_{3i} sites, thus allowing electric dipole transitions. Intense additional satellites were indeed observed: for instance, the low temperature $^4I_{9/2}(1) \rightarrow$ $^4F_{3/2}(1, 2)$ absorption spectrum of Nd:Sc$_2$O$_3$ ceramic shows, besides the normal lines N

at 897.1 nm and 877.4 nm, two additional inhomogeneous broadened asymmetrical lines at 885.3 and 874.9 nm; at high amounts of HfO_2 sintering aid the intensities of these lines become very large, almost similar to the normal centers. The shift of these new lines from the normal lines in other transitions could be very large, for instance, in the transition $^4I_{9/2}(1) \rightarrow {}^2P_{1/2}$ the shift is 230 cm^{-1}. A similar situation was reported [45] for the Nd:Y_2O_3 ceramic, where besides the normal lines at 876.8 nm and 892.2 nm new lines are observed at 875 nm, 884 nm and 889 nm. The effect of the sintering aid is felt by other ions too, for instance in the case of Yb:Sc_2O_3 a new sharp and fairly intense line at 969.55 nm shows in the vicinity of the $^2F_{7/2}(1) \rightarrow {}^2F_{7/2}(1)$ line at 974.8 nm of the C_2 center, whereas in Yb:Y_2O_3 a new line at 973.4 nm shows near the normal 975.4 nm absorption line at 10 K. The characteristics and the origin of these new centers have not been fully explored. Besides the substitutional and interstitial Hf^{4+} models, the possibility of the doping RE^{3+} ion entering in the HfO_2 non-diluted in sesquioxide has also been discussed. Because of the similarity of the ionic radii of Hf^{4+} and Zr^{4+}, the perturbing effects of these two ions would be expected to be quite similar. The need for charge compensation for the Sa^{4+} ions can also induce a reduction in the valence state of the doping laser ion, particularly in the case of Yb-doped ceramics. Quite large concentrations of Yb^{2+} can be present in these ceramics after the final sintering stage and the trivalent state can be restored by heat treatment in an oxidizing atmosphere. The fairly large amount of sintering aid in the case of sesquioxides can reduce the heat conductivity: comparison of the atomic masses indicates that the substitutional Hf^{4+} ion would have little influence in the case of Lu_2O_3, whereas Zr^{4+} has the smallest effect for Y_2O_3 or Sc_2O_3.

10.1.2.2 Multicenter structures in charge compensated systems

Substitution of the host cations with aliovalent rare earth ions requires charge compensation and at low doping concentrations this could be done either by cationic vacancies or with substitutional anions, or by co-doping with charge compensation ions. The attractive electrostatic force between the compensated ion and compensator can determine the correlated placement of the latter with respect to the doping ion. The various modalities of placement of the charge compensator at the available lattice sites or vacancies close to the doping ion induce a chain of discrete crystal field perturbations at the doping ion site. Since these perturbations are determined by the difference in electric charge between the compensator and the substituted site, they are strong, and can induce resolved effects in the optical spectra, resulting in specific multicenter structures, whereas the unresolved effect of the farther compensators induces strong and asymmetric inhomogeneous broadening of the lines. The relative intensities of these centers depend on the packing of the lattice sites and on the degree of correlation of placement, which can be influenced by the thermodynamic conditions of the fabrication technology. Typical examples of cubic systems with charge compensation are the alkaline earth fluoride MeF_2 crystals (with Me = Ca, Sr, Ba, Pb) doped with RE^{3+} ions [46], where the charge compensation at low doping can be made either by interstitial F^- ions, or by co-doping with monovalent alkali ions. However, the resulting multicenter structure of the absorption and emission spectra is not very favorable for efficient laser emission. In the case of high doping of the systems with charge compensation, severe rearrangement of the crystalline lattice could take place, which could enable intrinsic compensation: an example of such a case is the highly doped Yb^{3+}:CaF_2 crystals, where a unique type of hexametric ensemble of Yb^{3+} ions is formed [47].

These complex charge compensation processes in RE^{3+}-doped MeF_2 crystals require good starting homogeneity of the material before crystallization and high mobility of the ions at crystallization. Such conditions would be difficult to achieve in the case of RE^{3+}-doped alkaline earth fluoride ceramics. Preparation of highly doped (to 10% Yb) CaF_2 powders by mechanical mixing or by wet chemistry, based on solutions of nitrates of Ca and Yb and of HF, followed by precipitation or the inverse micelle method was reported [48, 49], and transparent ceramics from these powders were produced by sintering at 900 °C, followed by hot isostatic pressing. High resolution high angle annular dark field scanning transmission electron microscopy (HAADF-STEM) investigation of these samples revealed that at high concentrations (above 1 at.%) the Yb ions are organized in clusters rather than as isolated ions, but the structure of these clusters was not firmly established. The presence of oxygen rich regions at grain boundaries was also revealed.

An alternative technique for the production of transparent fluoride ceramics with controled doping is to use material fabricated initially as a single crystal and to transform it into polycrystalline material by uniaxial hot pressing ("hot forming"). This method preserves the advantages of the melt-growth method for controlling the doping, it reduces the oxygen contamination and gives increased resistance to cleavage. This method was demonstrated for the mixed fluoride CaF_2–SrF_2 doped with 5% YbF_3 [50] and the spectroscopic properties of the ceramics were similar to those of single crystals.

10.1.2.3 Compositionally disordered materials

Compositional disorder, in which specific types of sites can be occupied by ions of different species or valence, can preserve the global crystallographic structure type and symmetry of the crystal but influences the local composition dependent structure and properties, such as the crystal field at the site of the doping ions. This then determines the specific multicenter structure of the optical spectra and inhomogeneous line broadening. *Accidental compositional disordering* of the laser material can be determined by the doping and by charge compensation as well as by technology dependent defects.

Accidental disordering suggests the possibility of producing manageable effects on the spectra of the doping ions (especially the absorption and emission linewidths) by suitable control of the distribution of the crystal field in compositionally disordered host crystalline lattices (*controlled host disordering*). Very relevant in this respect is *controlled host disordering by solid-solution mixing* based on control of the distribution of the cationic species by selection of the species and proportions of the components in solid solutions. The undoped systems with some of the cationic sites occupied by several species of cations of different valence selected to maintain global charge neutrality are cases of *natural* (*intrinsic*) *disorder*. Finally, there are systems with *combined intrinsic-controlled disorder* based on solid solutions of naturally disordered materials or on selection of the cationic species. The controlled compositional disordering of the host laser material and proper selection of materials with intrinsic disorder can then be an important means for tailoring new laser materials and laser emission regimes.

10.1.2.3.1 Disordered materials obtained by solid-solution mixing Disordered mixed materials can be obtained from solid solutions of similar compounds of various isovalent cationic species: when these compounds are isomorphic, such as cubic sequioxides or garnets, the whole scale of proportion from 0% to 100% is possible in many cases. The original structure of a compound can also be preserved by mixing, within certain limits, with compounds of different symmetry, for example cubic sesquioxides mixed with a small amount of the lower symmetry sesquioxides Gd_2O_3 or La_2O_3. The RE^{3+} doping of the ordered materials has a disordering effect similar to solid-solution mixing, although cases exist when the doping ion is not a compound of the same structure as the host (for example Nd:YAG). The disorder of the cationic coordination spheres is influenced by their structure: for moderate packing of the cationic sites, such as in garnets, quite a high proportion of each of the mixing species would be necessary to obtain the desired effects on the RE^{3+} spectra, whereas in the case of tight packing of sesquioxides, much lower mixing could be sufficient.

A major issue in characterization of the disorder in solid-solution materials is the variety (crystal field) and structure (composition of the nearest coordination spheres) of the sites offered for substitution with laser ions. Many papers consider that in a solid solution of two isomorphous compounds (cations A and B) there will be only two types of sites, namely those corresponding to the original sites in the individual compounds. Actually, in such two-component solid solutions the crystallographic sites will be occupied at random by the two species of cations and thus each site will be surrounded by cationic coordination spheres whose composition is determined by the statistics of random occupancy. For a coordination sphere with m sites, the probability that n of these are occupied by cations B and $(m - n)$ by cations A is described by Eq. (2.24), where C and $(1 - C)$ are the relative concentrations of the two cations and thus $(m + 1)$ different compositions would be obtained. For instance, when $m = 4$, the composition of this coordination sphere could be 4A with probability C_A^4, (3A + B) with probability $4C_B C_A^3$, (2A + 2B) with probability $6C_B^2 C_A^2$, (A + 3B) with probability $4C_B^3 C_A$ and 4B with probability C_B^4. The largest occurrence probability would be determined by the ratio of the concentrations C_A and C_B: when these are equal, the center (2A + 2B) will have the largest probability, whereas the centers corresponding to the original compounds, 4A and 4B, have the smallest occurrence. The lines corresponding to 4A or 4B coordination in these materials could be slightly displaced from their positions in the individual A and B compounds, due to modification of the lattice parameter. Although the probabilities for the various centers are symmetrical around that of the center (2A + 2B), the positions of their lines can not be symmetrical around the line of this center and the most perturbed lines can be placed outside the range of the original 4A and 4B lines.

The crystal field perturbations in these systems are based on size-difference effects, so their strength is limited: in most cases, particularly at room temperature and with a high degree of mixing, the homogeneous and inhomogeneous broadening precludes the spectral resolution of the multicenter structure, resulting in broad, possibly asymmetric bands, which could show shoulders or bumps.

A large variety of disordered garnets can be obtained using solid solutions of garnets differing in composition of any or of several types of cationic sites; nevertheless, these systems should be characterized by the explicit composition. Strong broadening effects have been reported for many mixed garnets such as $Y_3ScAl_4O_{12}$ [51, 52], $Lu_3ScAl_4O_{12}$ [53], $(YGd_2)Sc_2(Al_2Ga)O_{12}$ [54, 55], and so on. For instance, the width of the 1 μm emission line of Nd^{3+} in $(YGd_2)Sc_2(Al_2Ga)O_{12}$, is 5.3 nm, much larger than 1 nm and respectively 1.3 nm in the garnets involved in mixing, $Y_3Sc_2Ga_3O_{12}$ and $Gd_3Sc_2Al_3O_{12}$. In the case of sesquioxides, the solid-solution method has been used especially for mixed Y–La sesquioxides [56, 57]; mixed CaF_2–SrF_2 ceramics are also reported [50].

10.1.2.3.2 Materials with intrinsic disorder Stronger perturbation of the crystal field by the difference in electric charge can be exploited in materials with intrinsic disorder characterized by multiple occupancy of the sites of a given cationic coordination sphere by cations of different valences. In the case of the garnet structures $A_3B_2C_3O_{12}$ such situations can arise when A is a divalent ion (Ca^{2+}) and C is basically trivalent (Ga^{3+}): the charge compensation imposes an effective average charge of the octahedral *a*-sites (ions B) equal to 9 and this can be tentatively achieved by a mixture of five-valence ions, such as Nb^{5+} and trivalent ions (Ga^{3+}), resulting in the calcium–niobium–gallium garnet (CNGG) [58, 59]. However, the stoichiometric compound with 75%Nb–25%Ga in the octahedral sites does not melt congruently, whereas the congruent-melting compound $Ca_3Nb_{1.68}Ga_{3.2}O_{12}$ is non-stoichiometric and implies cationic vacancies; additional co-doping with Li^+ in the octahedral *a*-sites leads to the stoichiometric garnet $Ca_3Li_yNb_{(1.5+y)}Ga_{(3.5-2y)}O_{12}$ (CLNGG) [59], the best optical quality being obtained for $y = 0.275$. The melting temperature of these materials is low (∼1480 °C), and good quality crystals can be grown by the Czochralski technique; recently RE (Nd, Er, Tm)-doped CLNGG ceramics have also been produced [60], with spectroscopic properties similar to the single crystals [59, 61]. In CLNGG the RE^{3+} ions substitute the Ca^{2+} ions and the charge compensation is accomplished by proper modification of the proportion of Nb and Ga. The various modalities of placement of the cations with different electric charge in the cationic sublattices determines a chain of discrete crystal field potentials at the RE^{3+} site, with differences much stronger than the ionic size effects, leading to multicenter spectral structures of strongly inhomogeneous broadened lines, which usually result in broad, multi-peaked bands.

The probabilities of occurrence of the various structural centers in these systems can be calculated, similar to the case of solid solutions. The absorption bands $^4I_{9/2}(1) \rightarrow {}^4F_{3/2}(1)$ of the 1 at.% Nd:CLNGG and $^2F_{7/2}(1) \rightarrow {}^2F_{5/2}(1)$ of the 4.3 and 10 at.% Yb:CLNGG crystals and ceramics are broad and multi-peaked and at low temperatures can be decomposed into several components whose relative intensities indicate that the most probable composition of the first octahedral coordination sphere is 4Nb, followed by 3Nb1Li. The shape and composition of these bands [60] differ from those of CNGG [62]. The $^2F_{7/2}(1) \rightarrow {}^2F_{5/2}(1)$ absorption band in Yb:CLNGG is centered around 973 nm of $\sim$2.6 nm FWHM at 300 K and 2.3 nm at 10 K and the strongest 300 K emission line, at $\sim$ 1027 nm ($\sim$20 nm FWHM) can be assigned to the $^2F_{5/2}(1) \rightarrow {}^2F_{7/2}(3)$ transition and remains broad at low temperatures ($\sim$14 nm at 10 K). The consistency of the description of the relative intensities of the lines corresponding to the various structural centers with the statistics of random occupancy of the first octahedral coordination sphere and with the concentrations of the Nd^{5+}, Li^+ and Ga^{3+} similar to the average composition of the material indicates a distribution of these ions which is close to random in the whole volume. The broad optical bands of the RE^{3+}-doped CLNGG crystals and ceramics in a large temperature range, including cryogenic temperatures, is determined by the fact that these are envelopes of homogeneously and inhomogeneously broadened spectra of several structural RE^{3+} centers differing in the composition of the cationic coordination spheres.

As mentioned above, crystal field disordering caused by a difference in electric charge, leading to a multicenter structure of optical spectra, can also be observed in the case of sesquioxide ceramics fabricated with large concentrations of sintering aid SaO_2. This could indicate that quite large concentrations of four-valence ions from the sintering aid (Zr^{4+}, Hf^{4+}, Th^{4+}) enter into the crystalline lattice of these ceramics.

Other classes of RE^{3+}-doped transparent ceramics with intrinsic disorder are the lead lanthanum zirconate titanate (Pb,La)(Zr,Ti)O_3 materials (PLZT) [63, 64] and Ba(Zr,Mg,Ta)O_3 [65]. Strong inhomogeneous broadening was observed for various doping RE^{3+} and the emission cross-section was kept within reasonable limits, offering a prospective material for short pulse laser emission.

10.1.2.3.3 Combined disordering by solid solution of intrinsic disordered materials
The widths of the spectral lines can be further increased by combining the intrinsic disorder with disordering by solid solution. An example is the calcium–lithium–niobium–tantalum–gallium garnet, CLNTGG [60], where half of the Nb^{5+} ions of CLNGG are replaced by Ta^{5+}: the various combinations of Nb^{5+} and Ta^{5+} ions in octahedral sites induce additional broadening of the RE^{3+} absorption and emission lines. Compared with CLNGG, the 1027 nm emission band of Yb^{3+} is broadened to 24 nm at 300 K and to 16 nm at 10 K.

Another example of combined broadening in ceramics is given by mixed solutions of sesquioxides in the presence of a large amount of the sintering aid SaO_2. It was found that with solid solutions of Y_2O_3 with 25% of Sc_2O_3 or Lu_2O_3 in the presence of 6 mol% ZrO_2 sintering aid, the 1 μm emission line of Nd^{3+} can be broadened to 40 nm [66].

These examples show that the distribution of the crystal field potential in compositionally disordered laser materials can induce a multicenter spectral structure and inhomogeneous broadening of lines that combines with the temperature dependent broadening to determine broad absorption and emission bands whose shape is determined by the ratio of these processes. Such materials could be of importance for special laser regimes, particularly

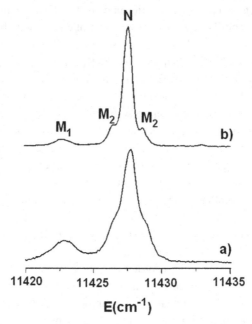

Figure 10.6 The satellite structure in the $^4I_{9/2} - {}^4F_{3/2}(1)$ 10 K absorption of Nd:YAG ceramic: (a) 6.6 at.% Nd; (b) 1 at. % Nd.

for ultra-short pulse generation. Complete use of these characteristics would then impose careful selection of the pump radiation (wavelength, bandwidth) in order to excite all the doping centers and of the laser regime to make full use of the broad emission.

10.1.2.4 Satellites dependent on the doping concentration

10.1.2.4.1 Systems with unique doping species The doping concentration (C_{RE}) dependent satellites are linked with the perturbations inside the ensembles of doping ions. The presence of doping ions in near lattice sites induces satellites M_i whose relative intensities depend on C_{RE}, while the more distant ions determine C_{RE} dependent inhomogeneous broadening. In the case of Nd-doped garnet (YAG or GSGG) ceramics, two types of such satellites have been observed in low temperature (10 K) absorption spectra, M_1 and M_2 [6–10, 25, 67] and their shift from the main line N and the relative intensities are identical with those measured for melt-grown and flux-grown crystals. The picture of the M_i satellites depends on the electronic transition, the M_1 satellite shows as a unique line, whereas in several electronic transitions the line of the satellite M_2 is split into two components of equal intensities, as shown in Figure 10.6 for the transition $^4I_{9/2}(1) \rightarrow {}^4F_{3/2}(1)$ of Nd:YAG ceramics with 1 and 6.6 at.% Nd. The shift of these satellites depends on the electronic transition and is larger for the satellite M_1 (up to 6 cm^{-1} in the case of YAG) than for M_2 (up to 2 cm^{-1}) and is larger in YAG than in GSGG. Their spectral resolution is additionally influenced by the broadening determined by farther Nd ions, as well as by homogeneous

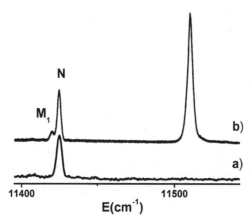

Figure 10.7 10 K spectra of 2.4 at.% Nd:YAG: (a) $^4F_{3/2}(R_i) \rightarrow {}^4I_{9/2}(Z_1)$ emission under non-selective excitation and (b) $^4I_{9/2}(Z_1) \rightarrow {}^4F_{3/2}$ (R_i) absorption.

broadening: in the case of Nd:YAG both satellites are resolved at very low temperatures in the transition $^4I_{9/2}(1) \rightarrow {}^4F_{3/2}(1)$ but none in the transition $^4I_{9/2}(1) \rightarrow {}^4F_{3/2}(2)$.

At low Nd concentrations, the ratio of the global intensity of satellite M_2 to that of M_1 is close to 2 and this can be described by Eq. (2.24) for random distribution, with $n = 1$ and $m = 8$ (second order or n.n.n. pairs according to Table 10.1) and respectively 4 (first order or n.n. pairs). However, Eq. (2.24) gives the relative concentration of the perturbing centers in each coordination sphere whereas, by definition, the measured satellites M_2 correspond to perturbations coming from the n.n.n. sphere only when no intervening doping ions are present, in the n.n. sphere. Thus, good agreement can be obtained only at low doping concentrations, but at higher C_{Nd} the relative intensity of M_2 would become gradually lower than predicted by Eq. (2.24), due to increased occurrence of centers with perturbing ions in both the n.n. and n.n.n. coordination spheres. Additionally, the probability of centers with $n > 1$ in these spheres becomes sizable. At high C_{Nd} the satellite M_1 shows a tendency to splitting (below 1 cm^{-1}), which could be a sign of exchange interactions inside the n.n. Nd pairs. Moreover, new satellites T with stronger C_{Nd} dependence become apparent, which could be connected with triads of Nd ions in nearest lattice sites. The lower spectral shift of the M_i satellites compared with satellites P_i and the similarity of the perturbations induced by all the Nd ions from a coordination sphere is consistent with the smaller dimensional mismatch between the dodecahedral Nd^{3+} and Y^{3+} ions than between the octahedral Y^{3+} and Al^{3+} ions. Spectrally resolved concentration dependent satellites have been observed in the absorption spectra of other large RE^{3+} ions in garnets, such as Pr^{3+} or Sm^{3+}; however, the resolution is much reduced for the smaller ion Er^{3+} or Yb^{3+} owing to the weaker spectral shift.

An interesting feature of the non-selectively continuous wave excited emission is the absence of emission from the M_1 centers, as shown in Figure 10.7 for the 2.4 at.% Nd:YAG ceramics: this is indicative of very low emission quantum efficiency for these centers.

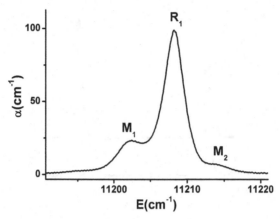

Figure 10.8 The Nd–Nd pair lines M_i in the $^4I_{9/2}(1) - {}^4F_{3/2}(1)$ 10 K absorption of (3 at. %)Nd: Y_2O_3 ceramic.

The low temperature high resolution absorption spectra of Nd^{3+} in the sesquioxide ceramics [17, 25, 68, 69] show two concentration dependent satellites, M_1 (more shifted) and M_2, such as for the transition $^4I_{9/2}(1) \rightarrow {}^4F_{3/2}(1)$ of Nd:Y_2O_3 shown in Figure 10.8. At low doping the relative intensities can be described with Eq. (2.24) assuming $n = 1$ and $m = 12$, respectively 6. According to Table 10.2, this corresponds to perturbations induced by Nd ions present in the first and respectively the second coordination spheres around the C_2 site, assuming random distribution of the doping ions and equal perturbations from the Nd ions in the centers C_2 and C_{3i}. The spectral shift of the satellites M_i in sesquioxides depends on the ionic radius of the host cation, being much larger in the case of Sc_2O_3 (from $10\,cm^{-1}$ to more than $40\,cm^{-1}$, depending on the transition) than for Y_2O_3 (to $20\,cm^{-1}$).

The relative intensities of the pair satellites can also be useful for assessing the distribution of doping in the early stages of the fabrication technology. It was thus found that the distribution of Nd in Nd:YAG powder prepared from alkoxide precursor is more homogeneous than the distribution of Nd in powder prepared from glycolate precursors; strong segregation and agglomeration of Nd in the ceramics prepared using the latter powders takes place [70].

10.1.2.4.2 Systems with several doping species The laser materials with sensitized emission imply doping with two species of ions, the sensitizer (S) and the activator (A) which can occupy the same type or different crystallographic sites. The crystal field perturbations caused by each of these species at the other species' site can induce additional spectral satellites dependent on the concentration of the perturbing ion and these perturbations are felt in a different way by the two ions. Several earlier works on sensitized systems revealed such satellites unambiguously. For instance co-doping of Nd^{3+} or Tm^{3+} (in the *c*-sites) with Cr^{3+} (in the *a*-sites) or of Tm^{3+} with Fe^{3+} (in tetrahedral *d*-sites) in garnets induces perturbations at the RE^{3+} site [71–73], whereas co-doping with Yb induces specific satellites

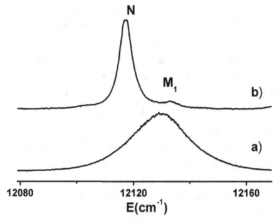

Figure 10.9 Co-doping effects on the $^4I_{9/2}(1) - {}^4F_{5/2}(1)$ Nd^{3+} absorption at 10 K of Sc_2O_3 ceramic with (a) 0.5 at.% Nd, 10 at.% Yb and (b) 0.5 at.% Nd.

in the absorption spectra of Pr^{3+} or Tm^{3+} [74]. The dependence of the relative intensities of these satellites on the doping concentrations can give information on the distribution of the perturbing ions.

In the case of heavy co-doping, such as for Nd in sesquioxide ceramics co-doped with Yb, the perturbed centers can become the dominant feature of the spectra [17, 53] and the systems behave as disordered solid solutions. Figure 10.9 shows the absorption spectrum $^4I_{9/2}(1) \rightarrow {}^4F_{3/2}(1)$ of 0.5 at.% Nd in the Sc_2O_3 ceramics: in the absence of Yb, only the satellites M_1 and M_2 of Nd are seen, whereas co-doping with 10 at.% Yb introduces in all transitions a broad absorption line, peaking between the original N line and the M_1 satellite. At such high Yb concentration (10 at.%) Eq. (2.24) is no longer valid, but it can be estimated that the probability of isolated N centers becomes very small and most of the Nd^{3+} ions are involved in n.n. Nd–Yb pairs and in a variety of more complex assemblies that determine a broad and unresolved distribution of crystal field perturbations.

10.2 The quantum states of the doping ions

10.2.1 Rare earth doped ceramics

10.2.1.1 Ce-doped ceramics

The trivalent Ce^{3+} ion has a very simple electronic structure: the ground configuration is $4f^1$, with a unique spectral term ^2F, which is split by the spin–orbit interaction in two manifolds, $^2F_{5/2}$ (ground state) and $^2F_{7/2}$, placed ~ 2500 cm^{-1} higher. The nearest excited electronic configuration is $5d^1$, with a unique spectral term ^2D. The Ce^{3+} ion enters in dodecahedral c-sites and the local symmetry D_2 determines the splitting of the $^2F_{5/2}$ and $^2F_{7/2}$ manifolds of the ground configuration 4f into three and respectively four Stark components (Kramers doublets), whereas the ^2D term of the excited configuration 5d is split into five singlets. Optical spectra for Ce^{3+} in YAG are reported for melt-grown crystals, ceramics and nanomaterials and the absorption lines are broad and peak at the average values 21 870 cm^{-1}, 29 390 cm^{-1}, 44 470 cm^{-1} and 48 880 cm^{-1}; it was also inferred that the unresolved fifth absorption band could be placed between the two last observed bands [75, 76]. An additional absorption band around 37 000 cm^{-1}, observed in crystals but not in ceramics, was connected with the Y^{3+} (a) antisites. The low temperature absorption spectra show that the zero phonon line to the lowest $5d^1$ crystal field level lies at 20 140 cm^{-1}. The room temperature emission spectrum contains a strong and asymmetric broad band originating from this level and peaking around 530 nm; this band is the convolution of two emission bands to the two manifolds $^2F_{5/2}$ and $^2F_{7/2}$ of the ground configuration, and at low temperatures this band is resolved in emission lines to the Stark levels of these manifolds. The most efficient excitation band for this emission is the 21 868 cm^{-1} (457 nm) band. Both this excitation band and the emission correspond to allowed electric dipole inter-configurational transitions and are very intense and the emission lifetime is very short and depends on Ce concentration and temperature; at 300 K it is in the region of 100 ns.

10.2.1.2 Pr-doped ceramics

Despite the difficulties of doping garnet crystals with Pr^{3+}, determined by the very small segregation coefficient (near to zero in the case of YAG), high doping concentrations (up to 4.5–5 at.%) of Pr^{3+} are reported in several garnet transparent ceramics such as YAG [77] or LuAG [78]. Despite the variety of emission lines in the visible range from the level $^3P_0$, the difficulties of pumping this level have tempered the interest in using these materials for laser emission and they have mostly been investigated as scintillators. A recent investigation of Pr^{3+} in YSAG ceramics claims that the material shows potential for 1.3 μm laser emission [79]. In sesquioxide ceramics Pr^{3+} was reported only in Sc_2O_3 [10], with spectroscopic properties of the centers of low symmetry C_2 similar to that of single crystals. A rich absorption spectrum from the ground level $^3H_4$ to the group of levels $^3P_{0,1,2}$ and $^1I_6$, in the region 19 500–23 000 cm^{-1}, with several lines suitable for excitation (absorption cross-sections of 2 to 3×10^{-20} cm^2) was observed; however, similar to the crystals, no emission from the level $^3P_0$ was obtained. The only visible emission originates from the level $^1D_2$, but this level could not be excited efficiently by pumping into the upper levels and its absorption cross-section is very weak.

In the case of Pr-doped SrF_2 ceramic produced by hot pressing of precursor crystalline material, the broad 444 nm absorption proved suitable for diode laser pumping to provide room temperature laser emission at 639 nm from the transition $^3P_0 \rightarrow {}^3F_2$ [80].

10.2.1.3 Nd-doped ceramics

10.2.1.3.1 Nd in garnet ceramics Investigation of Nd-doped YAG translucent ceramics [81, 82] and of highly transparent coarse-grained [6–8, 26, 27, 83–87] or fine-grained laser ceramics [88, 89] revealed the similarity of the optical spectra with those of single crystals

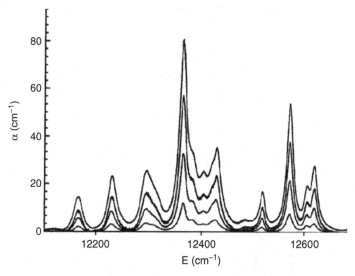

Figure 10.10 The absorption spectra of several Nd:YAG (1, 3.4, 6.6, 8 at.% Nd, from bottom to top) ceramic samples in the $^4I_{9/2} - {}^4F_{5/2}$ range at 300 K.

over the range of similar Nd concentrations, showing that the great majority of the doping ions in the ceramics reside in similar sites to those occupied in crystals. The only marked difference is the absence of the P_i satellites, which can also determine slightly enhanced peak intensity of the N centers. The absorption spectra for doping concentrations to 9 at.% in ceramics, unavailable in crystals, revealed several specific effects of C_{Nd} on the absorption lines: modification of the shape and increased (by up to $\sim$10% at 300 K) FWHM, leading to corresponding sub-linear dependence of the peak cross-section on C_{Nd}; a weak shift of the peaks is also observed, and this could be determined both by the asymmetric perturbations and by the expansion of the crystalline lattice. These effects are dependent on the electronic transition: the room temperature absorption spectrum $^4I_{9/2} \rightarrow ({}^4F_{5/2}, {}^2H_{9/2})$ for various values of C_{Nd} is shown in Figure 10.10, and part of the absorption spectrum $^4I_{9/2} \rightarrow {}^4F_{3/2}$ is shown in Figure 10.11. The effects of Nd concentration are also observed in the emission lines, by lineshape distortion with asymmetric inhomogeneous broadening and shifting of the apparent peak emission wavelength; moreover, the non-uniform variation of the emission cross-section and of the emission quantum efficiency across the broadened line influences the shape of the luminescent lines.

Most spectroscopic studies on Nd:YAG ceramics confirm the similarity of the spectroscopic properties to those of single crystals over the Nd concentration range available in both materials, with some spread that, undoubtedly, could be influenced by the experimental details of the measurement. Unexplained large differences are reported, however, between some of the spectroscopic properties of diluted Nd:YAG fine-grained ceramics and crystals (see Table 5 of reference [90]). Variation of the linewidth of emission lines and of the crystal field splitting of the emitting manifold $^4F_{3/2}$ and of the manifold $^4I_{13/2}$ with

Table 10.4 *The Stark components of the manifolds involved in the laser emission of Nd:YAG*

Manifold	Stark components (cm^{-1})
$^4I_{9/2}$	0, 133, 198, 312, 857
$^4I_{11/2}$	2000.5, 2028.5, 2111, 2146, 2460, 2513
$^4I_{13/2}$	3921, 3930, 4039, 4046, 4435, 4445, 4498
$^4I_{15/2}$	5760, 5816, 5936, 5970, 6571, 6586, 6641, 6740
$^4F_{3/2}$	11425.5, 11509

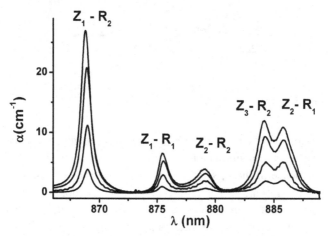

Figure 10.11 Partial absorption spectra of Nd:YAG (1, 3.4, 6.6, 8 at.% Nd, from bottom to top): ceramic samples in the $^4I_{9/2} - {^4F_{3/2}}$ range at 300 K.

concentration was also reported, the effects being differentiated for single crystals, fine- and coarse-grained ceramics; these changes were attributed exclusively to the variation of the emission branching ratios with concentration [91].

The Stark structure of the manifolds involved in the laser emission of low concentrated Nd:YAG (the emitting manifold $^4F_{3/2}$ and the terminal manifolds $^4I_{9/2}$ for emission in the 0.9 μm range, $^4I_{11/2}$ for 1 μm emission, and $^4I_{13/2}$ for 1.3 μm emission), are given in Table 10.4. The values given in this and other tables are compiled from a large volume of data found in the literature, sometimes with a quite wide spread, selected and supplemented based on data measured by the authors.

The Judd–Ofelt parameters and the radiative lifetime for the Nd:YAG ceramics correspond to those determined in reference [92] for Nd:YAG single crystals, and the calculated radiative lifetime τ_{rad} is 260 μs. The spectroscopic quality factor $X = \Omega_4/\Omega_6 = 0.43$, the branching ratios for the luminescence emission to the ^4I_j manifolds are $\beta_{3/2,9/2} = 0.34$, $\beta_{3/2,11/2} = 0.54$, $\beta_{3/2,13/2} = 0.11$ and $\beta_{3/2,15/2} = 0.01$ and thus the emission to $^4I_{11/2}$

dominates. There are many transitions between the Stark levels of $^4F_{3/2}$ and $^4I_{11/2}$ that could be suitable for laser emission: the stimulated emission spectra indicate that the strongest emission line peaks at $\sim$1.0641 μm and contains contributions from two inter-Stark transitions ($^4F_{3/2}(1) \rightarrow {}^4I_{11/2}(2)$ and $^4F_{3/2}(2) \rightarrow {}^4I_{11/2}(3)$) in a variable proportion that depends on temperature [93]. The room temperature peak emission cross-section at this wavelength was calculated from the spectroscopic data and measured by various methods: the early data gave a very large spread, from $\sim$2 to 8.8×10^{-19} cm^2, but the more recent studies give values concentrated in the range 2.6–3×10^{-19} cm^2. Several other transitions, such as $R_2 \rightarrow Y_6$ (1112 nm), $R_1 \rightarrow Y_5$ (1116 nm) or $R_1 \rightarrow Y_6$ (1123 nm) have room temperature emission cross-sections in the region of 3×10^{-20} cm^2 and show potential for efficient four-level laser emission. In the $^4F_{3/2} \rightarrow {}^4I_{9/2}$ emission the largest lines are at 946 nm ($^4F_{3/2}(1) \rightarrow {}^4I_{9/2}(5)$) and 938 nm ($^4F_{3/2}(2) \rightarrow {}^4I_{9/2}(5)$), with emission cross-sections 3×10^{-20} cm^2. The absorption lines are quite sharp and the width depends on temperature and on C_{Nd}: at 300 K the line $^4I_{9/2}(1) \rightarrow {}^4F_{5/2}(1)$, the strongest from its group (Figure 10.10), is quite sharp ($<$1 nm) and its peak absorption cross-section for 1 at.% Nd is 8.2×10^{-20} cm^2, whereas the stronger absorption line in the manifold $^4F_{3/2}$ is $^4I_{9/2}(1) \rightarrow {}^4F_{3/2}(1)$, with peak cross-section 2.85×10^{-20} cm^2. Quite intense absorption lines, suitable for lamp pumping, can be seen in the regions of $\sim$530 nm (absorption in the manifolds $^4G_{9/2}$, $^2K_{13/2}$, $^4G_{7/2}$), $\sim$600 nm ($^4G_{5/2}$ and $^4G_{7/2}$), and $\sim$760 nm ($^4F_{7/2}$, $^4S_{3/2}$).

Properties similar to those of single crystals have been observed for Nd:GSGG ceramics too [67], and the absence of P_i satellites in the spectra of crystals makes the comparison more straightforward. The larger lattice of GSGG compared with YAG determines weaker crystal field effects and oscillator strengths; moreover, the optical lines are broader and the peak absorption and emission cross-section are smaller by a factor of $\sim$2. The satellites M_i show as shoulders on the main lines N even at low temperatures. An interesting particularity of Nd:GSGG is the almost perfect coincidence of the gaps Z_2–R_1 and Z_3–R_2.

Spectroscopic investigation of Nd-doped $Y_3ScAl_4O_{12}$ ceramic with up to 5 at.% Nd shows the effect of the inhomogeneous broadening on reduction of absorption and emission cross-sections. By contrast to Nd:YAG, the room temperature luminescence line at 1061 nm ($\sim$5.5 nm FWHM) that originates from the first Stark level of $^4F_{3/2}$ has higher peak cross-section than the line at 1063 nm from the second Stark level [94].

10.2.1.3.2 Nd-doped sesquioxide ceramics Although the Nd^{3+} ions can enter in both sites of the cubic sesquioxides R_2O_3, very few optical transitions satisfy the selection rules for magnetic dipoles and thus the spectra are dominated by the low symmetry center C_2. The doping Nd concentrations in R_2O_3 ceramics depend on the ionic radius of R^{3+} and are larger than in the case of crystals. The large density of cationic sites in sesquioxides determines higher densities of doping ions than in garnets for a given relative doping concentration C_{Nd}. The optical spectra of the Nd-doped sesquioxide ceramics are similar to those reported for single crystals. At low C_{Nd} the spectral lines are sharp, at 10 K the homogeneous broadening is weak, due to the low phonon energies, the pair satellites M_1 and M_2 are well resolved and the main lines N show inhomogeneous broadening. With increasing temperature the

Table 10.5 *The energy levels of interest for laser emission for the C_2 symmetry center in Nd:Sc$_2$O$_3$ ceramics*

Manifold	Stark components (cm^{-1})
$^4I_{9/2}$	0, 37, 352, 534, 790
$^4I_{11/2}$	1905, 1944, 2240, 2359, 2443, 2475
$^4I_{13/2}$	3803, 3830, 4180, 4265, 4380, 4417, 4440
$^4F_{3/2}$	11147, 11398

lines broaden and at 300 K the width is dominated by homogeneous broadening; however, the lineshapes can be distorted by the unresolved pair satellites.

In the case of sesquioxides the nephelauxetic effect lowers the spectral term ^4F and its manifolds; combined with the strong axial crystal field effects, this shifts the lines of interest for laser emission or diode laser pumping toward the infrared compared with garnets. In the case of the Nd-doped Sc$_2$O$_3$ ceramics the energy levels of the center C_2, of interest for laser emission [17, 95, 96], are given in Table 10.5.

The strongest absorption line in the 300 K absorption spectrum $^4I_{9/2} \rightarrow (^4F_{5/2}, ^2F_{9/2})$ of Nd:Sc$_2$O$_3$ lies at 825.7 nm (effective absorption cross-section $\sim 7 \times 10^{-20}$ cm^2) followed by the line at 808.3 nm ($\sim 4 \times 10^{-20}$ cm^2), both corresponding to transitions from the lowest Stark level Z_1 of the ground manifold $^4I_{9/2}$, whereas in the absorption spectrum $^4I_{9/2} \rightarrow {}^4F_{3/2}$ the strongest lines are at 897.1 nm ($\sim 2 \times 10^{-20}$ cm^2), which corresponds to the transition $Z_1 \rightarrow R_2$, and the hot-band $Z_2 \rightarrow R_1$ at 880.2 nm ($\sim 2.1 \times 10^{-20}$ cm^2). The quality factor $X = 1.53$ determines branching ratios $\beta_{3/2,9/2} \approx \beta_{3/2,11/2} = 0.45$ and $\beta_{3/2,13/2} = 0.1$, whereas $\beta_{3/2,15/2}$ is very small. Due to the large crystal field splitting of $^4F_{3/2}$ (249 cm^{-1}) the strongest emission lines at 300 K originate from the lowest Stark level R_1 (fractional thermal population coefficient $f_l = 0.77$).

Due to the much larger incorporation of Nd and to the larger availability of the host material, the system Nd:Y$_2$O$_3$ has been much investigated. The positions of the lines in the optical spectra of Nd:Y$_2$O$_3$ ceramics [68, 69, 97–99] are similar to those of single crystals [100, 101]. As in the case of Nd:Sc$_2$O$_3$ ceramics, the absorption spectra of Nd:Y$_2$O$_3$ ceramics are dominated by the C_2 centers, although several additional lines in the absorption $^4I_{9/2} \rightarrow {}^4G_{7/2}$ could be attributed to magnetic dipole lines in the center C_{3i}[63]. The positions of the Stark components for the center C_2 correspond fairly well to those reported for single crystals (Table 10.6).

The most intense Nd^{3+} absorption lines in Y$_2$O$_3$ of interest for diode laser pumping correspond to transitions starting from the ground level $^4I_{9/2}(Z_1)$ to $^4F_{5/2}$ at ~ 820 nm and ~ 807 nm, and to $^4F_{3/2}(R_1)$ at ~ 892 nm [69]. The lines in the optical spectra of Nd:Y$_2$O$_3$ ceramics at low C_{Nd} become very narrow at low temperatures, as shown in Figure 10.12 for the $^4I_{9/2}(1) \rightarrow {}^4F_{3/2}(1,2)$ absorption; Hb in this figure denotes the hot-band absorption from the second Stark level Z_2 of the ground manifold $^4I_{9/2}$ which is thermally populated

Table 10.6 *The energy levels of interest for laser emission for the C_2 symmetry center in $Nd:Y_2O_3$ ceramics*

Manifold	Stark components (cm^{-1})
$^4I_{9/2}$	0, 27, 267, 446, 642
$^4I_{11/2}$	1897, 1935, 2147, 2271, 2331, 2359
$^4I_{13/2}$	3814, 3840, 4093, 4200, 4280, 4305, 4329
$^4F_{3/2}$	11208, 11406

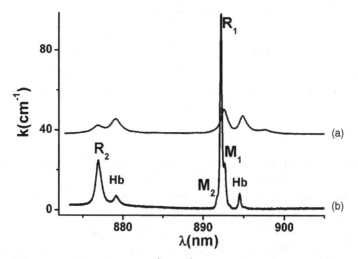

Figure 10.12 The highest energy part of the $^4I_{9/2} - {}^4F_{3/2}$ absorption spectrum of Nd (3 at.%):Y_2O_3 ceramic: (a) 10 K and (b) 300 K. R_1 and R_2 denote the transitions $^4I_{9/2}(1) \rightarrow {}^4F_{3/2}(1,2)$ and Hb the hot bands $^4I_{9/2}(2) \rightarrow {}^4F_{3/2}(1,2)$.

even at low temperatures. The strong temperature dependence of the linewidth, together with the variation of the fractional thermal population factor f for the Stark level $^4I_{9/2}(1)$, induces a strong temperature dependence of the peak absorption cross-section, as illustrated in Figure 10.13. This behavior in the Nd:Y_2O_3 ceramics [69] diverges from the unusual data reported in reference[101] for single crystals. The lines shift to longer wavelengths by about 0.4 nm from 80 K to 300 K. The large crystal field splitting of $^4F_{3/2}$ determines a large fractional population factor f_1 for the first Stark component, and even at room temperature the emission from this level dominates. The Judd–Ofelt parameters predict a quality factor $X \approx 1.55$, so the branching ratio $\beta_{3/2,9/2} = 0.49$ is larger than $\beta_{3/2,11/2} = 0.42$. The emission cross-sections reported by various authors vary widely; the average value for the 946.8 nm $R_1 \rightarrow Z_3$ line is 3.8×10^{-20} cm^2, whereas for the 1078 nm $^4F_{3/2}(1) \rightarrow {}^4I_{11/1}(2)$ line it is 6.8×10^{-20} cm^2.

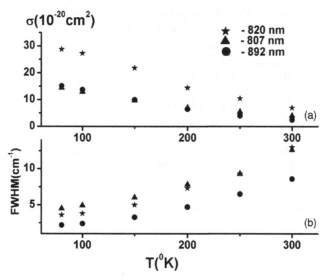

Figure 10.13 Temperature variation of the Nd^{3+} absorption in Y_2O_3 ceramics: (a) peak cross-sections and (b) linewidths (FWHM) for 820 nm , 807 nm and 892 nm lines.

10.2.1.3.3 Other Nd-doped transparent ceramics

In the transparent PLZT ceramics the Nd^{3+} ions enter in the large A sites and the density at 1 at.% Nd doping equals 2.8×10^{20} ions/cm^{-3}. The random distribution of the Pb^{2+} and La^{3+} ions in the A sites and of the Zr^{4+} and Ti^{4+} ions in the B sites determines strong inhomogeneous broadening of the Nd^{3+} absorption and emission lines, which reduces the peak cross-sections. The absorption line at 802 nm has 16 nm FWHM and 1.5×10^{-20} cm^2 peak absorption cross-section, whereas the 1064 nm emission line is 36.2 nm broad and has 3.5×10^{-20} cm^2 peak cross-section. The emission lifetime at very low Nd concentrations is 142 μs, close to the Judd–Ofelt radiative lifetime, but with increasing concentration it shortens and the decay becomes non-exponential. Measurements at low pump intensity indicate that the self-quenching of emission is caused by cross-relaxation energy transfer determined by dipole–dipole interaction with the C_{DA} parameter smaller by more than an order of magnitude than for Nd:YAG. Thus, despite the larger concentration of acceptor sites, the reduction of emission quantum efficiency is smaller and quite high Nd doping concentrations can be used to give good absorption efficiency [63, 64, 102, 103]. In principle, the very broad emission lines in this system would enable ultra-short pulse laser emission; however, the large width of the absorption line as an envelope of broadened lines of a multicenter structure can make pumping with conventional diode lasers (2–3 nm linewidth) quasi-selective, for restricted and aleatory varieties of centers.

Despite the high density of Nd^{3+} ions (1.32×10^{22} cm^{-3}), the emission lifetime of transparent ceramic of neodymium zirconate $Nd_2Zr_2O_7$, with cubic pyrochlore structure, is quite large, 480 μs. The Nd^{3+} ions occupy sites of D_{3d} symmetry and show broad absorption bands around 800 and 900 nm. The quite broad ($\sim$20 nm) 1054.5 nm emission line could be suitable for short pulse laser emission [104].

The optical spectra of 0.5 at.% Nd^{3+} in SrF_2 transparent ceramics are similar to those of single crystals and are dominated by the tetragonal (1,0,0) center L and by the Nd–Nd pairs (center M) [105]. The strongest absorption line in the transition $^4I_{9/2} \to {}^4F_{3/2}$ peaks at 796 nm and includes contributions from both centers. The time-resolved luminescence spectra $^4F_{3/2} \to {}^4I_{11/2}$ of the tetragonal center L are dominated by the lines at 1037 and 1044 nm, whereas the M centers give broad emission with two peaks at 1045 and 1060 nm.

The 2 at.% Nd-doped transparent ceramics of the anisotropic strontium fluoroapatite S-FAP, produced by slip casting under a magnetic field of 1.4 T to orient the single crystal grains, followed by sintering in air and by hot isostatic pressing, has an 807.5 nm absorption line with 1.7 nm FWHM and absorption coefficients of 36.2 cm^{-1} and 28.6 cm^{-1} for π and σ polarization. The emission line at 1063 nm has 1.1 nm FWHM, and the emission cross-section is larger for π than for σ polarization [106].

10.2.1.4 Sm³⁺-doped ceramics

Trivalent samarium has a long lived manifold $^4G_{5/2}$ in the 17 500 cm^{-1} range, separated from the nearest lower manifold $^6F_{11/2}$ by about 7000 cm^{-1}. A very dense packing of Stark energy levels originating from the manifolds $^6H_{5/2,7/2,9/2,11/2,13.2,15/2}$ and $^6F_{1/2,3/2,5/2,7/2,9/2,11/2}$ is jammed from 0 to about 10 800 cm^{-1} and emission to these levels in the visible and near infrared range can take place. There are many energy levels above $^4G_{5/2}$ that could be used for pumping, but the absorption cross-sections are quite weak. At low Sm concentrations in YAG, the lifetime of the $^4G_{5/2}$ level is close to 2 ms, but the very strong concentration quenching precludes utilization of high doping concentrations. Sm^{3+}-doped YAG ceramic was proposed as a suppressor of ASE in Nd:YAG lasers due to negligible absorption in the 808 nm range and good room temperature absorption $^6H_{5/2}(1) \to {}^6F_{9/2}(5)$ for the 1064 nm Nd:YAG emission [107]. Further high resolution investigation [108] indicates that this transition peaks at 1065.4 nm and the 1064 nm Nd emission is absorbed in a tail of this transition; at low temperature this tail resolves in a satellite that corresponds to the n.n. M_1 pair and thus good absorption would require high Sm concentrations.

10.2.1.5 Dy³⁺-doped ceramics

Spectroscopic investigation of Dy^{3+}-doped YAG ceramics [109] revealed properties similar to those reported for single crystals and provided a more complete energy level diagram. In these materials Dy^{3+} has a unique emitting level in the visible, $^4F_{9/2}$, that could give emission in blue ($\sim$487 nm) on the quasi-three-level scheme $^4F_{9/2} \to {}^6H_{15/2}$ transition, or in yellow ($\sim$582.7 nm) on the four-level scheme $^4F_{9/2} \to {}^6H_{13/2}$. The moderate emission cross-sections of these emission transitions together with the long lifetime ($\sim$1.3 ms at low concentrations) recommend this material for Q-switched laser emission. The emission can be excited by GaN diode lasers at various wavelengths in blue or violet; however, a major shortcoming of this material is the weak absorption of these lines. Unfortunately, the quite

Table 10.7 *The Stark energy levels of the $^5I_7$ and $^5I_8$ manifolds of Ho:YAG*

Manifold	Stark components (cm^{-1})
$^5I_8$	0, 4, 41, 51, 141, 144, 150, 162, 389, 418, 448, 596, 520, 531, 535
$^5I_7$	5229, 5232, 5243, 5250, 5303, 5312, 5320, 5341, 5352, 5375, 5395, 5404, 5418

strong concentration self-quenching cross-relaxation processes favored by the energy level scheme prevent utilization of highly concentrated materials and further research to improve the pump absorption is necessary.

10.2.1.6 Ho^{3+}-doped ceramics

Holmium-doped YAG crystal gives 2 μm laser emission on the transition $^5I_7 \rightarrow {}^5I_8$. In the D_2 symmetry of the c-site of YAG the energy manifolds of Ho^{3+} are completely split into singlets and a very dense structure of Stark levels results [110, 111], as given in Table 10.7.

The room temperature 2 μm emission spectrum of Ho:YAG crystals and ceramics contains two main lines, 2120 nm (cross-section 0.5×10^{-20} cm^2) which includes contributions of transitions from the first two Stark components of $^5I_7$ to the higher Stark levels of $^5I_8$, and 2090 nm (1.2×10^{-20} cm^2) with contributions of transitions from several lowest components of $^5I_7$ to Stark levels of $^5I_8$ in the range 448 to 520 cm^{-1}. This emission was difficult to excite with lamps. However, it was found that very efficient energy transfer from Tm^{3+} can take place in Tm–Ho co-doped materials, although additional upconversion loss can occur; this can be avoided by direct pumping in the strong 1907 nm absorption line $^5I_8(1) \rightarrow {}^5I_{7/2}(3)$ with Tm lasers [112]. Recent development of efficient InP diode lasers in the 1.9 μm range can open the way to highly efficient, low quantum defect infrared Ho lasers.

Spectroscopic investigation of 0.2 at.% Ho^{3+} in transparent Sc$_2$O$_3$ ceramic [113] enabled the construction of a detailed energy level diagram for the low symmetry centers C_2: all energy manifolds are split into singlets, resulting in a very rich energy level scheme. The Judd–Ofelt parameters evaluated from the absorption spectra were used to calculate the emission cross-sections and branching ratios for the emission transitions and of the radiative lifetimes. These parameters indicate that several emission transitions, such as $^5I_7 \rightarrow {}^5I_8$ in the 2 μm range and $^5I_6 \rightarrow {}^5I_8$ around 1.2 μm as well as the visible emission from ($^5S_2$, $^5F_4$) are suitable for laser emission. The absorption cross-section and the emission cross-section calculated by the reciprocity method for the transition $^5I_7 \rightarrow {}^5I_8$ are given in Figure 10.14. The most intense emission is at 2110 nm, with emission cross-section 0.26×10^{-20} cm^2, whereas that of the strongest line at 1226 nm in the transition $^5I_6 \rightarrow {}^5I_8$ equals 0.75×10^{-20} cm^2. The calculated radiative lifetimes for levels $^5I_7$ and $^5I_6$ are 8.7 ms and respectively 3.9 ms. It is thus evident that both these emission transitions are suitable for quasi-three-level laser emission and the performance could be improved by operating these lasers at cryogenic temperatures.

Table 10.8 *The Stark components of the* $^4I_{15/2}$,
$^4I_{13/2}$ *and* $^4I_{11/2}$ *manifolds of Er:YAG ceramics*

Manifold	Stark components (cm^{-1})
$^4I_{15/2}$	0, 22, 61, 79, 416, 430, 526, 573
$^4I_{13/2}$	6549, 6599, 6605, 6786, 6805, 6823, 6885
$^4I_{11/2}$	10256, 10287, 10362, 10373, 10414, 10419

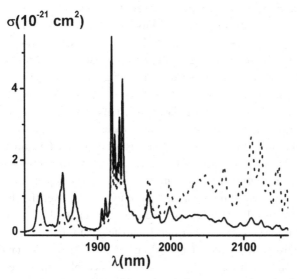

Figure 10.14 The room temperature absorption and emission cross-sections (reciprocity method) of the 0.2 at.% Ho $^5I_7 \leftrightarrow {}^5I_8$ transition.

The spectra of Ho^{3+} in the Y$_2$O$_3$ nanoceramics confirm the potential of the $^5I_7 \rightarrow {}^5I_8$ transition for 2 μm laser emission; in this case, the radiative lifetime of level $^5I_7$, calculated with the integrated absorption coefficient, is 14.9 ms [114]. These radiative lifetimes are close to those measured for Ho-doped Sc$_2$O$_3$ and respectively Y$_2$O$_3$ crystals [95].

10.2.1.7 Er^{3+}-doped ceramics

The Er^{3+} ion can give laser emission in the visible or infrared: of particular interest are the $^4I_{11/2} \rightarrow {}^4I_{13/2}$ emission in the 2.7–2.95 μm range and the $^4I_{13/2} \rightarrow {}^4I_{15/2}$ emission at 1.5–1.6 μm. The energy levels and transition probabilities of Er-doped YAG transparent ceramics [115, 116] are similar to those of the corresponding single crystals. The Stark components of the energy manifolds implied in these infrared emission processes are given in Table 10.8.

Raman spectroscopy of Er:YAG ceramic revealed phonon properties similar to those of the single crystals [117]; it was also shown that the Judd–Ofelt parameters for 50 at.%

Er-doped YAG ceramics and crystals are similar. The radiative lifetime of the level $^4I_{13/2}$, inferred from comparison of the emission cross-section calculated by reciprocity and the Fuchtbauer–Ladenburg relation, was 6.2 ms, close to the measured value for diluted samples and to that reported for single crystals. The Judd–Ofelt analysis gives almost equal ($\sim$150 s^{-1}) radiative de-excitation rates for the levels $^4I_{11/2}$ and $^4I_{13/2}$ [118, 119]. However, due to the large phonon energies (to 857 cm^{-1}) the electron–phonon interaction reduces the luminescence lifetime of the $^4I_{11/2}$ level to $\sim$110 µs, but it has little influence on $^4I_{13/2}$.

The most intense emission line in the 3 µm range for Er:YAG is $^4I_{11/2}(2) \rightarrow {}^4I_{13/2}(8)$ at 2.939 µm, with room temperature cross-section around 4.5×10^{-20} cm^2. However, the long luminescence lifetime of the terminal level $^4I_{13/2}$ can saturate this transition rapidly, and efficient laser emission was obtained only in concentrated Er:YAG crystals, where upconversion $(^4I_{13/2}, {}^4I_{13/2}) \rightarrow (^4I_{15/2}, {}^4I_{9/2})$, followed by the fast electron–phonon relaxation $^4I_{9/2} \rightarrow {}^4I_{11/2}$, was able to shorten considerably this lifetime and to recycle part of the excitation accumulated in this level back to the emitting level. It was also found that a second upconversion process $(^4I_{11/2}, {}^4I_{11/2}) \rightarrow (^4I_{15/2}, {}^4F_{7/2})$ can depopulate the emitting level $^4I_{11/2}$ and that the characteristics of the 3 µm laser emission (efficiency, wavelength) depend on the ratio of the rates of these two upconversion processes [120]. At low Er concentration these two electron transfer upconversion processes are dominated by direct donor–acceptor energy transfer, which is proportional to the Er concentration. However, with increased Er concentration the migration-assisted transfer, with rates dependent on the square of concentration, become more important and gain dominance over the direct transfer. The best 3 µm CW laser results with Er:YAG were in fact reported for Er concentrations of 50%, where the migration-assisted transfer dominates, with the rates 1.3×10^{-15} cm^3 s^{-1} for upconversion from $^4I_{13/2}$ and 3.1×10^{-15} cm^3 s^{-1} for $^4I_{11/2}$. Excitation with 960 nm diode laser radiation into the level $^4I_{11/2}$ causes strong visible luminescence that confirms efficient energy transfer upconversion from this level. At Er concentrations above several percent, the increased migration could contribute to efficient energy transfer to accidental impurities too.

The upconversion from the emitting level $^4I_{13/2}$ and from the upper level $^4I_{11/2}$ have a deleterious effect on the 1.5–1.6 µm laser emission $^4I_{13/2} \rightarrow {}^4I_{15/2}$ and this limitation imposes utilization of dilute Er materials; this is also important for limitation of reabsorption in the quasi-three-level scheme of these lasers. Traditionally, the emission of the $^4I_{13/2}$ level was obtained by diode laser pumping around 970 nm into the upper level $^4I_{11/2}$ (absorption cross-section 0.5×10^{-20} cm^2). Sensitization with the strongly absorbing Yb^{3+} was effective in increasing the pumping efficiency: the absorbed excitation was then transferred to the emitting level by electron–phonon interaction, however, a large part of the excitation could be lost by energy transfer or ESA upconversion. In order to reduce the lifetime of the level $^4I_{11/2}$, cross-relaxation with other doping ions, such as Ce^{3+} in which the final state of the donor is the manifold $^4I_{13/2}$, could be effective. Nevertheless, these pumping schemes determine a quite large quantum defect which reduces the laser emission efficiency and increases heat generation. A very tempting solution for these lasers could be direct pumping into the $^4I_{13/2}$ level, resulting in a very small quantum defect. Such pumping was investigated

Table 10.9 *The Stark components for the lowest energy manifolds of Er^{3+} in Y_2O_3 ceramics*

Center	Manifold	Stark components (cm^{-1})
C_2	$^4I_{15/2}$	0, 39, 76, 89, 161, 263, 493, 506
	$^4I_{13/2}$	6511, 6544, 6589, 6595, 6685, 6841, 6868
	$^4I_{11/2}$	10192, 10213, 10243, 10263, 10360, 10378
C_{3i}	$^4I_{15/2}$	0, 42, 81
	$^4I_{13/2}$	6460, 4469, 6534, 6570, 6647, 6801, 6862

for Er-doped laser materials using Er-doped fiber lasers and confirmed the efficiency of the process. The recent strong development of diode lasers in the absorption range of the manifold $^4I_{13/2}$ gives high expectation for the development of highly efficient, low quantum defect 1.5 µm Er lasers. As seen from Table 10.8, the Stark levels in Er:YAG form two well-separated groups, each of four closely packed levels in the $^4I_{15/2}$ ground manifold and three and four levels in the excited manifold $^4I_{13/2}$. The transitions involving levels from the lowest groups are sharper than those involving levels from the second group. The strongest lines in the 300 K $^4I_{15/2} \rightarrow {}^4I_{13/2}$ absorption spectrum are at 1532 nm, with absorption cross-section 2.3×10^{-20} cm^2 and at 1475 nm (1.8×10^{-20} cm^2), whereas the emission cross-section, determined by the reciprocity method for the strongest lines 1617 nm ($^4I_{13/2}(2) \rightarrow {}^4I_{15/2}(5)$) and 1645 nm ($^4I_{13/2}(2) \rightarrow {}^4I_{15/2}(7)$) is 0.67×10^{-20} cm^2 and 0.59×10^{-20} cm^2, respectively [115]. The radiative lifetime of $^4I_{13/2}$, inferred from comparison of the emission cross-section values calculated by reciprocity and the Fuchtbauer–Ladenburg relation, is 6.2 ms, close to the measured value for diluted samples and to that reported for single crystals. It is worth mentioning that the strong 1532 nm absorption line in Er:YAG corresponds in fact to the superposition of the transitions $^4I_{15/2}(2) \rightarrow {}^4I_{13/2}(1)$ (the most intense) and $^4I_{15/2}(4) \rightarrow {}^4I_{13/2}(3)$ and this could contribute to low quantum defect 1.6 µm laser emission in diluted Er:YAG materials. At 77 K the absorption lines narrow considerably, leading to strong enhancement of peak cross-sections: for the 1532.3 nm absorption line (FWHM ∼0.03 nm) the cross-section becomes ∼7.6×10^{-19} cm^2, and for the (4 → 2) 1534 nm absorption (0.18 nm) the cross-section is ∼2.1×10^{-20} cm^2 [121].

The optical spectra of the Er-doped cubic sesquioxide ceramics are similar to those of single crystals: Er^{3+} ions can enter in both sites, but the optical spectra are dominated by those entering in the low symmetry C_2 center, whereas the ions in the inversion C_{3i} sites can have only magnetic dipole allowed transitions such as $^4I_{15/2} \leftrightarrow {}^4I_{13/2}$ [122]. The Stark components of the lowest energy manifolds of Er^{3+} in Y_2O_3 ceramic are given in Table 10.9.

The strongest 300 K absorption line in the manifold $^4I_{11/2}$ of Er^{3+} in Y_2O_3 is around 974 nm. Due to the lower phonon energies, the non-radiative de-excitation of $^4I_{11/2}$ at low Er concentrations is not very strong, and shows moderate temperature dependence,

Table 10.10 *Stark components of the lowest energy manifolds of the Er^{3+} C_2 center in Sc_2O_3 ceramics*

Manifold	Stark components (cm^{-1})
$^4I_{15/2}$	0, 39, 78, 90, 277, 588, 604
$^4I_{13/2}$	6514, 6548, 6594, 6604, 6696, 6906, 6935
$^4I_{11/2}$	10191, 10214, 10251, 10266, 10393, 10416

2.4 ms at 300 K and 4.2 ms at 77 K. The emission spectrum $^4I_{11/2} \rightarrow {}^4I_{13/2}$ contains several intense emission lines, such as 2707 nm ($^4I_{11/2}(6) \rightarrow {}^4I_{13/2}(5)$) with emission cross-section 0.98×10^{-20} cm^2 at 300 K and 5.1×10^{-20} cm^2 at 77 K), 2715 nm ($1 \rightarrow 1$, 0.91×10^{-20} cm^2 and respectively 5.46×10^{-20} cm^2), 2725 nm ($2 \rightarrow 2$, 0.78×10^{-20} cm^2 and 1.2×10^{-20} cm^2), and 2740 nm ($3 \rightarrow 4$, 0.74×10^{-20} cm^2 and 2.53×10^{-20} cm^2) [123, 124]. The 77 K absorption spectrum $^4I_{15/2} \rightarrow {}^4I_{13/2}$ is dominated by the $1 \rightarrow 1$ line at 1535.7 nm, and the emission cross-sections of the various $^4I_{13/2} \rightarrow {}^4I_{15/2}$ transitions are smaller by almost an order of magnitude than for the $^4I_{11/2} \rightarrow {}^4I_{13/2}$ transitions; the calculated radiative lifetime of the $^4I_{13/2}$ is 7.6 ms.

The optical spectra of the Er-doped Sc_2O_3 ceramics [125–127] are similar to those reported for single crystals; however, a more complete energy level scheme for the Er^{3+} ions residing in the low symmetry sites C_2 can be identified (Table 10.10).

The calculated radiative lifetimes for several manifolds are 4.5 ms for $^4I_{13/2}$, 4.97 ms for $^4I_{11/2}$, 2.6 ms for $^4I_{9/2}$, 0.391 ms for $^4F_{9/2}$, 0.42 ms for $^SI_{3/2}$, 0.09 ms for $^2H_{11/2}$ and 0.17 ms for $^4F_{7/2}$. However, the measured lifetimes in Er:Sc_2O_3 ceramic can be influenced by non-radiative effects (electron–phonon interaction, energy transfer) or reabsorption: 16 μs for the emission of the thermallized levels ($^4S_{3/2}$, $^2H_{11/2}$), ∼1 μs for $^4F_{9/2}$, less than 0.1 μs for $^4I_{9/2}$, and 340 μs for $^4I_{11/2}$, 5–6 ms for the level $^4I_{13/2}$, in fair agreement with the values reported previously for single crystals. The emission cross-sections calculated by the reciprocity method, together with these lifetimes, indicate that many of the energy levels of Er^{3+} show potential for laser emission, such as $^4I_{13/2} \rightarrow {}^4I_{15/2}$, whose 300 K effective stimulated emission spectrum, calculated by the reciprocity method, is shown in Figure 10.15, together with the absorption cross-section spectrum. The largest emission cross-section in this range (∼1.8×10^{-20} cm^2, about 20% larger than for Nd:YAG) corresponds to the transition $^4I_{13/2}(1) \rightarrow {}^4I_{15/2}(1)$ at 1535 nm, ∼1. The laser emission on this transition would determine a pure three-level laser scheme, while the cross-sections of the transitions to higher Stark levels of $^4I_{15/2}$, such as $^4I_{13/2}(1) \rightarrow {}^4I_{15/2}(4)$ at 1535 nm, $^4I_{13/2}(1) \rightarrow {}^4I_{15/2}(5)$ at 1581 nm or $^4I_{13/2}(1) \rightarrow {}^4I_{15/2}(6)$ at 1603 nm could be enough for a quasi-three-level laser scheme. The reabsorption could be reduced at low temperatures, where a simultaneous increase in the effective emission cross-section is obtained [128]. Moreover, efficient quasi-three-level Er laser emission with very low quantum defect can be obtained by pumping resonantly the $^4I_{15/2}$ level with an Er fiber laser or with diode lasers.

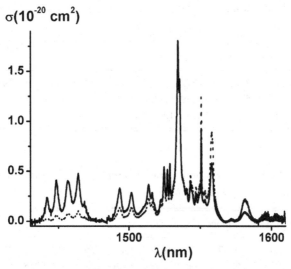

Figure 10.15 The room temperature absorption (continuous line) and emission (reciprocity method, dotted line) spectra of Er: Sc_2O_3 $^4I_{13/2} \leftrightarrow {}^4I_{15/2}$ transition.

Several manifolds of Er^{3+} in Sc_2O_3 ceramic can give efficient luminescence in the visible, such as the green 550 nm $(^2H_{11/2}, {}^4S_{3/2}) \rightarrow {}^4I_{15/2}$ transition or the red 660 nm $^4F_{9/2} \rightarrow {}^4I_{15/2}$ transition. A major difficulty in this case is excitation of the emission by direct pumping in the visible, although upconversion of infrared excitation radiation might prove efficient in intrinsic or sensitized schemes. Practically all the emitting levels of Er^{3+} show strong concentration quenching by energy transfer upconversion (in the case of the manifolds $^4I_{13/2}$ and $^4I_{11/2}$) or down-conversion, that dominates the de-excitation of the upper manifolds.

10.2.1.8 Tm^{3+}-doped ceramics

The degeneracy of the energy manifolds of Tm^{3+} in the dodecahedral site of garnets is completely removed by the crystal field interaction, resulting in a rich Stark level structure; several emission transitions in the visible and infrared can be useful for laser emission. Particularly important is the emission $^3F_4 \rightarrow {}^3H_6$ in the 2 μm range. Table 10.11 gives the Stark structure of the Tm:YAG manifolds involved in the infrared emission [129].

This energy level scheme favors the cross-relaxation process $(^3H_4, {}^3H_6) \rightarrow ({}^3F_4, {}^3F_4)$ that would enable 2 μm emission from the level $^3F_4$ under diode laser pumping in the 780–800 nm range: by this process, a quantum of pumping radiation at 800 nm will be transformed into two 2 μm emission quanta. Generally the optical spectra of the ceramic Tm:YAG materials [130, 131] are similar to those of single crystals, although differences in the high energy side of the emission spectrum $^3F_4 \rightarrow {}^3H_6$ have been noticed. The absorption cross-section at 785 nm in the manifold $^3H_4$ is 0.87×10^{-20} cm^{-1}, whereas the strongest

Table 10.11 *Stark levels of Tm:YAG manifolds involved in infrared emission*

Manifold	Stark components (cm^{-1})
$^3H_6$	0, 27, 216, 240, 257, 300, 450, 588, 610, 650, 690, 730, (798)
$^3F_4$	5555, 5764, 5832, 5901, 6042, 6111, 6170, (6175), 6199
$^3H_4$	12607, 12644, 12732, 12747, 12824, (12973), 13036, 13112, 13152

Table 10.12 *Stark levels of the lowest manifolds of the Tm^{3+} center of C$_2$ symmetry in Sc$_2$O$_3$ ceramics*

Manifold	Stark components (cm^{-1})
$^3H_6$	0, 50, 89, 278, 315, 370, 403, 420, 545, 566, 621, 655, 771, 991
$^3F_4$	5635, 5687, 5821, 6091, 6110, 6170, 6203, 6241, 6258
$^3H_5$	8276, 8328, 8340, 8498, 8512, 8600, 8815, 8851
$^3F_4$	12551, 12665, 12686, 12723, 12845, 12963, 12987, 13155, 13220

emission lines in the 2 μm range are at 2013, 1967, 1887 and 1792 nm, with emission cross-sections around 0.5×10^{-20} cm^{-1}; the emission lifetime was 10.5 ms.

The cross-relaxation $(^3H_4, {}^3H_6) \rightarrow (^3F_4, {}^3F_4)$ could be favored by hosts with high density of cationic sites and low lattice parameters, such as the sesquioxides. The laser transition $^3F_4 \rightarrow {}^3H_6$ is magnetic dipole forbidden so the Tm^{3+} ions in the C_{3i} sites of sesquioxides are inactive. In the low symmetry center C_2 in sesquioxides all the energy manifolds of Tm^{3+} are split into singlets, resulting in a rich Stark energy level scheme. Spectroscopic investigation of Tm^{3+}:Sc$_2$O$_3$ ceramics with up to 5 at.% Tm [132] enabled a more complete energy level diagram of the low symmetry C_2 center, as given in Table 10.12.

In the 5 at.% Tm:Sc$_2$O$_3$ ceramic the emission of $^3H_4$ is almost completely quenched, revealing the high efficiency of the cross-relaxation. For Tm concentration up to ~3 at.% the ion–ion cross-relaxation process can be described by assuming mixed multipole interaction (dipole–dipole, dipole–quadrupole, quadrupole–quadrupole) [132], although at higher Tm concentration additional de-excitation processes with features corresponding to cooperative de-excitation became apparent [133]. The Judd–Ofelt analysis of these ceramics enabled calculation of the radiative lifetime of the level $^3H_4$, $\tau_{rad} = 2.77$ ms, close to the fluorescence lifetime at low Tm concentrations (2.62 ms). The 300 K emission cross-sections calculated with the reciprocity method show that the strongest emission lines are at 1970 nm (emission cross-section 0.97×10^{-20} cm^2) and 1989 nm (0.91×10^{-20} cm^2), and several lines in the range 2100–2150 nm have emission cross-sections close to 0.2×10^{-20} cm^2. This indicates that Tm:Sc$_2$O$_3$ ceramic could be useful for tunable emission in these wavelength ranges [134].

Table 10.13 *The Stark levels for Yb:YAG (the asterisks denote the measured positions of electron–phonon resonantly shifted Stark levels of the excited manifold)*

Manifold	Stark components (cm^{-1})
$^2F_{7/2}$	0, 565, 612, 785
$^2F_{5/2}$	10327, 10623*, 10921*

Tm^{3+} in very fine ($\sim$500 nm) grained Lu_2O_3 ceramics shows two absorption lines, 769 nm (cross-section 0.38×10^{-20} cm^2) and 811 nm (0.32×10^{-20} cm^2), which can be pumped with diode lasers and the strongest emission lines in the transition $^3F_4 \rightarrow {}^3H_6$ are at 1942, 1963 and 2066 nm [135].

10.2.1.9 Yb^{3+}-doped ceramics

10.2.1.9.1 Yb-doped garnet ceramics The most investigated Yb-doped laser garnet material is Yb:YAG. Although Yb:LuAG may have superior qualities (higher cross-sections, smaller reduction of heat conduction at high Yb concentration), the cost and availability of the raw material have contributed to this choice. Highly transparent Yb:YAG ceramics with high Yb concentrations C_{Yb} can be fabricated, with spectroscopic properties similar to those of single crystals [136]. Despite the simple electronic structure of Yb^{3+}, the absorption and emission spectra are quite complex: the pure electronic lines are accompanied by sharp and intense vibronic satellites, and the accidental resonance of the gaps of the Stark levels with the phonons can induce a shift in the electronic transition and enhancement of the intensity of the resonant vibronic satellite. These circumstances make the interpretation of spectra and localization of the Stark levels very difficult [137, 138]. The Stark levels for Yb:YAG are given in Table 10.13.

At 300 K, Yb:YAG can give laser emission from the transitions $^2F_{5/2}(1) \rightarrow {}^2F_{7/2}(3)$ and $^2F_{5/2}(1) \rightarrow {}^2F_{7/2}(4)$, in a quasi-three-level scheme. The spectroscopic parameters of interest for pumping and laser emission of Yb:YAG are given in Table 10.14.

The fractional thermal population coefficients of the terminal levels ($f_{t3} = 0.046$ and $f_{t4} = 0.0192$ at 300 K) for laser emission and reabsorption can be reduced by cooling to 77 K ($f_{t3} = 1.05\times10^{-5}$ and $f_{t4} = 4.1\times10^{-7}$), transforming these laser schemes into almost pure four-level emission. The modification of the thermal coefficients of the Stark levels from which the absorption and emission transitions originate or homogeneous broadening when lowering the temperature to 77 K modifies the spectroscopic parameters of the transitions of interest for laser emission, as shown in Table 10.14.

The sharp absorption line $^2F_{7/2}(1) \rightarrow {}^2F_{5/2}(1)$ at 968 nm is difficult to pump, especially at low temperatures and with conventional diode lasers (FWHM 2–3 nm and with thermal shift), although spectrally narrowed and stabilized diodes could eliminate this problem.

Table 10.14 *Spectroscopic parameters of interet for laser emission of Yb:YAG*

Process	Transition	Temperature (K)	Wavelength (nm)	FWHM (nm)	Peak effective cross-section (10^{-20} cm^2)
Absorption	$^2F_{7/2}(1) \rightarrow {}^2F_{5/2}(1)$	300	968.8	2.8	0.7
		77		0.1–02	>13
	$^2F_{7/2}(1) \rightarrow {}^2F_{5/2}(2)$	300	940.6	18	0.82
		77		13	1.7
Emission	$^2F_{7/2}(1) \rightarrow {}^2F_{5/2}(3)$	300	1030.1	6	2.1
		77		1.5	13.2
	$^2F_{7/2}(1) \rightarrow {}^2F_{5/2}(4)$	300	1048	10	0.38
		77			0.75

Nevertheless, the pumping can be also done into the broader but weaker absorption line $^2F_{7/2}(1) \rightarrow {}^2F_{5/2}(2)$. The low absorption cross-sections of Yb:YAG impose high C_{Yb} in the bulk or thin-disk lasers. The emission radiative lifetime of $^2F_{5/2}$ is quite long (0.95 ms) and, despite the increased electron–phonon interaction of Yb^{3+}, the large $^2F_{7/2}$–$^2F_{5/2}$ energy gap renders the non-radiative de-excitation quite small. The lifetime measurements for diluted samples confirm this value; however, with increasing C_{Yb}, special care must be taken to avoid artificial lengthening of the measured lifetime by reabsorption. Additionally, for C_{Yb} typically above 10 at.% Yb, a quite pronounced reduction in lifetime was observed and this was attributed to cooperative upconversion processes up to the charge transfer and to the conduction bands or to migration-assisted energy transfer to accidental impurities, especially Er or Tm. The particularities of fabrication techniques can determine differences between the spectroscopic, dynamic and laser emission properties of heavily doped Yb:YAG crystals and ceramics, as will be discussed in Section 11.4.

In the Yb:LuAG ceramics prepared by solid-state reaction, the room temperature absorption cross-sections of transitions $^2F_{7/2}(1) \rightarrow {}^2F_{5/2}(1)$ at 969 nm and $^2F_{7/2}(1) \rightarrow {}^2F_{5/2}(2)$ at 937 nm are 0.43×10^{-20} cm^2 and respectively 0.66×10^{-20} cm^2, and the cross-section of the strongest transition $^2F_{5/2}(1) \rightarrow {}^2F_{7/2}(3)$ is 0.66×10^{-20} cm^2 [139].

The ability of Yb laser materials to generate ultra-short pulses is determined by the emission linewidth: in the ordered garnets (YAG, LuAG, and so on) the absorption and emission lines are mainly homogeneously broadened by electron–phonon interaction and sharpen considerably when cooling to cryogenic temperatures. As shown in Section 10.1.2.3, the width of the optical lines of Yb^{3+} can be increased controllably in a large temperature range by using disordered garnets based on solid-solution mixing of ordered garnets (such as Y$_3$ScAl$_4$O$_{12}$ [51], Lu$_3$ScAl$_4$O$_{12}$ [53], (YGd$_2$)Sc$_2$(Al$_2$Ga)O$_{12}$ [55]), or intrinsic disordered garnets with mixed occupancy of one or more cationic sublattices with cations of different valence, such as the CNGG–CLNGG family, or by solid-solution mixing of intrinsic disordered garnets, such as CLNTGG. The 300 K linewidth of the most intense

Table 10.15 *The Stark components of the*
Yb^{3+} C$_2$ center in cubic sesquioxides

Material	Manifold	Stark components
Y$_2$O$_3$	$^2F_{7/2}$	0, 350, 534, 946
	$^2F_{5/2}$	10242, 10518, 11027
Lu$_2$O$_3$	$^2F_{7/2}$	0, 505, 558, 975
	$^2F_{5/2}$	10245, 10530, 11020
Sc$_2$O$_3$	$^2F_{7/2}$	0, 580, 650, 1110
	$^2F_{5/2}$	10258, 10620, 11194

$^2F_{5/2}(1) \rightarrow {}^2F_{7/2}(3)$ emission line of Yb^{3+} in garnets is around 10 nm in ordered garnets, it increases ~20–40% in solid-solution garnets, and practically doubles in the intrinsic disordered garnets and becomes ~2.5 times larger in the systems with mixed disorder. Moreover, whereas in Yb:YAG the width of this line decreases by ~6.5 times when cooling from 300 K to 77 K, in the case of CLNTGG it decreases by only 25–30%. The $^2F_{7/2}(1) \rightarrow {}^2F_{5/2}(1)$ absorption in the intrinsic disordered garnets (CLNGG, CLNTGG) shows as a broad (~2.6 nm at 300 K, 2.3 nm at 77 K), multi-peaked band that could be used for diode laser pumping.

Besides infrared absorption and emission, the Yb:YAG crystals and ceramics show emission bands at 340 nm and 460 nm from the charge transfer band to the $^2F_{7/2}$ and $^2F_{5/2}$ levels of Yb^{3+}. The very short, Yb concentration dependent lifetime (several nanoseconds) makes the Yb:YAG ceramics of interest for fast scintillators [140, 141].

10.2.1.9.2 Yb-doped cubic sesquioxide ceramics High quality transparent ceramics of Yb^{3+}-doped cubic sesquioxides can be produced by solid-state and wet chemistry methods for all the sesquioxides [17, 43, 142–144]. The positions of the spectral lines are similar to those reported for single crystals [15, 16, 69, 101, 145–148] and the quite large differences in the relative intensities reported in various papers are most likely caused by the resolution of the experiment, since some of the lines are very sharp. Yb^{3+} can occupy both the C_2 and C_{3i} symmetry centers in these materials, but whereas in the first center both the electric and magnetic dipole transitions are allowed, in the second center these are restricted to the magnetic dipole transition. Due to the higher abundance of the C_2 center with electric dipole transitions and strong electron–phonon interaction, the optical spectra of Yb^{3+} are dominated by this center and its Stark levels are given in Table 10.15.

The optical lines of Yb^{3+} in cubic sequioxide ceramics show strong temperature dependent homogeneous broadening, as illustrated in Figure 10.16 for Y$_2$O$_3$ ceramic. The presence of the C_{3i} symmetry center becomes evident in the sharp $^2F_{7/2}(1) \leftrightarrow {}^2F_{5/2}(1)$ absorption and in the non-selectively excited emission line at low temperatures and is placed ~10–11 cm^{-1} higher in energy than the line corresponding to the C_2 center. The peak intensity of the absorption line $^2F_{7/2}(1) \rightarrow {}^2F_{5/2}(1)$ for the center C_{3i} is about 15 times smaller than

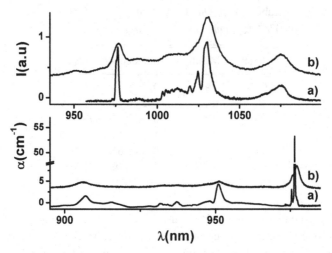

Figure 10.16 The Yb^{3+} (1 at.%) absorption (lower layer) and emission (upper layer) in Y$_2$O$_3$ ceramic at: (a) 10 K and (b) 300 K. There is a break in the intensity scale of the 10 K absorption spectrum.

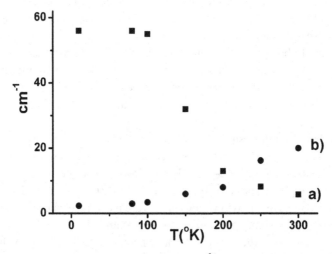

Figure 10.17 Temperature dependence of the 976 nm Yb^{3+} absorption of the C_2 center in 1 at.% Yb:Y$_2$O$_3$ ceramic: (a) peak absorption coefficient and (b) linewidth (FWHM).

for the center C_2, whereas an accurate comparison in emission is difficult because of the stronger reabsorption of the emission of the second center. The temperature dependence of the peak absorption coefficient and of the linewidth is shown in Figure 10.17. The peak absorption coefficient of the line $^2F_{7/2}(1) \rightarrow {}^2F_{5/2}(1)$ (976 nm) at 10 K in Yb:Y$_2$O$_3$ ceramic is about an order of magnitude larger than for any other absorption line. Similar to garnets, the optical spectra of Yb-doped sesquioxides reveal two absorption bands suitable for diode laser pumping, the sharp and intense transition $^2F_{7/2}(1) \rightarrow {}^2F_{5/2}(1)$ and the broader but less

Table 10.16 *Spectroscopic characteristics of interest for laser emission of Yb^{3+} in cubic sesquioxides, at 300 K*

Material	Process	Transition	Wavelength (nm)	FWHM (nm)	Peak effective cross-section (10^{-20} cm^2)	Lifetime (ms)
Y_2O_3	Absorption	$^2F_{7/2}(1) \to {}^2F_{5/2}(1)$	976	2	2.8	
		$^2F_{7/2}(1) \to {}^2F_{5/2}(2)$	951	14.5	0.8	
	Emission	$^2F_{5/2}(1) \to {}^2F_{7/2}(3)$	1031	14	1	0.85
		$^2F_{5/2}(1) \to {}^2F_{7/2}(4)$	1076	16	0.45	
Lu_2O_3	Absorption	$^2F_{7/2}(1) \to {}^2F_{5/2}(1)$	976	2.2	3.1	
		$^2F_{7/2}(1) \to {}^2F_{5/2}(2)$	949	10	0.95	
	Emission	$^2F_{5/2}(1) \to {}^2F_{7/2}(3)$	1033	10.5	1.1	0.80
		$^2F_{5/2}(1) \to {}^2F_{7/2}(4)$	1080	15	0.4	
Sc_2O_3	Absorption	$^2F_{7/2}(1) \to {}^2F_{5/2}(1)$	975	2	4.3	
		$^2F_{7/2}(1) \to {}^2F_{5/2}(2)$	940	10	0.85	
	Emission	$^2F_{5/2}(1) \to {}^2F_{7/2}(3)$	1041	14	1.3	0.73
		$^2F_{5/2}(1) \to {}^2F_{7/2}(4)$	1094	18	0.25	

intense transition $^2F_{7/2}(1) \to {}^2F_{5/2}(2)$, and two emission lines, $^2F_{5/2}(1) \to {}^2F_{7/2}(3)$ and $^2F_{5/2}(1) \to {}^2F_{7/2}(4)$.

The room temperature spectroscopic characteristics of the transitions of interest for laser emission of Yb^{3+} in cubic sesquioxides (absorption and emission wavelengths, linewidths and cross-sections, lifetimes) are given in Table 10.16.

At room temperature, both emission transitions $^2F_{5/2}(1) \to {}^2F_{7/2}(3)$ and $^2F_{5/2}(1) \to {}^2F_{7/2}(4)$ determine a quasi-three-level laser scheme. Owing to the quite high energy and good separation of the level $^2F_{7/2}(4)$, its thermal population is quite low ($f_t(4) < 1\%$), the reabsorption of the laser radiation is weak and the pump intensity necessary to reach transparency can be lower than for the more intense $^2F_{5/2}(1) \to {}^2F_{7/2}(3)$ emission ($f_t(3)$ around 6%). However, the larger quantum defect inherent in this transition favors increased heat generation. The more intense, lower quantum defect transition $^2F_{5/2}(1) \to {}^2F_{7/2}(3)$ would be favored by cooling to cryogenic temperatures (80 K), where $f_t(3)$ decreases by about three orders of magnitude, transforming the laser emission to an almost pure four-level scheme; this will determine also the increase in the thermal population of the emitting level $^2F_{5/2}(1)$ as well as enhancement of the peak emission cross-section due to the narrowing of lines.

The widths of the emission lines in the Yb-doped sesquioxides are fairly large and can be further increased by structural disorder in solid solutions with other cubic or lower symmetry sesquioxides or by doping with aliovalent cations. Disordering the cubic Y_2O_3 by solid-solution mixing with 10% of the much larger La^{3+} ion induces strong broadening (to 6.9 nm) of the absorption line $^2F_{7/2}(1) \to {}^2F_{5/2}(1)$ as well as ~50% broadening of the emission $^2F_{5/2}(1) \rightarrow {}^2F_{7/2}(3)$ [149]. Similar broadening was observed in Yb:Y_2O_3 ceramic with 3% ZrO_2 sintering aid, and it was claimed that the emission cross-section was larger than in $(La,Y)_2O_3$ ceramic. Enhanced broadening could be obtained by combining the size disordering by solid-solution mixing with electric charge disordering caused by large amounts of SaO_2 sintering aid: transparent Yb-doped $(Sc_{0.25}Y_{0.75})_2O_3$ ceramics with 6 wt.% ZrO_2 were produced and strong broadening of the Yb emission was observed [150].

10.2.1.9.3 Yb in alkaline earth fluoride ceramics The occurrence of a unique structural center involving hexametric ensembles of Yb ions in heavily doped CaF_2 crystals [47] renewed the prospect of using this material for high power laser emission. The Stark levels of this center are 0, 50, 100, and 530 cm^{-1} for the manifold $^2F_{7/2}$ and 10200, 10300 and 10580 cm^{-1} for the $^2F_{5/2}$ manifold. The strongest absorption line is $^2F_{7/2}(1) \rightarrow {}^2F_{5/2}(1)$ at 980.5 nm and the low energy of the second and third Stark levels of $^2F_{7/2}$ could make them of interest for hot-band pumping. All these transitions could give low quantum defect laser emission; pumping in the broad $^2F_{7/1}(1) \rightarrow {}^2F_{5/2}(2)$ could also be possible. The emission spectrum is broad, with the $^2F_{5/2}(1) \rightarrow {}^2F_{7/2}(4)$ line at 1049 nm, with cross-section 0.16×10^{-20} cm^2 and 2.4 ms lifetime.

Transparent CaF_2 ceramics with high (up to 10%) Yb doping were produced by reacting mixed solutions of Ca and Yb nitrates with fluoric acid, followed by precipitation, vacuum sintering and hot isostatic compression [48, 49]. High resolution electron microscopy revealed the presence of Yb clusters, without clear identification of their nature. As mentioned before, energy dispersive X-ray investigation of these ceramics indicates a strong (to 50%) increase in Yb concentration at the ceramic grain boundaries. Accidental poisoning with oxygen during the technological process was revealed.

The precipitation of the wet-reacted material was also used to fabricate 5 at.% Yb-doped CaF_2–SrF_2 mixed cubic fluoride ceramics and it was claimed that the difficult to control segregation can hamper uniform composition of the mixed fluorides and the distribution of the doping ion [50, 151]. To avoid these difficulties, Bridgman crystals of Yb-doped CaF_2 or (CaF_2–SrF_2) mixed fluoride were hot pressed, resulting in highly transparent and homogeneous ceramics, with spectroscopic properties similar to the corresponding single crystals. The fracture toughness was almost double, making these materials more suitable for high power laser emission. The peak of the zero phonon $^2F_{7/2}(1) \rightarrow {}^2F_{5/2}(1)$ absorption line is shifted to longer wavelengths compared with CaF_2 and SrF_2 and in the case of the 65%CaF_2–30%SrF_2–5%YbF_3 ceramic it lies at 972 nm and the emission lifetime is 2.1 ms. Due to the structural solid-solution disordering, these mixed ceramics could be more suitable for short pulse generation than Yb-doped CaF_2.

10.2.2 Transition metal doped ceramics

10.2.2.1 Cr^{2+}-doped ceramics

Transparent Cr^{2+}-doped ZnSe ceramics with spectroscopic characteristics similar to those of single crystals can be produced by uniaxial hot (1400–1500 K) pressing (350 MPa). The strong and broad vibronic transitions between the Stark levels $^5D_2$ (ground) and ^5E of the ground spectral term ^5D in the tetrahedral T_d crystal field of ZnSe are suitable for efficient broadly tunable or short pulse laser emission or for Q-switching of several RE^{3+} infrared lasers. These ceramics show a very strong ($\sigma_a = 1.1 \times 10^{-18}$ cm^2) and broad (350 nm) absorption band peaking around 1770 nm and the emission band (860 nm) peaks at 2450 nm, with emission cross-section $\sigma_e = 1.1 \times 10^{-18}$ cm^2. The 300 K luminescence lifetime at very low Cr concentrations is 5.4 μs, practically equal to the radiative lifetime, but drops severely at concentrations above 10^{19} cm^{-3} and at higher temperatures [152, 153].

10.2.2.2 Cr^{3+}-doped ceramics

Trivalent chromium can substitute Al^{3+}, Ga^{3+} or Sc^{3+} in the octahedral a-sites, and the relative positions of the excited levels ^2E and $^4T_2$ are determined by the strength of the crystal field, i.e. the size of the substituted site [154, 155]. In Al garnets the crystal field acting on the Cr^{3+} ions that substitute Al^{3+} in octahedral sites is strong, the $^4T_2$ level is placed well above ^2E and at 300 K, despite a weak thermallization, the emission is dominated by the spin forbidden, long lifetime (millisecond range) red emission from ^2E to the ground level $^4A_2$. By contrast, when Cr^{3+} substitutes large cations such as Sc^{3+}, the energy of the level $^4T_2$ is lowered in the vicinity of ^2E, a strong thermallization takes place and the spin forbidden criterion is relaxed: the emission becomes broad and extends into the near infrared range, its lifetime is shortened and the cross-section is increased. However, the spectroscopic properties of the Cr^{3+}-doped garnets, particularly the involvement of ESA, are not very suitable for efficient laser emission. The success of other materials, such as the colquirites LiCaAlF$_6$ (LiCAF) or LiSrAlF$_6$ (LiSAF) [156] in granting weak crystal field sites for the Cr^{3+} ions raised hopes for alternative materials with very broadband emission under diode laser pumping that could be used for ultra-short pulse generation with high global optical–electric efficiency. Unfortunately, their thermomechanical properties are not very good, growth of high quality large single crystals is not easy and the low symmetry precludes fabrication of transparent ceramics.

Spectroscopic investigation of Cr-doped Sc$_2$O$_3$ single crystals revealed [157, 158] two centers, the center I, attributed to Cr^{3+} in the C_{3i} site, whose strong and broad absorption and emission vibronic bands could be suitable for broadband laser emission in the near infrared under visible pumping, and the center II, of unidentified origin, with absorption in the emission range of center I and with short-lived emission in the infrared. The quite large difference between the six-fold coordinated ionic radii of Cr^{3+} and Sc^{3+} (Table 2.1) determines very small segregation coefficients for Cr in the melt growth of Sc$_2$O$_3$ crystals and limits severely the maximum doping concentration. Recent studies revealed that Cr can be introduced in larger concentration in Sc$_2$O$_3$ ceramics and that the doping is uniform over the whole body of the ceramic [159]. The spectroscopic properties of Cr-doped Sc$_2$O$_3$ ceramics are similar to those reported previously for single crystals, and detailed investigation of these properties in ceramics revealed additional information on the electronic structure of the emitting centers, particularly the lower position of the $^4T_2$ level, below ^2E in center I. The emission can be excited by visible diode laser (670 nm range) and the lifetime of emission is fairly large, $\sim$70 µs at low concentration. The nature of center II was not clearly established, it could be Cr^{3+} in the C_2 site or Cr in another valence state and it was found that it could contribute to de-excitation of center I. Nevertheless, the spectroscopic properties of center I demonstrate that weak crystal field centers of Cr^{3+} can exist in Sc-based compounds and that such centers could give broad emission under diode laser pumping.

10.2.2.3 Cr^{4+}-doped ceramics

According to the ionic radius, Cr^{4+} could substitute the Al^{3+} or Ga^{3+} ions in both six-fold and four-fold coordinated sites of garnets; obviously, this substitution would require charge compensation, such as co-doping with Ca^{2+} or Mg^{2+} that could enter in the dodecahedral c-sites. Spectroscopic investigation [160–164] showed that the optical spectra are dominated by the Cr^{4+} ions in the tetrahedral d-sites of distorted tetrahedral local symmetry D_{2d}. The electronic structure of the Cr^{4+} in these sites is complex and the intense and broad near infrared ($\sim$750 nm to 1150 nm) vibronic absorption band is dominated by the spin-allowed electric dipole transitions from the ground level $^3B_1$ to various excited crystal field levels ($^3A_2$, $^3B_2$, ^3E), all originating from the lowest spectral term ^3F of the ground configuration $3d^2$ and with contribution (around 1.1–1.2 µm) of the spin forbidden absorption in the level ^1E, originating from the first excited spectral term ^1D. The emission spectrum corresponds to the transition $^3B_2 \rightarrow {}^3B_1$: at 300 K the emission is broad ($\sim$230 nm) and strong, with peak emission cross-section (at 1380 nm) equal to 3.3×10^{-19} cm^2, whereas at 10 K it shows sharp lines at 1278.8 and 1275.2 nm, corresponding to zero phonon emission from the spin–orbit split (28 cm^{-1}) level $^3B_2(^3T_2)$. The lifetime of emission depends strongly on temperature, 30.1 µs at 15 K and 4.1 µs at 300 K; compared with the calculated radiative lifetime of 48 µs, this indicates quite low quantum efficiency even at very low temperatures. The Cr^{4+}:YAG ceramics show strong and broad infrared absorption and emission lines [165], similar to those of single crystals.

The strong and broad near infrared absorption band shows that Cr^{4+}:YAG can serve as a saturable absorber for passive Q-switching of Nd or Yb 1 µm lasers ($\sigma_{a,1064nm} = 4.4 \times 10^{-18}$ cm^2) or as a suppresser of ASE for the 1 µm luminescence. The concentration of Cr^{4+} in YAG is typically of the order of a tenth of a percent and the initial transmission of the saturable absorber can be controlled by the Cr^{4+} concentration and by size. The intense and broad absorption band, coupled with the high cross-section and large width of the emission band suggest the possibility of use for laser emission, particularly for ultra-short pulse generation; such a laser could be pumped by the fundamental radiation of Nd lasers or with diode lasers. A major drawback for both laser emission and Q-switching is the excited state absorption in this range ($\sigma_{ESA,1064nm} = 4.4 \times 10^{-18}$ cm^2) to short lived ($<$0.1 ns) levels in the visible. The saturated ESA at 1064 nm would then enable mode-locking. The spectroscopic properties depend on polarization, but the properties of Cr^{4+}:YAG ceramics in unpolarized light are similar to those of single crystals. Mixed doped (Nd^{3+}–Cr^{4+} or

Yb^{3+}–Cr^{4+}) components enable self Q-switching, or self-mode-locking. In large-lattice garnets such as YSGG, the emission band is shifted to the infrared and the relative excited state absorption is weaker; however, no reports on ceramics from these materials have been published.

The utilization of Cr^{4+}:YAG ceramics as saturable absorbers [166–168] is increasing strongly. Since the duration of the passively Q-switched pulse depends on the length of resonator, including the intra-cavity components, a modality to reduce it would be the bonding of the active material and of SA, provided they have similar refractive index, such as for the Cr^{4+}-, Nd^{3+}- or Yb^{3+}-doped YAG. In the case of ceramics the technical problems connected with the bonding could be avoided by fabrication of composite components with a sharp interface between the laser active part and the SA part. Alternatively, the active and SA ions could be uniformly co-doped [169, 170]: this approach could be efficient for pumping wavelengths poorly absorbed by SA, such as for 809 nm diode laser pumping of Nd, but can induce additional problems in the case of diode pumping of Yb materials since the absorption at the 940 nm pump wavelength is about 70% at the 1030 nm Yb^{3+} emission wavelength. Moreover, increased Cr^{4+} concentration reduces the Yb emission lifetime.

10.3 Radiative and non-radiative de-excitation processes

10.3.1 Emission decay in doped laser ceramics

10.3.1.1 The effect of non-radiative processes on emission decay

Non-radiative de-excitation can influence the emission decay in doped materials by processes which are identical for all doping ions, such as the electron–phonon interaction, which does not change the exponential character of the decay but reduces its lifetime to τ_D (Eq. (2.38)), and by processes involving the ensemble of doping ions, such as the energy transfer, which modify the shape of the decay according to Eq. (2.48) which can be transcribed as

$$I(t) = I(0) \exp\left(-\frac{t}{\tau_D}\right) \exp[-P'(t)], \qquad (10.1)$$

where the global transfer function $P'(t) = P(t) + \bar{W}$ includes the contribution of the direct and of migration-assisted transfer. The experimental energy transfer function $P'(t)$ can be obtained by subtracting from the normalized global decay $I(t)/I(0)$ the decay corresponding to the lifetime τ_D, i.e. $P'(t) = -\ln[(I(t)/I(0)] - t/\tau_D$. The Nd:doped YAG crystals and ceramics illustrate this very well.

In Czochralski grown Nd:YAG laser crystals C_{Nd} was traditionally limited to 1–1.2 at.% Nd. In the early studies on Nd:YAG crystals the emission decay $^4F_{3/2}$ was considered exponential, although small departures at early times were noticed. The lifetime measured after a certain lapse of time from the beginning of decay in 1 at.% Nd:YAG crystals was smaller (230–235 μs) than the calculated radiative lifetime of 260 μs [92]. It was observed

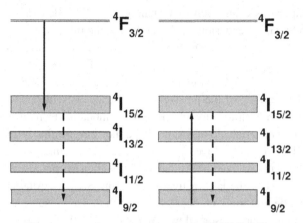

Figure 10.18 Scheme of down-conversion cross-relaxation in Nd:YAG: full arrows, cross-relaxation; dashed arrows, subsequent electron–phonon relaxation.

[171] that the departure from exponential shows systematic dependence on C_{Nd}, and this was linked with the energy transfer inside the system of Nd ions.

The Nd^{3+} energy level scheme can favor conversion of excitation by energy transfer.

(1) *The down-conversion of excitation* by interaction between an excited and a non-excited Nd^{3+} ion, according to the cross-relaxation processes $(^4F_{3/2}, {}^4I_{9/2}) \rightarrow (^4I_{15/2}, {}^4I_{15/2})$ and $(^4F_{3/2}, {}^4I_{9/2}) \rightarrow (^4I_{13/2}, {}^4I_{15/2})$: at room temperature the first of these, shown schematically in Figure 10.18, dominates [3]. In Nd:YAG the terminal levels of the donor and the acceptor are de-excited by electron–phonon interaction and thus all the excitation of the initial level of the donor is wasted as heat. Down-conversion by cross-relaxation was observed for many emitting levels of other ions, such as Er^{3+} $((^4S_{3/2}, {}^4I_{15/2}) \rightarrow (^4I_{9/2}, {}^4I_{13/2}))$, Tm^{3+} $((^3H_4, {}^3H_6) \rightarrow (^3F_4, {}^3F_4))$, and so on. For some of these the final level of the acceptor can give laser emission.

(2) *The upconversion of excitation* by interaction of two Nd^{3+} ions excited in the level $^4F_{3/2}$ according to the cross-relaxation mechanisms $(^4F_{3/2}, {}^4F_{3/2}) \rightarrow (^4I_{15/2}, {}^4G_{5/2})$, $(^4F_{3/2}, {}^4F_{3/2}) \rightarrow (^4I_{13/2}, {}^4G_{7/2})$, $(^4F_{3/2}, {}^4F_{3/2}) \rightarrow (^4I_{11/2}, {}^4G_{9/2})$ or $(^4F_{3/2}, {}^4F_{3/2}) \rightarrow (^4I_{9/2}, {}^2P_{1/2})$. Due to the dense packing of levels between all these final states of the acceptor and the emitting level, almost all the upconverted excitation relaxes non-radiatively to the emitting level $^4F_{3/2}$. Thus, in contrast to down-conversion where each individual act of energy transfer combined with the electron–phonon interaction transforms completely the excitation of the donor ion into heat, each upconversion act transforms only one of the two initial excitations into heat, since the acceptor returns to its initial excited state. The upconversion processes are shown schematically in Figure 10.19; the final result is similar for all four processes mentioned above. The energy transfer rates for these four upconversion processes could be very different; however, because of the similar final effect of the joint upconversion–(electron–phonon) processes, and because it is impossible to separate their effects, they are accounted for in the rate equation

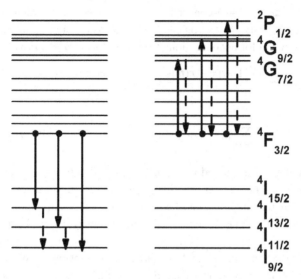

Figure 10.19 Scheme of energy transfer upconversion $(^4F_{3/2}, {}^4F_{3/2}) \rightarrow (^4I_{13/2}, {}^4G_{7/2})$, $(^4F_{3/2},$ ${}^4F_{3/2}) \rightarrow (^4I_{11/2}, {}^4G_{9/2})$, $(^4F_{3/2}, {}^4F_{3/2}) \rightarrow (^4I_{9/2}, {}^2P_{1/2})$: full arrows, cross-relaxation; dashed arrows, subsequent electron–phonon relaxation.

by a common transfer rate. Upconversion by energy transfer has been observed in the de-excitation of several energy levels of other RE^{3+} ions too.

In several laser materials, self-de-excitation by energy transfer involves only one of these processes, such as down-conversion in the case of the level $^3H_4$ of Tm^{3+} or upconversion in the case of the levels $^4I_{13/2}$ and $^4I_{11/2}$ of Er^{3+}. However, in other cases, such as the $^4F_{3/2}$ level of Nd^{3+}, these processes act simultaneously and at each moment of time the balance is determined by the instantaneous fraction $r(t)$ of Nd ions excited into the level $^4F_{3/2}$. The temporal dependence of $r(t)$ is similar to that of the emission intensity $I(t)$, i.e.

$$r(t) = r(0) \exp\left(-\frac{t}{\tau_D}\right) \exp[-P'(t)], \tag{10.2}$$

the fraction of excited ions at the beginning of decay $r(0)$ being determined by the excitation intensity. Due to the simultaneous presence of down-conversion and upconversion, the energy transfer in the case of the $^4F_{3/2}$ level of Nd^{3+} corresponds to the presence of two types of acceptors, Nd^{3+} ions in the ground state, of relative concentration $(1 - r(t))C_{Nd}$, which act as acceptors in down-conversion, and excited Nd^{3+} ions of relative concentration $r(t)C_{Nd}$ which act as acceptors in upconversion [7, 172]. Since these two classes of acceptors can occupy the same type of lattice sites i, the direct energy transfer function can be written

$$P(t) = \sum_i \ln\left\{1 - C_{Nd} + [1 - r(t)]C_{Nd} \exp\left(-W_i^{dw}t\right) + r(t)C_{Nd} \exp\left(-W_i^{up}t\right)\right\},$$

$$\tag{10.3}$$

where W_i^{dw} and W_i^{up} are the down-conversion and respectively upconversion energy transfer rates. At low C_{Nd}, this transfer function can be written as the sum of the transfer functions for down-conversion and for upconversion, $P(t) \approx P^{dw}(t) + P^{up}(t)$, with

$$P^{dw}(t) = \sum_i \ln \left\{ 1 - [1 - r(t)]C_{Nd} + [1 - r(t)]C_{Nd} \exp\left(-W_i^{dw}t\right) \right\} \quad (10.4)$$

and

$$P^{up}(t) = \sum_i \ln \left\{ 1 - r(t)C_{Nd} + r(t)C_{Nd} \exp\left(-W_i^{up}t\right) \right\} \quad (10.5)$$

and these two equations are coupled via $r(t)$. Due to the temporal dependence of $r(t)$, the evolution of decay with the transfer function given by Eq. (10.3) cannot be described by a closed analytical function and must be calculated numerically. An important characteristic of $P(t)$, Eq. (10.3), is the dependence on the pump intensity. However, at very low pump intensities $r(0)$ is very small (below a percent) and the upconversion can be safely neglected.

The migration-assisted transfer at high Nd doping concentrations determines additional self-quenching down-conversion and upconversion. In the case of down-conversion the relative concentrations of acceptors for migration and for the donor–acceptor process are $C_A^{(DD)} = C_A^{(DA)} = [1 - r(t)]C_{Nd}$ and the averaged migration-assisted transfer rate becomes

$$\bar{W}^{dw} = \bar{W}_0^{dw}[1 - r(t)]^2 C_{Nd}^2, \quad (10.6)$$

whereas in the case of upconversion $C_A^{(DD)} = [1 - r(t)]C_{Nd}$ and $C_A^{DA} = r(t)C_{Nd}$ and

$$\bar{W}^{up} = \bar{W}_0^{up} r(t)[1 - r(t)]C_{Nd}^2. \quad (10.7)$$

Thus, due to the decrease in $r(t)$ during decay, $\bar{W}^{dw}$ increases steadily in time, whereas $\bar{W}^{up}$ is small at very large $r(t)$, then increases to a maximum value for $r(t)/r(0) = 0.5$ and decreases again for smaller $r(t)$. The effects are more pronounced for large fraction $r(0)$.

10.3.1.2 Emission decay in Nd-doped materials at low excitation intensities

By refining successively the measurements and interpretation of the emission decay of Nd:YAG crystals at low excitation intensity ($r(0)$ close to zero), it was found [173] that the experimental transfer function $P(t)$ corresponds to down-conversion self-quenching, assuming random discrete distribution of the Nd ions at the available lattice sites (Eqs. (2.42) with $C_A \approx C_{Nd}$). The transfer function for Nd:YAG crystals with C_{Nd} from 0.1 to 2.5 at.% Nd, measured with high temporal resolution after very short (10 ns) pulse excitation is very complex; however it shows several temporal intervals of distinct behavior.

- A very sharp drop in emission that ends practically within the first 2 μs of decay, which can pass unnoticed at reduced temporal resolution of detection (microsecond), but becomes obvious with a detection resolution of tens of nanoseconds. This drop

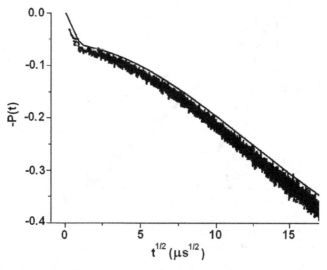

Figure 10.20 Experimental and calculated emission decay for 1.5 at.% Nd:YAG.

in emission corresponds well to $4C_{Nd}$ and, according to Section 2.2.1.8.2, it can be attributed to energy transfer inside the first (n.n.) Nd ion pairs, governed by very strong, short-distance interaction between the Nd ions, most probably superexchange, which determines very fast quenching of emission of these pairs.

- A quasi-linear temporal dependence with the rate W_{lin}, corresponding to the linear approximation of the transfer function $P(t)$, given by Eq. (2.44).
- A portion that can be described by a $t^{1/2}$ dependence, similar to the continuous distribution in the case of transfer determined by dipole–dipole interaction, which corresponds to the transfer function given by Eq. (2.40) and this enables estimation of the energy transfer microparameter. Using the absolute concentration of Nd as the density of acceptors, the value $C_{DA} = 1.8 \times 10^{-40}$ cm^6 s^{-1} was obtained for Nd:YAG. With this transfer microparameter, the rate W_{lin} calculated with Eq. (2.44) by summing over all the lattice c-sites available to the Nd ions around the c-site of the donor is about four times larger than the value measured experimentally. However, agreement can be restored by excluding the nearest neighbors from the summation and assuming that the short range interaction inside the n.n. Nd pair has a very small but finite effect on the n.n.n. pairs too. The exclusion of the first four acceptor sites from the summation in W_{lin} was justified by the fact that the transfer to these ions is practically exhausted in the fast initial $4C_{Nd}$ drop in emission. The fast initial decay enables evaluation of the superexchange parameters and the global transfer function (Eq. (2.42)) calculated with these superexchange parameters and with the electric dipole–dipole interaction parameter C_{DA} as well as the measured $P(t)$ is illustrated in Figure 10.20. Limitation of the multipole interactions between the Nd ions to the dipolar interaction is consistent with the selection rules for the transitions involved in this cross-relaxation, which exclude the quadrupole interactions.

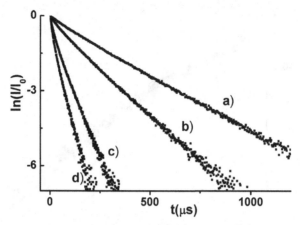

Figure 10.21 The concentration dependence of 300 K $^4F_{3/2}$ decays of: (a) 1 at.%, (b) 3.4 at.%, (c) 6.6 at.% and (d) 8.2 at.% Nd in YAG ceramic.

- A final portion with quasi-linear dependence on time, corresponding to migration-assisted transfer with the rate $\bar{W}$. Whereas the approximate transfer functions for the first three steps described above correspond to the direct donor–acceptor transfer and are linear in C_{Nd}, the final portion shows quadratic dependence on concentration, $\bar{W} = \bar{W}_0 C_{Nd}^2$, with $\bar{W}_0 = 240$ s^{-1} (at.% Nd)^2. Such dependence on time and concentration corresponds to a hopping mechanism of migration and to dipole–dipole interaction between the excited and the unexcited donors. Obviously, the passage from one portion to the next is gradual.

The emission decay of the various spectral satellites in Nd:YAG crystals under selective excitation confirms [4] the conclusion from global decay: the decay of the P_i satellites is similar to that of the unperturbed center N, whereas the decay of satellites M_i is much faster. For the pair satellite M_1 the de-excitation rate is $\sim 2.5 \times 10^6$ s^{-1} and is almost independent of C_{Nd}, whereas the emission decay of center M_2 is nearly exponential, with concentration-dependent lifetime of about 85 µs for 0.5 at.% Nd and 80 µs for 1 at.% Nd. The emission decay of Nd:YAG ceramics is similar to that observed in single crystals with the same Nd concentration. Moreover, extension of the studies to higher Nd concentrations which is made possible using ceramics [6, 8, 174] shows that the decay accelerates strongly (Figure 10.21) and can be described very well up to $\sim$6–7 at.% Nd with the transfer parameters inferred from the decay in diluted Nd:YAG crystals. However, at even higher Nd concentrations a slightly stronger acceleration of decay is observed, which could be attributed to the onset of new energy transfer processes.

The possibility of describing emission decay in Nd:YAG ceramics over a large range of concentrations with the same set of energy transfer parameters, assuming random distribution of the Nd^{3+} ions in the YAG lattice, suggests that although regions of enhanced Nd concentrations exist at the grain boundaries, their effect on the global decay is not large enough to alter its characteristics sizably and the possible effects could be hidden by the experiment. As discussed before, a similar conclusion can be drawn from the investigation of the relative intensities of the pair satellites M_i. A similar situation is observed in the case of Nd-doped GSGG [67] and CLNGG [61] crystals and ceramics: since the Nd concentration in these crystals can be much larger than in YAG, the comparison with ceramics can be extended to higher concentrations and it was found that the decay for these two types of materials is similar.

In the case of Nd-doped cubic sesquioxides, no detailed investigation of the self-quenching energy transfer processes has been reported for crystals. However, the dependence on Nd concentration of decay in Nd-doped cubic sesquioxide ceramics Y_2O_3 [68] and Sc_2O_3 [17] at low excitation densities shows similar features to the case of garnets, although the acceleration of decay by energy transfer is more evident. This can be determined by the higher density and tighter packing of the cationic sites and by larger energy transfer microparameters C_{DA} due to a more favorable energy level structure. As in the case of garnets, it was found that at low Nd concentrations the direct donor–acceptor down-conversion process is determined by mixed superexchange and dipole–dipole ion–ion interactions. However, the selection rules for the transitions involved in the cross-relaxation exclude the Nd^{3+} ions in the inversion sites of C_{3i} symmetry from acting as acceptors and thus only the ions in the C_2 sites are involved. The dipole–dipole energy transfer microparameters C_{DA} determined from the emission decay are $\sim 2.7 \times 10^{-39}$ cm^2 for Nd:Sc_2O_3 and $\sim 1.75 \times 10^{-39}$ cm^2 for Nd:Y_2O_3. With increased Nd concentrations the migration-assisted processes become evident and above 3 at.% in Y_2O_3 an additional static down-conversion mechanism becomes active. This was tentatively attributed to ion-pair cooperative interaction [132], as will be described later.

10.3.1.3 Emission decay in Nd-doped materials at high excitation intensities

At high excitation densities a stronger acceleration of the emission decay in Nd^{3+}-doped materials, including Nd:YAG single crystals [175, 176] and thin films [177], was observed at the early stage and this was attributed to upconversion. However, no specific investigation of upconversion processes in Nd-doped YAG ceramics was reported. In most of the papers dealing with upconversion in Nd-doped crystals or glasses, its effect on decay was accounted for assuming that this is the only cause for departure from exponential behavior, whereas the effect of down-conversion was either neglected or included as a constant contribution to the reduction in lifetime. Moreover, the upconversion was characterized by a transfer rate W^{up}, constant over all the decay. The evolution of the population of the level $^4F_{3/2}$ was described with a Bernoulli rate equation considering the transfer rate either to be identical to that for any acceptor ion (as in the case of the average-distance model), or as an average that could include the effect of direct and migration-assisted processes too,

$$\frac{dn_D}{dt} = -\frac{t}{\tau_D} - W^{up} n_D^2. \tag{10.8}$$

This equation is attractive, since it has an analytical solution and enables a simple evaluation of the upconversion rate. Equation (10.8) predicts the acceleration of decay at any moment of time. The upconversion rates inferred from these studies are quite large: analysis of the experimental data on thin films of Nd:YAG with Nd concentration up to 6 at.% indicates a global upconversion rate $W^{up} \approx 6.9 \times 10^3 \ \text{s}^{-1}$ for 1 at.% Nd:YAG [177], which is smaller by about 5.6 times than the value reported previously for single crystals [175]. It was also determined that the global upconversion rates are 5 to 16 times larger than the down-conversion rates.

Although Eq. (10.8) can provide an apparent fit of the experimental data for Nd:YAG, it does not reflect the actual picture of interactions and the upconversion transfer rate W^{up} has no clear physical meaning. As discussed above, owing to the complex energy level structure of Nd^{3+}, the effect on decay of upconversion cannot be treated independently from the effect of down-conversion. Both these processes involve direct donor–acceptor and migration-assisted transfer characterized by specific dependence on the excitation intensity and on time and, except for migration-assisted down-conversion at very low excitation intensity, none of these could be characterized by a constant rate. Moreover, since the concentration of acceptors for upconversion decreases, whereas that of acceptors for down-conversion and for migration increases during decay, the relative contribution of upconversion to de-excitation decreases, whereas that of down-conversion increases, and thus the balance is modified permanently in favor of down-conversion. In the case of the direct transfer processes, these circumstances can be accounted for with the rate equation (2.49) and with the transfer function (10.3) that accounts for the joint action of upconversion and down-conversion.

Even for systems where only upconversion is present, the existence of a particular configuration of acceptor ions at different distances around each donor would make the utilization of a unique transfer rate inadequate and this means that the rate equation (10.8) must be replaced by the rate equation (2.49) with the transfer function $P^{up}(t)$ (Eq. (10.5)). As discussed above, the instantaneous values of the acceptors for down-conversion and for upconversion are modified during decay and the corresponding transfer functions of the two processes have a different evolution. However, under the common action of both processes the global transfer function $P(t)$ (Eq. (10.3)) will be an ever increasing function of time since the upconversion, in conjunction with the electron–phonon interaction, prepares the system for down-conversion by returning the acceptor to its initial excited state and by increasing the concentration of Nd^{3+} ions in the ground state. This transfer function depends on the fraction of excited ions $r(0)$ and on the ratio of the energy transfer rates for upconversion and down-conversion $R_i = W_i^{up} / W_i^{dn}$. If $R_i < 1$ the transfer function in the presence of upconversion and down-conversion will have smaller values than when only down-conversion is present, and the opposite holds for $R_i > 1$. The migration-assisted transfer can be accounted for by including in the rate equation (2.49) the additional terms $(-\bar{W}^{dw} n_D - \bar{W}^{up} n_D)$, with $\bar{W}^{dw}$ and $\bar{W}^{up}$ given by Eqs. (10.6) and (10.7).

The rate equations that account for the joint action of upconversion and down-conversion do not have analytical solutions and can only be solved numerically, although on definite temporal intervals, such as the beginning and the end of decay, simplified approximate transfer functions can be used. It was thus found [172] that the decays reported in reference

[177] can be well described by numerical solution of the rate equation with the transfer function (2.46) assuming individual upconversion rates W_i^{up} about 20–30 times larger than the down-conversion rates W_i^{dw}. The increased upconversion rate compared with the down-conversion reflects a much larger transfer microparameter that could be determined by the higher superposition of the emission of the donor and absorption of the acceptor and by the larger variety of upconversion schemes. Nevertheless, despite the much higher upconversion rates, a consistent discussion of upconversion in the case of Nd^{3+} should account for the simultaneous presence of down-conversion [7].

10.3.2 Emission quantum efficiency

By definition, the emission quantum efficiency η_{qe} expresses the fraction of the excitation reaching the emitting level that is found in the luminescence emission, and is influenced by the joint effect of all non-radiative processes (electron–phonon relaxation, energy transfer). The emission quantum efficiency can be measured by comparing either the amount of luminescence or of the non-radiative processes, mainly heat generation, with the amount of exciting radiation absorbed by the system of active ions in the absence of laser emission. It can also be estimated by comparing the heat generation in the absence and in the presence of lasing, or from its effect on the laser emission threshold. As discussed in Chapter 2, all these measurements in fact provide the product $\eta_q = \eta_p \eta_{qe}$ referred to usually as the quantum efficiency; thus, η_{qe} can be extracted from η_q only if the pump level efficiency η_p is known and vice versa. It is then important to estimate these two efficiencies from independent measurements or considerations. The emission quantum efficiency can be calculated if the characteristics of the de-excitation processes are known from other types of experiments, such as the emission decay. On the other hand, the dense packing of energy levels above the emitting level $^4F_{3/2}$ of Nd^{3+}, which enables successive de-excitation by low order electron–phonon interaction, does not offer any physical reason to suspect lower than unity pump level efficiency η_p. In fact, until the advent of the model of dead sites [20–22], practically all studies on quantum efficiency in Nd-doped materials assumed $\eta_p = 1$. Measurement of η_p is difficult, however, it can be deduced from its effect on the pump saturation [178] or by comparing the measured η_q when pumping in various energy levels with the value obtained by pumping into the emitting level: no obvious evidence for $\eta_p < 1$ was reported.

An additional complication in measurement of the emission quantum efficiency can be introduced by accidental absorbing centers, such as the color centers or impurities, which can reduce the amount of excitation acting on the active ions; moreover, all excitation absorbed in these parasitic centers can be transformed completely into heat and this can alter the evaluation of the absorption and de-excitation processes of the active laser ions.

10.3.2.1 Measurement of emission quantum efficiency

The emission quantum efficiency of Nd:YAG has been measured by a very large variety of methods based either on direct measurement of the amount of emitted radiation [20, 22,

179, 180], of the rise in temperature under pumping [181, 182] or by various effects of heating (thermomechanical, thermo-optic and so on) by calorimetric interferometry [183], photoacoustic spectrometry [184], thermal lensing [185, 186], laser thermal depolarization [187] and thermal line broadening [188]. Most measurements refer to 1 at.% Nd:YAG crystals and a quite large spread of data has been reported, from 0.47 to almost 1, and there are cases when measurements on the same sample by different methods have provided quite different values. Obviously, such measurements could be influenced by model, experimental and calibration error or by uncertainty of various parameters used in interpretation of the data. Several authors [182, 183, 187, 188] report data for different Nd concentrations in YAG and η_q for 1 at.% Nd was around 0.8.

10.3.2.2 Calculation of emission quantum efficiency from the effect of non-radiative de-excitation on the emission decay

A parameter of major importance in evaluation of η_{qe} from the emission decay is the radiative lifetime τ_{rad} of the emitting level. This cannot be measured directly, but it can be calculated either from spectroscopic data (Eq. (2.12)) or, more accurately, by using the Judd–Ofelt parameters, according to Eq. (2.28). Calculation of τ_{rad} with Eq. (2.12) for the Nd:YAG crystal predicted 411 μs [179], and the measurements [189] on powder samples indeed provided a lifetime of 420 μs, but only 280 μs in crystals; the difference was attributed to defects introduced by the crystal growth process. However, it was shown [190] that the emission lifetime of Nd in YAG or Y_2O_3 powders depends on the size of powder, which determines the effective refractive index matching of the material with surrounding medium: it could be very long for measurement on fine powders in air (as large as 510 μs in YAG), but equal to the value in crystals for the powders immersed in index-matching liquid. The radiative lifetime for Nd:YAG calculated from the Judd–Ofelt analysis, 259 ± 20 μs [92], was confirmed by the low temperature luminescence lifetime of very diluted Nd:YAG crystals [171].

10.3.2.2.1 The effect of multiphonon de-excitation According to Eq. (2.34) the emission decay in the presence of multiphonon de-excitation with rate W^{nr} remains exponential, with the lifetime $\tau_f = \tau_{rad} [1 + \tau_{rad} W^{nr}]^{-1}$. The emission lifetime of level $^4F_{3/2}$ of Nd:YAG at different temperatures indicates W^{nr} values of several tens per second at 300 K and thus the emission quantum efficiency of Nd:YAG at very low C_{Nd} in the presence of multiphonon relaxation, $\eta_{qe}^{mp} = (1 + \tau_{rad} W^{nr})^{-1}$, is higher than 0.99.

10.3.2.2.2 The effect of energy transfer self-quenching: Nd:YAG emission quantum efficiency at low excitation intensity As discussed before, at early stages of investigation of Nd:YAG crystals the emission decay was considered nearly exponential and the lifetime measured after a certain lapse of time from the beginning of decay was around 230–235 μs for 1 at.% Nd. By using the radiative lifetime of 411 μs calculated from the spectroscopic properties of Nd:YAG a quantum efficiency of 0.56 was evaluated [189]. However, this

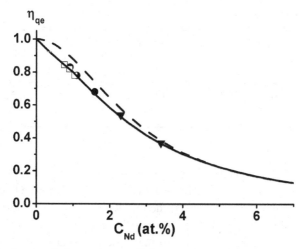

Figure 10.22 Calculated emission quantum efficiency at low excitation in Nd:YAG with the data derived from the emission decay: the model of discrete random distribution (thick line) and experimental data from interferometric calorimetry (squares), heat loading (circles) and thermal depolarization (triangles). The dashed line represents calculation using Eq. (2.53a).

value was later adjusted to ∼0.9 by using the 260 μs radiative lifetime calculated from the Judd–Ofelt analysis, and this new value is still largely accepted. The range of C_{Nd} in Nd:YAG was then extended using flux-grown crystals or thin films and the measured lifetimes were fitted with Eq. (2.54) corresponding to the average-distance model, although the critical Nd concentration N_0 reported by different authors varies in a wide range, from 3.8 to 10×10^{20} cm^{-3}. However, as shown above, none of these models describes correctly the emission decay of Nd:YAG.

The Nd:YAG emission quantum efficiency for low Nd concentrations (to ∼1.2 at.%) calculated using the model of discrete random distribution assuming only direct donor–acceptor transfer (Eq. (2.56)) with the energy transfer microparameters inferred from the emission decay of Nd:YAG crystals and with the lattice sum covering 102 coordination spheres l, indicates a parameter $b \approx 0.2$, which would correspond to $\eta_{qe} \approx 0.82$ for 1 at.% Nd. Inclusion of the migration-assisted transfer (Eq. (2.57)) leads to the emission quantum efficiency presented in Figure 10.22: for 1 at.% Nd, $\eta_{qe} \approx 0.8$, leading to an effective lifetime $\tau_{eff} \approx 208$ μs. For the diluted samples (to ∼ 1.2 at.% Nd) these data can still be described fairly well with the simplified Eq. (2.56) and with the effective parameter $b_{eff} \approx 0.223$. Figure 10.22 also includes the experimental data for η_q measured by interferometric calorimetry [183], heat load coefficient [182] and thermal depolarization [184]. The data obtained from pump induced thermal line broadening effects [188] correspond very well to the calculated values given in Figure 10.22. The very good fit of these experimental data for η_q with the calculated values of η_{qe} can then be explained assuming a pump level efficiency $\eta_p \approx 1$ and thus $\eta_q \cong \eta_{qe}$. The numerically calculated η_{qe} of Nd:YAG for C_{Nd} to 8 at.% is almost identical to that calculated with Eq. (2.58a). Figure 10.22 also

Table 10.17 *Emission quantum efficiency of Nd:YAG and Nd:GSGG at low excitation*

	C_{Nd} (%)										
	0.5	1	1.5	2	2.5	3	3.5	4	5	6	7
Nd:YAG	0.89	0.80	0.68	0.58	0.49	0.41	0.35	0.30	0.22	0.17	0.13
Nd:GSGG	0.91	0.82	0.73	0.65	0.57	0.50	0.44	0.38	0.30	0.24	0.20

shows the concentration dependence of η_{qe} calculated with Eq. (2.58b) using the critical concentration $C_0 \approx 4$ at.% Nd, determined by the migration-assisted parameter $\bar{W}_0 = 240\,\text{s}^{-1}$: it is obvious that this equation only describes the emission quantum efficiency well at high Nd concentrations, where the migration-assisted transfer dominates. The calculated values of emission quantum efficiency (to the closest second decimal) for several Nd concentrations in Nd:YAG and in Nd:GSGG under weak pump intensity are given in Table 10.17.

The emission quantum efficiency discussed so far refers to the whole system of doping Nd ions. However, it can also be calculated for the various structural centers corresponding to the resolved spectral satellites. Thus, in the case of Nd:YAG, η_{qe} for the n.n. Nd pair corresponding to the M_1 satellite is very low, $\sim 1.5 \times 10^{-3}$, and practically independent of C_{Nd}, whereas for the n.n.n. Nd pairs (the M_2 centers) it shows a slight dependence on concentration and is ~ 0.3 for 1 at.% Nd. At the same time, η_{qe} for the centers P_i is similar to that of the unperturbed centers N. These differences in η_{qe} between the various structural centers in Nd:YAG explain in a consistent way the differences between the high resolution absorption and emission spectra under non-selective excitation, particularly the absence of emission lines corresponding to the n.n. Nd pairs in emission under CW excitation.

The values of η_{qe} calculated with the energy transfer parameters derived from decay enable estimation of the thermal load coefficient in the absence of laser emission, $\eta_h^{(f)} = 1 - \eta_p \eta_{qe} \eta_{qd}^{(f)}$. For 1 at.% Nd and 808 nm pumping, using $\eta_{qe} = 0.8$ and assuming $\eta_p \approx 1$, the calculated $\eta_h^{(f)} = 0.377$ agrees well with the data measured in reference [182]. This shows that the heating of Nd:YAG under 808 nm pumping can be completely explained by self-quenching, and the value of the pump level efficiency $\eta_p \approx 0.9$, i.e. a fraction of 10% of Nd^{3+} ions in dead sites in 1 at.% Nd:YAG, inferred in [182], is caused by the use of the improper value $\eta_{qe} \approx 0.9$.

10.3.2.2.3 The effect of energy transfer self-quenching: Nd:YAG emission quantum efficiency at high excitation intensity The complex emission decay of Nd:YAG at high excitation, with permanent competition between upconversion and down-conversion in de-excitation of the level $^4F_{3/2}$, means that in this case it is not possible to find an analytical solution for the emission quantum efficiency. Numerically calculated values of η_{qe} for 1 at.% Nd:YAG using the model of discrete random distribution of Nd^{3+} ions for several

Table 10.18 *Calculated emission quantum efficiency of 1 at.% Nd:YAG for various fractions of excited ions* r*(0) and ratios* R *between the upconversion and down-conversion rates*

	$r(0)$					
R	~0	0.2	0.4	0.6	0.8	1.0
0, dw	0.802	0.812	0.822	0.832	0.843	0.854
10, up	1.000	0.960	0.924	0.893	0.863	0.834
30, up	1.000	0.932	0.874	0.828	0.791	0.756
1, dw+up	0.802	0.802	0.802	0.802	0.802	0.802
10, dw+up	0.802	0.780	0.761	0.745	0.729	0.713
30, dw+up	0.802	0.760	0.727	0.698	0.671	0.648

fractions of initially excited ions $r(0)$ and for various ratios $R = W_i^{up}/W_i^{dw}$ between the energy transfer rates corresponding to upconversion and those for down-conversion are given in Table 10.18.

Calculation of η_{qe} for Nd:YAG under strong excitation shows the following trends.

- In the absence of upconversion (down-conversion only, dw) η_{qe} increases with increasing initial excitation fraction $r(0)$; this is caused by the reduction of the concentration of acceptors for the down-conversion process (unexcited Nd^{3+} ions) at high excitation.
- In the absence of down-conversion (upconversion only, up) η_{qe} is close to unity for weak $r(0)$ (weak acceptor concentrations for upconversion), and decreases with increasing $r(0)$; this is more accentuated for higher values of upconversion rates. In the presence of upconversion only, η_{qe} could remain high at quite large upconversion rates and large $r(0)$: for Nd:YAG it would be larger, even at quite high R ratios (to ~20–30, depending on C_{Nd}), than the value corresponding to down-conversion only up to quite high $r(0)$ (0.5 to 0.8).
- When the upconversion and down-conversion transfer rates are equal ($R = 1$), η_{qe} in the presence of both processes is independent of $r(0)$ and equals the value calculated for the case of down-conversion only under weak excitation ($r(0) \approx 0$).
- For $R = W_i^{up}/W_i^{dc} > 1$, η_{qe} in the presence of both processes takes the value corresponding to down-conversion only at $r(0) \approx 0$ and decreases with increasing $r(0)$. The effect is accentuated for higher ratios R and higher C_{Nd}. For low C_{Nd} the reduction in η_{qe} is almost linear in $r(0)$, i.e. in excitation power, but becomes super-linear for higher concentrations.

The data from Table 10.18 show that consideration of upconversion processes only in the self-quenching of Nd can be misleading. Even for the ideal case of $r(0) = 1$, and for ratios of the transfer rates $R = 30$, the calculated reduction in emission quantum efficiency $(1 - \eta_{qe})$ for 1 at.% Nd and 1.5 at.% Nd inYAG is 0.352 and respectively 0.483, compared to 0.20

and 0.32 in the absence of upconversion; for realistic values of $r(0)$ this reduction will be considerably smaller. This points out that the reduction in η_{qe} in Nd-doped materials cannot be attributed exclusively to upconversion and that the upconversion parameters inferred in such an approach could be unreliable since they do not reflect a physical reality; this stresses the need for joint consideration of both processes of self-quenching. The strong effect of down-conversion in reducing the emission quantum efficiency in these laser materials is linked, as explained before, to the wastage of the unique absorbed quantum in each act of down-conversion, whereas the upconversion wastes only one of the two excitations of the interacting ions, since the acceptors return practically to their initial state.

10.4 Distribution of the doping ions in ceramics

The granular structure of ceramics determines the manifestation of the various compositional and crystallographic properties and of many physical properties at the level of each particular grain. Some of these properties can show quite non-uniform distribution inside the grains; moreover, some of them can show natural or induced dependence on orientation even in the case of cubic material. Of particular relevance is the segregation of the foreign ions at the grain boundaries, dependent on the difference in ionic radius from the host ions; obviously, this segregation is accompanied by depletion of these ions in the rest of the grain. The segregation will influence the variety and relative concentrations of ensembles of doping ions in near lattice sites. These ensembles should involve ions in defective sites too, and this could be expected to influence the structure of spectral satellites (variety, relative intensities) and the emission decay of the doping ions and the local distribution of the refractive index.

Although in the global spectroscopic measurements the averaging of the properties of the more concentrated and of the depleted regions could obscure such differences to a certain extent, specific manifestation of the high-degree ensembles at high segregation should still be evident and this would introduce differences compared with crystals. However, the satellite structures in the high resolution global absorption spectra of Nd:YAG ceramics, as well as the emission decay did not reveal [6–8, 25] obvious departure from the predictions of a global random distribution model up to high Nd concentrations and this could be indicative of the limited global extent of segregation. Moreover, the laser performance and the global heat generation of the ceramics did not prove inferior to that of the corresponding crystals. Nevertheless, such departures from random distribution could be hidden by the inaccuracies of the global measurements; the possibility that the variety of centers determined by segregation would determine a broad, less structured background of the lines should not be excluded, although this has not been shown. Such a situation requires high resolution spatially resolved spectroscopic investigation in conjuction with microscopic methods of elemental analysis and consistent correlation with the data of global investigation.

The investigation of coarse-grained ceramics by scanning confocal microscopic Raman and luminescence spectroscopy [31, 36, 191] revealed sharp differences between the intensity and structure of the 869 nm emission line $^4F_{3/2}(2) \rightarrow {}^4I_{9/2}(1)$ of Nd^{3+} near the grain boundaries and far from the grain boundaries. The minimum of emission was observed at the grain boundaries and was dependent on C_{Nd} (reduction by $\sim$3% compared with the regions far from the boundaries for 1 at.% Nd) and the width of the drop in intensity profile was $\sim$2 μm, much larger than the size of the exciting spot (335 nm) [191]. This drop in emission intensity was attributed to decreased η_{qe} by enhanced self-quenching determined by higher C_{Nd} at grain boundaries: for the 1 at.% Nd ceramic this would correspond to enhancement by $\sim$0.05 at.% Nd. The difference between the profile of the emission line close or far from the grain boundary is manifested by the presence of several spectral satellites ($+21.15$ cm^{-1}, $+6.6$ cm^{-1} and -9.3 cm^{-1} from the line peak at 11505 cm^{-1} for transition $^4F_{3/2}(2) \rightarrow {}^4I_{9/2}(1)$), whose relative intensities increase with doping. The most shifted satellite was attributed [191] to the n.n. pair satellites M_1 and the other two to the n.n.n. pair satellites M_2 and it was concluded that they reflect increased C_{Nd} at the grain boundaries. Observation of a reduction in emission lifetime by $\sim$10 μs, i.e. by $\sim$5%, in this region was used as an additional argument. From these data it was concluded that about 20% of the volume of the ceramic grains in the coarse-grained ceramics is affected by Nd segregation. However, no variations in concentration of this extent were detected by elemental analysis with energy dispersive spectrometry (EDS), secondary ion mass spectroscopy (SIMS) and electron probe microanalysis (EPMA) [191], although further investigation by EDS [36] reports C_{Nd} increased by 25% in a 33 nm wide area at grain boundaries. Similar local modifications of emission were observed in fine-grained Nd:YAG ceramics, but to a much smaller extent.

Despite the observed local modification of spectroscopic properties at the grain boundaries in Nd:YAG ceramics, these data raise several points of concern. Thus, the region influenced by increased Nd concentration evidenced by luminescence is much larger, and the maximum modification of the emission intensity is much smaller than for the usual impurity segregation processes in ceramics. This could be influenced by the resolution of the experiment or by variation of the refractive index, as well as by possible back-diffusion of Nd. The shifts of the spectral satellites from the main line observed in emission are much larger than the usual shift of the pair satellites for Nd:YAG, and do not correspond to the structure of satellites in the absorption line $^4I_{9/2}(1) \rightarrow {}^4F_{3/2}(2)$. Moreover, as discussed in Section 10.3.2.2.2, the emission quantum efficiency for the n.n. pair satellite M_1 is close to zero so it practically does not show in the CW excited emission. Thus it is tempting to attribute at least part of the observed satellites to defective sites or to Nd centers perturbed by the presence of structural defects near the Nd ion, such as increased concentrations of Si from the sintering aid at the grain boundaries as revealed by SIMS [192] or EDX [193].

Increased doping concentration at grain boundaries due to segregation was also observed by imaging confocal spectroscopy (ICS) and EDS in Ce-doped $(Y,Gd)_3Al_5O_{12}$ [194] and YAG ceramics [195]. In contrast to the Nd-doped ceramics, where the reduction of emission at grain boundaries was attributed to increased self-quenching, in the case of Ce^{3+}, which does not show self-quenching, the emission intensity in ICS is larger at the grain boundaries as a result of the segregation confirmed by EDS. Investigation of Yb-doped YAG and Y_2O_3

ceramics [196] by these techniques did not reveal differences between the grain boundaries and the rest of the grain: it was thus inferred that the segregation behavior of different RE^{3+} ions (Ce^{3+}, Nd^{3+}, Yb^{3+}) parallels the segregation coefficient in the melt growth of doped garnets. However, energy dispersive X-ray spectroscopy revealed a 50% increase in Yb concentration at the grain boundaries in $Yb:CaF_2$ ceramics [49].

10.5 Conversion of excitation in doped ceramics

Besides phonon relaxation, the excitation of the excited levels can be converted into the excitation of other levels by radiative processes (upconversion excited state absorption, or down-conversion by cascade emission) or by energy transfer inside the system of absorbing ions or to other species of ions. There is a large variety of manifestations of these processes and their management imposes specific requirements on the laser material.

10.5.1 Sensitized conventional emission

Sensitization of emission can be an effective means of excitation of weakly absorbing ions by energy transfer from strongly absorbing ions. Efficient sensitization assumes good packing of the activator sites around the sensitizer and large transfer microparameters, i.e. good superposition of donor emission and acceptor absorption. The efficiency of the direct transfer increases at high acceptor concentrations, and at high donor concentrations it can be supplemented by migration-assisted transfer. A problem of concern is the back transfer from the activator to the sensitizer, determined by the particularities of their optical spectra. The enhanced compositional and doping capabilities of ceramic materials enable optimization of the conditions for controlled sensitization.

10.5.1.1 Sensitization of Nd emission

Sensitization of RE^{3+} emission, with fairly weak and narrow absorption lines in the visible, by energy transfer from strongly absorbing 3d ions, was considered for a long time as a way of increasing the efficiency of traditional lamp pumping. The advent of strong and efficient CW diode lasers that enable resonant pumping in levels close to the emitting level reduced the relevance of CW lamp pumping, although it can still be important for high power pulsed pumping. Another area of relevance for absorption of visible radiation is solar pumping. Utilization of visible pumping for these infrared lasers would require laser materials with high heat conduction and which can be produced in large sizes, and in this respect garnet ceramics could be the materials of choice.

Trivalent chromium enters in the octahedral sites of garnets and has strong and broad absorption bands in the visible dominated by the spin-allowed transitions from the singlet state $^4A_2$ to the triplet states $^4T_2$ and $^4T_1$. In YAG the crystal field acting on Cr^{3+} is strong and places ^2E as the lowest excited level, at 616 cm^{-1} below $^4T_2$. At low temperatures its emission spectrum is dominated by the weak, spin-forbidden transition $^2E \rightarrow {}^4A_2$, and it consists mainly of the sharp R_1 (687.3 nm) line with phonon side-bands at 704 nm and 723 nm. This emission has very poor superposition on the absorption in the high energy side of the levels ($^4F_{7/2}$, $^4S_{3/2}$) of Nd^{3+}, determining a small Cr → Nd energy transfer microparameter C_{DA}. Increasing the temperature determines thermallization of the levels ^2E and $^4T_2$, leading to temperature dependent Cr^{3+} emission: the lifetime shortens and a broad emission band develops at longer wavelengths (to the 800 nm range), increasing the superposition with the Nd ($^4F_{7/2}$, $^4S_{3/2}$) and ($^4F_{5/2}$, $^2H_{9/2}$) absorption bands [197]. The microparameter C_{DA} is larger in Cr–Nd co-doped garnets with larger sites available to Cr, such as the Sc garnets GSGG or YSGG: the $^4T_2$ level is lowered close to ^2E and their thermallization is much stronger, leading to more intense broad-band emission of Cr, with strong superposition on the Nd absorption. Efficient lamp pumped (Cr, Nd):GSGG lasers were demonstrated [198]; unfortunately, the smaller heat conduction of this material limits its practical use.

The Cr^{3+} and Nd^{3+} concentrations in YAG crystals are usually below 1 at.% and Cr → Nd ET is restricted to the low efficiency direct process. The emission decay suggests that the ion–ion coupling is mainly d-d, although the faster decay at early times could be indicative of superexchange coupling too. Increasing C_{Nd} to enhance the transfer efficiency is impractical since it accelerates self-quenching. However, investigation of (Cr,Nd):YAG crystals with higher Cr concentrations revealed the onset of migration-assisted transfer [199]. The high Cr concentrations available in the YAG ceramics could favor this type of transfer, renewing the prospect of efficient sensitization of Nd emission [200, 201].

The emission decay of Cr^{3+} in YAG at very small (0.1 at.%) Cr concentration is exponential, with ~1.8 ms lifetime at 300 K, much longer than that of Nd^{3+}. Co-doping with Nd accelerates the decay of Cr; however, for 1 at.% Cr and 1 at.% Nd the effective emission lifetime of Cr^{3+} is still longer than that of directly excited Nd^{3+} emission. Thus, according to Section 2.2.1.8, the sensitized decay of Nd^{3+} after short pulse excitation in Cr^{3+} will show a rising part, determined by the intrinsic lifetime of Nd^{3+}, the maximum of emission is reached at quite a long time after excitation, and the decay part corresponds to the residual lifetime of Cr^{3+}. Such behavior is typical for incomplete donor–acceptor transfer; however, even in this situation laser performance can be superior to the direct pumping in Nd^{3+}, but this is caused by the strongly increased pump absorption in Cr. Nevertheless, complete use of the radiation absorbed by Cr^{3+} for laser emission would require optimization of the transfer, particularly of the Nd and Cr concentrations. Increased transfer efficiency by migration assistance for high Cr concentrations was demonstrated for (5 at.% Cr, 1 at.% Nd):YAG, where the Nd emission rise-time was shortened dramatically and the final decay part became very close to that of YAG doped with Nd only [199]. Enhanced Cr^{3+} → Nd^{3+} energy transfer at higher Cr concentrations was also shown by increased Nd luminescence intensity [202].

Another possibility for sensitization of Nd emission under short wavelength pumping is co-doping with Ce^{3+} whose emission in the range 500–650 nm shows good superposition with the intense Nd^{3+} absorption lines $^4I_{9/2}$ → ($^4G_{7/2}$, $^2G_{9/2}$, $^2K_{13/2}$), grouped around 530 nm, and $^4I_{9/2}$ → ($^2G_{7/2}$, $^4G_{9/2}$) around 590 nm [203–205]. The very strong absorption bands around 460 nm and 240 nm limit the Ce^{3+} concentrations necessary for efficient pump absorption to several tenths of a percent. Although in this system the energy transfer rate must be very high to compete with the very fast Ce^{3+} intrinsic de-excitation, very efficient $Ce \rightarrow Nd$ energy transfer has been demonstrated. A further approach, of use in the case of broad-band pumping sources, could be sensitization of Nd^{3+} emission by joint (Cr^{3+}, Ce^{3+}) co-doping [205].

10.5.1.2 Sensitization of Er emission

Sensitization with Cr^{3+} was a way of exciting the Er^{3+} infrared emitting levels $^4I_{11/2}$ and $^4I_{13/2}$ under lamp pumping before the advent of diode lasers in the 950 nm range. The direct absorption of this radiation by Er^{3+} is weak; however, sensitization with the strongly absorbing Yb^{3+} ion, whose unique excited manifold lies very close to the $Er^{3+} \, ^4I_{11/2}$ manifold, could populate this level [206]. At high Yb concentrations the efficiency of transfer is increased by migration among the Yb^{3+} ions. Spectroscopic investigation of (Yb,Er) co-doped YAG or Y_2O_3 ceramics, reveals [207–209] properties similar to those of single crystals. In the case of solid solution mixed $(Y,La)_2O_3$ ceramics, considerable broadening of the 1.5 μm emission was observed [209], with potential for short pulse laser emission.

Due to the fairly long lifetime of the level $^4I_{11/2}$ of Er^{3+} in many materials, sensitization with Yb could have two major drawbacks: back transfer from Er to Yb and sensitized upconversion from this level. This can limit the utility of this sensitization scheme for the 3 μm $^4I_{11/2} \rightarrow \, ^4I_{13/2}$ laser emission. However, in the case of the 1.5 μm $^4I_{13/2} \rightarrow \, ^4I_{15/2}$ laser emission, efficient de-excitation of $^4I_{11/2}$ to $^4I_{13/2}$ can be obtained by co-doping with Ce^{3+} and using the phonon-assisted cross-relaxation $(^4I_{11/2}(Er), \, ^2F_{5/2}\,(Ce)) \rightarrow (^4I_{15/2}(Er), \, ^2F_{7/2}\,(Ce))$. Such an approach could be difficult in the case of hosts with small cationic sites such as the sesquioxides, and this would recommend direct diode laser pumping into the level $^4I_{13/2}$.

10.5.1.3 Sensitization of Yb emission

A shortcoming of Yb^{3+} is the lack of absorption in the visible, where strong pulse pumping sources exist, and the quite low absorption in the infrared (cross-sections $\sim 10^{-20}$ cm^2 on sharp peaks). A possible sensitizer for Yb^{3+} is the better absorbing Nd^{3+} ion, by the energy transfer $[(^4F_{3/2})_{Nd}, (^2F_{7/2})_{Yb}] \rightarrow [(^4I_{9/2})_{Nd}, (^2F_{5/2})_{Yb}]$. $Nd \rightarrow Yb$ energy transfer has been reported in different materials, but the efficiency was usually below 70%, with few exceptions going to higher values, such as 85%. The spectroscopic properties recommend the garnets and sesquioxides as materials with potential for efficient $Nd \rightarrow Yb$ transfer and the ceramic techniques can offer the possibility of optimizing the laser material [17, 69, 99, 210].

The dependence of the superposition of Nd emission with the Yb absorption of the host is shown in Figure 10.23. In YAG, only two Nd emission transitions are involved in transfer, in Y_2O_3 there are five transitions and in Sc_2O_3 there are eight. The increased superposition in this group of hosts determines a corresponding increase in the transfer microparameter

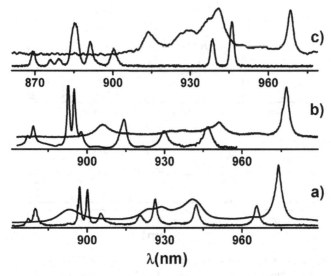

Figure 10.23 Overlap of the Yb^{3+} absorption (upper curves in every layer) with $Nd^{3+}\,{}^4F_{3/2} - {}^4I_{9/2}$ emission at 300 K in (Nd, Yb) co-doped samples of (a) Sc_2O_3, (b) Y_2O_3 and (c) YAG ceramic.

C_{DA}. In these materials the lowest Stark level of the ${}^4F_{3/2}$ manifold of Nd^{3+} is higher in energy than the lowest Stark level of the $Yb^{3+}\,{}^2F_{5/2}$ manifold and this reduces the back Nd-to-Yb transfer.

The Nd → Yb energy transfer at 300 K is confirmed by the Yb^{3+} steady state emission under excitation in Nd^{3+}, as illustrated in Figure 10.24 for (0.1 at.% Nd, x at.% Yb):Sc_2O_3 and (1 at.% Nd, x at.% Yb):Y_2O_3 ceramics with $x = 0$, 1 and 3. With increasing C_{Yb}, the emission of Nd weakens while that of Yb is enhanced; the transfer diminishes, but it is still strong at 10 K. A similar situation holds in (Nd,Yb)-doped YAG ceramic. In contrast to YAG, where the Nd^{3+} and Yb^{3+} ions occupy a unique crystallographic site, in sesquioxides they can occupy sites of C_2 and C_{3i} symmetry: whereas the excitation in the Nd^{3+} ions is made almost exclusively in the C_2 centers, the Yb^{3+} absorption is electric dipole allowed in the C_2 centers and magnetic dipole allowed in both centers and thus the energy transfer from the center C_2 of Nd^{3+} can take place to both these Yb^{3+} centers. This was revealed by the emission of both Yb^{3+} centers under excitation in Nd^{3+} in (Nd,Yb):sesquioxide ceramics at 10 K.

The Nd → Yb transfer is confirmed by the C_{Yb} dependent acceleration of Nd decay after short pulse (10 ns) excitation at 532 nm, as shown in Figure 10.25 for Nd:YAG and by the specific evolution of the emission of the Yb ions. The Nd → Yb transfer competes with the Nd self-quenching, and analysis of the decay showed that the direct Nd → Yb transfer microparameter $C_{DA(Nd \to Yb)}$ increases along the series YAG–Y_2O_3–Sc_2O_3 and is larger than $C_{DA(Nd \to Nd)}$ by ∼12 times in YAG and ∼4 times in sesquioxides. Since in these materials C_{Yb} was much larger than C_{Nd}, the Nd → Yb transfer dominates the self-quenching; its efficiency is 96% and >98% in (0.1% Nd, x% Yb):Sc_2O_3 with $x = 3$

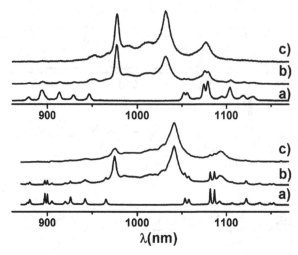

Figure 10.24 The 300 K emission under non-selective excitation of Nd^{3+} (in the visible) of (0.3 at.% Nd, x at.% Yb):Sc_2O_3 ceramic (lower layer) and (1 at.% Nd, x at.% Yb):Y_2O_3 ceramic (upper layer), with variable Yb^{3+} concentrations x: (a) $x = 0$, (b) $x = 1$, (c) $x = 3$ at.%.

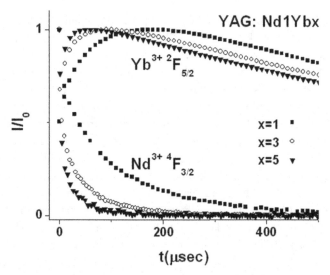

Figure 10.25 300 K $Nd^{3+}\,^4F_{3/2}$ (lower curves) and $Yb^{3+}\,^2F_{5/2}$ (upper curves) emission decays of Nd_1Yb_x ($x = 1, 3, 5$):YAG under 532 nm Nd^{3+} excitation.

and 4, 93% and 98% in (1% Nd, x% Yb):Y_2O_3 with $x = 3$ or 5, and 87% and 95% in (1% Nd, x% Yb):YAG with $x = 3$ or 5.

Sensitization of Yb emission with broad-band absorbing ions such as Cr^{3+} is very tempting for the generation of high energy short or ultra-short pulses under flashlamp excitation. Such sensitization would assume emission of the sensitizer in the near infrared range, i.e.

materials with weak crystal field at the Cr^{3+} site. Efficient transfer was demonstrated from the Cr^{3+} center I in Sc_2O_3 ceramic [211], but in this case the absorption of center II in the region of Yb emission could introduce large losses. Alternatively, sensitization of Yb emission could be obtained by two-step energy transfer in (Cr, Nd, Yb)-doped materials.

10.5.2 Down-conversion and quantum splitting of excitation

Since the down-conversion by cross-relaxation on intermediate levels transforms the excitation of the excited donor level into two excitations, the final state of the donor and that of the acceptor, the process is often called quantum cutting (splitting) of excitation, especially when these final states are metastable or de-excite subsequently to metastable states and can give emission of two lower energy quanta. The acceptor can be another ion or an ensemble (mostly pairs) of strongly coupled ions in near sites, and in this case the acceptor act of transfer involves cooperative optical processes by which the energy lost by the donor determines simultaneous excitation of both ions from the acceptor pair. Whereas in d-d ion–ion transfer the transfer function $P(t)$ shows a linear dependence on the acceptor concentration and $t^{1/2}$ dependence at long times, for cooperative ion-pair transfer $P(t)$ shows a squared dependence on acceptor concentration and $t^{1/3}$ dependence [212].

10.5.2.1 Energy transfer down-conversion in Nd^{3+}-doped ceramics

The down-conversion of excitation by energy transfer from the emitting level $^4F_{3/2}$ of Nd^{3+} in garnet or cubic sesquioxide ceramics was described in Section 10.3 and it was shown that it leads to specific C_{Nd} dependent loss of excitation and enhancement of heat generation. At low to medium C_{Nd} this down-conversion is determined by ion–ion electric dipole and superexchange interactions, with increasing contribution of the migration-assisted processes. However, the emission decay at high C_{Nd} (larger than $\sim$7 at.% Nd in YAG [8, 174] and $\sim$3 at.% Nd in Y_2O_3 [68]) shows faster acceleration than predicted by the ion–ion transfer and reveals additional energy transfer paths of de-excitation. The analysis of these effects [133] suggests that this additional de-excitation could be the cooperative energy transfer processes $[^4F_{3/2}, (^4I_{9/2}, {}^4I_{9/2})] \rightarrow [^4I_{13/2}, (^4I_{13/2}, {}^4I_{13/2})]$ or $[^4I_{9/2}, (^4I_{15/2}, {}^4I_{15/2})]$, and in the case of sesquioxides only the Nd^{3+} ions in the C_2 sites can act as acceptors.

10.5.2.2 Energy transfer down-conversion in Er^{3+}-doped ceramics

The green emission $(^4S_{3/2}, {}^2H_{11/2}) \rightarrow {}^4I_{15/2}$ of Er^{3+} concentration quenching is attributed to the ion–ion down-conversion cross-relaxation $[(^4S_{3/2}, {}^2H_{11/2}), {}^4I_{9/2}] \rightarrow (^4I_{9/2}, {}^4I_{13/2})$. However, in highly concentrated Er:YAG crystals [213] a new de-excitation becomes active and the additional term in the transfer function shows temporal and concentration dependence specific to cooperative cross-relaxation, such as $[^2H_{11/2}, (^4I_{15/2}, {}^4I_{15/2})] \rightarrow [^4I_{13/2}, (^4I_{13/2}, {}^4I_{13/2})]$. A similar situation was observed in the case of Er:Sc_2O_3 ceramics [133], whose emission decay shows strong concentration dependence (Figure 10.26). Although

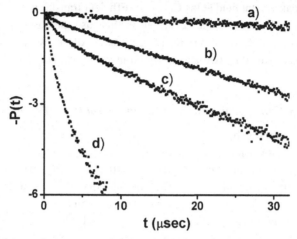

Figure 10.26 The $^4S_{3/2}$ Er^{3+} transfer function in Sc$_2$O$_3$ for various concentrations: (a) 0.3 at.%, (b) 1 at.%, (c) 3 at.% and (d) 10 at.% Er.

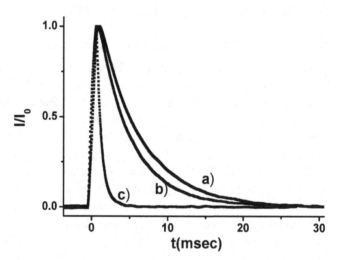

Figure 10.27 Room temperature emission dynamics of $^4I_{13/2}$ emission of (a) 1 at.% Er^{3+}, (b) 3 at.% Er^{3+} and (c) 10 at.% Er^{3+} in Sc$_2$O$_3$ ceramics under pumping in the $^4S_{3/2}$ level.

the pumping into the ($^4S_{3/2}$, $^2H_{11/2}$) levels only excites the ions in C_2 sites of the Sc$_2$O$_3$ lattice, ions from both C_2 and C_{3i} sites can act as acceptors in ion–ion as well as in ion-pair cooperative processes. The efficient energy transfer is shown by the temporal evolution of emission from the $^4I_{13/2}$ level under short pulse excitation at 532 nm (Figure 10.27): the rise part is determined by the de-excitation of the ($^4S_{3/2}$, $^2H_{11/2}$) levels, whereas the decay part is determined by the de-excitation of the longer lived $^4I_{13/2}$ level. The shortening of the rise-time of emission with the Er concentration reveals the increase in transfer efficiency, whereas the acceleration of the decay part is caused by the increased efficiency of upconversion from the $^4I_{13/2}$ level.

10.5.2.3 Energy transfer down-conversion in Tm^{3+}-doped ceramics

The 2 μm laser emission $^3F_4 \rightarrow {}^3H_6$ of Tm^{3+} under visible or near infrared 800 nm excitation in the level $^3H_4$ is based on the quantum splitting ion–ion cross-relaxation process $(^3H_4, {}^3H_6) \rightarrow (^3F_4, {}^3F_4)$. By this process an excitation in the level $^3H_4$ is split into two excitations of the 2 μm laser emitting level $^3F_4$ and this process was used for laser emission in Tm:YAG crystals and recently in ceramics [129]. Tm-doped cubic sesquioxides offer conditions for efficient quantum splitting due to the high density and tight packing of the cationic sites. However, only the Tm^{3+} ions in the C_2 sites can be excited by absorption in the $^3H_4$ level, and only ions in these sites can act as acceptors in the cross-relaxation process, similar to Nd^{3+}. The emission spectra and decay of 1 at.% $Tm:Sc_2O_3$ ceramics by pumping into or above $^3H_4$ revealed that although after a certain lapse of time from the beginning of decay the energy transfer function corresponds to d-d interaction and has $t^{1/2}$ dependence, at early times the d-q interaction is also active and determines fast de-excitation [130]. On increasing the Tm concentration, the decay accelerates faster than is predicted based on ion–ion cross-relaxation, the difference showing $t^{1/3}$ and Nd concentration squared dependence; this was tentatively attributed [133] to the onset of a cooperative ion-pair cross-relaxation process $[^3H_4, (^3H_6, {}^3H_6)] \rightarrow [^3H_6, (^3F_4, {}^3F_4)]$.

10.5.2.4 Sensitized energy transfer quantum cutting in ceramics

Cooperative quantum cutting in systems with ions emitting in the green was observed in several ceramic materials by the acceleration of emission decay when co-doping with Yb^{3+}. The acceptors in this process are strongly coupled pairs of Yb^{3+} ions in nearest lattice sites; at the high Yb concentrations required for large concentrations of pairs, migration on the Yb ions could also favor the inverse process, cooperative sensitization from Yb. In the case of (Ce^{3+}, Yb^{3+}) co-doped YAG ceramics, infrared emission of Yb was observed by pumping in the ultraviolet [214, 215] as a result of cooperative energy transfer from the $5d_1$ level of Ce^{3+} to a strongly coupled pair of Yb^{3+} ions, $[(5d_1)_{Ce}, (^2F_{7/2}, {}^2F_{7/2})_{Yb}] \rightarrow [(^2F_{5/2})_{Ce}, (^2F_{5/2}, {}^2F_{5/2})_{Yb}]$; the inverse process, cooperative sensitization of Ce^{3+} by Yb^{3+}, was also observed.

The emission $^5D_4 \rightarrow {}^4F_6$ of Tb^{3+} in the blue–green region shows good superposition with the convolution of the Yb^{3+} absorption spectrum, suggesting the possibility of cooperative energy transfer, $[(^5D_4)_{Tb}, (^2F_{7/2}, {}^2F_{7/2})_{Yb}] \rightarrow [(^5F_6)_{Tb}, (^2F_{5/2}, {}^2F_{5/2})_{Yb}]$. The tight packing of the cationic sites in sesquioxides could favor such cooperative processes and indeed they were observed in (Tb, Yb) co-doped Y_2O_3 phosphors [216] and Sc_2O_3 ceramics [133]. However, similar to the (Ce^{3+}, Yb^{3+}) systems, this cooperative down-conversion process competes with the cooperative upconversion-sensitized emission of Tb.

10.5.3 Upconversion

The upconversion can involve one or several (n) steps, so the final state can be much higher in energy than the exciting quantum and the power of the short-wavelength emission from this state is non-linear with respect to the excitation absorbed from the pump radiation, $P_{em} \propto (P_{abs})^n$; emission can also be generated from all intermediate (reservoir) levels. However, in other cases, such as the one-step energy transfer upconversion from the level $^4F_{3/2}$ of Nd^{3+}, discussed above, the final state of upconversion can be de-excited almost completely back to the reservoir level by electron–phonon interaction. The upconversion can be caused by ESA, or by various types of intrinsic or sensitized energy transfer, such as step-by-step transfer, cooperative upconversion and cooperative-sensitized excitation, or by a combination of such processes as in the case of photon avalanche. Whereas the ESA can be at the same or at a different wavelength with respect to the radiation that prepares the reservoir state, the energy transfer processes require a unique wavelength of external excitation. In many cases the ESA and energy transfer upconversion can co-exist and compete. Owing to the rich energy level structures, with many coincident energy gaps and with several metastable states, the RE^{3+} ions can show a variety of upconversion processes. The most investigated are Ho^{3+}, Er^{3+} and Tm^{3+} which can give emission in the blue, green or red [217].

10.5.3.1 Upconversion in Ho-doped ceramics

Ho^{3+} can give emission in the visible but, because of quite poor absorption, direct excitation is difficult. However, the emitting levels can be excited by Yb sensitized upconversion. In a first step, Yb^{3+} can sensitize the level $^5I_6$, from which the excitation can be upconverted by another step of transfer or by ESA to upper levels. By pumping at 980 nm in $(Yb, Ho)_2O_3$, bright green emission ($^5F_4, {}^5S_2) \to {}^5I_8$ at 548 nm was observed, accompanied by emission at 753 nm on transitions ($^5F_4, {}^5S_2) \to {}^5I_7$ and $^5F_5 \to {}^5I_8$ and by much weaker emission lines at higher energies [218]. Ultraviolet and violet upconversion in Ho-doped Y_2O_3 ceramic was observed by 532 nm excitation in the ($^5F_4, {}^5S_2$) levels [219].

10.5.3.2 Upconversion in Er-doped ceramics

In addition to the energy transfer upconversion from the levels $^4I_{13/2}$ and $^4I_{11/2}$ discussed before, the energy level scheme of Er^{3+} can accommodate a variety of energy transfer (step-by-step or cooperative) and/or ESA processes that can upconvert the excitation to levels placed in the ultraviolet. Strong emission in the green around 523 nm ($^2H_{11/2} \to {}^4I_{15/2}$) and 559 nm ($^4S_{3/2} \to {}^4I_{15/2}$) and in the red, 670 nm ($^4F_{9/2} \to {}^4I_{15/2}$), was observed by pumping in the 940–980 nm range in Er- or (Yb, Er)-doped YAG [220, 221] or sesquioxide (Y_2O_3 [222], Sc_2O_3 [223]) ceramics. The balance between the green and red emission can be changed by increasing C_{Er} which modifies the balance between the energy transfer down-conversion processes that de-populate ($^2H_{11/2}, {}^4S_{3/2}$) and those that populate the $^4F_{9/2}$ level from upper levels. Pumping at various wavelengths around 800 nm in $^4I_{9/2}$ ceramic revealed the effect of wavelength on the competition of ESA and ETU in $Er:Sc_2O_3$

ceramic: although the excitation from $^4I_{9/2}$ relaxes rapidly to $^4I_{11/2}$, quite strong, wavelength dependent ESA to the level $^2H_{9/2}$ from which it relaxes to levels ($^2H_{11/2}$, $^4S_{3/2}$) takes place, and this can compete with the ETU ($^4I_{11/2}$, $^4I_{11/2}$) → ($^4I_{15/2}$, $^4F_{7/2}$) in exciting the green emission. Upconversion to very high energy levels can take place by ETU from the ($^2H_{11/2}$, $^4S_{3/2}$) levels and the effect can be enhanced by additional ESA under 532 nm pumping. The efficient ET excitation of the $^4I_{11/2}$ level of Er^{3+} by ET from Yb can be raised to high energy Er levels [224] by further transfer from Yb^{3+}, by step-by-step (up to five steps) or cooperative sensitization, especially at high C_{Yb}.

10.5.3.3 Upconversion in Tm-doped ceramics

Direct pumping of the excited level $^1G_4$ of Tm^{3+} which gives bright blue emission in the 480 nm range from the transition $^1G_4 \rightarrow {}^3H_6$ is difficult, and the energy level structure of Tm^{3+} does not allow efficient ETU. Sensitization of this emission by co-doping with Yb was observed in various materials under 940–980 nm pumping, including Y$_2$O$_3$ [225] and Sc$_2$O$_3$ [136] ceramics, and is based on three successive steps of transfer, [($^2F_{5/2}$)$_{Yb}$, ($^3H_6$)$_{Tm}$] → [($^2F_{7/2}$)$_{Yb}$, ($^3H_5$)$_{Tm}$], followed by relaxation to $^3F_4$, [($^2F_{5/2}$)$_{Yb}$, ($^3F_4$)$_{Tm}$] → [($^2F_{7/2}$)$_{Yb}$, ($^3F_2$, $^3F_3$)$_{Tm}$], followed by relaxation to $^3H_4$, and [($^2F_{5/2}$)$_{Yb}$, ($^3H_4$)$_{Tm}$] → [($^2F_{7/2}$)$_{Yb}$, ($^1G_4$)$_{Tm}$]. By a fourth step of transfer, [($^2F_{5/2}$)$_{Yb}$, ($^1G_4$)$_{Tm}$] → [($^2F_{7/2}$)$_{Yb}$, ($^1D_2$)$_{Tm}$], 360 nm emission from the $^1D_2$ level of Tm^{3+} under infrared pumping in Yb was observed in (Tm, Yb):Y$_2$O$_3$ ceramic. In the case of Sc$_2$O$_3$ ceramic, co-doping with Nd^{3+} and pumping in the 800 nm range increases considerably the intensity of emission compared with Yb pumping for the same excitation power, similar to the previous observations on crystals or glasses. This enhancement is based on the increased pump absorption by Nd^{3+} and on the very efficient energy transfer to Yb^{3+}, as well as on the additional sensitization of the Tm upconversion steps by energy transfer from Nd^{3+}, [($^4F_{3/2}$)$_{Nd}$, ($^3F_4$)$_{Tm}$] → [($^4I_{11/2}$)$_{Nd}$, ($^3F_2$, $^3F_3$)$_{Tm}$], and [($^4F_{3/2}$)$_{Nd}$, ($^3H_4$)$_{Tm}$] → [($^4I_{9/2}$)$_{Nd}$, ($^1G_4$)$_{Tm}$].

The high efficiency of ET from Yb to various RE^{3+} ions, even when these are accidental impurities, can contribute to parasitic de-excitation of Yb laser materials, especially at high C_{Yb} values that favor migration-assisted transfer, and the upconversion excited emission from these ions is a sign of their presence.

10.5.3.4 Cooperative upconversion processes in ceramics

Cooperative optical processes in dimers of Yb^{3+} ions have been reported for several Yb-doped transparent ceramics such as YAG [10.226], Y$_2$O$_3$ [10.227] and Sc$_2$O$_3$ [228 229]. The intensity of cooperative absorption of Yb^{3+} is very weak, 10^{-4}–10^{-5} of that of single-ion absorption and in many cases it can be obscured by absorption of trace RE impurities, especially Tm. The cooperative emission of Yb^{3+} shows in the blue–green and corresponds to the convolution of the isolated ion infrared emission spectrum of Yb. Emission decay of the cooperative emission starts immediately after the excitation pulse, without rise-time, and its lifetime is half that of the isolated Yb^{3+} ions. The cooperative emission of Yb^{3+} can interfere with the sensitized upconversion in trace impurity ions, especially Tm (on the high-energy side) or Er (on the low-energy side). As discussed above, the cooperative processes in dimers of RE^{3+} ions can contribute to intrinsic or sensitized down-conversion or upconversion energy transfer.

10.5.4 Non-linear effects in cubic oxide ceramics

Cubic materials can have non-zero third order optical susceptibility, resulting in an efficient spontaneous Raman shift of exciting external radiation. Such processes have been observed in fine grained ceramics of ordered (YAG, the strongest shifted line at 370 cm^{-1}) or disordered ($\{YGd_2\}Sc_2(Al_2Ga)O_{12}$, 358 cm^{-1}) garnets and sesquioxides Y_2O_3 (378 cm^{-1}), Lu_2O_3 (392 cm^{-1}) and Sc_2O_3 (419 cm^{-1}) [154, 230–232]. The shifted lines could be useful for laser emission at new wavelengths, or for control of the linewidth for short-pulse generation.

10.6 Conclusions from high resolution optical spectroscopy of laser ceramics

The results obtained from the high resolution spectroscopy investigation of transparent laser ceramics show that this could be a major instrument for characterization of these materials. The multicenter structure revealed by the satellite structure and the lineshape of the spectral lines, as well as the dynamic properties of the various centers can provide important and sometimes unique data regarding the microstructure of the material, the distribution of the doping ions at the available lattice sites, the relation between the host and the doping ion, and the effect of the fabrication process. These data are necessary for a meaningful comparison of ceramics with the corresponding single crystals or for characterization of materials that are not available (composition, doping concentrations) as single crystals. At the same time, the spectroscopic investigation provides the spectroscopic parameters necessary for modeling the laser emission process and heat generation and for the selection of suitable materials, pumping conditions and laser design for optimization of the laser emission properties and power scaling. Moreover, these studies enable identification of new quantum electronics processes that could contribute to the diversification of coherent photon sources.

Similar to the corresponding single crystals, the global optical spectra of the RE^{3+} in the laser ceramics are dominated by the ions residing in the specific available lattice sites of the basic crystallographic structure of the material, and their quantum states (energy levels, transition probabilities) are similar to those found in the single crystals. No additional lines that could be attributed to defective centers at the grain boundaries were observed in coarse-grained ceramics, most probably due to the low (surface/volume) ratio. The satellite structure of perturbed centers P_i induced by departures from stoichiometry in the melt-grown YAG crystals is almost completely absent in ceramics, most likely due to the lower fabrication temperature and to the mechanism of formation of the crystalline grains, and indicates that the composition of the ceramic YAG is much closer to the ideal garnet formula. By contrast, the structures of satellites induced by the statistical ensembles of doping ions in the global absorption spectra are similar in ceramics and crystals and correspond quite well to random distribution of these ions at the available sites. Similar to single crystals, no sizable structure of spectral satellites has been found to confirm the perturbing effects expected in the presence of high concentrations of "dead sites" or of inversion of sites assumed by several authors, suggesting that these models do not have physical reality. No evidence of obvious perturbing caused by the Si^{4+} ions from the sintering aid in solid-state synthesized Nd:YAG ceramics was found, probably due to the low relative concentration of such centers. By contrast, in the sesquioxide ceramics, where the amount of sintering could be much higher, new strongly perturbed centers of quite high intensity were found.

The emission decay of laser ceramics is similar to that of the corresponding laser crystals and shows that the energy transfer processes that determine self-quenching of emission are the same. Moreover, the emission decay in Nd:YAG ceramics at high Nd concentrations can be described using the energy transfer parameters determined from the emission decay in the more diluted crystals. The effect of self-quenching on the emission decay can be modeled to a very good approximation assuming random distribution of the doping ions; this is in agreement with the findings from investigation of the M_i satellites. Accurate knowledge of the emission decay then enables evaluation of the effect of the doping concentration and pump intensity on the emission quantum efficiency.

Nevertheless, the specific processes connected with grain growth in ceramics, particularly the segregation of foreign ions at grain boundaries, could determine specific local departures from this statistics, which would manifest at grain level and require investigation with spatially resolved microscopic methods. Scanning confocal Raman or luminescence investigation of YAG ceramics doped with large RE^{3+} ions indicates extended regions with enhanced concentrations of ions at ceramic grain boundaries, although microscopic X-ray analysis and the theory of dopant segregation indicate that the regions affected by segregation are much more confined. This calls for further effort to correlate the data and to explain the differences obtained with different methods of investigation. The apparent small influence of these local irregularities on the global spectroscopic properties and emission decay or on laser emission is, most probably, connected with the low (surface/volume) ratio in laser ceramics. The granular structure of the ceramics could influence the transverse structure of the laser beams, but the manifestation of grain orientation-dependent thermal birefringence even in the case of cubic materials could largely contribute to this.

The extended compositional versatility and doping (concentration, distribution) control in ceramics enables a more complete characterization of the spectroscopic properties and of the flow of excitation inside the laser material. The composition, doping and temperature dependence of these properties would then determine the possibility of embedding the laser emission scheme within the actual electronic structure of the laser material and will provide invaluable selection and design criteria for the laser material, pumping system and laser resonator. The data on the energy level structure, on the absorption and emission cross-sections and on the physical mechanisms that control the homogeneous and inhomogeneous broadening of lines, together with data on non-radiative de-excitation provide information on the potential pump and laser emission characteristics and contribute to optimization of the pumping system and of laser design to grant maximum utilization of the absorbed pump radiation for laser emission and to reduce heat generation. By offering data for optimization or extension of existing laser materials and emission schemes, as well as by the identification and characterization of new quantum electronics processes, the spectroscopic characterization of ceramic laser material could be of major relevance for diversification of laser emission schemes, for extension of the range of laser emission wavelengths, for optimization of laser efficiency, for diversification of temporal regimes of emission, for control of the laser beam characteristics and for power or energy scaling. The other general characteristics of the ceramic materials, particularly the increased compositional and doping capability, improvement in thermomechanical properties and fabrication of large-size bodies would be the practical demonstration of this potential. It is then obvious that the ceramic materials could not only replace the corresponding crystalline materials in the construction of solid-state lasers, but they also represent a major innovative factor in extension of the range of variety, performance and applications of these lasers.

References

[1] M. K. Ashurov, Y. K. Voronko, V. V. Osiko, A. A. Sobol, and M. I. Timoshechkin, Spectroscopic study of stoichiometry deviation in crystals with garnet structure, *Phys. Staus Solidi A,* **42** (1977) 101–110.

[2] V. V. Osiko, Yu. K. Voronko, and A. A. Sobol, Spectroscopic investigation of defect sructures and structural transformations in ionic crystals, in *Crystals 10,* Springer, Heidelberg (1984), pp. 38–86.

[3] A. Lupei, V. Lupei, S. Georgescu, and W. M. Yen, Mechanism of energy transfer between Nd^{3+} ions in YAG, *J. Lumin.* **39** (1987) 35–43.

[4] V. Lupei, A. Lupei, C. Tiseanu, S. Georgescu, C. Stoicescu, and P. M. Nanau, High resolution optical spectroscopy of Nd in YAG – a test for structural and distribution models, *Phys. Rev. B.* **95** (1995) 8–17.

[5] A. Lupei, V. Lupei, and E. Osiac, Spectral and dynamical effects of octahedral impurities on RE^{3+} in garnets, *J. Phys. Condens. Mater.* **10** (1998) 9701–9710.

[6] V. Lupei, A. Lupei, S. Georgescu, T. Taira, Y. Sato, and A. Ikesue, The effect of Nd concentration on the spectroscopic and emission decay properties of highly-doped Nd:YAG ceramics, *Phys. Rev. B* **64** (2001) 092102.

[7] V. Lupei, RE^{3+} emission in garnets: multisites, energy transfer and quantum efficiency, *Opt. Mater.* **19** (2002) 95–107.

[8] V. Lupei, A. Lupei, S. Georgescu, B. Diaconescu, T. Taira, Y. Sato, S. Kurimura, and A. Ikesue, High-resolution spectroscopy and emission decay in concentrated Nd:YAG ceramics, *J. Opt. Soc. Am. B* **19** (2002) 360–368.

[9] V. Lupei, A. Lupei, and A. Ikesue, Single crystal and transparent ceramic Nd-doped oxide laser materials: a comparative spectroscopic investigation, *J. Alloys. Comp.* **380** (2004) 61–70.

[10] V. Lupei, A. Lupei, and A. Ikesue, Transparent polycrystalline ceramic laser materials, *Opt. Mater.* **130** (2008) 1781–1786.

[11] U. Aschauer and P. Brown, Atomistic modelling study of surface segregation in Nd:YAG, *J. Am. Ceram. Soc.* **89** (2006) 3812–3816.

[12] U. Aschauer, P. Brown, and S. C. Parker, Surface and mirror twin grain boundary segregation in Nd:YAG: an atomistic study, *J. Am. Ceram. Soc.* **91** (2008) 2698–2705.

[13] N. I. Agladze, H. S. Bagdasarov, E. A. Vinogradov, V. I. Zhekov, T. M. Murina, M. N. Popova, and E. A. Fedorov, Shape of the spectral lines in $(Y_{1-x}Er_x)_3Al_5O_{12}$ garnets, *Kristallografiya* **33** (1988) 912–919 (in Russian).

[14] J. B. Gruber, R. P. Leavitt, C. A. Morrison, and N. C. Chang, Optical spectra, energy levels, and crystal field analysis of tripositive rare earth ions. IV. C_{3i} sites, *J. Chem. Phys.* **82** (1985) 5373–5378.

[15] K. Petermann, G. Huber, L. Fornasiero, S. Kuch, E. Mix, V. Peters, and S. A. Basun, Rare earth doped sesquioxides, *J. Lumin.* **87–89** (2000) 973–975.

[16] L. Fornasiero, E. Mix, V. Peters, K. Petermann, and G. Huber, Czochralski growth and laser parameters of RE^{3+}-doped Y_2O_3 and Sc_2O_3, *Ceram. Int.* **26** (2000) 589–592.

[17] V. Lupei, A. Lupei, and A. Ikesue, Transparent Nd and (Nd, Yb)-doped Sc_2O_3 ceramics as potential new laser materials, *Appl. Phys. Lett.* **86** (2005) 111118.

[18] L. Rogobete, A. Lupei, V. Lupei, A. Petraru, and B. Diaconescu, On the inhomogeneous broadening by point defects on RE^{3+} optical lines in YAG, *Proc. SPIE* **4430** (2001) 97–105.

[19] J. Dong and K. W. Lu, Noncubic symmetry in garnet structures studied using extended X-ray absorption fine structure spectra, *Phys. Rev. B* **43** (1991) 8808–8821.

[20] D. P. Devor and L. G. DeShazer, Evidence of Nd:YAG quantum efficiency on nonequivalent crystal field effects, *Opt. Commun.* **46** (1983) 97–102.

[21] D. P. Devor, L. G. DeShazer, and R. C. Pastor, Hydroxyl impurity effects in YAG $(Y_3Al_5O_{12})$, *J. Chem. Phys.* **81** (1984) 4104–4117.

[22] D. P. Devor, L. G. DeShazer, and R. C. Pastor, Nd:YAG quantum efficiency and related radiative properties, *IEEE J. Quantum Electron.* **25** (1989) 1863–1873.

[23] H. Yagi, T. Yanagitani, and K. Ueda, $Nd^{3+}:Y_3Al_5O_{12}$ laser ceramics: flashlamp pumped laser operation with a UV cut filter, *J. Alloys Comp.* **421** (2006) 195–199.

[24] A. J. Stevenson, B. C. Bittel, C. G. Leh, X. Li, E. C. Dickey, P. M. Lenahan, and G. L. Messing, Color center formation in vacuum sintered $Nd_{3x}Y_{3-3x}Al_5O_{12}$ transparent ceramic, *Appl. Phys. Lett.* **98** (2011) 051906.

[25] V. Lupei, A. Lupei, and A. Ikesue, Multicenter structure and dynamical processes in the rare earth doped garnet and sesquioxide laser crystals and ceramics, *Spectrosc. Lett.* **43** (2010) 357–372.

[26] A. Ikesue, I. Furusato, and K. Yamada, Fabrication of polycrystalline transparent YAG ceramics by a solid-state reactive method, *J. Am. Ceram. Soc.* **78** (1995) 225–228.

[27] A. Ikesue, T. Kinoshita, K. Kamata, and K. Yoshida, Fabrication and optical properties of high-performance polycrystalline Nd:YAG ceramics for solid-state lasers, *J. Am. Ceram. Soc.* **78** (1995) 1033–1040.

[28] A. Ikesue, K. Yoshida, T. Yamamoto, and I. Yamaga, Optical scattering centers in polycrystalline Nd:YAG laser, *J. Am. Ceram. Soc.* **80** (1997) 1517–1552.

[29] S. H. Lee, S. Kochawattana, G. L. Messing, J. Q. Dumm, G. Quarles, and V. Castillo, Solid-state reactive sintering of transparent polycrystalline Nd:YAG ceramics, *J. Am. Ceram. Soc.* **89** (2006) 1945–1950.

[30] A. Ikesue and Y. L. Aung, Sythesis and performance of advanced ceramic lasers, *J. Am. Ceram. Soc.* **89** (2006) 1936–1944.

[31] S. Kochawattana, A. Stevenson, S. H. Lee, M. Ramirez, V. Gopalan, J. Dumm, V. K. Castillo, G. J. Quarles, and G. L. Messing, Sintering and grain growth profiles in SiO_2 doped Nd:YAG, *J. Eur. Ceram. Soc.* **28** (2008) 1527–1534.

[32] A. Maitre, Ch. Salle, R. Boulesteix, J-F. Boumard, and Y. Rabinovitch, Effect of silica on the reactive sintering of polycrystalline Nd:YAG ceramics, *J. Am. Ceram. Soc.* **91** (2008) 406–413.

[33] R. Boulesteix, A. Maitre, J.-F. Baumard, C. Salle, and Y. Rabinovitch, Mechanism of the liquid-phase sintering for Nd:YAG ceramics, *Opt. Mater.* **31** (2009) 711–715.

[34] J. P. Hollingsworth, J. D. Kantz, and T. F. Soules, Neodymium diffusion during sintering of Nd:YAG transparent ceramics, *J. Phys. D: Appl. Phys.* **42** (2009) 052001.

[35] W. Liu, B. Jiang, W. Zhang, J. Li, J. Zhou, D. Zhang, Y. Pan, and J. Guo, Influence of the heating rate on optical properties of Nd:YAG laser ceramics, *Ceram. Int.* **36** (2010) 2197–2201.

[36] A. J. Stevenson, X. Li, M. A. Martinez, J. M. Anderson, D. L. Suchi, E. R. Kupp, E. C. Dickey, K. T. Mueller, and G. L. Messing, Effect of SiO_2 on densification and microstructure development in Nd:YAG transparent ceramics, *J. Am. Ceram. Soc.* **94** (2011) 1380–1387.

[37] M. M. Kuklja, Defects in yttrium aluminum perovskite and garnet crystals: atomistic study, *J. Phys. Condens. Mater.* **12** (2000) 2953–2967.

[38] Y. Kuru, E. O. Savasir, S. Z. Nergiz, C. Oncel, and M. A. Gulgun, Enhanced co-solubilities of Ca and Si in YAG ($Y_3Al5O12$), *Phys. Status Solidi C* **5** (2008) 3383–3386.

[39] Y. Li, S. Zhou, H. Lin, X. Hou, W. Li, H. Teng, and T. Jia, Fabrication of Nd:YAG transparent ceramics with TEOS, MgO and compound additives as sintering aids, *J. Alloys Comp.* **502** (2010) 225–230.

[40] A. J. Stevenson, E. R. Kupp, and G. L. Messing, Low temperature transient liquid phase sintering of $B_2]_3$-SiO_2 doped Nd:YAG transparent ceramic, *J. Mater. Res.* **26** (2011) 1151–1158.

[41] C. Greskovich and J. P. Chernoch, Polycrystalline ceramic lasers, *J. Appl. Phys.* **44** (1973) 4599–4606.

[42] C. Greskovich and J. P. Chernoch, Improved polycrystalline ceramic laser, *J. Appl. Phys.* **45** (1974) 4495–4502.

[43] A. Ikesue, K. Kamata, and K. Yoshida, Synthesis of transparent Nd-doped HfO_2–Y_2O_3 ceramics using HIP, *J. Am. Ceram. Soc.* **79** (1996) 359–365.

[44] W. Li, S. Zhou, H. Lin, H. Teng, N. Liu, J. Li, X. Hou, and T. Jia, Controlling of grain size with different additives in Tm^{3+}:Y_2O_3 transparent ceramics, *J. Am. Ceram. Soc.* **93** (2010) 3819–3822.

[45] S. N. Ushakov, M. A. Uslamina, and E. V. Zharikov, Spectral properties of Nd^{3+} ions in samples of transparent Y_2O_3 ceramics, *Opt. Spectrosc.* **106** (2009) 549–555.

[46] M. Siebold, S. Bock, U. Schramm, B. Xu, J. L. Doualan, P. Camy, and R. Moncorge, Yb:CaF_2 – a new old crystal, *Appl. Phys. B* **97** (2009) 327–338.

[47] V. Petit, P. Camy, J.-L. Doualan, X. Portier, and R. Moncorge, Spectroscopy of Yb^{3+}:CaF_2: from isolated centers to clusters, *Phys. Rev. B* **78** (2008) 085131.

[48] P. Aubry, A. Bensalah, P. Gredin, G. Patriarche, D. Vivien, and M. Mortier, Synthesis and optical characterization of Yb-doped CaF_2 ceramics, *Opt. Mater.* **31** (2009) 750–753.

[49] A. Lyberis, G. Patriarche, P. Gredin, D. Vivien, and M. Mortier, Origin of light scattering in ytterbium doped calcium fluoride transparent ceramics for high power lasers, *J. Eur. Ceram. Soc.* **31** (2011) 1619–1630.

[50] T. T. Basiev, M. E. Doshenko, P. P. Feotilov, V. A. Konyushkin, S. V. Kuznetsov, V. V. Osiko, and M. Sh. Akchurin, Efficient laser based on CaF_2–SrF_2–YbF_3 nanoceramics, *Opt. Lett.* **33** (2008) 521–523.

[51] J. Saikawa, Y. Sato, T. Taira, and A. Ikesue, Absorption, emission spectrum properties, and efficient laser performances of $Yb:Y_3ScAl_4O_{12}$ ceramics, *Appl. Phys. Lett.* **85** (2004) 1898–1900.

[52] Y. Sato, J. Saikava, T. Taira, and A. Ikesue, Characteristics of Nd^{3+}-doped $Y_3ScAl_4O_{12}$ ceramic laser, *Opt. Mater.* **29** (2007) 1277–1288.

[53] S. Cheng, X. Xu, D. Li, D. Zhou, F. Wu, Z. Zhao, and J. Xu, Spectral properties of Yb^{3+} ions in $Lu_3ScAl_4O_{12}$ single crystal, *J. Alloys Comp.* **506** (2010) 513–515.

[54] H. Okada, M. Tanaka, H. Kiriyama, Y. Nakai, Y. Ochi, A. Sugiyama, H. Daido, T. Kimura, T. Yanagitani, H. Yagi, and N. Meichin, Laser ceramic materials for sub-picosecond solid-state lasers using Nd^{3+}-doped mixed scandium garnets, *Opt. Lett.* **35** (2010) 3048–3050.

[55] M. Tokuragawa, H. Kurokawa, A. Shirakawa, K. Ueda, H. Yagi, T. Yanagitani, and A. A. Kaminskii, Continuous-wave and mode-locked lasers on the base of partially disordered crystalline Yb^{3+}: $\{YGd_2\}[Sc_2](Al_2Ga)O_{12}$ ceramics, *Opt. Express* **18** (2010) 4390–4395.

[56] J. Ding, Q. Xing, Z. Tang, J. Xu, and L. Su, Investigation of the spectroscopic properties of $(Y_{0.92-x}La_{0.08}Nd_x)_2O_3$ transparent ceramic, *J. Opt. Soc. Am. B* **24** (2007) 681–684.

[57] Q. Yang, S. Lu, B. Zhang, H. Zhang, J. Zhou, Z. Yum, Y. Qi, and Q. Lou, Properties and laser performances of Nd-doped yttrium lanthanum oxide transparent ceramics, *Opt. Mater.* **33** (2011) 692–694.

[58] A. A. Kaminskii, E. L. Belokoneva, A. V. Butashin, A. A. Markosian, B. V. Mill, and O. Nikolskaia, Crystal structure and spectral luminescence properties of the cation-deficient garnet $Ca_3(Nb, Ga)_2Ga_3O_{12}$–Nd^{3+}, *Inorg. Mater.* **22** (1986) 1061–1071.

[59] Y. K. Voronko, A. A. Sobol, A. Y. Karasik, N. A. Eskov, P. A. Ryabochkina, and S. N. Ushakov, Calcium niobium gallium and calcium lithium niobium gallium garnets doped with rare earth ions – effective laser materials, *Opt. Mater.* **20** (2002) 197–209.

[60] V. Lupei, A. Lupei, C. Gheorghe, L. Gheorghe, A. Achim, and A. Ikesue, Crystal field disorder effects in the optical spectra of Nd^{3+} and Yb^{3+}-doped calcium lithium niobium gallium garnet laser crystals and ceramics, *J. Appl. Phys.* **112** (2012) 063110.

[61] Y. K. Voronko, A. V. Popov, A. A. Sobol, and S. N. Ushakov, Yb^{3+} ions in calcium niobium gallium garnet crystals: nearest neighbor environment and optical spectra, *Inorg. Mater.* **42** (2006) 1133–1137.

[62] A. Lupei, V. Lupei, L. Gheorghe, L. Rogobete, E Osiac, and A. Petraru, The nature of nonequivalent Nd^{3+} centers in CNGG and CLNGG, *Opt. Mater.* **16** (2001) 403–411.

[63] A. S. S. de Camargo, L. A. De O. Nunes, I. A. Santos, D. Garcia, and J. A. Eiras, Structural and spectroscopic properties of rare-earth (Nd^{3+}, Er^{3+}, and Yb^{3+}) doped transparent lead lanthanum zirconate titanate ceramics, *J. Appl. Phys.* **95** (2004) 2135–2137.

[64] A. S. S. de Camargo, C. Jacinto, L. A. O. Nunes, T. Catunda, D. Garcia, E. R. Botero, and J. A. Eiras, Effect of Nd^{3+} concentration quenching in highly doped lead lanthanum zirconate titanate, *J. Appl. Phys.* **101** (2007) 053111.

[65] A. A. Kaminskii, H. Kurokawa, A. Shirakawa, K. Ueda, N. Tanaka, P. Becker, L. Bohaty, M. Akchurin, M. Tokurgawa, S. Kuretake, Y. Kintaka, K. Kageyama, and H. Takagi, $Ba(Mg,Zr,Ta)O_3$ fine-grained ceramic: a novel laser gain material with disordered structure for high-power laser systems, *Laser Phys. Lett.* **6** (2009) 304–310.

[66] V. V. Osipov, O. L. Khasanov, V. I. Solomonov, V. A. Shidov, A. N. Orlov, V. V. Platonov, A. V. Spirina, K. L. Luk'yashin, and E. S. Dvilis, Highly transparent ceramics with disordered structure. $Nd:Y_2O_3$, *Russ. Phys. J.* **53** (2010) 263–269.

[67] V. Lupei, A. Lupei, C. Gheorghe, and A. Ikesue, Comparative high resolution spectroscopy and emission dynamics of Nd-doped GSGG crystals and transparent ceramics, *J. Lumin.* **128** (2008) 885–887.

[68] A. Lupei, V. Lupei, T. Taira, Y. Sato, A. Ikesue, and C. Gheorghe, Energy transfer processes of Nd^{3+} in Y_2O_3 ceramics, *J. Lumin.* **102–103** (2003) 72–76.

[69] A. Lupei, V. Lupei, A. Ikesue, and C. Gheorghe, Spectroscopic and energy transfer investigation of Nd/Yb in Y_2O_3 transparent ceramics, *J. Opt. Soc. Am. B* **27** (2010) 1002–1010.

[70] S. Mathur, H. Shen, M. Veith, R. Rapalaviciute, and T. Agne, Structural and optical properties of highly Nd-doped yttrium aluminium garnet ceramics from alkoxide and glycolate precursors, *J. Am. Ceram. Soc.* **89** (2006) 2027–2033.

[71] J. Mares, W. Nie, and G. Boulon, Multisites and energy transfer in Cr^{3+}-Nd^{3+} codoped YAG and $YAlo_3$ crystals, *J. Phys.* **51** (1990) 1655–1669.

[72] A. Lupei, C. Tiseanu, and V. Lupei, Correlation between spectral and structural data of $YAG:Tm^{3+}$ and $YAG:Cr^{3+}, Tm^{3+}$, *Phys. Rev. B* **47** (1993) 14084–14092.

[73] V. Lupei, G. Boulon, A. Lupei, M. J. Elejalde, A. Brenier, and C. Pedrini, Energy transfer from tetrahedral Fe^{3+} to Tm^{3+} in garnets, *Phys. Rev. B* **49** (1994) 7076–7079.

[74] A. Lupei, V. Lupei, and A. Petraru, Spectroscopy of Yb^{3+} co-doped $Pr^{3+}:YAG$ laser crystals, *Proc. SPIE* **3405** (1998) 563–569.

[75] P. A. Tanner, L. Fu, L. Ning, B. M. Cheng, and M. G. Brik, Soft synthesis and vacuum ultraviolet spectra of $YAG:Ce^{3+}$ nanocrystals: reassignment of Ce^{3+} energy levels, *J. Phys. Condens. Mater.* **19** (2007) 216213.

[76] Yu. Zorenko, T. Voznyak, V. Gorbenko, E. Zych, S. Nizankovski, A. Danko, and V. Puzikov, Luminescence properties of $Y_3Al_5O_{12}:Ce$ nanoceramics, *J. Lumin.* **131** (2010) 17–21.

[77] A. Ikesue, and Y. Sato, Synthesis of Pr heavily doped transparent YAG ceramics, *J. Ceram. Soc. Jpn.* **109** (2001) 640–642.

[78] T. Yanagida, A. Yoshikawa, A. Ikesue, K. Kamada, and Y. Yokota, Basic properties of ceramic Pr:LuAG scintillators, *IEEE Trans. Nucl. Sci.* **56** (2009) 2955–2959.

[79] T. Feng, J. Shi, and D. Jiang, Optical properties of transparent Pr:YSAG ceramic, *Ceram. Int.* **35** (2009) 427–431.

[80] T. T. Basiev, V. A. Konyushkin, D. V. Konyushkin, M. E. Doroshenko, G. Huber, F. Reichert, N.-O. Hansen, and M. Futuro, First ceramic laser in the visible spectral range, *Opt. Mater. Express* **1** (2011) 1511–1514.

[81] M. Sekita, H. Haneda, T. Yanagitani, and S. Shivasaki, Induced emission cross-section of $Nd:Y_3Al_5O_{12}$ ceramic, *J. Appl. Phys.* **67** (1990) 453–458.

[82] M. Sekita, H. Haneda, S. Shivasaki, and T. Yanagitani, Optical spectra of undoped and rare-earth (Pr, Nd, Eu and Er) doped transparent ceramic $Y_3Al_5O_{12}$, *J. Appl. Phys.* **69** (1991) 3709–3718.

[83] I. Shoji, S. Kurimura, Y. Sato, T. Taira, A. Ikesue, and K. Yoshida, Optical properties and laser characteristics of highly Nd^{3+}-doped $Y_3Al_5O_{12}$ ceramics, *Appl. Phys. Lett.* **77** (2000) 939–941.

[84] V. Lupei, A. Lupei, N. Pavel, I. Shoji, T. Taira, and A. Ikesue, Laser emission under thermally-activated resonant pump in concentrated Nd:YAG ceramics, *Appl. Phys. Lett.* **79** (2001) 590–592.

[85] V. Lupei, A. Lupei, N. Pavel, T. Taira, and A. Ikesue, Comparative investigation of spectroscopic and laser emission characteristics under direct 885-nm pump of concentrated Nd:YAG ceramics and crystals, *Appl. Phys. B* **73** (2001) 757–762.

[86] V. Lupei, T. Taira, A. Lupei, N. Pavel, I. Shoji, and A. Ikesue, Spectroscopy and laser emission under hot band resonant pump in highly-doped Nd:YAG ceramics, *Opt. Commun.* **195** (2001) 225–232.

[87] V. Lupei, N. Pavel, and T. Taira, Efficient laser emission in concentrated Nd laser materials under pumping into the emitting level, *IEEE J. Quantum Electron.* **38** (2002) 240–245.

[88] J. Lu, M. Prabhu, J. C. Song, C. Li, J. Xu, K. Ueda, A. A. Kaminskii, H. Yagi, and T. Yanagitani, Optical properties and highly efficient laser oscillation of Nd:YAG ceramics, *Appl. Phys. B* **71** (2000) 469–473.

[89] J. Lu, K. Ueda, H. Yagi, T. Yanagitani, Y. Akiyama, and A. A. Kaminskii, Neodymium doped yttrium aluminum garnet ($Y_3Al_5O_{12}$) nanocrystalline ceramics – a new generation of solid state laser and optical materials, *J. Alloys Comp.* **341** (2002) 220–225.

[90] G. A. Kumar, J. Lu, A. A. Kaminskii, K. Ueda, H. Yagi, T. Yanagitani, and N. V. Unnikrishnan, Spectroscopic and stimulated emission characteristics of Nd^{3+} in transparent YAG Ceramics, *IEEE J. Quantum Electron.* **40** (2004) 747–758.

[91] Y. Sato and T. Taira, Variation of the stimulated emission cross-section in Nd:YAG caused by the structural changes of Russell–Saunders manifolds, *Opt. Mater. Express* **1** (2011) 514–522.

[92] W. F. Krupke, Radiative transition probabilities within the $4f^3$ ground configuration of Nd:YAG, *IEEE J. Quantum Electron.* **7** (1971) 153–159.

[93] A. A. Kaminskii, S. N. Bagaev, K. Ueda, A. Sirakawa, T. Tokurakawa, H. Yagi, T. Yanagitani, and J. Dong, Stimulated emission spectroscopy of fine-grained garnet ceramics Nd:YAG between 77–650 K, *Laser Phys. Lett.* **6** (2009) 682–687.

[94] Y. Sato, J. Saikava, T. Taira, and A. Ikesue, Characteristics of Nd^{3+}-doped $Y_3ScAl_4O_{12}$ ceramic laser, *Opt. Mater.* **29** (2007) 1277–1288.

[95] L. Fornasiero, E. Mix, V. Peters, K. Petermann, and G. Huber, Czochralski growth and laser parameters of RE^{3+}-doped Y_2O_3 and Sc_2O_3, *Ceram. Int.* **26** (2000) 589–592.

[96] L. Fornasiero, E. Mix, V. Peters, E. Heumann, K. Petermann, and G. Huber, Efficient laser operation of Nd:Sc_2O_3 at 966 nm, 1082 nm and 1486 nm, *OSA Trends Opt. Photon.* **26** (1999) 249–254.

[97] J. Lu, T. Murai, K. Takaichi, T. Uematsu, K. Ueda, H. Yagi, T. Yanagitani, and A. A. Kaminskii, Nd^{3+}:Y_2O_3 ceramic laser, *Jpn. J. Appl. Phys.* **40** (2001) L1277–L1279.

[98] J. B. Gruber, D. K. Sardar, K. L. Nash, and R. M. Yow, Comparative study of the crystal field splitting of trivalent neodymium energy levels in polycrystalline ceramic and nanocrystalline yttrium oxide, *J. Appl. Phys.* **102** (2007) 023103.

[99] V. Lupei, A. Lupei, C. Gheorghe, S. Hau, and A. Ikesue, Efficient sensitization of Yb^{3+} emission by Nd^{3+} in Y_2O_3 transparent ceramics and the prospect for high energy Yb lasers, *Opt. Lett.* **34** (2009) 2141–2143.

[100] N. C. Chang, J. B. Gruber, R. P. Leavitt, and C. A. Morrison, Optical spectra, energy levels, and crystal field analysis of tripositive rare earth ions in Y_2O_3. 1. Kramers ions in C_2 sites, *J. Chem. Phys.* **76** (1982) 3877–3889.

[101] B. M. Walsh, J. M. McMahon, W. C. Edwards, N. P. Barnes, R. W. Equal, and R. L. Hutchinson, Spectroscopic characterization of Nd: Y_2O_3: application toward a differential absorption lidar system for remote sensing of ozone, *J. Opt. Soc. Am. B* **19** (2002) 2893–2903.

[102] A. S. S. Camargo, C. Jacinto, T. Catunda, L. A. O. Nunes, D. Garcia, and J. A. Eiras, Thermal lens and Auger upconversion losses effect on the efficiency of Nd^{3+}-doped lead lanthanum zirconate titanate transparent ceramic, *J. Opt. Soc. Am. B* **23** (2006) 2097–2106.

[103] T. B. de Queiroz, D. Mohr, H. Eckert, and A. S. S. De Camargo, Preparation and structural characterization of rare-earth doped lead lanthanum zirconate titanate, *Solid State Sci.* **11** (2009) 1363–1369.

[104] T. Feng, D. R. Clarke, D. Jiang, J. Xia, and J. Shi, Neodymium zirconate ($Nd_2Zr_2O_7$) transparent ceramic as solid state laser material, *Appl. Phys. Lett.* **98** (2011) 151105.

[105] T. T. Basiev, M. E. Doroshenko, V. A. Konyushkin, and V. V. Osiko, SrF_2:Nd^{3+} laser fluoride ceramics, *Opt. Lett.* **35** (2010) 4009–4011.

[106] J. Akiyama, Y. Sato, and T. Taira, Laser ceramics with rare-earth-doped anisotropic materials, *Opt. Lett.* **35** (2010) 3598–3600.

[107] H. Yagi, J. F. Bisson, K. Ueda, and T. Yanagitani, $Y_3Al_5O_{12}$ ceramic absorbers for the suppression of parasitic oscillation in high-power Nd:YAG lasers, *J. Lumin.* **121** (2006) 88–94.

[108] A. Lupei, V. Lupei, C. Gheorghe, and A. Ikesue, Spectroscopic investigation of Sm^{3+} in YAG ceramic, *Rom. Rep. Phys.* **63** (2011) 817–822.

[109] A. Lupei, V. Lupei, C. Gheorghe, A. Ikesue, and M. Enculescu, Spectroscopic characteristics of Dy^{3+} doped $Y_3Al_5O_{12}$ transparent ceramics, *J. Appl. Phys.* **110** (2011) 083120.

[110] S. A. Payne, L. L. Chase, L. K. Smith, W. L. Kway, and W. F. Krupke, Infrared cross-section measurements for crystals doped with Er^{3+}, Tm^{3+} and Ho^{3+}, *IEEE J. Quantum Electron.* **28** (1992) 2619–2630.

[111] J. B. Gruber, M. D. Selzer, V. J. Pugh, and F. S. Richardson, Electronic energy level structure of trivalent holmium in yttrium aluminum garnet, *J. Appl. Phys.* **77** (1995) 5882–5901.

[112] W. X. Zhang, J. Zhou, W. B. Liu, J. Li, L. Wong, B. X. Jiang, Y. B. Pan, X. J. Cheng, and J. Q. Liu, Fabrication, properties and laser performance of Ho:YAG transparent ceramic, *J. Alloys Comp.* **506** (2010) 745–748.

[113] C. Gheorghe, A. Lupei, V Lupei, and A. Ikesue, Spectroscopic properties of Ho^{3+} doped Sc_2O_3 transparent ceramic for laser materials, *J. Appl. Phys.* **105** (2009) 123110.

[114] M. Galceran, M. C. Pujol, P. Gluchowski, W. Strek, J. J. Carvajal, X. Mateos, M. Aguilo, and F. Diaz, A promising $Lu_{2-x}Ho_xO_3$ laser nanoceramic: synthesis and characterization, *J. Am. Ceram. Soc.* **93** (2010) 3764–3772.

[115] J. B. Gruber, A. S. Nijjar, D. K. Sardar, R. M. Yow, C. C. Russel, T. H. Aliik, and B. Zhandi, Spectral analysis and energy-level structure of Er in polycrystaline ceramic garnet $Y_3Al_5O_{12}$, *J. Appl. Phys.* **97** (2005) 063519.

[116] D. Sardar, C. C. Russell, J. B. Gruber, and T. Allik, Absorption intensities and emission cross sections of principal intermanifold and inter-Stark transitions of Er^{3+} in polycrystalline ceramic garnet $Y_3Al_5O_{12}$, *J. Appl. Phys.* **97** (2005) 123501.

[117] U. Hömmerich, C. Hanley, E. Brown, S. B. Trivedi, and J. M. Zavadac, Spectroscopic studies of the 1.5 μm ($^4I_{15/2} \rightarrow {}^4I_{13/2}$) emission from polycrystalline ceramic Er:YAG and Er:KPb$_2$Cl$_5$, *J. Alloys Comp.* **488** (2009) 624–627.

[118] G. Qin, J. Lu, J. F. Bisson, Y. Feng, K. Ueda, H. Yagi, and T. Yanagitani, Upconversion luminescence of Er^{3+} in highly transparent YAG ceramics, *Solid State Commun.* **132** (2004) 103–106.

[119] S. Georgescu, V. Lupei, A. Lupei, V. I. Zhekov, T. M. Murina, and A. M. Prokhorov, Concentration effects on the upconversion from $^4I_{13/2}$ level of Er^{3+} in YAG, *Opt. Commun.* **81** (1991) 186–192.

[120] V. Lupei, S. Georgescu, and V. Florea, On the dynamics of population inversion for three-micron Er^{3+} lasers, *IEEE J. Quantum Electron.* **29** (1993) 426–434.

[121] N. Ter-Gabrielyan, V. Fromzel, L. D. Merkle, and M. Dubinskii, Resonant in-band pumping of cryo-cooled Er^{3+}:YAG at 1532, 1534 and 1546 nm: a comparative study, *Opt. Mater. Express* **1** (2011) 223–233.

[122] J. B. Gruber, K. L. Nash, D. K. Sardar, U. V. Valiev, N. Ter-Gabrielyan, and L. D. Merkle, Modeling optical transitions of Er^{3+} in C_2 and C_{3i} sites in polycrystalline Y$_2$O$_3$, *J. Appl. Phys.* **104** (2008) 023101.

[123] T. Sanamyan, I. Simmons, and M. Dubinskii, Er^{3+} doped Y$_2$O$_3$ ceramic laser at ~2.7 μm with direct pumping of the upper laser level, *Laser Phys. Lett.* **7** (2010) 206–209.

[124] T. Sanamyan, M. Kanskar, Y. Xiao, D. Kodlaya, and M. Dubinskii, High power, diode pumped 2.7 μm Er^{3+} Y$_2$O$_3$ laser with near quantum defect-limited efficiency, *Opt. Express* **19** (2011) A1082–A1087.

[125] V. Lupei, A. Lupei, C. Gheorghe, and A. Ikesue, Ceramic laser materials for high-performance solid-state lasers. *Proc. SPIE* **6552** (2007) 655251.

[126] A. Lupei, V. Lupei, C. Gheorghe, and A. Ikesue, Excited state dynamics of Er^{3+} in Sc$_2$O$_3$ ceramic, *J. Lumin.* **128** (2008) 918–920.

[127] C. Gheorghe, S. Georgescu, V. Lupei, A Lupei, and A. Ikesue, Absorption intensities and emission cross section of Er^{3+} in Sc$_2$O$_3$ transparent ceramic, *J. Appl. Phys.* **103** (2008) 083116.

[128] N. Ter-Gabrielian, L. D. Merkle, A. Ikesue, and M. Dubinskii, Ultralow quantum-defect eye safe Er:Sc$_2$O$_3$ laser, *Opt. Lett.* **33** (2008) 1524–1526.

[129] C. Tiseanu, A. Lupei, and V. Lupei, Energy levels of Tm^{3+} in yttrium aluminum garnet, *J. Phys. Condens. Mater.* **7** (1975) 8477–8486.

[130] Y. W. Zhou, Y. D. Zhang, X. Zhong, Z. Y.Wei, W. X. Zhang, B. X. Jiang, and Y. B. Pan, Efficient Tm:YAG ceramic laser at 2 μm, *Chin. Phys. Lett.* **27** (2010) 074213.

[131] S. Zhang, M. Wang, L. Xu, Y. Wang, Y. Tang, X. Chen, W. Chen, J. Xu, B. Jiang, and Y. Pan, Efficient Q-switched Tm:YAG ceramic slab laser, *Opt. Express* **19** (2011) 727–732.

[132] V. Lupei, A. Lupei, C. Gheoghe, and A. Ikesue, Spectroscopic characteristics of Tm in Tm and Tm, Nd, Yb: Sc$_2$O$_3$ ceramic, *J. Lumin.* **128** (2008) 901–905.

[133] V. Lupei, A. Lupei, C. Gheorghe, A. Ikesue, and E. Osiac, Energy transfer driven infrared emission processes in rare earth doped Sc$_2$O$_3$ ceramics, *J. Lumin.* **129** (2009) 1862–1865.

[134] C. Gheorghe, A. Lupei, V. Lupei, A. Ikesue, and M. Enculescu, Intensity parameters of Tm^{3+} doped Sc$_2$O$_3$ ceramic transparent laser materials, *Opt. Mater.* **33** (2011) 501–505.

[135] O. L. Antipov, S. Yu. Golovkin, O. N. Gorshkov, N. G. Zakharov, A. P. Zinoviev, A. P. Kasatkin, M. V. Kruglyov, M. O. Marychev, A. A. Novikov, N. V. Sakharov,

and E. V. Chuprunov, Structure, optical and spectroscopic properties and efficient two-micron lasing of new Tm^{3+}:Lu_2O_3 ceramics, *Quantum Electron.* **41** (2011) 863–866.

[136] K. Takaichi, H. Yagi, J. Lu, A. Shirakawa, K. Ueda, and T. Yanagitani, Yb^{3+}-doped $Y_3Al_5O_{12}$ ceramics – a new solid-state laser material, *Phys. Status Solidi A* **200** (2003) R5–R8.

[137] A. Lupei, V. Enaki, V. Lupei, C. Presura, and A. Petraru, Resonant electron-photon coupling of Yb^{3+} in YAG, *J. Alloys Comp.* **275–277** (1998) 196–199.

[138] A. Lupei, V. Lupei, C. Presura, V. N. Enaki, and A. Petraru, electron–phonon coupling effects on Yb^{3+} spectra in several laser crystals, *J. Phys. Condens. Mater.* **11** (1999) 3769–3778.

[139] C. W. Xu, D. W. Luo, J. Zhang, H. Yang, X. P. Qin, W. D. Tan, and D. Y. Tang, Diode pumped highly efficient Yb:$Lu_3Al_5O_{12}$ ceramic laser, *Laser Phys. Lett.* **9** (2012) 30–34.

[140] I. Kamenskikh, M. Chugunova, S. T. Fredrich-Thornton, C. Pedrini, K. Petermann, A. Vasil'ev, U. Wolters, and H. Yagi, Potentiality of ceramic scintillators: general considerations and YAG-Yb optical ceramic performance, *IEEE Trans. Nucl. Sci.* **57** (2010) 1211–1217.

[141] M. M. Chugunova, I. A. Kamenskih, V. V. Mikhailin, and S. A. Ushakov, Luminescence properties of transparent ceramics $Y_3Al_5O_{12}$:Yb, *Opt. Spectrosc.* **109** (2010) 887-892.

[142] J. Lu, K. Takaichi, T. Uematsu, A. Shirakawa, M. Musha, K. Ueda, H. Yagi, T. Yanagitani, and A. A. Kaminskii, Yb^{3+}:Y_2O_3 ceramic – a novel solid state material, *Jpn. J. Appl. Phys.* **41** (2002) L1373–L1375.

[143] J. Lu, J. F. Bisson, K. Takaichi, T. Uematsu, A. Shirakawa, M. Musha, K. Ueda, H. Yagi, T. Yanagitani, and A. A. Kaminskii, Yb^{3+}:Sc_2O_3 ceramic laser, *Appl. Phys. Lett.* **83** (2003) 1101–1103.

[144] K. Takaichi, H. Yagi, A. Shirakawa, K. Ueda, S. Hosokawa, and A. A. Kaminskii, Lu_2O_3:Yb^{3+} ceramic – a novel gain material for high power solid state lasers, *Phys. Status Solidi A* **202** (2005) R1–R3.

[145] L. Laversenne, Y. Guyot, C. Goutaudier, M. T. Cohen-Adad, and G. Boulon, Optimization of spectroscopic properties of Yb^{3+}-doped refractory sesquioxides: cubic Y_2O_3, Lu_2O_3 and monoclinic Gd_2O_3, *Opt. Mater.* **16** (2001) 475–483.

[146] G. Boulon and V. Lupei, Energy transfer and cooperative processes in Yb^{3+}-doped cubic sesquioxide laser ceramics and crystals, *J. Lumin.* **125** (2007) 45–54.

[147] G. L. Bourdet, O. Casagrande, N. Deguil-Robin, and B. Le Garrec, Performances of cryogenic cooled laser based on ytterbium doped sesquioxide ceramics, *J. Phys. Conf. Ser.* **112** (2008) 032054.

[148] L. D. Merkle, G. A. Newburgh, N. Ter-Gabrielyan, A. Michael, and M. Dubinskii, Temperature-dependent lasing and spectroscopy of Yb:Y_2O_3 and Yb:Sc_2O_3, *Opt. Commun.* **281** (2008) 5855–5861.

[149] Q. H. Yang, J. Ding, H. W. Zhang, and J. Xu, Investigation of the spectroscopic properties of Yb^{3+}-doped yttrium lanthanum oxide transparent ceramic, *Opt. Commun.* **273** (2007) 238–241.

[150] V. V. Osipov, K. E. Luk'yashina, V. I. Solomonov, V. A. Shitov, A. N. Orlov, V. V. Platonov, and A. V. Spirin, Effect of iso- and heterovalent additives on characteristics of highly transparent Nd (Yb):Y_2O_3 ceramics, *Bull. Lebedev Phys. Inst.* **34** (2009) 347–349.

[151] T. T. Basiev, M. E. Doroshenko, and V. A. Konyushkin, Nd^{3+} and Yb^{3+} doped fluoride laser ceramics, *Techn. Dig. OSA Adv. Opt. Mater. Conf AIOM*, Istanbul (2011), paper AIThA3.

[152] A. Gallian, V. V. Fedorov, S. B. Mirov, V. V. Badikov, S. N. Galkin, E. F. Voronkin, and A. I. Lalayants, Hot pressed ceramic Cr^{2+}:ZnSe gain-switched laser, *Opt. Express* **14** (2006) 11694–11701.

[153] S. Mirov, V. Fedorov, I. Moskalev, D. Martyshkin, and C. Kim, Progress in Cr^{2+} and Fe^{2+} doped mid-IR laser materials, *Laser Photon Rev.* **4** (2010) 21–41.

[154] B. Henderson and R. H. Bartram, *Crystal Field Engineering of Solid-State Laser Materials*, Cambridge University Press (2000).

[155] A. A. Kaminskii, Laser crystals and ceramics: recent advances, *Laser Photon. Rev.* **1** (2007) 93–177.

[156] S. A. Payne, L. L. Chase, and G. D. Wilke, Optical spectroscopy of the new laser material, $LiSrAlF_6$:Cr^{3+} and $LiCaAlF_6$:Cr^{3+}, *J. Lumin.* **44** (1989) 167–176.

[157] G. Huber, S. A. Payne, L. L. Chase, and W.F. Krupke, Optical spectroscopy of Cr^{3+} in ScF_3 and Sc_2O_3, *J. Lumin.* **39** (1988) 259–268.

[158] S. Kuck, L. Fornasiero, E. Mix, and G. Huber, Spectroscopic properties of Cr-doped Sc_2O_3, *J. Lumin.* **87–89** (2000) 1122–1125.

[159] V. Lupei, A. Lupei, A. Ikesue, and S. Florea, Spectroscopic properties of chromium doped Sc_2O_3 ceramics, *CLEO-EQEC*, Munchen, Germany (2009).

[160] H. Eilers, U. Hommerich, S. M. Jacobsen, W. M. Yen, K. R. Hoffman, and W. Jia, Spectroscopy and dynamics of Cr^{4+}:$Y_3Al_5O_{12}$, *Phys. Rev. B* **49** (1994) 15505–15513.

[161] S. Kuck, K. Petermann, U. Pohlmann, and G. Huber, Near-infrared emission of Cr^{4+}-doped garnets: lifetimes, quantum efficiencies and emission cross-sections, *Phys. Rev. B* **51** (1995) 17323–17331.

[162] S. Kuck, Laser-related spectroscopy of ion-doped crystals for tunable solid-state lasers, *Appl. Phys.* **72** (2001) 515–562.

[163] Y. Kalisky, Cr^{4+} doped crystals: their use as lasers and passive Q-switches, *Prog. Quantum Electron.* **26** (2004) 249–303.

[164] E. A. Khazanov and A. M. Sergeev, Concept study of a 100 PW femtosecond laser based on laser ceramics doped with chromium ions, *Laser Phys.* **17** (2007) 1398–1403.

[165] A. Ikesue, K. Yoshida, and K. Kamata, Transparent Cr^{4+} doped YAG ceramics for tunable lasers, *J. Am. Ceram. Soc.* **78** (1995) 2545–2547.

[166] Y. Feng, J. Lu, T. Takaichi, K. Ueda, H. Yagi, T. Yanagitani, and A. A. Kaminskii, Passively Q-switched ceramic Nd^{3+}:YAG/Cr^{4+}:YAG lasers, *Appl. Opt.* **43** (2004) 2944–2947.

[167] H. Yagi, K. Takaichi, K. Ueda, T. Yanagitani, and A. A. Kaminskii, Influence of annealing conditions on the optical properties of chromium-doped $Y_3Al_5O_{12}$, *Opt. Mater.* **29** (2006) 392–396.

[168] J. Dong, A. Shirakawa, K. Takaichi, K. Ueda, H. Yagi, T. Yanagitani, and A. A. Kaminskii, All-ceramic passive Q-switched Yb:YAG/Cr^{4+}:YAG microchip laser, *Electron. Lett.* **42** (2006) 1154–1155.

[169] J. Li, Y. S. Wu, Y. P. Pan, and J. Guo, Fabrication of Cr^{4+},Nd^{3+}:YAG transparent ceramics for self-Q-switched lasers, *J. Non-Cryst. Solids* **325** (2006) 2404–2407.

[170] J. Dong, K. Ueda, H. Yagi, and A. A. Kaminskii, Laser-diode pumped self-Q-switched microchip lasers, *Opt. Rev.* **15** (2008) 57–74.

[171] V. Lupei, A. Lupei, S. Georgescu, and C. Ionescu, Energy transfer between Nd^{3+} ions in YAG, *Opt. Commun.* **60** (1986) 59–63.

[172] V. Lupei, B. Diaconescu, and A. Lupei, The effects of upconversion on the decay of the $^4F_{3/2}$ level of Nd^{3+} in YAG at high pump intensities, *Proc. SPIE* **4430** (2001) 88-96.

[173] V. Lupei and A. Lupei, Emission dynamics of of $^4F_{3/2}$ level of Nd^{3+} in YAG at low pump intensities, *Phys. Rev. B* **61** (2000) 8087–8098.

[174] L. D. Merkle, M. Dubinskii, K. L. Schepler, and S. M. Hedge, Concentration quenching in fine-grained ceramic Nd:YAG, *Opt. Express* **14** (2006) 3893–3903.

[175] Y. Guyot, H. Manaa, J. Y. Rivoire, R. Moncorge, N. Garniers, E. Descroix, M. Bon, and P. Laporte, Excited state absorption and upconversion studies of Nd-doped single crystals $Y_3Al_5O_{12}$, $YLiF_4$ and La $MgAl_{11}O_{11}$, *Phys. Rev. B* **51** (1995) 784 799.

[176] M. Pollnau, P. J. Hardman, M. A. Kern, W. A. Clarkson, and D. C. Hanna, Upconversion-induced heat generation and thermal lensing in Nd:YLF and Nd: YAG, *Phys. Rev. B* **58** (1998) 16076–16092.

[177] S. Guy, C. L. Bonner, D. P. Shepherd, D. C. Hanna, A. C. Tropper, and B. Ferrand, High inversion intensities in Nd:YAG upconversion and bleaching, *IEEE J. Quantum Electron.* **34** (1998) 900–909.

[178] Y. Sato, T. Taira, N. Pavel, and V. Lupei, Laser operation with near quantum-defect slope efficiency in $Nd:YVO_4$ under direct pump in the emitting level, *Appl. Phys. Lett.* **82** (2003) 844–847.

[179] S. Singh, R. G. Smith, and L. G. Van Uitert, Stimulated emission cross section and fluorescent quantum efficiency of Nd in yttrium aluminium garnet at room temperature, *Phys. Rev. B* **10** (1974) 2566–2572.

[180] E. M. Dianov, A. Ya. Karasik, V. B. Neustruev, A. M. Prokhorov, and I. A. Shcherbakov, Direct measurements of fluorescent quantum yield from the metastable $^4F_{3/2}$ state of Nd^{3+} in $Y_3Al_5O_{12}$ crystals, *Sov. Phys. Doklady* **20** (1976) 622–628.

[181] C. J. Kennedy and J. D. Barry, New evidence on the quantum efficiency of Nd:YAG, *Appl. Phys. Lett.* **31** (1977) 91–93.

[182] T. Y. Fan, Heat generation in Nd:YAG and Yb:YAG, *IEEE J. Quantum Electron.* **29** (1993) 1457–1459.

[183] K. K. Deb, R. G. Buser, and J. Paul, Decay kinetics of $^4F_{3/2}$ fluorescence of Nd^{3+} in YAG at room temperature, *Appl. Opt.* **20** (1981) 1203–1206.

[184] A. Rosencwaig and E. A. Hildum, Nd^{3+} fluorescence quantum-efficiency measurements with photoacoustics, *Phys. Rev. B* **23** (1981) 3301–3307.

[185] C. Jacinto, A. A. Andrade, T. Catunda, S. M. Lima, and M. L. Baeso, Thermal lens spectroscopy of Nd-YAG, *Appl. Phys. Lett.* **86** (2005) 034104.

[186] C. Jacinto, T. Catunda, D. Jaque, L. E. Bausa, and J. Garcia-Sole, Thermal lens and heat generation of Nd:YAG lasers operating at 1.064 and 1.34 μm, *Opt. Express* **16** (2008) 6317–6323.

[187] I. Shoji, Y. Sato, S. Kurimura, V. Lupei, T. Taira, A. Ikesue, and K. Yoshida, Thermal-birefringence-induced depolarization in Nd:YAG ceramics, *Opt. Lett.* **27** (2002) 234–236.

[188] A. Benayas, D. Jaque, C. Jacinto, and A. A. Kaminskii, Luminescence quantum efficiency of $Nd^{3+}:Y_3Al_4O_{12}$ garnet laser ceramics determined by pump induced line broadening, *IEEE J. Quantum Electron.* **46** (2010) 1870–1876.

[189] S. Singh, W. A. Buser, W. H. Grodkiewicz, M. Grosso, and L. G. van Uitert, Nd-doped yttrium aluminum garnet with improved fluorescence lifetime of the $^4F_{3/2}$ state, *Appl. Phys. Lett.* **29** (1976) 343–345.

[190] H. P. Christensen, D. R. Gabbe, and H. P. Jenssen, Fluorescence lifetimes for neodymium-doped yttrium aluminum garnet and yttrium oxide powders, *Phys. Rev. B* **25** (1982) 1467–1473.

[191] M. O. Ramirez, J. Wisdom, H. Li, Y. L. Aung, J. Stitt, G. L. Messing, V. Dierolf, Z. Liu, A. Ikesue, R. L. Byer, and V. Gopalan, Three-dimensional grain boundary spectroscopy in transparent high power ceramic laser materials, *Opt. Express* **16** (2008) 5965–5873.

[192] H. Haneda, Role of diffusion phenomena in the processing of ceramics, *J. Ceram. Soc. Jpn.* **111** (2003) 439–447.

[193] J. Li, Y. Wu, Y. Pan, W. Liu, Y. Zhu, and J. Guo, Solid-state reactive fabrication of Cr, Nd:YAG transparent ceramics: the influence of raw material, *J. Ceram. Soc. Jpn.* **116** (2008) 572–577.

[194] W. Zhao, C. Mancini, D. Amans, G. Boulon, T. Epicier, Y. Min, H. Yagi, T. Yanagitani, T. Yanagida, and A. Yoshikawa, Evidence of the inhomogeneous Ce^{3+} distribution across grain boundaries in transparent polycrystalline Ce^{3+}-doped $(Gd,Y)_3Al_5O_{12}$ garnet optical ceramics, *Jpn. J. Appl. Phys.* **42** (2010) 022602.

[195] W. Zhao, S. Anghel, C. Mancini, D. Amans, G. Boulon, T. Epicier, Y. Shi, X. Q. Feng, Y. B. Pan, V. Chani, and A. Yoshikawa, Ce^{3+} dopant segregation in $Y_3Al_5O_{12}$ optical ceramics, *Opt. Mater.* **33** (2011) 684–687.

[196] V. I. Chani, G. Boulon, W. Zhao, T. Yanagida, and A. Yoshikawa, Correlation between segregation of the rare earth dopants in melt crystal growth and ceramic processing for optical applications, *Jpn. J. Appl. Phys.* **49** (2010) 075601.

[197] P. Hong, X. X. Zhang, C. W. Struck, and B. Di Bartolo, Luminescence of Cr^{3+} and energy transfer between Cr^{3+} and Nd^{3+} ions in yttrium aluminium garnet, *J. Appl. Phys.* **78** (1995) 4659–4667.

[198] E. V. Zharikov, V. A. Zhitniuk, G. M. Zverev, S. P. Kalitin, I. I. Kuratev, V. V. Laptev, A. M. Onishkenko, V. V. Osiko, A. S. Pimenov, A. M. Prokhorov, V. A. Smirnov, M. F. Stel'makh, A. V. Shestakov, and I. A. Shcherbakov, Active media for high-efficiency neodymium lasers with non-selective pumping, *Sov. J. Quantum Electron.* **12** (1982) 1652–1653.

[199] A. Lupei, V. Lupei, A. Petraru, and M. Petrache, Energy transfer processes in Cr^{3+}, Nd^{3+}YAG, *Proc. SPIE* **3405** (1997) 587–595.

[200] A. Ikesue, K. Kamata, and K. Yoshida, Synthesis of Nd^{3+}, Cr^{3+}-codoped YAG ceramics for high efficiency solid state lasers, *J. Am. Ceram. Soc.* **78** (1995) 2545–2547.

[201] H. Yagi, T. Yanagitani, H. Yoshida, M. Nakatsuka, and K. Ueda, The optical properties and laser characteristics of Cr^{3+} and Nd^{3+} co-doped $Y_3Al_5O_{12}$ ceramic, *Opt. Laser Technol.* **39** (2007) 1295–1300.

[202] K. Fujioka, T. Saiki, S. Motokoshi, Y. Fujimoto, H. Fujita, and M. Nakatsuka, Luminescence properties of highly Cr co-doped Nd:YAG powder produced by sol-gel method, *J. Lumin.* **130** (2010) 455–459.

[203] J. Mares, Energy transfer in YAG:Nd codoped with Ce, *Czech. J. Phys. B* **35** (1985) 883–891.

[204] Y. Li, S. Zhou, H. Lin, X. Hou, and W. Li, Intense 1064 nm emission by the efficient energy transfer from Ce^{3+} to Nd^{3+} in Ce/Nd codoped YAG transparent ceramic, *Opt. Mater.* **32** (2010) 1223–1226.

[205] P. Samuel, T. Yanagitani, H. Yagi, H. Nakao, K. Ueda, and S. M. Baby, Efficient energy transfer between Ce^{3+} and Nd^{3+} in cerium co-doped Nd:YAG laser quality transparent ceramics, *J. Alloys Comp.* **507** (2010) 475–478.

[206] T. Schweizer, T. Jensen, E. Heumann, and G. Huber, Spectroscopic properties and diode pumped 1.6 µm laser performance in Yb-codoped Er:$Y_3Al_5O_{12}$ and Er:Y_2SiO_5, *Opt. Commun.* **118** (1995) 557–561.

[207] H. Eilers, Effect of particle/grain size on the optical properties of Y_2O_3: Er, Yb, *J. Alloys. Comp.* **474** (2009) 569–572.

[208] J. Zhou, W. Zhang, T. Huang, L. Wang, J. Li, W. Liu, B. Jiang, Y. Fan, and J. Guo, Optical properties of Er, Yb co-doped YAG transparent ceramics, *Ceram. Int.* **37** (2011) 513–519.

[209] S. Lu, Q. Yang, B. Zhang, and H. Zhang, Upconversion and infrared luminescence in Er^{3+}/Yb^{3+} codoped Y_2O_3 and $(Y_{0.9}La_{0.1})_2O_3$ transparent ceramic, *Opt. Mater.* **33** (2011) 746–749.

[210] V. Lupei, A. Lupei, C. Gheorghe, and A. Ikesue, Sensitization of Yb^{3+} emission in (Nd,Yb):$Y_3Al_5O_{12}$ transparent ceramics, *J. Appl. Phys.* **108** (2010) 123112.

[211] V. Lupei, A. Lupei, C. Gheorghe, and A. Ikesue, Spectroscopic properties of Cr^{3+} and energy transfer to rare earth ions in Sc_2O_3 ceramics (unpublished).

[212] T. T. Basiev, K. K. Pukhov, and I. T. Basieva, Cooperative quenching kinetics: computer simulation and analytical solution, *Chem. Phys. Lett.* **432** (2006) 367–370.

[213] A. Lupei, V. Lupei, S. Georgescu, I. Ursu, V. I. Zhekov, T. M. Murina, and A. M. Prokhorov, Many body energy transfer processes between Er^{3+} ions in YAG, *Phys. Rev. B* **41** (1990) 10923.

[214] J. Ueda and S. Tanabe, Visible to near infrared conversion in Ce^{3+}-Yb^{3+} codoped YAG ceramics, *J. Appl. Phys.* **106** (2009) 043101.

[215] H. Lin, S. Zhou, H. Tang, Y. Li, X. Hou, and T. Jia, Near infrared quantum cutting in heavy doped $Ce_{0.03}Yb_{3x}Y_{(2.97-x)}Al_5O_{12}$ transparent ceramics for crystalline silicon soloar cells, *J. Appl. Phys.* **107** (2010) 043107.

[216] J. L. Yuan, X. Y. Zeng, J. T. Zhao, Z. J. Zhang, H. H. Chen, and X. X. Yang, Energy transfer mechanisms in Tb^{3+},Yb^{3+} codoped Y_2O_3 downconversion phosphor, *J. Phys. D: Appl. Phys.* **41** (2008) 105406.

[217] F. Auzel, Upconversion and anti-Stokes processes with f and d ions in solids, *Chem. Rev.* **104** (2004) 139–173.

[218] L. C. An, J. Zhang, M. Liu, and S. W. Wang, Upconversion properties of Yb^{3+}, Ho^{3+}:Lu_2O_3 sintered ceramic, *J. Lumin.* **122/123** (2007) 125–127.

[219] F. Qin, Y. Zheng, Y. Yu, Z. Cheng, P. S. Tayebi, W. Cao, and Z. Zhang, Ultraviolet and violet upconversion luminescence in Ho^{3+}-doped Y_2O_3 ceramic induced by 532 nm CW laser, *J. Alloys Comp.* **509** (2011) 1115–1118.

[220] G. Qin, J. Lu, J. F. Bisson, Y. Feng, K. Ueda, H. Yagi, and T. Yanagitani, Upconversion luminescence of Er^{3+} in highly transparent YAG ceramics, *Solid State Commun.* **132** (2004) 103–106.

[221] M. Liu, S. Wang, J. Zhang, L. An, and L. Chen, Preparation and upconversion luminescence of $Y_3Al_5O_{12}$:Yb^{3+}, Er^{3+} transparent ceramics, *J. Rare Earths* **24** (2006) 732–735.

[222] J. Zhang, S. Wang, L. An, M. Liu, and L. Chen, Infrared to visible upconversion of Er: Y_2O_3 transparent ceramics, *J. Lumin.* **122–123** (2007) 8–10.

[223] X. Hu, S. Zhou, W. Li, Y. Li, H. Liu, H. Teng, and T. Jia, Investigation of upconversion in Er^{3+}/Yb^{3+} codoped yttria ceramics, *J. Am. Ceram. Soc.* **93** (2010) 2779–2782.

[224] A. Lupei, V. Lupei, C. Gheorghe, A. Ikesue, and E. Osiac, Upconversion emission of RE^{3+} in Sc_2O_3 ceramic under 800 nm pumping, *Opt. Mater.* **31** (2009) 744–749.

[225] X. Hou, S. Zhou, H. Lin, H. Teng, Y. Li, W. Li, and T. Jia, Violet and blue up-conversion luminescence in Tm/Yb codoped Y_2O_3 transparent ceramic, *J. Appl. Phys.* **107** (2010) 083101.

[226] X. Xu, Z. Zhao, P. Song, B. Jiang, G. Zhou, J. Xu, P. Deng, G. Bourdet, J. C. Chanteloup, J. P. Zou, and A. Fulop, Upconversion luminescence in Yb^{3+}-doped YAG ceramics, *Physica B* **357** (2005) 365–369.

[227] C. G. Dou, Q. H. Yang, X. M. Hu, and J. Xu, Cooperative upconversion luminescence of ytterbium doped yttrium lanthanum oxide ceramic, *Opt. Commun.* **281** (2008) 692–695.

[228] G. Boulon and V. Lupei, Energy transfer and cooperative processes in Yb3+-doped cubic sesquioxide laser ceramics and crystals, *J. Lumin.* **125** (2007) 45–54.

[229] V. Lupei, A. Lupei, G. Boulon, A. Jouini, and A. Ikesue, Assessment of the distribution of the Yb3+ ions in Sc_2O_3 ceramics from cooperative absorption and emission, *J. Alloys Comp.* **451** (2008) 179–181.

[230] A. A. Kaminskii, K. Ueda, H. J. Eichler, S. N. Bagaev, K. Takaichi, J. Lu, A. Shirakawa, H. Yagi, and T. Yanagitani, and Observation of nonlinear lasing chi((3))-effects in highly transpatrent nanocrystallline Y_2O_3 and $Y_3Al_5O_{12}$ ceramics, *Laser Phys. Lett.* **1** (2004) 6–11.

[231] A. A. Kaminskii, S. N. Bagaev, H. J. Eichler, K. Ueda, K. Takaichi, A. Shirakawa, H. Yagi, T. Yanagitani, and H. Rhee, Observation of high order Stokes and anti-Stokes-(3)- generation in highly transparent laser-host Lu_2O_3 ceramics, *Laser Phys. Lett.* **3** (2006) 310–313.

[232] A. A. Kaminskii, S. N. Bagaev, K. Ueda, H. Yagi, H. J. Eichler, A. Shirakawa, M. Tokurakawa, H. Rhee, K. Takaichi, and T. Yanagitani, Nonlinear-laser-effects in novel garnet-type fine-grained ceramic host $\{YGd_2\}[Sc_2](Al_2Ga)O_{12}$ for Ln^{3+} lasants, *Laser Phys. Lett.* **6** (2009) 671–677.

11

Ceramic lasers

Since the demonstration of the first transparent ceramics of laser quality [1], which enabled laser emission performance similar to that of single crystals, research and engineering in the field of laser materials, processes, devices and applications has developed explosively, as reflected by numerous review papers [2–13]. As discussed in Chapter 2, the performance of solid-state lasers is determined by a complex interplay between the laser material, the pumping process and the laser design in determining the optimum utilization of the pump radiation for efficient laser emission in the desired wavelength range, temporal regime and power range, with suitable beam characteristics. A major task of this correlated approach is the maximum limitation of the non-radiative de-excitation processes, which lead to wastage of excitation and to the generation of heat, as well as the proper control of the thermal field inside the laser material and of the thermomechanical and thermo-optical processes. The undesired effects of the residual luminescence emission, such as an amplified spontaneous emission, must also be avoided. These restrictions impose specific requirements on the three main components (material, pumping, design) as well as on the whole system in a correlated approach. The spectroscopic properties of the ceramic laser materials, discussed in Chapter 10, offer a good basis for selection of the laser material and pumping conditions and indicate that the new possibilities offered by ceramic laser materials can contribute to the extension of laser performance and of the areas of application.

11.1 Pumping schemes

In principle, the pumping schemes must fulfill two major tasks. A high degree of absorption of pump radiation and optimal utilization of the absorbed radiation for laser emission are required, while limiting the possibility for manifestation of the de-excitation processes which are not involved in the laser emission scheme. In addition to the absorption coefficient of the laser material, the pump absorption is essentially influenced by the superposition of the pump radiation spectrum with the absorption spectrum of the laser material. From a practical point of view, this would require the development of pumping sources correlated with the absorption properties (spectrum peak position and FWHM, cross-sections) of the laser material as well as the selection of laser material parameters (doping concentration, size) and laser design to give complete absorption and proper distribution of the absorbed power.

The efficiency of utilization of the absorbed excitation in the laser process as well as the limitation of parasitic heat generation is essentially connected with the quantum defect between the pump and laser emission. Since for each laser emission scheme the quantum of the laser emission determined by the position of the emitting and of the terminal levels, as well as the position of the terminal level relative to the ground state are fixed, the only way of influencing the quantum defect would be via the wavelength of the pump by reducing or eliminating the upper quantum defect between the pump energy level and the emitting laser level and/or by optimal use of the thermallization of the energy levels with the neighboring levels.

Generally, pumping sources with broad-band or multiline emission (flashlamps, solar radiation) show poor superposition with the absorption lines of RE^{3+} laser ions, and thus a large part of the pump radiation travels inside the laser material without being absorbed by the laser active ions. Moreover, this modality of pumping involves a variety of absorption transitions, each leading to a different quantum defect; however, an average quantum defect can be defined and usually it is quite high. The development of efficient and high power diode lasers in various wavelength ranges could enable resonant excitation of an individual absorption line with suitable absorption properties. However, diode laser pumping raises several points of concern. Traditionally, the width of the emission line of diode lasers is a few nanometers, which could be too broad for pumping several intense but narrow absorption bands of the RE^{3+} ions in many laser materials, particularly at low temperatures, and thus part of the pump radiation will not be absorbed. An additional complication can be introduced by the thermal shift of the emission line of these diode lasers and by the thermal shift of the absorption lines of the RE^{3+} ions. Although modern solutions for very narrow-emitting and wavelength stabilized diodes are available, their utilization is not very straight forward because of the thermal shift of the narrow absorption lines of the material. Special care must be taken in the case of laser materials with multicenter structure designed to give a broad emission spectrum, since these systems show broad and sometimes multi-peaked absorption lines, corresponding to the various structural centers. In such cases the emission spectrum of the pump source should have complete superposition on the whole inhomogeneously broadened absorption line to grant excitation of all these centers in order to make complete use of the broad emission for short pulse laser generation.

Resonant pumping as close as possible to the emitting laser level, preferably direct pumping into this level, can be very important for low quantum defect laser schemes, resulting in increased laser parameters and reduced heat generation. Pumping on transitions originating from thermally populated upper Stark levels of the ground manifold can reduce the quantum defect further, and accidental spectral degeneracy of several such transitions can be useful for increasing the pumping efficiency. By its effect on the intracavity laser intensity I_ω at the fundamental frequency ω, direct pumping into the emitting level can contribute to marked enhancement of the efficiency of the non-linear $n\omega$ processes since $I_{n\omega} \propto (I_\omega)^n$ [5, 14]. In this case, an additional contribution to the global efficiency is given by the reduction of thermal effects on the beam quality.

11.1.1 Nd ceramic materials

Owing to the strongly absorbing levels in the visible and infrared, Nd laser ceramic materials can be pumped with a very large variety of pumping sources (lamps, diode lasers, solar radiation). The poor utilization of the pump emission spectrum and the large quantum defect in the case of lamp pumping makes resonant pumping with diode lasers in levels close to the emitting level of major importance for high performance laser emission. The first experiments on diode pumped Nd:YAG lasers used direct pumping into the level $^4F_{3/2}$ in the 870 nm range [15, 16], but the low power of the available diodes and the weak absorption in this level at the traditional 1 at.% Nd doping (room temperature peak absorption coefficient $\sim$4 cm^{-1}) determined the change to pumping into the strong $^4I_{9/2} \rightarrow {}^4F_{5/2}$ absorption around 809 nm (absorption coefficient $\sim$11.4 cm^{-1}) with the more efficient and powerful AlGaAs diode lasers. Pumping at 809 nm introduces a relative quantum defect of $\sim$0.24 for the laser emission at 1.06 μm, and about 30% of this quantum defect represents the upper part (the difference between the pump level and the emitting level), which has no role in the laser emission itself.

As illustrated in Figure 10.13, at larger Nd concentrations the absorption coefficient for various transitions to $^4F_{3/2}$ becomes fairly large and, in principle, almost all transitions originating from the Stark components of the ground manifold can be used for pumping in this level. Obviously, the higher this Stark component is, the more the quantum defect will decrease, although the absorption can be hampered by the smaller fractional thermal population. Thus, the relative quantum defect for the 1064 nm laser emission ranges from 18.3% when pumping in the transition $^4I_{9/2}(1) \rightarrow {}^4F_{3/2}(2)$ to 11% for the transition $^4I_{9/2}(5) \rightarrow {}^4F_{3/2}(1)$. As mentioned, for Nd:YAG the largest effective absorption coefficient at room temperature in level $^4F_{3/2}$ corresponds to the sharp (below 1 nm) 869 nm transition $^4I_{9/2}(1) \rightarrow {}^4F_{3/2}(2)$; however, the thermally activated (hot-band) transitions $^4I_{9/2}(2) \rightarrow {}^4F_{3/2}(1)$ and $^4I_{9/2}(3) \rightarrow {}^4F_{3/2}(2)$ show quasi-degeneracy, resulting in a fairly broad (around 3 nm) two-peaked (885.7 nm and 884.4 nm) band centered at 885 nm that can be used for diode laser pumping. The transitions corresponding to these absorption lines are shown in Figure 11.1. The room temperature average absorption coefficient of this band for 1 at.% Nd is around 1.65 cm^{-1} and increases slightly below linear at high C_{Nd} due to the broadening of lines, as shown in Figure 11.2. The temperature dependence in the 300–400 K range of the fractional thermal coefficients f of the Stark levels of the ground manifold $^4I_{9/2}$, which influence the effective absorption cross-sections at various temperatures, is shown in Figure 11.3. In this temperature range f decreases by $\sim$15.4% for the first Stark level Z_1 (from 0.465 to 0.405), it remains almost constant for Z_2, then increases gradually for the upper Stark levels: the sum of the coefficients f for levels Z_2 and Z_3 corresponding to the 885 nm band increases by only 5.2% (from 0.424 to 0.446), which

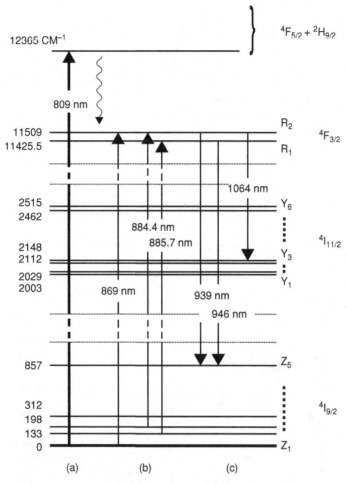

Figure 11.1 Schemes of pumping and laser emission in Nd:YAG: (a) pumping into the level $^4F_{5/2}$ at 809 nm; (b) pumping into the level $^4F_{3/2}$; (c) the most intense emission lines in the 0.9 and 1 μm range.

partially compensates for the thermal decrease of the peak absorption and emission cross-sections. This would give much higher thermal stability of pump absorption and laser emission for the 885 nm than the traditional 809 nm pumping; moreover, it will reduce the relative quantum defect by ~30%, from 0.24 to 0.168 in the case of 1064 nm laser emission, and by 55.5%, from 0.145 to 0.065 for 946 nm laser emission.

Quasi-degeneracy of two thermally activated transitions can occur in several other Nd laser materials, including garnets. A very interesting case is Nd:GSGG where the hot bands $^4I_{9/2}(2) \rightarrow {}^4F_{3/2}(1)$ and $^4I_{9/2}(3) \rightarrow {}^4F_{3/2}(2)$ are practically coincident [17]. Moreover, the sum of thermal population coefficients $f_2 + f_3$ is larger than f_1 even at room temperature (0.444 compared to 0.425) and the difference increases at 400 K (0.458 versus 0.374).

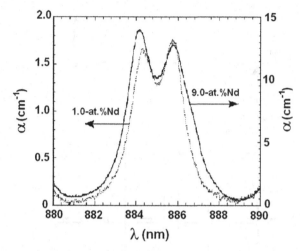

Figure 11.2 Detail of the room temperature absorption spectrum of Nd:YAG in the region of the $Z_2 \rightarrow R_1$ and $Z_3 \rightarrow R_2$ transitions.

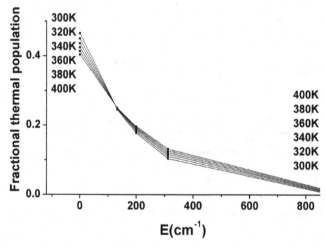

Figure 11.3 Temperature dependence of the thermal population coefficients of the Stark levels of the ground manifold $^4I_{9/2}$ of Nd in YAG.

The strong development (extension of wavelength range, increased power) of diode lasers in recent decades has led to renewed interest in pumping the Nd lasers into the emitting level, and several results for ceramic lasers will be discussed in Section 11.4.1. The effect of the low absorption cross-sections of the 869 nm and 885 nm bands in the $^4I_{9/2} \rightarrow {}^4F_{3/2}$ spectrum of Nd:YAG on the pump absorption efficiency can be managed by controlling the path of the pump radiation inside the laser material, for example by using long laser rods with conventional 1 at.% Nd doping in end-pumped configuration [18, 19], zig-zag configurations or multipass pumping [5] and/or by higher doping concentrations [20–23]. Nevertheless, all of these approaches can have undesired consequences: the long laser rods lead to increased residual optical losses, multi-pass pumping is technically complex, whereas high Nd doping concentrations can favor self-quenching of the emission.

Laser emission when pumping at 946 nm in the $^4I_{9/2}(5) \rightarrow {}^4F_{3/2}(1)$ transition, which gives the lowest quantum defect, was reported for Nd:YAG crystal [24], but the efficiency relative to the incident pump power was low because of the low thermal population of the Z_5 level at room temperature. Raising the temperature could improve this absorption to a certain extent, but it must be handled with care because of thermallization of the emitting level with the level $^4F_{5/2}$.

The pump diode lasers are operated in CW or QCW mode, but not in the high energy pulse regime, and pumping of Nd ceramic lasers with flashlamps could be still necessary to generate high energy pulses. Attempts to use pumping of Nd:YAG ceramics with lamps with the ultraviolet part cut by special Sm-doped filters to prevent absorption in the stable or transient color centers proved successful [25] and showed that these materials could replace Nd:glass materials in high energy lasers, with the obvious advantage of better thermomechanical properties. Since a large part of the lamp emission is not absorbed by the Nd^{3+} emission, co-doping with strongly absorbing ions such as Cr^{3+} that can transfer the excitation to Nd^{3+} was attempted [26, 27]; this approach also proved useful in the case of solar-pumped Nd:YAG lasers [28].

For diode laser pumping, Nd-doped sesquioxides offer several lines in the near infrared, the strongest line in the transition $^4I_{9/2} \rightarrow {}^4F_{5/2}$ being in the region of 820–825 nm, with room temperature absorption cross-sections of the order of 5–7×10^{-20} cm^2, followed by the absorption around 808 nm with absorption cross-section $\sim 4 \times 10^{-20}$ cm^2. Pumping in the transition $^4I_{9/2} \rightarrow {}^4F_{3/2}$ can be done on the $Z_1 \rightarrow R_2$ line around 877 nm or on the hot-band transition $Z_2 \rightarrow R_1$, both with cross-sections around 2×10^{-20} cm^2; no spectrally degenerate hot bands are available. These ceramics can also be pumped with flashlamps.

11.1.2 Yb-doped ceramics

Because of their simple electronic structure, with the excited manifold around 10 000 cm^{-1}, Yb-doped laser materials can only be pumped by infrared diode lasers. The weak absorption of the pump radiation imposes the utilization of high Yb concentrations or laser design with a long pump radiation path inside the laser material. However, very high Yb concentrations can lead to the onset of parasitic de-excitation by cooperative processes inside the system of Yb ions or by migration-assisted energy transfer to accidental impurities. Because of the quasi-three-level nature of the Yb laser emission, utilization of large laser rods is not recommended and the long absorption path can be realized by multi-pass pump radiation in a thin laser material or by using long fibers doped at very low concentration. The reabsorption at 300 K of the laser emission on the strongest emission line $^2F_{5/2}(1) \rightarrow {}^2F_{5/2}(3)$ of Yb^{3+}

recommends utilization of low temperatures to reduce the fractional thermal population of the terminal laser level.

As discussed in Section 10.2.1.9, the strongest absorption line of Yb^{3+} corresponds to the transition $^2F_{7/2}(1) \rightarrow {}^2F_{5/2}(1)$ in the 965–980 nm range, but the reduced linewidth, especially at low temperatures, makes the pumping process difficult and the broader but much less intense transition $^2F_{7/2}(1) \rightarrow {}^2F_{5/2}(2)$ in the region of 940 nm is often preferred. Sensitization of Yb^{3+} emission by energy transfer from Nd^{3+} with transfer efficiency exceeding 96% in the case of sesquioxides and YAG was demonstrated [29–31], but laser emission in the case of cubic materials has so far only been reported for heavily doped (Nd, Yb):CaF_2 single crystals [32].

11.1.3 Other RE^{3+} doped ceramics

The well developed diode lasers in the 800 and 940–980 nm ranges enable excitation of infrared (1.5–3 μm) emitting levels for other RE^{3+} ions. Thus, the 3 μm emitting level $^4I_{11/2}$ and the 1.5 μm emitting level $^4I_{13/2}$ of Er^{3+} can be excited by 940–980 nm pumping into the level $^4I_{11/2}$ (directly or by sensitization with Yb^{3+}), whereas the 2 μm emission of the level $^3F_4$ of Tm^{3+} can be excited by pumping around 790–800 nm in the level $^3H_4$, followed by the quantum cutting $(^3H_4,^3H_6) \rightarrow (^3F_4,^3F_4)$ cross-relaxation. In both these cases the quantum defect determines heat generation. Recent development of efficient diode lasers in the infrared (1.5 and 2 μm ranges) makes resonant pumping into the emitting level of quasi-three-level infrared emitting RE^{3+} very attractive. Such a modality of pumping would allow low quantum defect laser emission and efficient utilization of the absorbed radiation, and has already been demonstrated for ions such as Ho^{3+}, Er^{3+} or Tm^{3+} in ceramic materials, as will be described later in this chapter.

11.2 Radiative and non-radiative processes in ceramics

As discussed in Section 2.3, in pumped laser material placed in a laser resonator, the excited ions can de-excite by various radiative (laser or luminescence) or non-radiative (electron–phonon interaction, energy transfer) processes and the amount of ions participating in laser emission can be delineated by the laser emission efficiency η_l. The laser emission is given by the excited ions inside the laser mode volume and pumped above threshold, whereas the excited ions inside the pumped volume but outside the laser mode volume, as well as those that form the inversion of population at the laser threshold, can de-excite by luminescence and non-radiative processes that manifest in heat generation and generally $\eta_l = \eta_v(1 - P_{th}/P)$. The relevance of the superposition integral of the laser mode and the pumped volume η_v is determined by the regime of laser emission. The material factors that influence the laser threshold and thus the laser emission efficiency η_l in the case of CW emission are given explicitly in Eq. (2.85). Several of these factors, such as the emission quantum efficiency and the effective lifetime, can be influenced by the conditions of the

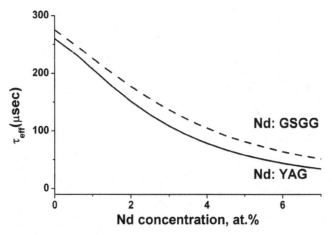

Figure 11.4 Calculated effective lifetime for Nd:YAG and Nd:GSGG.

experiment (concentration of the doping ions, temperature), whereas the quantum defect ratio $\eta_{qd}^{(l)}$ is influenced by the pump wavelength.

11.2.1 De-excitation of the emitting level in laser ceramics

As discussed above, control of the pump absorption can require increased Nd concentrations. In the case of Nd^{3+} ions, self-quenching of the emission by down-conversion or upconversion inside the system of doping ions can reduce the emission quantum efficiency η_{qe} which influences the emission threshold and the generation of heat. The effect of C_{Nd} and of pump intensity on η_{qe} in Nd-doped laser materials was discussed in Section 10.3.2.2 and is illustrated in Table 10.17 for Nd-doped YAG and GSGG under weak excitation (down-conversion only). The calculated concentration dependence of the effective lifetime $\tau_{eff} = \tau_f \eta_{qe}$ for these two materials is given in Figure 11.4, whereas the effect of the pump intensity (down-conversion and upconversion) on η_{qe} in Nd:YAG is illustrated in Table 10.18.

The increased self-quenching of emission was long used as an argument against the highly doped Nd laser materials. However, it was argued [21–24] that the effect of C_{Nd} must be judged by the joint effect of the pump absorption efficiency η_a and the emission quantum efficiency η_{qe} on the laser parameters expressed in the incident pump power: the slope efficiency is not influenced by η_{qe} but is proportional to η_a, whereas the laser threshold is inversely proportional to the product $\eta_a \eta_{qe}$, which can be considered as a figure of merit for the effect of concentration on threshold. Thus, although the decrease in η_{qe} determined by the increased C_{Nd} enhances the threshold expressed in the absorbed power $P_{th}^{(a)}$, in certain concentration ranges the increased absorption can compensate for this effect, leading to a lower laser threshold expressed in the incident power $P_{th}^{(in)}$. Obviously, when using a given pumping source, the absorption efficiency of a laser material is influenced by

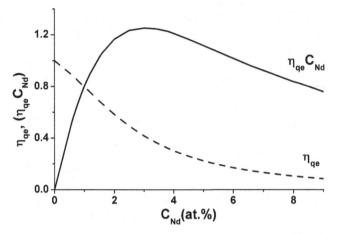

Figure 11.5 Calculated dependence on concentration of the product $C_{Nd}\eta_{qe}$ for Nd:YAG.

its particular geometry (shape, size); however, the product $C_{Nd}\eta_{qe}$ can be considered as an approximate material-only figure of merit for the effect of C_{Nd} on the laser threshold $P_{th}^{(in)}$ in a given material in similar conditions.

The calculated product $C_{Nd}\eta_{qe}$ for Nd:YAG is shown in Figure 11.5 and suggests that the global effect of quite high C_{Nd} (3–4 at.% Nd range) on threshold could be favorable; moreover the increased η_a at high C_{Nd} contributes to a larger slope efficiency $\eta_{sl}^{(i)}$. This indicates that the higher C_{Nd} necessary to grant efficient pump absorption in the case of 885 nm pumping of Nd:YAG would not have strong negative global effects on the laser emission parameters; moreover, in this case the additional positive effect on the quantum effect (Stokes) ratio $\eta_{qd}^{(l)}$ must also be taken into account. More accurate evaluation of the effect of C_{Nd} should account for the actual value of η_a and this can give information on the optimal concentration and size of the laser material. In the case of high excitation intensities that favor enhanced upconversion self-quenching, η_{qe} depends on the pump intensity (Table 10.18) and the data from Figure 11.5 will no longer be valid; however, even in this case increased C_{Nd} could prove useful.

The Nd concentration can influence in a selective manner the lineshape and width of the various absorption and emission lines via the unresolved perturbing crystal field effect inside the ensembles of Nd ions in near lattice sites. In the case of the $^4I_{9/2} \rightarrow ^4F_{3/2}$ absorption of Nd:YAG ceramics, the linewidth can increase by up to $\sim 10\%$, depending on transition, in the C_{Nd} range from 1 to 9 at.% Nd, with corresponding reduction of the peak cross-sections. The positions of the lines can change slightly with the Nd concentration, both because of the asymmetric perturbation effects and because of the slight modification of the crystal lattice parameter. Since the concentration induced broadening is based on the crystal field perturbations produced by the other doping ions, the emission quantum efficiency can vary non-uniformly across the emission band and thus the emission lineshape of a transition can differ from that of the absorption.

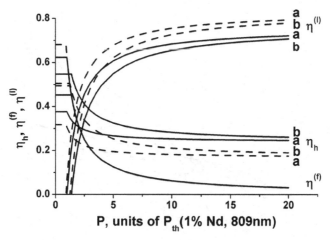

Figure 11.6 Laser, fluorescence and heat generation efficiencies in the 1064 nm Nd:YAG laser at 809 nm (solid line) and 885 nm (dotted line) pumping: (a) 1at.% Nd, (b) 2 at.% Nd.

Besides its effect on absorption, the temperature can also influence the emission properties by changing the fractional coefficients of the two Stark levels R_1 and R_2 of the emitting manifold ${}^4F_{3/2}$ and the emission cross-sections σ_e, and thus their effective emission cross-sections $\sigma_{eff} = \sigma_e f_l$. As discussed in Section 10.2.1.3.1, the crystal field splitting of this manifold can range from 0 to $\sim$250 cm^{-1}, depending on the host. In the case of Nd:YAG it amounts to $\sim$84 cm^{-1} and strong emission lines can be generated from both Stark components, in proportion dependent on temperature. The main emission line at 1064 nm contains contributions from transitions from both these levels: at room temperature the fractional thermal coefficients of levels R_1 and R_2 are 0.6 and respectively 0.4 and despite the lower value for the latter its emission will dominate due to the larger absolute emission cross-section σ_e. Lowering the temperature will determine a crossing point between the effective emission cross-sections $\sigma_{eff} = \sigma_e f_l$ of these two transitions, below which the emission from level R_1 will dominate. The detailed analysis of the effective emission cross-sections from these two Stark levels in a wide temperature range (77 to 650 K) [33] reveals the modification of the proportion of their contributions.

11.2.2 The global balance of de-excitation processes

The dependence of the global fractional laser emission efficiency $\eta^{(l)}$ defined by Eq. (2.64), of the luminescence emission efficiency $\eta^{(f)}$ (Eq. (2.66)) and of the heat load parameter η_h (Eq. (2.68)) on the operating point P/P_{th} in absorbed power for 1 at.% and 2 at.% Nd:YAG is given in Figure 11.6 for pumping at 809 nm and 885 nm assuming $\eta_v = 1$. In order to compare the effect of the concentration and of the pump wavelength, the unit of power for all curves is the threshold for 1 at.% Nd:YAG pumped at 809 nm. Figure 11.6 shows that 885 nm pumping determines higher $\eta^{(l)}$ and lower η_h than 809 nm pumping,

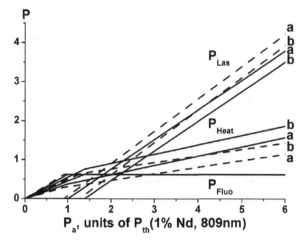

Figure 11.7 Intracavity laser power, thermal power and fluorescence power for CW Nd:YAG laser under 809 nm (solid line) and 885 nm (dotted line) pumping: (a) 1 at.% Nd, (b) 2 at.% Nd.

and above the specific threshold for each pump wavelength and Nd concentration, the dependence of $\eta^{(f)}$ on the operating point is the same regardless of pump wavelength and C_{Nd}.

The power dissipated in each de-excitation process, $P_{Las} = \eta^{(l)} P^{(a)}$, $P_{Fluo} = \eta^{(f)} P^{(a)}$ and $P_{Heat} = \eta_h P^{(a)}$, is represented in Figure 11.7. From this figure it is evident that P_{Fluo} increases linearly with the absorbed power up to the specific threshold of each situation then remains constant and with the same value for all cases. At the same time, P_{Heat} increases up to the threshold value with the slope given by Eq. (2.69) for each case, then changes to the slope given by Eq. (2.70) which depends on the pumping wavelength but is the same for all Nd concentrations.

The variation of the CW laser emission efficiency η_l with the pump-laser volume super-position efficiency η_v, illustrated in Figure 2.5 for Nd:YAG, determines the dependence of the efficiencies of de-excitation processes on η_v. According to Eqs. (2.95) and (2.96), poor volume superposition will increase the laser threshold and reduce the laser slope efficiency; at the same time, it will increase the heat loading parameter η_h, as illustrated in Figure 11.8 for $\eta_v = 1$ (lines a), 0.9(b), 0.8(c) and 0.7(d).

The variation of η_h with η_v will determine the change in slope of the thermal power P_{Heat} in Figure 11.7 above the laser threshold, $\eta_{h,ath}$, according to Eq. (2.107). Thus, in the case of 1 at.% Nd:YAG the heat load parameter $\eta_h^{(f)}$ in the absence of laser emission, which determines the slope $\eta_{h,bth}$ of P_{Heat} below the threshold in Figure 11.7 (Eq. (2.106)), calculated with the emission quantum efficiency $\eta_{qe} = 0.8$ (Table 10.17) is ~0.377. Above the threshold, for CW 1064 nm laser emission pumped at 809 nm, the slope $\eta_{h,ath}$ takes the values 0.24, 0.254, 0.267 and 0.281 for $\eta_v = 1, 0.9, 0.8$ and 0.7 respectively. Generally, the measured heat dissipation should be correlated with the laser emission parameters: a slope

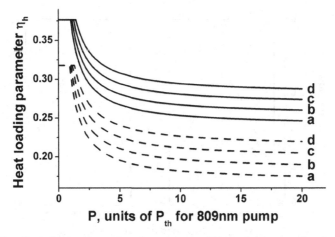

Figure 11.8 The effect of pump-to-mode volume superposition on heat generation in a 1064 nm CW Nd:YAG laser for 1 at.% Nd: solid lines $\lambda_p = 809$ nm, dashed lines $\lambda_p = 885$ nm.

efficiency lower than predicted by the measured residual losses, the outcoupling losses and the quantum defect could be indicative of poor pump-to-laser volume superposition.

11.2.3 The spatial distribution of the de-excitation processes

The dependences of the efficiencies of the de-excitation processes on the operating point presented above are global; however, as discussed in Section 2.5, the spatial distribution of the absorbed power and of the CW laser mode determines the spatial distribution of the de-excitation processes. Knowledge of this distribution is important because in many cases the maximum local rather than the global values of the efficiencies of the de-excitation processes can determine the limits of pump power. For instance, thermal stress fracture can occur when the heat generation in certain regions of the pumped laser material, and not the value averaged over the whole volume, rises above the fracture limit. Such distributions can be calculated for particular cases. Figure 11.9 shows the calculated radial distribution of the efficiencies $\eta^{(l)}$ (solid line), $\eta^{(f)}$ (dashed line) and η_h (dotted line) in the transverse section of a uniformly pumped cylindrical Nd:YAG laser rod that gives single-mode 1064 nm laser emission with the waist equal to 0.6 of the rod diameter. The data are calculated for 1 at.% and 2 at.% Nd, and the pumping is at 809 nm (a) or 885 nm (b) and the ratio $(P_{th}/P) = 0.1$ for each C_{Nd} and pump wavelength.

Figure 11.9 shows that $\eta^{(l)}$ is similar for both concentrations for a given wavelength of pump, with maximal value larger in the case of 885 nm compared with 809 nm pumping, but the heat generation and the luminescence emission have minimal values on the rod axis and increase with distance from the axis in a manner dependent on C_{Nd} and on pump wavelength. Obviously, for any radial point r, the relation $\eta^{(l)} + \eta^{(f)} + \eta_h = 1$ holds. The ratio of heat generation to luminescence increases with C_{Nd}, and is smaller in the case of

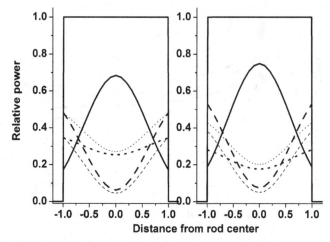

Figure 11.9 Radial distribution of the luminescence (dashed lines) and heat generation (dotted lines) for Gaussian laser emission (solid line) in 1 at.% Nd (thick lines) and 2 at.% Nd (thin lines) in YAG under 809 nm (left) and 885 nm (right) pumping.

885 nm compared with 809 nm pumping. Since the excited ions below the laser threshold de-excite by luminescence and non-radiative processes, these global distributions will be modified by changing the power operation point (P/P_{th}).

These data show clearly that in the presence of laser emission the distribution of heat generation in CW Nd lasers does not follow the distribution of the absorbed power, and that both the pump-mode volume superposition and the operating point can contribute to this. Cooling of the laser rods modifies the thermal field inside the pumped laser material and it was calculated that in the case of uniform generation of heat in a cylindrical rod, uniform lateral cooling induces a parabolic thermal field, with the maximum on the rod axis [34], and this determines a spherical thermal lens effect. However, it is evident that for a laser rod with heat generation distribution as indicated in Figure 11.9, the lateral cooling will produce a distribution of thermal field which is different from parabolic, leading to aspherical aberrations that could determine an apparent dioptric power lower than predicted by Eqs. (2.110): evaluating the heat load parameter from the measured dioptric power with this equation could then provide underestimated values.

Knowledge of the distribution of luminescence emission is also necessary in order to find a suitable solution for limiting the amplified spontaneous emission. It is obvious that cladding the rods with materials that absorb the luminescence escaping from the rod and transform it into heat would introduce an additional annular contribution to the generation of heat.

At higher C_{Nd}, the departures of the distributions of the non-laser de-excitation processes from the distribution of pumping in the case of $\eta_v < 1$ are strongly accentuated, as shown in Figure 11.10 for the radial distribution of the heat generation (solid lines) in a section

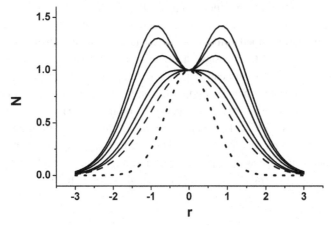

Figure 11.10 Radial distribution of heat generation (thick lines) for a Gaussian pumping profile (dashed line) and Gaussian laser mode (dotted line) with $w_p/w_l = 1.6$, for different Nd concentrations, in ascending order 0.5, 1, 2, 3, 4 at.% Nd in a Nd:YAG laser with 809 nm pumping.

transverse to the laser axis in Nd:YAG pumped with a Gaussian beam at 809 nm and with Gaussian CW laser emission at 1064 nm, with $(w_l/w_p) = 0.6$. The data are normalized relative to the value of heat generation at the center of the rod for each concentration. The shapes of the pump and laser modes are also given, with the maxima normalized to 1 and pumping high above threshold $((P_{th}/P) \approx 0)$ is assumed. The departure of heat generation from the distribution of the pump radiation is obvious and it increases with C_{Nd}. Even for quite small mismatch between the pumped volume and the laser mode volume, the distribution of heat generation will broaden compared with a Gaussian pump distribution. At high pump intensities the heat generation is increased due to the reduction of η_{qe} by upconversion and the departure of the profile of heat generation from the profile of absorbed power in the case of lasers with $\eta_v < 1$ increases.

The distribution of the de-excitation processes presented in Figure 11.9, which corresponds to a uniformly transverse pumped laser rod, is also valid for the situation in a transverse plane at a given distance inside the material in the case of end-top-hat pumping, in a plane-wave approximation. However, in such an end-pumping case, as well as for the case represented in Figure 11.10, the pump radiation attenuates due to absorption along its path inside the laser material and this modifies the operating point (P/P_{th}) for any transverse plane along this path, with an effect on the global distribution of the de-excitation processes.

These data illustrate the very important role of the laser emission efficiency η_l on the distribution of the de-excitation processes and show that proper management of the pump and laser mode volumes is crucial for high CW laser efficiency and for reduction of the parasitic luminescence and heat generation processes which can lead to loss of excitation by amplified spontaneous emission, distortion of the beam and mechanical fracture [35]. Such information is also important for the design of the cooling system and for evaluation of material characteristics such as the emission quantum efficiency from thermal lensing data.

In the case of end-pumped uniformly doped laser rods, the strongest absorption at the entrance face can lead to excessive heating in this region, which limits severely the maximum power of the laser and can impose a reduction in the doping concentration and lengthening of the rod, with negative influence on the residual optical losses. In order to mitigate such effects and to optimize the process of cooling of the rod, it was calculated that (quasi-) uniform distribution of heat generation along the rod and elimination of regions with excessive absorption and heating, without influencing the global absorbed power, could be obtained using rods with a profiled distribution of Nd concentration along the rod or using multi-segmented rods composed of parts with different uniform Nd concentrations [36, 37].

11.2.4 Limitation of heat generation and extension of power scaling

Since heat generation is the main limiting factor for power scaling of Nd lasers, a generalized figure of merit for the effect of pump wavelength, Nd concentration and operating point on scaling could be the ratio between the fractional laser efficiency and the heat load parameter, $(\eta^{(l)}/\eta_h)$ [13, 35]. Pumping on transitions that give a low quantum defect leads to simultaneous enhancement of the laser efficiency and reduction of heat generation, and this can be accomplished in the case of Nd lasers by direct pumping into the emitting manifold. As discussed above, the 885 nm band collecting the transitions $Z_2 \rightarrow R_1$ and $Z_3 \rightarrow R_2$ is very suitable for pumping Nd:YAG lasers at room temperature. The dependence of the generalized figure of merit $(\eta^{(l)}/\eta_h)$ on these factors for Nd:YAG is shown in Figure 11.11 for pumping with 809 nm and 885 nm diode lasers. The units for the absorbed power are the same in all cases and correspond to the threshold for 1 at.% Nd:YAG pumped at 809 nm. It is thus obvious that the highest figure of merit expressed in absorbed power corresponds to the 885 nm pumped weakly doped material, where high values, in the range of 4.6, can be obtained pumping high above threshold, whereas under 809 nm pumping this is limited to ~3. The ratio between the generalized figures of merit for 885 nm and 809 nm pumping high above threshold is in the region of 1.55: this implies that for the same amount of heat generation the power that can be extracted from the laser pumped at 885 nm can be up to ~55% larger than that extracted from 809 nm diode laser pumping. Obviously, in practice the effect of the incident pump radiation is more relevant: the situation can approach the data from Figure 11.11 if absorption efficiencies close to 1 can be achieved for all these concentrations by suitable tailoring of the absorption efficiency (doping concentration, path of pump radiation inside the laser material). Such a situation can be met in the case of 885 nm pumped, weakly doped YAG only for large laser rods, and in normal cases use of more concentrated materials will be necessary. Calculation for various situations indicates that enhanced power scaling under 885 nm pumping compared with 809 nm pumping can be achieved for most cases of practical interest [13, 35].

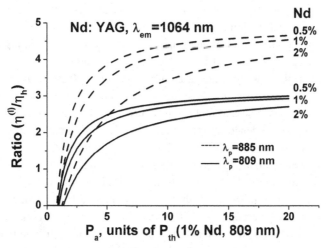

Figure 11.11 Generalized figure of merit ($\eta^{(l)}/\eta_h$) for the 1064 nm Nd:YAG CW laser under 809 nm (solid line) and 885 nm (dotted line) pumping.

In the case of Yb ceramic laser materials these processes are much less severe. However, the necessity for high doping concentrations can lead to very strong migration of excitation and easy access to minute amounts of accidental impurities as well as cooperative processes. All these processes can induce heating effects much above, sometimes twice, the limits imposed by the quantum defect even at moderate Yb concentration and can limit the range of useful Yb concentrations.

11.3 Ceramic laser materials and components

Single crystal and glass laser components doped with a unique species of laser active ion, as a rule, have uniform composition over the whole body of the material. In the case of melt-grown crystals, the segregation coefficient can induce variation of the doping concentration along the crystal and this can limit the useful length from which the laser rods can be processed. Moreover, when doping these crystals with two species of ions, as is the case for sensitized laser materials, the differences in the segregation coefficients of these ions can influence their relative distribution and the sensitization process along the grown crystal. By contrast, ceramic techniques offer the possibility to control the spatial doping distribution and to produce ceramic materials with a controlled profile of doping or of monolithic multifunctional materials, either as composite materials consisting of parts with definite function or as materials co-doped uniformly with different ions with definite functions [38–47]. As discussed in the previous chapters of this book, most laser ceramic materials refer to active laser materials; however, ceramic materials with saturable absorption properties, particularly Cr^{4+}-doped YAG, have also been investigated for passive Q-switching and/or mode-locking of solid-state lasers.

A large variety of garnet and sesquioxide ceramic laser rods with uniform RE^{3+} doping have been produced and, in principle, the size of the ceramic material is limited mainly by the capacity of the technological equipment. In the case of Nd:YAG, high quality rods with uniform doping in the standard 1 at.% Nd range with diameter to ~10 mm and length to 150–200 mm, or plates with surface area of the order of 100 cm^2, have been produced, and extension to larger size seems possible. Such ceramic laser rods have been used in the construction of transverse and end-pumped lasers, in zig-zag, thin-disk, corner cube pumped and in brazing-incidence lasers. A major advantage of ceramic techniques, spatially controlled doping could be very instrumental in controlling the distribution of the absorbed power or of heat generation along the laser rods under longitudinal pumping.

One of the main thermomechanical effects that determines thermal lensing is the bulging of the rods due to the non-uniform distribution of the thermal field inside the pumped and externally cooled laser material. A method of reducing this effect is bonding to both ends of the doped rod of non-doped segments of laser material that contribute to dissipation of heat and to the mechanical restriction of bulging. In the case of crystalline rods, such end caps have been thermally bonded or mechanically clamped to the rod and perfect contact is not easy to achieve. However, it was shown that deposition of thin SiO_2 layers onto the surfaces to be joined and heating at 1200 °C without any pressure allows perfect bonding of crystalline and ceramic YAG [48]. In the case of ceramics, monolithic composite laser materials with undoped caps can be produced directly. The undoped material can also be clad around the laser active material of any shape and can contribute to dissipation of heat and/or to guiding the laser radiation. Ceramic techniques enable the cladding or capping of crystalline materials with undoped ceramic parts. The diffusion of doping ions into the undoped part has been investigated by various methods [47, 49–51] and it was found that a diffusion layer, which can extend to a few tens of micrometers depending on the fabrication process, exists in the case of Nd:YAG.

Very complex laser materials which are suitable for multiple total reflection active mirrors (TRAM) can be produced by ceramic techniques [52]. Composite multifunctional monolithic components consisting of the active material and a saturable absorber (Nd:YAG/Cr^{4+}:YAG) acting as Q-switch can also be produced: such composite materials avoid the necessity for bonding and will give high strength. Mixed ceramic materials doped uniformly with laser ions and SA ions can also be produced. However, in this case the SA ions can influence the losses that contribute to the threshold. Moreover, although the doping concentration of Cr^{4+} ions in these materials is low, below a tenth of a percent, energy transfer from the laser ions can take place, especially at the high concentrations of these ions that favor efficient migration-assisted energy transfer. Ceramic techniques also allow the fabrication of monolithic composite multifunctional materials consisting of an active laser part and a part with the role of suppressor for amplified spontaneous emission, such as (Nd:YAG/Sm:YAG).

Ceramic techniques prove useful for the fabrication of high performance Faraday rotators for isolation of laser amplifiers and birefringence compensation in two-pass high power laser systems. The main requirements for such components are a high Verdet constant, high optical quality, high laser damage threshold, thermal strength, and size scalability. Terbium-doped garnet crystals have very good functional parameters; however, ceramic techniques could be useful for size scaling and for improving the thermomechanical properties. The Verdet constants for ceramic terbium gallium garnet (TGG) [53, 54] and terbium aluminum garnet (TAG) [55] are similar to those of the single crystal materials.

Ceramic materials can be produced in shapes close to those required for the final product and are then processed by mechanical or combined mechanical and thermochemical techniques and a very thorough cleaning is then necessary. Due to the granular structure of ceramics, processing of flat surfaces is not easy, and irregular surfaces can contribute to scattering of light. As a final step, advanced methods for strengthening the surfaces of the processed rods by thermochemical methods have been investigated [56].

11.4 Ceramic lasers

Despite the short period of time since the advent of the first ceramic laser, the variety of lasers based on these materials is already extremely large. This is only normal, since ceramic lasers can use the very rich experience accumulated with single crystal lasers and accompanying techniques. They also fill many hopes that could not be put in practice with crystal lasers. The ceramic laser materials allow optimization of design solutions for the various classes of lasers, from microlasers to huge systems, according to their specific requirements and make possible tailoring of solid-state lasers to match the requirements of the applications. Moreover, the extended capabilities of ceramic materials can offer conditions for new processes and allow considerable extension of the variety and characteristics of solid-state lasers.

11.4.1 Nd-doped ceramic lasers

As discussed in Chapter 10, the spectroscopic and emission dynamics properties of the Nd^{3+} ion allow laser emission in a large variety of temporal regimes from CW to very short (below picosecond) pulses. There are several transitions of quite large emission cross-section in the 0.9, 1 and 1.3 μm range, originating from the two Stark levels of the manifold $^4F_{3/2}$ to various Stark levels of the manifolds $^4I_{9/2}$, $^4I_{11/2}$ and $^4I_{13/2}$, respectively. Laser emission of Nd was intensively investigated in various cubic crystalline materials such as garnets and sesquioxides, and this stimulated interest in the production of polycrystalline materials of these compounds by ceramic techniques.

11.4.1.1 Free generation (CW, QCW) Nd ceramic lasers

11.4.1.1.1 Nd-doped garnets

A. Nd garnet lasers at 1 μm The main activity in the field of Nd laser ceramics was directed at Nd:YAG. At room temperature the transitions $^4F_{3/2}$ (R_i) $\rightarrow$ $^4I_{11/2}$ (Y_j) contain several lines of high emission cross-section that could give laser emission: the most intense and most utilized emission line is at 1064.15 nm, with effective emission cross-section of $\sim(2.8\text{–}3)\times10^{-19}$ cm^2, based mainly on transition $2\rightarrow3$, with a small contribution from transition $1\rightarrow2$. Other lines in this transition [33] that could give laser emission are $2\rightarrow1$ at 1052.1 nm (room temperature effective emission cross-section $\sim0.94\times10^{-19}$ cm^2), $1\rightarrow1$ at 1061.5 nm ($\sim0.94\times10^{-19}$ cm^2), $2\rightarrow4$ at 1068.2 nm ($\sim0.6\times10^{-19}$ cm^2), $1\rightarrow3$ at 1073.7 nm ($\sim1.64\times10^{-19}$ cm^2), $1\rightarrow4$ at 1077.9 nm ($\sim0.76\times10^{-19}$ cm^2), $2\rightarrow6$ at 1111.9 nm ($\sim0.36\times10^{-19}$ cm^2), $1\rightarrow5$ at 1115.8 nm ($\sim0.41\times10^{-19}$ cm^2) and $1\rightarrow1$ at 1122.5 nm ($\sim0.4\times10^{-19}$ cm^2).

The first Nd:YAG ceramic laser demonstrated CW laser emission in 1.1 at.% doped material produced by solid-state reactive sintering, with efficiency similar to a corresponding crystal laser under diode laser pumping at 809 nm [1]. The strong development of these materials and the introduction of new fabrication techniques based on wet synthesis [57] led to a tremendous volume of activity on 1064 nm laser emission under Ti:sapphire or 809 nm diode laser pumping, with continuously improving performances. The range of Nd concentrations for the CW laser regime was soon extended to 4.6 at.% Nd [58] and 6.9 at.% Nd [24], whereas the development of large fine-grained ceramic laser rods enabled demonstration of 1.5 kW range CW laser emission diode laser pumping [4]. Comparative investigation of ceramic and single crystal Nd:YAG lasers demonstrated similar and sometimes improved qualities [1–4, 22, 59]. Diode laser pumping at 809 nm in the energy level $^4F_{5/2}$ of conventional or composite ceramic Nd:YAG lasers demonstrated remarkable performances, such as 511 W from a composite 1 at.% rod with transverse pumping [60] or 144 W for end-pumped core-doped rod [41]. Composite ceramic rods consisting of an Nd-doped YAG core with Sm:YAG cladding manifested higher QCW laser parameters compared with uniformly doped rods or with rods with undoped YAG cladding [61]. The highest reported performances for diode laser pumped Nd:YAG lasers are in the 100 kW range, either from a system of seven coherently coupled MOPA lasers [62, 63] or in a low-duty regime on connecting six modules of pairs of thin Nd:YAG ceramic plates in a single-aperture power oscillator with the laser beam traveling in a zig-zag (ThinZAG concept) [64, 65]. High power (1020 W) 1064 nm laser emission was obtained in a QCW (1000 Hz) 808.5 nm transverse diode pumped MOPA architecture with three 1 at.% Nd:YAG ceramic rods [66]. When tested in single-rod oscillator configuration, each of these rods provided 426 W average power with optical–optical efficiency similar to that of single crystal rods. Very high power (67 kW) in a burst-mode high aperture heat capacity laser was obtained with a cascade of five $10\times10\times2$ cm^3 Nd:YAG ceramic plates [67] face-pumped with 808 nm diode lasers. Side-pumping of such large-aperture Nd:YAG ceramic plates clad with Sm:YAG ceramic in order to suppress ASE proved very efficient in improving the stability and the front of the laser beam [68].

The first Nd:YAG ceramic lasers directly pumped at 885 nm with Ti:sapphire or diode lasers [20–23, 69] demonstrated clearly the advantages of this modality of pumping by enhanced laser performance (lower threshold and higher slope efficiency) as illustrated in Figure 11.12 for uncoated laser rods. These performances were subsequently enhanced

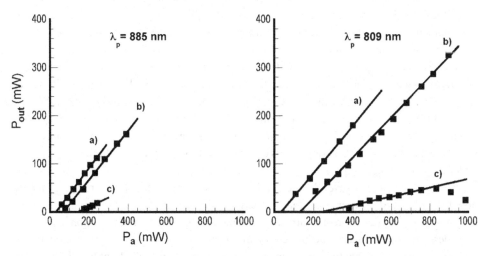

Figure 11.12 Earliest data on CW laser emission in Nd:YAG ceramics under 885 nm pumping (a) 1.0 at.%, 3 mm thick, efficiency 0.54, (b) 3.8 at.%, 2.5 mm thick, efficiency 0.51, (c) 6.8 at.%, 0.4 mm thick efficiency 0.19; and at 809 nm pumping (a) 1.0 at.%, 3 mm thick, efficiency 0.49, (b) 3.8 at.%, 2.5 mm thick, efficiency 0.42, (c) 6.8 at.%, 0.4 mm thick, efficiency 0.09.

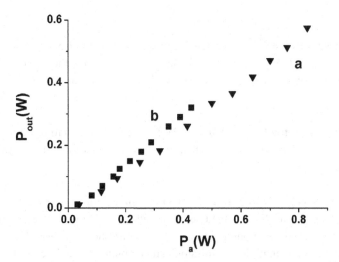

Figure 11.13 1064 nm CW laser emission of 1 at.% Nd:YAG under Ti:sapphire pumping (a) at 809 nm and (b) at 885 nm.

close to the limit imposed by the quantum defect, as shown in Figure 11.13 for AR coated higher quality ceramics pumped by a Ti:sapphire laser to enable comparison under similar conditions. The improvement of the CW laser parameters was generally larger than 9.5% predicted by the quantum defect ratio, owing to the additional contribution of the reduced heat generation and of the increased beam quality. Subsequently, a highly efficient

250 W laser under diode laser pumping was reported [70]. Nevertheless, the strong recent development of 885 nm laser diodes could contribute to considerable scaling of power under this wavelength of pump.

A flashlamp pumped 1064 nm Nd:YAG ceramic laser was operated in QCW free-generation regime under 5 ms pulse excitation at 20 Hz with a Xe lamp whose ultraviolet radiation was filtered with Sm-doped flow glass tube: an average power of 387 W was obtained from the 1.1 at.% Nd:YAG ceramic rod of 8.3 mm diameter and 152 mm length, slightly better than from a single crystal rod [71]. No solarization of the rod was observed, indicating that the Nd:YAG ceramic rods could be used for generation of high energy pulses under lamp pumping.

As discussed above, several other emission transitions in the 1 μm range could be possible in Nd:YAG. Laser emission on such transitions can be selected by using wavelength selecting optics (mirrors, intracavity elements – gratings, prisms, etalons, selective absorbers and so on) that avoid the competition from other transitions and by operating under conditions such that the competing transitions cannot reach the laser threshold. Quite efficient QCW laser (180 μs) emission with 1.1 kHz repetition rate (optical slope efficiency 24.7%) was thus obtained at 1123 nm in Nd:YAG ceramic under 1000 W diode laser pumping [72, 73] by using a strictly controlled coating to avoid competition from nearby transitions. Efficient CW emission of Nd:YAG ceramic at 1052 nm was also demonstrated [74].

B. Laser emission in the 1.3 μm range Laser emission in the 1.3 μm range in Nd:YAG ceramics with high slope efficiency (~35%) was reported on the quite intense (room temperature emission cross-section ~0.8×10^{-19} cm^2) transition $^4F_{3/2}$ (2) → $^4I_{13/2}$(1) at 1.319 μm [75, 76].

C. Laser emission in the 0.9 μm range CW laser emission at 946 nm on the transition $^4F_{3/2}(2)$ → $^4I_{9/2}(5)$ in Nd:YAG ceramics had parameters similar to those of crystal Nd:YAG lasers under the same conditions [77].

D. Multiple wavelength Nd ceramic laser emission Multiple laser emission at several discrete wavelengths can be obtained using broad-band or multiple dichroic resonator optics and under conditions of pumping that allow the threshold for several transitions to be reached. Multiple wavelength laser emission can be obtained in uni-center laser materials on emission lines inside the same inter-manifold transition or between different manifolds or in multicenter materials using emission lines from the various structural centers excited simultaneously by pumping. Laser emission at two or more wavelengths by transitions inside the same or different inter-manifold transitions was demonstrated in several Nd-doped crystals and ceramics. Thus, for the 1.3 μm range, simultaneous emission was obtained on the 1.319 μm and 1.338 μm emission lines in Nd:YAG ceramics [78], and further tri-wavelength laser emission including these transitions and the 1.064 μm line was reported [79]. The multi-wavelength lasers could be of interest for subsequent non-linear processes such as mixing or frequency multiplication, which would enable generation of radiation from the visible to the terahertz region.

Multiple wavelength laser emission can be obtained in disordered systems, where the multicenter structure can offer different emission wavelengths for the same transition from the various centers. Such dual wavelength emission was obtained in an Nd-doped solid-solution disordered $Y_3Sc_{1.5}Al_{3.5}O_{12}$ ceramic thin-disk laser [80]. Particularly useful for multi-wavelength emission could be the compositionally disordered garnets with intrinsic disorder such as the CNGG–CLNGG family where the multicenter structure of the spectra is more accentuated.

E. Sensitized CW laser emission in Nd-doped garnet ceramics The increased concentration and uniformity of doping in ceramics enable tailoring of laser materials with sensitized emission. Of particular importance are the (Cr^{3+}, Nd):YAG ceramics since Cr^{3+} has strong and broad absorption bands in the visible and thus these materials can be used for lamp or solar pumping. In the case of crystals, the segregation coefficients for these two species of doping ions are different, and thus the conditions of sensitization are non-uniform. The possibility of fabricating highly transparent (Cr^{3+}, Nd):YAG ceramics was demonstrated [26] soon after the first Nd:YAG transparent ceramic. Highly efficient laser emission and amplification qualities in such ceramics were demonstrated under lamp pumping [27, 81–84]. Under solar pumping using a 4 m^2 Fresnel lens collector, 80 W CW laser emission was obtained, with 4.3% solar–laser power conversion efficiency [28, 85]. These results show that the (Cr^{3+}, Nd):YAG ceramics have good potential for the construction of highly efficient lamp pumped Nd lasers and solar pumped lasers. The analysis of these results indicates further directions for optimization of the laser material to increase the energy transfer from Cr^{3+} to Nd^{3+}.

11.4.1.1.2 Nd sesquioxide ceramic lasers The first Nd-doped oxide ceramic laser reported was Y_2O_3 with 10% ThO_2 under lamp pumping [86, 87]. The laser was operated in free generation (150 μs pulses) with quite low efficiency, 0.14% at 40 J pumping, owing to high residual optical losses. Higher 1 μm laser performance was later reported for CW emission under diode laser pumping for $Nd:Y_2O_3$ [88] and $Nd:Lu_2O_3$ fine-grained ceramics [89]. Since the crystal field splitting is large in these materials, the transitions with strongest emission cross-sections at room temperature originate from the lowest Stark component of the emitting level. Moreover, the strongest of these transitions, with similar emission cross-sections, terminate on the lowest two Stark components of the manifold $^4I_{11/2}$ which are only 35 cm^{-1} apart from each other in both materials and thus dual CW laser emission at 1074.6 nm and 1078.6 nm in the case of Y_2O_3 and 1075.9 nm and 1080 nm for Lu_2O_3 ceramic was observed. Strong broadening of the emission line (to 5 nm) was observed in Nd-doped solid-solution $Y_{1.8}La_{0.2}O_3$ ceramic, leading to 1079 nm laser emission under diode laser pumping [90].

11.4.1.1.3 Nd in other ceramic laser materials Nd^{3+} in intrinsic disordered ceramics shows broad absorption and emission bands that could be used for multiple wavelength or tunable laser emission or for the generation of ultra-short pulses. Under 809 nm diode laser pumping, disordered transparent ceramics of the cubic perovskites $Ba(Zr, Mg, Ta)O_3$ can show single (1075 nm) or multiple wavelength emission in the range 1062–1075 nm, with FWHM that shows potential for short pulse mode-locked emission [91]. The 1 at.% Nd-doped electro-optic lead zirconate titanate ceramic PLZT (10/65/35) gives CW laser emission at 1064.4 nm under pulsed (100 ms) 805 nm CW diode laser pumping that shows promise for multifunctional laser emission [92].

Quasi-CW pulsed diode laser pumping of 0.5 at.% $Nd:SrF_2$ at 790 nm, which corresponds to the maximum of the $^4I_{9/2} \rightarrow {}^4F_{5/2}$ absorption of the isolated tetragonal centers L, produced double-wavelength laser emission, with a main line at 1037 nm and a weaker line at 1044 nm. The slope efficiency in absorbed power was 19%, slightly lower than in the corresponding single crystals. Pumping at 796 nm, which corresponds to the maximum of the composed absorption band for the center L and for the Nd-pair center M, increases the laser threshold and reduces the slope efficiency for the L centers but no emission from the M centers was obtained [93].

Although the studies on transparent ceramics concentrated on cubic laser materials, recent success in the fabrication of ceramics of anisotropic materials (strontium fluoroapatite S-FAP) with grains oriented in a magnetic field prior to sintering [94] have enabled QCW 1063.1 nm laser emission under 807.5 nm diode laser pumping [95]. Despite the low actual efficiency, these materials could contribute in the future to considerable extension of the variety and applications of ceramic lasers.

11.4.1.2 Q-switched Nd ceramic lasers

The Nd:YAG ceramic laser can be Q-switched by using active methods (electro-optic or acousto-optic switching) or passive methods based on saturable absorption.

11.4.1.2.1 Active Q-switching Active Q-switching of Nd:YAG ceramic lasers has been done with electro-optic and acousto-optic switches. High repetition rate (1000 Hz) 1064 nm laser emission pulses of 23 mJ and with duration under 12 ns have been obtained with electro-optic Q-switching under CW 808 nm diode laser pumping [96], whereas QCW 808 nm pumping at a repetition rate of 100 Hz enabled the generation of 50 mJ pulses with duration of 6 ns [97]. Over 40 W average power 1064 nm laser emission with high spatial quality was generated from an 808 nm diode pumped, acousto-optically Q-switched MOPA in bounce configuration using 2 at.% Nd:YAG ceramic, with 35–80 ns pulses in the repetition range of 10 to 150 kHz [98]. Diode laser pumped acousto-optically Q-switched laser emission at tens of kilohertz repetition rate from Nd:YAG ceramics was reported at 1319 nm [99] and 946 nm [100]. Generally, the results on actively Q-switched emission from ceramic Nd:YAG lasers compare well with the best results obtained with single crystals.

11.4.1.2.2 Passive Q-switching The saturable absorbers used for Q-switching Nd lasers are mostly based on transition metal (TM) doped transparent materials, although semi-conductor materials (GaAs) have also been tested. As discussed in Chapter 2, the main characteristics of the SA are the saturation fluence $E_s = h\nu/\sigma_{gs}$, the initial transmission and the residual absorption. The saturation fluence is a material parameter of the SA, whereas the initial transmission can be controlled by the concentration and size, and the residual absorption can also be influenced by technology.

The TM ions usually used for passive Q-switching of infrared solid-state lasers are Cr^{4+} (940–1100 nm range), V^{3+} (1.30–1.35 μm) and Co^{2+} (1.3–1.6 μm). Although all these ions can be doped in cubic materials (Cr^{4+} and V^{3+} in garnets, Co^{2+} in spinels), ceramic SA Q-switches have so far been reported for Cr^{4+}:YAG [101–103]. These SA materials can be used as distinct components placed in the cavity together with the Nd laser active material. However, when both the laser and Q-switch material are based on the same host material, as in the case of Nd:YAG lasers passively Q-switched with Cr^{4+}:YAG, the difference between their refractive indices will be very small and thus no reflection would be expected from the interface when these materials are intimately joined. When both these materials are single crystals, they can be diffusion bonded or mechanically clamped; however, when one or both the laser materials and the SA are ceramic materials, fabrication of monolithic composite materials becomes possible. Ceramic materials homogeneously co-doped with Nd^{3+} and Cr^{4+} can also be produced, enabling self-Q-switched emission. Both these types of monolithic materials (composite and co-doped) enable construction of efficient high brilliance microchip lasers with very short laser cavity, able to generate high quality (single-mode) pulses of short duration.

The use of saturable absorbers for passive Q-switching of solid-state lasers was investigated intensively theoretically and experimentally and criteria for increasing the energy and decreasing the duration of the pulse, such as low initial transmission of the SA, increased transmission of the outcoupling mirror, short laser resonator, have been established [104–106]. The pump intensity has little effect on the energy and duration of pulses but influences the repetition rate. Most experimental studies on Cr^{4+}:YAG SA passive Q-switching of Nd lasers refer to CW diode pumped lasers. Constantly increasing performance of 808 nm diode laser pumped, Cr^{4+}:YAG Q-switched 1064 nm Nd:YAG ceramic lasers have been reported, with pulse energies ranging from several microjoules to almost 200 μJ, duration ranging from several to tens of nanoseconds, and peak powers from below 100 W to tens of kilowatts [107–110]. The instability of passive Cr^{4+}:YAG Q-switched emission of end-pumped Nd:YAG ceramic rods was suppressed by cladding with ceramic Sm:YAG [61]. Moreover, for SA with low T_0 these rods enabled similar or higher pulse energies than undoped YAG clad Nd:YAG rods and than uniformly doped Nd:YAG rods. Efficient Cr^{4+}:YAG passively Q-switched emission of Nd:YAG ceramic laser with peak power exceeding 20 kW at two wavelengths (1052 nm and 1064 nm) was obtained, with the SA also acting as wavelength selector [71, 111].

Pulsed pumping (flashlamps, QCW diode lasers) of the passively Q-switched 1 μm Nd:YAG lasers enables generation of high energy (millijoule to tens of millijoules) pulses.

Flashlamp pumping (5 Hz) of passively Q-switched ceramic YAG was reported for a 1 at.% Nd doped rod co-doped with 0.1 at.% Cr^{3+} to increase absorption [112]. In the free generation regime this laser delivered pulses of 2.1 J. By inserting a high T_0 (90%) SA Cr^{4+}:YAG ceramic in the cavity, a 1.1 J burst (130 µs) of Q-switched pulses (each of $\sim$300 ns) was obtained for each pump pulse. By decreasing T_0 to 30% the number of pulses decreased to a single pulse (130 ns), the laser threshold increased and the slope efficiency decreased, whereas the pulse energy decreased to 80 mJ. The results for (Nd, Cr):YAG rod were considerably improved compared to those obtained under similar conditions with a rod doped only with Nd. Multiple pulse emission under strong pulse pumping in passively Q-switched Nd lasers is a common fact. However, it was found earlier that combined active–passive Q-switching of high energy pulsed pumped Nd:YAG crystal lasers transforms the multiple pulse emission into a single pulse with much higher energy and shorter duration and with controlled jitter [113]; this approach proved useful in the case of CW pumped Nd:YAG lasers too [114]. Such an approach might prove useful for the generation of giant pulses from flashlamp pumped (Cr, Nd):YAG ceramics. Recently, pulse emission with energy in the range $\sim$2.4 mJ and peak power of 2.8 MW at 5–100 Hz repetition rate from a passively Q-switched composite 1.1 at.% Nd:YAG/Cr^{4+}:YAG ceramic under QCW 807 nm diode laser pumping was reported [115]. The laser was pumped by three independent beams and generated three laser beams that could be focused to give breakdown at three points whose positions could be controlled to provide optimized ignition in a car engine.

The (Nd, Cr):YAG ceramic was also used as an active mirror amplifier under flashlamp, arc lamp and solar pumping of low repetition rate free generation (160 µs) or of Q-switched pulses from a ceramic lamp-pumped (Nd, Cr):YAG laser. The characteristics of amplification for the short and long pulses were similar, and the output energy was limited by the saturation of the amplified input laser at quite low fluence [84]. However, this amplifier pumped with a CW arc lamp proved very efficient in providing high optical conversion in the case of lasers with high repetition rate [116]. Passive Q-switching of the 1.3 µm bounce Nd:YAG ceramic laser with crystalline V^{3+}:YAG saturable absorber (ground-state absorption cross-section 7.2×10^{-18} cm^2 and negligible excited state absorption) provided pulses with 10 to 114 kHz repetition rate and 55 to 140 ns duration [117], whereas the diode end-pumped laser produced 128 ns pulses at 230 kHz repetition rate, with 9% slope efficiency [118].

The self-Q-switching properties of co-doped (Cr^{4+}, Nd):YAG crystals have been investigated in detail with regard to the construction of single-mode microchip lasers [119, 120]. However, despite the fabrication of high quality (Cr^{4+}, Nd):YAG ceramic materials [121], no data on laser emission properties are reported.

Although the band-gap of GaAs semiconductor (1.42 eV) is much larger than the energy of the 1 µm laser quantum, a defect-related level inside the band-gap shows saturable absorption in this wavelength range. Passive Q-switching of 1064 nm laser emission with GaAs SA of a diode laser pumped ceramic Nd:YAG laser provided 4.7 ns pulses of 6.25 µJ [122].

11.4.1.3 Mode-locked Nd ceramic lasers

11.4.1.3.1 Passive saturable absorption mode-locking Mode-locking of solid-state lasers is an efficient means of reducing the pulse duration. As discussed in Chapter 2, mode-locking can be achieved by active methods (acousto-optic modulators) or by passive methods, based on intracavity intensity-dependent modification of the losses in the resonator with real or effective saturable absorbers or of the reflectivity of mirrors by non-linear processes, particularly by SA in semiconductor materials.

Active mode-locking generation of stable, regular trains of picosecond pulses was demonstrated in various CW pumped Nd-doped crystal lasers, but it has not been reported so far for Nd ceramic materials. In the case of SA mode-locking the pulse duration τ_p is limited by the product $\tau_p \Delta \nu$, where $\Delta \nu$ is the FWHM of the emission of the laser active material in the frequency domain, and equals 0.44 or 0.315 in the case of Gaussian or sech^2 temporal dependence, respectively. Passive mode-locking would require relaxation times from the upper state of the absorber, populated by absorption of the laser emission, of the order of picoseconds.

As discussed above, the Cr^{4+} ion in YAG shows saturable GSA to its first excited state in the 1 μm range of Nd lasers The lifetime of this excited state is 4.2 μs and this precludes mode-locking at the intracavity intensities that bleach the absorber, although it enables efficient Q-switching. However, by increasing the intracavity intensity, strong ESA of the 1 μm laser emission from the first excited state of Cr to upper levels takes place and, since the lifetime of these upper levels is very short, under 0.1 ns, this ESA could be used for mode-locking, although its saturation intensity can be five orders of magnitude larger than for Q-switching. Normally, in this case the train of mode-locked pulses is modulated in intensity inside the Q-switch temporal pulse shape (QML regime). The QML regime enables ML pulses of higher intensity than in normal ML and could be of practical interest. This regime of emission using Cr^{4+}:YAG as SA was investigated for several Nd-doped crystal lasers, but not for ceramics, and criteria for optimization were issued [123]. Moreover, it was found that in the case of (Cr^{4+}, Nd^{3+}) co-doped YAG the saturation intensity for ESA could be much reduced, to about only two orders of magnitude larger than for Q-switching [124].

Recent investigation revealed that very suitable saturable absorbers for mode-locking of solid-state lasers could be nanosystems such as single walled carbon nanotubes (SWCNT) or graphenes. Only the semiconducting carbon nanotubes have suitable saturable absorption properties and thus the performance is influenced by the fabrication technique. By using a polymer extraction technique, material containing a very high (80%) proportion of semiconducting single-wall carbon nanotubes was obtained, with saturation intensity of $1.7 \, \text{MW cm}^{-2}$, modulation depth 6% and non-residual loss 5.5% that enabled generation of stable 8.3 ps pulses with repetition rate 90 MHz from a 0.5 at.% Nd:YAG ceramic pumped with a CW 808 nm diode laser [125]. However, the average power for stable mode-locked emission of this laser was limited to 130 mW. A single layer of graphenes can absorb ~2.3% of the incident light, is practically insensitive to the wavelength and has very fast recovery time. Since the number of layers of graphenes deposited onto a transparent substrate can be controlled, mode-locking saturable absorbers with desired transparency can be fabricated. In the case of a diode laser pumped Nd:YAG ceramic laser, utilization of graphene SA enabled the generation of 4 ps pulses with repetition rate 88 MHz and average power up to 100 mW [126].

11.4.1.3.2 Mode-locking by controlled mirror reflection A very fast modification of the reflectivity of the rear mirror of the laser from a low to a higher value can be an effective means of mode-locking. The reflectivity can be controlled by inserting in front of a 100% reflective mirror a non-linear element that changes its transmission under the action of the intracavity intensity. This element can be a conventional non-linear optical material or a semiconductor saturable absorber. In order to obtain higher stability and performance, another layer with slightly lower reflectivity is coated onto the semiconductor SA material, thus forming a Fabry–Perot resonator that enables tight control of the SA bleaching. The device thus obtained is called a semiconductor saturable absorption mirror (SESAM).

The SESAMs have very low residual losses (below 0.2%) and short ($\sim$20 ps) recovery time, and the modulation depth is usually around 1%. It was observed that under low CW diode laser pumping at 808 nm the 1064 nm Nd:YAG ceramic operated in the QML regime; however, at higher pump power, stable mode-locking was obtained, with 8.3 ps pulses at 130 MHz repetition rate, and 1.59 W average power with 21% optical efficiency [127]. Slightly larger output power (2.6 W) was subsequently reported for Nd:YAG ceramic, with 26 ps pulses at 78 MHz repetition rate [128]. It was also shown that the Q-switched instability of the SESAM mode-locked Nd:YAG ceramic laser could be suppressed by a proper choice of the modulation depth of SESAM and of the length of the laser resonator [129].

The inhomogeneously broadened emission lines in solid-solution or intrinsic disordered garnet materials offer conditions for shorter mode-locked pulses. Complete use of the inhomogeneously broadened emission lines requires pumping over the whole width of the absorption bands. Quite efficient (slope efficiency 49.5%) passive mode-locking (10 ps duration at 77 MHz) was reported in the case of Nd-doped $Y_3ScAl_4O_{12}$ ceramic laser pumped with CW Ti:sapphire laser [130] using a non-linear LBO controlled mirror. The prospect of using solid-solution ceramic GSAG–YSGG in the proportion 2:1, with emission linewidth of 5.3 nm, was also discussed [131]. Of interest for short pulse generation could also be Nd-doped garnets with intrinsic disordered structure from the CNGG family: 1061 nm SESAM mode-locked pulses of 900 fs duration, corresponding to a FWHM of 1.8 nm, with repetition rate 88 MHz were reported for Nd:CLNGG crystal [132], whereas a composite Nd:CLNGG/Nd:CNGG crystal had an effective FWHM of 2.2–2.3 nm and pulses of 534 fs [133]. Recent investigation of Nd-doped CLNGG ceramics revealed the similarity with the single crystals and indicates the potential of these materials for short pulse generation: moreover, the additional disordering in the case of CLNTGG could contribute to further broadening of the emission lines [134]. Short pulse (1.4 ps, 1.5 FWHM) SESAM mode-locked 1062 nm emission was also demonstrated in disordered Nd-doped Ba(Zr, Mg, Ta)O_3 ceramics [91].

11.4.1.3.3 Effective saturable absorption mode-locking The energy of pulses achievable in passive mode-locking using GaAs SA is limited by the strong two-photon absorption that can dominate linear absorption at high intracavity intensities. However, it was found that the reflectivity of thick GaAs wafers can be influenced by optical interference modulation. By placing a high reflectivity deposited GaAs wafer as the end mirror of the laser, its reflectivity can be modulated by the combined Kerr effect and optical interference, which act as an effective saturable absorber, leading to mode-locking of the laser beam. Using this approach, mode-locked pulses of 4.1 ps at 129 MHz repetition rate, with average power of 2.84 W and slope efficiency of 48%, even larger than for CW operation, were obtained from a 0.5% Nd:YAG ceramic under CW diode laser pumping at 806 nm [135].

11.4.2 Ho-doped ceramic lasers

Direct pumping in the emitting manifold $^5I_7$ of Ho^{3+} demonstrated efficient quasi-three-level CW laser emission in the 2 μm range: 1.91 μm pumping with a Tm:YLF laser of 1 at.% Ho:YAG ceramic produced 2.09 μm laser emission with slope efficiency around 44% [136], whereas 1907 nm Tm fiber laser pumping of 1.5 at.% Ho:YAG ceramic produced 21 W of 2097 nm CW laser radiation with 63.3% slope efficiency and 61% optical efficiency in incident power [137]. Resonant 1.93 μm diode laser pumping of Ho:Y_2O_3 ceramic produced CW emission at 2.12 μm with 35% slope efficiency [138].

11.4.3 Er-doped ceramic lasers

Investigation of the laser emission of Er-doped transparent ceramics concentrated on the infrared 1.5 μm $^4I_{13/2} \rightarrow {}^4I_{15/2}$ and 3 μm $^4I_{11/2} \rightarrow {}^4I_{13/2}$ transitions under resonant pumping into the emitting manifold to give a low quantum defect. A very convenient absorption band for InP diode laser pumping of the $^4I_{13/2}$ level of Er:YAG is the strong absorption band at 1532.2 nm; according to Table 10.9, this band collects the hot band absorption transitions $^4I_{15/2}(2) \rightarrow {}^4I_{13/2}(1)$ at 1532.1 nm and $^4I_{15/2}(4) \rightarrow {}^4I_{13/2}(3)$ at 1532.3 nm. Quasi-continuous-wave (10 ms pulses, 5% duty cycle) fiber laser pumping at 1532 nm of a tape-cast three-segment Er:YAG composite ceramic rod (0 at.%, 0.25 at.% and 0.5 at.% Er) produced 7 W laser emission at 1647 nm with 56.9% slope efficiency in absorbed power at room temperature [139], whereas 1532 nm volume Bragg grating stabilized (Er, Yb) fiber laser CW pumping of a 1 at.% Er:YAG ceramic rod produced 13.8 W of 1645 nm radiation with 54.5% slope efficiency in incident power [140]; similar pumping also produced CW 1617 nm emission with 51.7% slope efficiency in a 0.5 at.% Er:YAG ceramic [141].

The Er:Sc_2O_3 ceramic pumped with an Er fiber laser at 1535 nm and placed in a laser resonator with volume Bragg grating sharp-edge dichroic input mirror delivered 3.3 W of CW laser emission at 1558 nm with 45% slope efficiency [142]. Laser emission in the 3 μm range has focused on Er-doped Y_2O_3 ceramic since the lifetime of luminescence in this material is considerably longer than in YAG or Sc_2O_3 due to the less efficient electron–phonon interaction determined by the lower phonon quanta. Spectrally narrowed laser diode CW pumping at 974 nm of 2 at.% $Er:Y_2O_3$ ceramic produced laser emission dominated by the transition $^4I_{11/2}(1) \rightarrow {}^4I_{13/2}(1)$ at 2707 nm, and containing unstable admixture from other transitions (2715 nm, 2725 nm and 2740 nm), with slope efficiency around 15% at room temperature [143]. Cooling at cryogenic temperatures increased the maximum extractable power and the slope efficiency: on pumping with 974 nm single-emitter narrow-band (FWHM $\sim$0.3 nm) surface-emitting distributed feedback laser diodes the slope efficiency became 27.5%, close to the quantum defect limit, whereas utilization of spatially combined and fiber coupled two-dimensional arrays of such laser diodes enabled generation of high power (14 W) laser emission with $\sim$26% slope efficiency from 2 at.% $Er:Y_2O_3$ ceramics [144]. The laser generated stable CW and QCW radiation at 2715 nm, with unstable contributions from 2707 nm and 2740 nm emissions.

11.4.4 Tm-doped ceramic lasers

Laser emission in the 2 μm range of Tm^{3+} in YAG ceramics was investigated under Ti:sapphire or diode laser pumping into the manifold $^3H_4$, using the down-conversion process $(^3H_4, {}^3H_6) \rightarrow (^3F_4, {}^3F_4)$. In order to produce efficient energy transfer down-conversion, Tm concentrations of 6 at.% were typically used. Although the quantum defect inherent in this excitation process together with the efficiency of down-conversion and with the quasi-three-level scheme of laser emission limit the performance of the laser, 2012 nm or 2015 nm CW emission slope efficiencies in the range 20–42% have been reported [145–147]. With increased 782 nm diode laser pump power (over 50 W) dual-wavelength CW emission at 2016 nm and 2006 nm was obtained, with slope efficiency over 36%, and Q-switching with an acousto-optic modulator produced pulses in the repetition range 500 Hz–5 kHz: at 500 Hz the duration of pulses was 69 ns, with energy 20.4 mJ [148]. Resonant pumping of 4 and 6 at.% Tm:YAG ceramic with a 1617 nm Er:YAG laser produced 2015 nm CW laser emission with 62.3% slope efficiency [149]. Diode laser pumping at 796 and 811 nm of 2 at.% $Tm:Lu_2O_3$ fine-grained ($\sim$550 nm) ceramic produced 2066 nm CW emission with $\sim$40% slope efficiency [150].

11.4.5 Yb-doped ceramic lasers

As discussed in Chapter 10, the spectroscopic properties at room temperature of Yb-doped ceramic materials could favor almost pure four-level emission on the transition $^2F_{5/2}(1) \rightarrow {}^2F_{7/2}(4)$ and quasi-three-level laser emission on the much more intense transition $^2F_{5/2}(1) \rightarrow {}^2F_{7/2}(3)$. The competition between the laser emission on these two transitions can be influenced by the ambient temperature and by the internal rise in temperature determined by the generation of heat in the pumped laser material, the latter being influenced by the absorbed pump power and by the transmission of the output mirror that influences the intracavity intensity. At cryogenic temperatures the coefficient $f_t(3)$ becomes negligibly small and the laser can practically generate in a four-level scheme on the transition $^2F_{5/2}(1) \rightarrow {}^2F_{7/2}(3)$.

A shortcoming of Yb laser materials is the low absorption cross-section of the broad absorption band $^2F_{7/2}(1) \rightarrow {}^2F_{5/2}(2)$ around 940 nm used traditionally for diode laser pumping. In ordered laser materials the $^2F_{7/2}(1) \rightarrow {}^2F_{5/2}(1)$ absorption line, with larger peak absorption cross-section which could allow low quantum defect laser emission, is very narrow, especially at low temperatures, and diode laser pumping is difficult. However, the recent advent of stabilized diode lasers has enabled pumping in this transition in several room temperature Yb lasers. As spectroscopic investigations show, this absorption line can be broadened inhomogeneously using disordered laser materials, although this will reduce the peak absorption cross-section. The low absorption cross-sections impose increased Yb concentration and/or lengthening the path of pump radiation inside the laser material. Although the simple electronic structure of Yb^{3+} precludes de-excitation processes such as the concentration dependent self-quenching by down-conversion or upconversion cross-relaxation, a reduction in the emission lifetime and in emission quantum efficiency as well as cooperative processes determined by interaction of two excited Yb^{3+} ions were observed at high Yb concentrations. These parasitic de-excitation processes are accompanied by the generation of up to twice as much heat than is predicted by the quantum defect [151]. This outlines the need for good matching of the pump and mode volume in Yb lasers, similar to Nd lasers.

The moderate emission cross-section, the long lifetime of the emission and the quite broad emission lines recommend Yb-doped materials for a large variety of laser regimes, from CW to ultra-short (femtosecond range) pulses, as well as for tunable emission. Such lasers can cover a broad range of sizes and power, from microchip lasers to high power lasers or high energy-peak power amplifiers of interest for nuclear physics and nuclear fusion.

11.4.5.1 Free generation (CW, QCW) lasers

11.4.5.1.1 Yb-doped garnet ceramic lasers
A. Room temperature Yb-doped garnet ceramic lasers The first room temperature continuous wave Yb:YAG ceramic laser used a low doped material (1 at.% Yb, 1.5 mm thick) whose optical quality, together with the non-optimized laser setup limited the slope efficiency to 26% [152] under 940 nm diode laser pumping; the transmission T of the output coupler was ~5% at 1030 nm laser wavelength. These performances were increased by using Yb:YAG ceramics of improved quality and higher Yb doping (9.8 at.%), enabling a higher absorption efficiency and a reduction in the thickness of material (1 mm). For $T = 5\%$, the low power pumping determined 1030 nm laser emission, but with increased pumping competition from the additional emission at 1049 nm appeared and became dominant at high power. However, for $T = 10\%$ or 20%, the 1030 nm emission remained dominant [153]. The performances of these end-pumped bulk Yb:YAG ceramics have been improved subsequently to tens of watts and the laser slope efficiency has been raised above 70% [154–159]. A diode laser at 968 nm proved beneficial [157]. Laser emission in a front face diode laser (970 nm) pumped compact active mirror configuration was also reported [160].

The reabsorption that limits the 1030 nm emission of Yb:YAG lasers can be diminished by reducing the thickness of the sample. This could make conventional end pumping inefficient. Two approaches can be used to circumvent this shortcoming: side pumping of a thin laser material whose diameter is large enough to grant efficient pump absorption and a thin-disk approach based on face pumping and repeated passes of the pump radiation inside the laser material by multiple reflection. In both these approaches the end mirror is coated on the laser material and the removal of heat or cooling occurs through this face. The heat flow in this case is along the laser axis and in principle no thermal lensing is present.

Increased power was obtained by 940 nm diode laser side-pumping of a thin (0.2 mm) 10% Yb:YAG ceramic disk (3.7 mm diameter) surrounded by undoped YAG ceramic that acted as a pump light pipe [161]: such a configuration enabled increased pump absorption efficiency and simultaneous reduction of reabsorption, resulting in 414 W CW (slope efficiency 47%) and 520 W QCW (10 ms, 10 Hz, slope efficiency 52%) 1030 nm laser emission. The output power density of these lasers was 3.9 kW cm^{-2} and respectively 0.19 MW cm^{-3}. The thin-disk approach was developed for crystalline Yb:YAG lasers and proved very successful [162]. The aperture of these lasers is 1–2 cm^2 and deleterious ASE and heating can limit the scalability. It was recently shown [44] that bonding an undoped cap to the Yb-doped disk could reduce ASE. By using a 9 at.% Yb:YAG ceramic thin disk (0.2 mm) bonded with a 1 mm thick undoped YAG ceramic cap, stable emission with very high power (6.5 kW, slope efficiency 57%) 1030 nm multimode laser emission was obtained [163] in single thin-disk configuration; the output density was 2.5 kW cm^{-2} under a pump density of 5 kW cm^{-2} and the power scaling was limited by the heating of the laser material. These results are better than those obtained with crystalline Yb:YAG. This result shows the promise of thin-disk Yb:YAG ceramics for the generation of very high CW power in optimized laser configurations with several disks coupled in series.

Broadly tunable laser emission with wavelength range dependent on transmission of the output mirror was obtained for 9.8 at.% Yb:YAG ceramic under 940 nm diode laser pumping: for $T = 1\%$ the laser was continuously tunable between 992.5 nm and 1098 nm, and extended to 992.5–1110.83 nm for $T = 0.1\%$, although the maximum extractable power decreased [164, 165]. The large tuning range suggests the possibility of using Yb:YAG ceramic for ultra-short pulse generation.

The results described above were obtained with Yb:YAG ceramics with C_{Yb} up to 10 at.% and the main limitation of power scaling was the difficulty in dissipating efficiently the heat generated in the pumped laser material. Heat removal in the face-cooled Yb laser materials demands thinner laser elements and thus the optimization of pump absorption and power scaling would impose high C_{Yb} and high excitation densities. Comparative investigation of the 1030 nm laser emission of 20% Yb:YAG ceramics and single crystals in minilaser double pump-pass configuration (1 mm plates mounted on a copper heat sink, diode laser pumping at 940 nm) [166] showed considerably higher maximum extractable power and linearity with the absorbed 940 nm pump power for ceramics than for crystals;

the red shift of emission wavelength at low T or at increased pump power was attributed to the increased intracavity intensity.

This comparative investigation of Yb:YAG ceramics and crystals was later extended to a larger C_{Yb} range [167, 168], for absorbed 940 nm pump power focused in a spot of ~170 µm on the laser material in the range 5–5.6 W and using a similar two pump-pass laser configuration. The experimental data show a certain spread that reflects variations in the quality of the samples under investigation and/or in the conditions of the experiment. It was thus found [168] that at concentrations around 10 at.% both types of materials have high laser performance, higher for single crystals. However, with increasing Yb concentrations the performances drop; the effect is stronger for crystals and at 20% Yb the ceramics show better performance. Thus, in the case of Yb:YAG crystal the optical–optical efficiency in absorbed power dropped from 61% for $T = 5\%$ and 58% for $T = 10\%$ and 10 at.% Yb to 33.5% and respectively 25% for 20 at.% Yb, whereas in the case of Yb:YAG ceramic this drop was from 45% for $T = 5\%$ and 49% for $T = 10\%$ and 10 at.% Yb to 38% and respectively 33% for 20 at.% Yb.

This decrease in performance is manifested in modification of both the pump threshold and slope efficiency, sometimes in an inconsistent way, which can reflect the existence of a variety of processes that influence these parameters. Among these processes, those that can influence the slope efficiency, such as optical losses (including reabsorption of the laser emission) or the pump-mode volume superposition, must be considered. The comparative investigation of 2 mm thick 10% at.% Yb and 1 mm 20 at.% Yb-doped YAG ceramics was extended by using higher 940 nm pumping power and selecting experimentally the optimal output mirror in a single-pass laser configuration [169]. For 1030 nm laser emission with $T = 6$, 12 or 18.5% the laser characteristics of the 10% and 20 at.% Yb ceramics were quite similar, and the output power at maximum pumping of 21 W corresponding to ~13 W absorbed power was 8.7 W and respectively 8.2 W. For the 1050 nm laser emission obtained with $T = 1.5\%$ the performances of the two ceramics were similar, slightly higher for 10 at.% Yb ceramic, and with the maximum output power in the range of 8 W. The behavior at high inversion population was tested with $T = 79\%$ and 97%. In the first case the laser emission was always stronger for the less doped sample; moreover, at a certain pump level the emission of the 20 at.% Yb-doped ceramic showed roll-over and stopped. For $T = 97\%$ the ceramic with 10 at.% Yb showed emission, whereas that with 20 at.% Yb did not lase at all. Generally, the beam quality of these lasers, as well as for the thin-disk lasers, is multimode, and deteriorates with increasing pump power.

Although several qualitative models to explain the drop in performance at higher Yb concentration and excitation intensity and the better performance of ceramics at high Yb concentrations have been suggested, no quantitative explanation of the experimental data has been given. Thus, in reference [168] the drop in performance at high Yb concentration and strong excitation was attributed to energy transfer to accidental impurities whose concentration increases with the Yb concentration. However, it must be taken into account that such energy transfer processes influence the lifetime of emission in the absence of laser emission, i.e. the emission quantum efficiency, and this in turn increases the emission threshold, but does not influence directly the slope efficiency of the CW laser. The better performance of ceramics at high doping was attributed [167] to increased phonon transmission because of segregation of the Yb ions at grain boundaries; here again it must be remarked that the segregation of Yb at grain boundaries in YAG ceramics is quite low [170] and a quantitative evaluation of the effect is necessary. On the other hand, the heavily doped Yb:YAG ceramics show strong photoconductivity increasing with the 2.5 power of absorbed pump power. Qualitative models involving upconversion of excitation to the charge transfer band and then to the conduction band of the material, including the involvement of transient or permanent Yb^{2+} centers, have been discussed [171–174] but a quantitative explanation of the observed effects has not yet been reported. Evaluation of the thermal effects for YAG ceramics with 10 and 20 at.% Yb at various intracavity intensities confirm the existence of an additional non-linear effect at high Yb concentrations and high excitation [169].

$Lu_3Al_5O_{12}$ is a tempting host for Yb since the doping has very little influence on thermal conductivity compared with YAG, although the ionic radii of Y^{3+} and Lu^{3+} are similar. Diode laser pumping at 940 nm of 5 at.% Yb:LuAG ceramic produced 1030 nm CW laser emission, with 63% slope efficiency, which was influenced by the residual absorption [175].

B. Cryogenic Yb-doped garnet ceramic lasers As discussed above, the thin-disk multi-pass configuration of Yb:YAG lasers enables generation of very high CW power that is limited mainly by the heating of the laser material, whose effects are accentuated by the three-level nature of the Yb lasers at room temperature. Moreover, the multi-pass techniques require complex advanced optical systems; another shortcoming of these lasers is the need for a high-reflection coating on the disk, which can reduce heat transfer to the coolant.

Many of the limitations of room temperature Yb lasers can be circumvented by lasing at cryogenic temperatures. Sound technical solutions for cooling the laser materials to cryogenic temperatures are now well established and the cost associated with cooling is much below the advantages gained in laser performance. The spectroscopic advantages (increased emission cross-section, reduced population of the terminal level that transforms the 1030 nm emission of the strong transition $^2F_{5/2}(1) \rightarrow {}^2F_{7/2}(3)$ into a four-level scheme) together with improvement of the material properties (increased heat conduction, reduced thermal expansion and thermo-optic coefficient) have been discussed in the literature and demonstrated with crystalline Yb:YAG lasers. In particular, these advantages enable utilization of thicker Yb:YAG plates and tailoring of the Yb concentration to optimize the laser performance. This approach was demonstrated in the case of Yb:YAG ceramics by using the total-reflection active-mirror (TRAM) technique that uses the total reflection on the bottom of the laser material plate instead of a high-reflection coating [176]; this technique has the additional advantage of separating the path of pump radiation from that of the laser beam. By using a composite laser material containing a 180 μm or 400 μm thick 9.8 at.% Yb:YAG ceramic plate with 40×40 mm^2 bottom surface bonded to undoped YAG ceramic and cooling by liquid nitrogen in direct contact with the bottom surface, an output power of 273 W was obtained under 940 nm diode laser pumping, with 72% slope efficiency and 65% optical–optical efficiency. The design of the TRAM laser was further extended by using a composite ceramic material with three active Yb:YAG plates placed in zig-zag configuration on an undoped YAG body [177]. The performance of this material was similar to that of the single active element composite and it was estimated that such an approach could enable scaling of Yb:YAG ceramic lasers to the range of 10 kW.

11.4.5.1.2 Yb-doped sesquioxide ceramic lasers

A. Room temperature Yb-doped sesquioxide lasers The sesquioxides are attractive host laser materials because of their high heat conduction, although doping with high concentrations of laser active materials can spoil this parameter considerably. Room temperature emission of Yb^{3+} was observed for transparent ceramics of all cubic sesquioxides (Y$_2$O$_3$, Lu$_2$O$_3$, Sc$_2$O$_3$). These ceramics can be pumped with diode lasers in the broad $^2F_{7/2}(1) \rightarrow {}^2F_{5/2}(2)$ absorption band in the region 937–950 nm or in the more intense but considerably sharper absorption $^2F_{7/2}(1) \rightarrow {}^2F_{5/2}(2)$ in the region 976–980 nm and, similar to garnets, concurrent room temperature emission can take place in the region 1030–1040 nm on the transition $^2F_{5/2}(1) \rightarrow {}^2F_{7/2}(3)$ in a quasi-three-level scheme or in the region 1077–1080 nm on the transition $^2F_{5/2}(1) \rightarrow {}^2F_{7/2}(4)$ in a four-level scheme. The emission wavelength can be selected by controlling the reabsorption by the size of the laser material and by the Yb concentration; utilization of selective optical components is also effective.

The performance of Yb-doped Y$_2$O$_3$ [178–181], Lu$_2$O$_3$ [182, 183, 184] and Sc$_2$O$_3$ [184, 185] ceramic lasers has shown continuous progress based on improved laser materials and laser design. Slope efficiencies in excess of 80% and optical–optical efficiencies larger than 70% for the transition $^2F_{5/2}(1) \rightarrow {}^2F_{7/2}(4)$, and somewhat lower for the transition $^2F_{5/2}(1) \rightarrow {}^2F_{7/2}(3)$, were reported. The emission lines were several nanometers broad, showing potential for short pulse emission. In the case of solid-solution (Y$_{0.9}$La$_{0.1}$)$_2$O$_3$ ceramic, efficient emission at 1080 nm was observed under diode pumping at 976 nm. Smooth tunable emission in the range 1018–1086 nm was observed under 940 nm pumping, with up to 30 nm broadband lasing, although under 976 nm pumping the tuning range narrowed to 1031–1083 nm [186].

B. Yb sesquioxide ceramic lasers at cryogenic temperature The improvement of the 1030 nm emission of the 10 at.% Yb-doped Y$_2$O$_3$, Lu$_2$O$_3$ and Sc$_2$O$_3$ ceramics at cryogenic temperature due to a reduction in reabsorption was investigated in the free generation regime under pulsed 940 nm diode laser pumping [187]. By using 600 μs pump pulses with energies up to 1 J at 10 Hz repetition rate the output energy increased by 3 to 6 times in the temperature interval from 80 to 320 K as a consequence of lowering the emission threshold and an enhancement of slope efficiency.

11.4.5.1.3 Yb-doped fluoride ceramic lasers Laser emission of diode laser pumped 1 at.% $Yb:CaF_2$ ceramic produced by sintering liquid synthesized nanopowders was influenced by the quite high residual optical losses [188]. However, the 3 at.% $Yb:CaF_2$ ceramics produced by hot pressing of Bridgman crystalline material had much better laser performance, with $\sim$30% slope efficiency in CW and $\sim$37% in QCW regime at room temperature [189]. Improved 1052 nm laser performance with 45% slope efficiency was reported for $65\%CaF_2$–$30\%SrF_2$–$5\%YbF_3$ hot pressed ceramic under 972 nm diode laser pumping, close to the value obtained with single crystals [190].

11.4.5.2 Q-switched emission of Yb-doped ceramics

Electro-optic Q-switching of Yb:YAG ceramics at cryogenic temperatures delivered 42 mJ, 200 ns pulses at 10 Hz repetition rate [191]. However, much shorter pulses were reported from ceramic Yb:YAG in the case of passive Q-switching with Cr^{4+}:YAG saturable absorbers. Thus, by using separate ceramic Yb:YAG active material and ceramic Cr^{4+}:YAG SA with initial transmission of 80%, Q-switched pulses as short as 335 ps with peak power above 150 kW at a repetition rate of 5 kHz were obtained, with high quality beam, close to the diffraction limit [192]. Utilization of composite 9.8 at.% Yb:YAG/0.1 at.% Cr^{4+}:YAG ceramics with 64% initial SA transmission enabled pulse energies of 125 µJ of 1.2 ns duration with peak power of 105 kW at 3.8 kHz repetition rate for 2.55 W absorbed 940 nm power [193]. By increasing the transmission of SA to 70%, in the case of an output coupler with $T = 50\%$, pulses of 172 µJ with duration 237 ps and peak power of 0.72 MW were obtained at a repetition rate of 3.5 kHz for an absorbed power of 3.53 W. No sign of saturation was observed at maximum power, indicating further potential for scaling [119].

11.4.5.3 Mode-locked emission

11.4.5.3.1 Mode-locked Yb-doped garnet ceramic lasers Mode-locking of Yb:YAG ceramic lasers under CW 940 nm or 968 nm diode laser pumping was investigated using SESAMs [194, 195] and the duration of the pulse was limited to several hundred femtoseconds because of the fairly narrow emission lines. The Yb concentration was usually in the 10 at.% range, and the emission wavelength (1030 nm or 1048 nm) was selected by the thickness of the laser material or by transmission of the output mirror; dual wavelength emission was also reported [196]. The average power was up to several watts and slope efficiencies up to 76% were reported.

SESAM mode-locked pulses of shorter duration were obtained based on the inhomogeneous broadening in Yb-doped solid-solution garnet systems. In the case of Yb-doped $Y_3ScAl_4O_{12}$ ceramic the emission linewidth at 1035.8 nm is 12.5 nm, larger by about 50% than in YAG, and this enabled generation of shorter (280 fs) mode-locked pulses [197]. Shorter pulses (69 fs) were obtained in the SESAM mode-locked laser based on the more disordered solid-solution $\{YGd_2\}[Sc](Al_2Ga)O_{12}$ ceramic [198]. The 1042 nm laser wavelength of this laser was shifted considerably from the value 1031 nm in CW emission.

The spectral linewidth of the mode-locked pulse was 22 nm, considerably larger than the luminescence FWHM (16 nm), and the broadening was attributed to participation of Kerr lens mode-locking.

11.4.5.3.2 Mode-locked Yb-doped sesquioxide ceramic lasers The emission cross-section and linewidth of Yb-doped cubic sesquioxide ceramics favors short pulse generation by mode-locking. Moreover, the high non-linear refractive index of these materials [199] can lead to strong Kerr lensing and self-phase modulation which can contribute to mode-locking when the length of the laser material is sufficient to allow high non-linear effects. Mode-locking by SESAM of Yb-doped sesquioxide ceramics produces pulses in the range of 200 fs to picoseconds [200–203]. Self-mode-locking by diffraction loss induced by thermal lensing in combination with Kerr self-focusing allows mode-locking of the 1078 nm emission in a heavily doped $Yb:Y_2O_3$ ceramic [204]. The Kerr effect in the presence of SESAM led to considerable shortening of the pulse duration for all the Yb-doped cubic sesquioxide ceramics: 92 fs for Sc_2O_3 [205], 65 fs for Lu_2O_3 [206] and 68 fs for Y_2O_3 [207]. Moreover, by using two ceramic gain elements ($Yb:Y_2O_3$ and $Yb:Sc_2O_3$) with carefully selected doping levels and thickness, pulses as short as 53 fs were obtained.

11.4.5.4 Yb ceramic amplifiers

According to their spectroscopic properties (long emission lifetime, moderate emission cross-section), Yb-doped ceramics satisfy the conditions for good storage of inversion and extraction of the stored energy in amplifiers. Investigation of the amplification characteristics of Yb:YAG ceramics at cryogenic temperatures enabled transformation of the laser scheme to a four-level scheme and an improvement in thermal properties. A MOPA system consisting of a fiber oscillator delivering 10 ns pulses of energy below 15 pJ at 100 Hz repetition rate was first amplified in a QCW (100 Hz, 1000 µs) 938 nm diode laser-pumped active mirror regenerative amplifier consisting of 9.8 at.% Yb:YAG ceramic to obtain 2.4 mJ pulses and finally amplified to more than 200 mJ in a four-pass main amplifier based on a diode laser (700 µs) pumped 5.8 at.% Yb:YAG ceramic [208]. The optical–optical efficiency of the main amplifier was 19% and the slope efficiency was 30%, and limitation of scaling attributed to saturation and ASE was observed. By improving the spatial beam coupling, the 10 Hz, 18 mJ pulses were amplified to $\sim$150 mJ with slope efficiency $\sim$44% and optical efficiency $\sim$30%, but above this energy strong saturation effects were evident due to enhanced ASE effects that became even stronger at higher repetition rates. Another approach to higher energy amplification made use of the TRAM design for the regenerative amplifier [209]. Besides the advantages of this design discussed already, the absence of HR coating can relax the limitations imposed by damage of this coating at high energies. By using a monolithic composite 20 at.% Yb:YAG/YAG ceramic, the 10.5 ns, 62 pJ pulses from a fiber laser were amplified to 5 mJ at repetition rates from 10 to 450 Hz. The energy extraction of the amplifier was 41% and the fluence reached 3.5 J cm^{-2}. These first experiments show the good potential of cooled Yb:YAG ceramics for the construction of high energy amplifiers.

11.4.6 Waveguide lasers

Short pulse laser [210–215] or particle beam [216] irradiation of doped ceramic plates can be very efficient in the creation of high quality three-dimensional photonic structures, particularly buried channel waveguides that can be used for laser emission. This process is based on controlled modification of the index of refraction by the electronic processes and thermally induced transformations associated with laser irradiation. Laser writing of such structures is mainly a one-step process that does not need special preparation, has good reproducibility, is a low-cost process and needs only a very short time. It could contribute to the fabrication of integrated photonic devices.

11.4.7 Non-linear ceramic lasers

The fundamental emission of ceramic lasers can be modified by various non-linear processes, such as frequency doubling with non-linear crystals or by using high order non-linear properties of the laser material. Thus, high power (104 W) 532 nm green radiation was obtained by frequency doubling of Q-switched composite ceramic Nd:YAG [217], and lasing effects at wavelengths determined by the third-order susceptibility $\chi^{(3)}$ were observed in various ceramics such as YAG, $\{YGd_2\}[Sc_2](Al_2Ga)O_{12}$ and the cubic sesquioxide ceramics [218–221].

11.4.8 The effect of granular structure on the laser beam properties

As discussed in Section 2.6.3, in a laser with a rod of optically isotropic crystalline material (such as the cubic garnets), the strain and stress caused by the thermal field inside the pumped laser material determine the difference between the radial and azimuthal refractive indices, i.e. thermally induced birefringence. The thermally induced birefringence depends on the orientation of the laser beam with respect to the cubic axes of the laser material and determines orientation dependent bifocal thermal lensing. At the same time it induces phase distortion leading to orientation dependent partial depolarization of a polarized beam traveling inside the laser material. This would indicate a complex depolarization behavior in the case of polycrystalline transparent ceramics, caused by the existence of crystalline grains of random orientation. Investigation of the global thermal depolarization in the case of Nd:YAG ceramics revealed behavior similar to that of (111) oriented single crystals [222, 223]. The effect increases with the absorbed pump power. In the absence of lasing, the depolarization in the Nd:YAG ceramics depends on the Nd concentration since the heating effects are directly linked with the concentration dependent emission quantum efficiency, as discussed in Section 10.3.2. The measured depolarization for Nd concentrations up to

3.4 at.% Nd corresponds precisely to the emission quantum efficiency calculated with Eq. (2.68). The depolarization under 885 nm diode laser pumping was smaller than in the case of 808 nm pumping because of the smaller quantum defect. Moreover, under the conditions of laser emission the depolarization is smaller than in the absence of lasing and corresponds to the tendency of the heat load parameter to hold a fixed value for each wavelength of pump regardless of the Nd concentration. Theoretical modeling and the experimental investigation of thermal depolarization in Nd:YAG ceramics indicated that because of the granular structure of the ceramics the polarized and de-polarized beams show transverse modulation with spacing corresponding to the ceramic grain size. The depth of modulation corresponds to the ratio between the thickness of the laser material and the grain size [224, 225]. The thermal small-scale modulation on the phase distortion in ceramics was confirmed by measurements on undoped CaF_2 ceramics heated by CO_2 laser [226].

The random distribution of the phase distortions caused by the random orientation of the ceramic grains could be expected to determine local modes of oscillation in lasers with small ratio of the thickness of the laser material to the grain size. Such a situation can be met in thin laser materials, such as used for microchip lasers. Indeed, in the case of a diode laser end-pumped highly doped Nd:YAG ceramic microchip laser with deposited mirrors, the global optical spectrum measured with a conventional (20 GHz resolution) wavemeter showed single longitudinal mode oscillation. However, measurement with a high-resolution (6 MHz) Fabry–Perot interferometer revealed that these laser emission modes consist of closely spaced multiple components that fluctuate in time and it was inferred that these modes correspond to the local modes emitted by the individual crystalline ceramic grains [227]. Further investigation revealed that the dynamics of modulation by coupled local modes is very complex and includes additional quasi-periodic and chaotic relaxation oscillations; periodic short-time quasi-Q-switched spikes were also observed and attributed to the existence of foreign inclusions with SA properties at the grain boundaries [228]. Measurements on 1 mm thick coarse-grained Nd:YAG ceramics with various Nd concentrations ranging from 1.1 at.% Nd (average grain size 51.85 μm) to 4.8 at.% Nd (average grain size 5.61 μm) revealed the presence of these local modes, whereas the fine-grained (1.16 μm) ceramics with 1.1 at.% Nd concentration and 5 at.% Yb:YAG ceramics (grain size 3.20 μm) generated single-frequency linearly polarized radiation [229].

Control of the birefringence and the polarization of the laser emission can be important for various applications of solid-state lasers and various techniques based on resonator configuration, insertion of intracavity elements (polarizers, grating mirrors, birefringent elements), direction and pump configuration and on exploitation of thermally induced modification of the refractive index can modify the stability of the laser resonator to obtain the desired state of beam polarization. Such techniques have been applied successfully in the case of ceramic Nd:YAG lasers for the control of birefringence [230] and for the generation of laser radiation with controlled polarization properties [231–234] or modal structure [229, 235–237], including beams that would enable creation of arrays of vortex lasers for optical tweezers or for cold atom trapping. Polarized laser emission was also reported for passively Q-switched [238] and mode-locked [239] Nd:YAG ceramic lasers.

11.5 Concluding remarks: the state of the art and directions of development of ceramic lasers

The ceramic lasers described in this chapter, as well as many other cases, show that ceramic lasers represent a vigorous sector of research in the field of lasers and their applications. Some of these lasers are already at a mature stage, but many others are still at the stage of investigation or development. The variety of new ceramic materials is leading to strong development of basic research in the field of doped materials (quantum states, radiative and non-radiative de-excitation and so on), interactions between the doping ions and energy transfer process, and new quantum electronic processes. Ceramic laser materials can contribute to major development of solid-state lasers, such as the extension of the wavelength range, improvement of efficiency, extension of the temporal regime to very short pulse generation, unprecedented scaling of the average or peak power or of energy and so on. Activity in the field of ceramic lasers can extend, benefit from, and contribute to the development of research on crystalline lasers and on other optical materials such as nanomaterials, phosphors and scintillators.

The unprecedented versatility of the ceramic materials (composition of the host material, doping concentration and distribution), which allows tailoring of the properties of the laser materials to the specific needs of the various applications, together with the ability to produce large active components with improved thermomechanical properties or multi-functional monolithic composite materials, gives confidence that research in this field will continue and expand. A crucial criterion for the success of these lasers is the maximum use of the radiation power of the excitation sources by simultaneous improvement of the pump absorption, laser emission efficiency and limitation of the parasitic de-excitation (luminescence, non-radiative processes) as well as of their effects (heat generation, amplified spontaneous emission). Since the thermomechanical and thermo-optic effects caused by heat generation are the main limiting factors for power or energy scaling of solid-state lasers, a tight control of the intensity and distribution of the thermal field and of its effects inside the laser material is crucial. The basis of this improvement resides in a correlated approach to all three main parts of the laser, the laser material, the pumping system and the laser design. The ability to tailor the properties of the laser material to the requirements imposed by this correlation is in turn determined by a deep understanding and exploitation of the intimate relation between composition–fabrication technology–structure–properties–functionality of these materials. With the opportunity to control the doping or to produce composite materials, ceramic materials offer unique possibilities.

The independent optimization of the various individual steps of the flow of excitation inside the laser material and of their consequences can generate contradictory results, so that the improvement of a step can reduce the efficiency of other steps. In this case figures of merit for any action must be introduced to characterize its global effect: typical examples are Nd laser materials where increasing the doping concentration in order to improve the pump absorption can enhance the self-quenching of emission. In this case the concentration must be chosen to give the optimal global effect. Ceramic materials allow much better control of this optimization compared with single crystals. These materials can also be useful in making possible laser emission from doping ions with favorable emission properties (new wavelengths, high emission cross-section, long emission lifetime) or new excitation schemes for which crystalline materials do not offer suitable conditions. The enhanced doping capabilities of ceramics (concentration and distribution) can also lead to improved excitation of emission by sensitization processes. The compositional versatility of transparent ceramics allows control of the inhomogeneous broadening of the emission lines in solid-solution mixed materials or as the effect of an incorporated sintering aid. This opens the possibility for tunable or multiple-wavelength emission or for the generation of short or ultra-short laser pulses.

Optimization of laser emission and minimization of heat generation recommend laser emission schemes with low quantum defect. In the case of resonant diode laser pumping, the maximum reduction can be obtained by pumping directly into the emitting level. Moreover, in addition to a reduction in the quantum defect, whenever possible, pumping in hot-bands of the absorption spectrum of the emitting level can allow use of part of the thermal energy of the material in laser emission. Significant examples of the benefits of direct diode laser pumping into the emitting level of RE^{3+}-doped ceramics have been described in this book. Such an approach could enable enhancement of performance or power scaling of various solid-state lasers, from microchip to very high power systems, and this can be facilitated by the vigorous recent development of high power diode lasers in various wavelength ranges. Owing to its effects on laser emission at the fundamental frequency and on heat generation, direct pumping could be very useful for increasing the global efficiency of the non-linear processes.

The possibility of controlling the laser parameters, the wavelength of emission, the temporal regime of emission, the beam quality (modal structure, state of polarization) and the power or energy range offers countless examples of possible applications of ceramic lasers, in industry, nuclear physics and energy, medicine and biology, information technology, communications, control of the environment, defense and security and so on. The performance of these lasers could be unprecedented, leading to solutions for problems that could not be solved otherwise.

Illustrative of the potential of ceramic lasers for new applications is the development of new sustainable energy sources, based on nuclear fusion or new modalities for using solar energy. Irradiation of pellets of deuterium–tritium mixture by high energy lasers was envisaged soon after the advent of the laser as a modality to supply enough energy to produce conditions for nuclear fusion [240]. The nuclear fusion reaction involves two steps, compression and ignition, which can be accomplished with a unique laser (central fusion) or with two different lasers (fast ignition). The first approach requires nanosecond laser pulse energies of several megajoules, whereas the second requires hundreds of kilojoule nanosecond pulses for compression and tens of kilojoule picosecond pulses for ignition. It was also determined that a nuclear fusion plant would require repetition rates for lasers larger than 10 Hz. The neutrons released in an inertial laser nuclear fusion facility can be used to heat a mantle and for the production of steam and electricity. Various solutions of the problem of laser nuclear fusion have been investigated [241]. A very tempting utilization of the neutrons released from (D–T) fusion would be the nuclear fission of fissile materials such as natural uranium or spent nuclear fuel from nuclear fission plants, leading to complete annihilation of these wastes (the LIFE project) and generation of heat [242].

Laser nuclear fusion was demonstrated less than a decade after the advent of lasers by observation of neutron emission from D–T mixtures. However, a very important step in demonstrating the practical utility of laser nuclear fusion would be demonstration of the break-even point, when the energy released becomes equal to the enormous energy spent creating fusion. Experiments on a variant of central fusion using high energy (192 large-aperture, 40×40 cm^2, beams totaling 1.8 MJ), lamp pumped frequency-tripled Nd:glass lasers (National Ignition Facility, NIF) [11. 243] now face this very important task. Similar characteristics are envisaged for the lasers for nuclear fusion in the project Laser Megajoule (LMJ) [244]. The repetition rate of these lasers is very low, one pulse in several hours, so the following steps would necessitate much higher repetition rates, which could not be reached with materials of low thermal conductivity and low thermal stress such as glasses, and with the high quantum defect pumping inherent to the lamp pumping.

In these circumstances, the most promising solution for practical production of electricity from inertial nuclear fusion would be utilization of large-aperture diode pumped ceramic lasers. Projects based on the development of diode laser pumped cooled (100–220 K) Yb-doped garnet or sesquioxide ceramics are in various stages of development [245–249]. Such lasers could fulfill the functional conditions (energy, pulse duration, repetition rate); at the same time, they can have very high optical–electrical efficiency, in excess of 10%, a very important condition for the energetic balance of a nuclear fusion plant.

The Yb^{3+}-doped (under direct or Nd-sensitized excitation) or the Cr^{4+}-doped ceramics could also be an alternative solution for the construction of high peak power lasers of interest in nuclear and atomic physics [250, 251]. The solar pumped lasers based on (Cr^{3+}, Nd)-doped YAG ceramics under Fresnel [85] or solar furnace [252] concentration can contribute to utilization of solar energy in a variety of applications such as the solar–magnesium–hydrogen cycle for the development of new mobile sources of energy [253], spatial communications and so on. The Q-switched multi-beam composite Nd:YAG/Cr^{4+}:YAG monolithic ceramics can enable the spatial control of pulses able to grant the optimal regime for ignition of automotive engines or of gas or liquid fuel furnaces.

From existing experience and from the many envisaged new applications, it is evident that the laser cannot be regarded as universal equipment with standard performance. The development, optimization and identification of new possible applications will require specialized lasers with specific characteristics, efficiency and functionality and in many cases such lasers could have almost unique application. Since the basic element of the laser is the active laser material, with the pumping system and the laser design allowing the maximal use of the properties of the laser material, a correlated approach to these main parts of the laser is the most efficient and realistic way to optimize laser performance. In many cases the complex, quantum nature of the laser material and of the excitation and emission processes requires a "back-to-basics" approach. The development of the basic, applied and technological research, the diversification and enhancement of performance and extension of the power range of ceramic lasers and the development of their applications can thus fully justify the assertion that the advent of transparent polycrystalline ceramic materials and lasers represents a revolutionary step in the laser field.

References

[1] A. Ikesue, T. Kinoshita, K. Kamata, and K. Yoshida, Fabrication and optical properties of high-performance polycrystalline Nd:YAG ceramics for solid state lasers, *J. Am. Ceram. Soc.* **78** (1995) 1033–1040.

[2] J. Lu, M. Prabhu, K. Ueda, H. Yagi, T. Yanagitani, A. Kudryashov, and A. A. Kaminskii, Potential of ceramic YAG lasers, *Laser Phys.* **11** (2001) 1053–1057.

[3] A. Ikesue, Polycrystalline Nd:YAG ceramic lasers, *Opt. Mater.* **19** (2002) 183–187.

[4] J. Lu, K. Ueda, H. Yagi, T. Yanagitani, Y. Akiyama, and A. A. Kaminskii, Neodymium doped yttrium aluminum garnet ($Y_3Al_5O_{12}$) nanocrystalline ceramics – a new generation of solid state laser and optical materials, *J. Alloys Comp.* **341** (2002) 220–225.

[5] V. Lupei, Efficiency enhancement and power scaling of Nd lasers, *Opt. Mater.* **24** (2003) 353–368.

[6] J. Wisdom, M. Digonnet, and R. L. Byer, Ceramic lasers: ready for action, *Photon. Spectra* (February 2004) 2–7.

[7] K. Ueda, J. F. Bisson, H. Yagi, K. Takaichi, A. Shirakawa, T. Yanagitani, and A. A. Kaminskii, Scalable ceramic lasers, *Laser Phys.* **15** (2005) 927–938.

[8] A. Ikesue and Y. L. Aung, Synthesis and performance of advanced ceramic lasers, *J. Am. Ceram. Soc.* **89** (2006) 1936–1944.

[9] A. Ikesue, Y. L. Aung, T. Taira, T. Kamimura, K. Yoshida, and G. L. Messing, Progress in ceramic lasers, *Annu. Rev. Mater. Res.* **36** (2006) 397–429.

[10] A. A. Kaminskii, Laser crystals and ceramics: recent advances, *Laser Photon. Rev.* **1** (2007) 93–177.

[11] T. Taira, RE^{3+} ion-doped YAG ceramic lasers, *IEEE J. Quantum Electron.* **13** (2007) 798–809.

[12] A. Ikesue and Y. L. Aung, Ceramic laser materials, *Nature Photonics*, **2** (2008) 721–727.

[13] V. Lupei, Ceramic laser materials and the prospect for high power lasers, *Opt. Mater.* **31** (2009) 701–706.

[14] V. Lupei, N. Pavel, and T. Taira, Basic enhancement of the overall optical efficiency of intracavity frequency-doubled devices for the 1 μm continuous-wave Nd:$Y_3Al_5O_{12}$ laser emission, *Appl. Phys. Lett.* **83** (2003) 3653–3655.

[15] M. Ross, YAG laser operation by semiconductor laser pumping, *Proc. IEEE* **56** (1968) 196–197.

[16] L. S. Rosenkrantz, GaAs diode-pumped Nd:YAG laser, *J. Appl. Phys.* **43** (1972) 4603–4605.

[17] V. Lupei, A. Lupei, C. Gheorghe, and A. Ikesue, Comparative high-resolution spectroscopy and emission dynamics of Nd-doped GSGG crystals and transparent ceramics, *J. Lumin.* **128** (2008) 885–887.

[18] R. Lavi, S. Jackel, Y. Tzouk, M. Wink, E. Lebiush, M. Katz, and I. Paiss, Efficient pumping scheme for neodymium-doped materials by direct excitation of the upper laser level, *Appl. Opt.* **38** (1999) 7382–7385.

[19] R. Lavi and S. Jackel, Thermally boosted pumping of neodymium lasers, *Appl. Opt.* **39** (2000) 3093–3098.

[20] V. Lupei, A. Lupei, S. Georgescu, T. Taira, Y. Sato, and A. Ikesue, The effect of Nd concentration on the spectroscopic and emission decay properties of highly-doped Nd:YAG ceramics, *Phys. Rev. B* **64** (2001) 092102.

[21] V. Lupei, T. Taira, A. Lupei, N. Pavel, I. Shoji, and A. Ikesue, Spectroscopy and laser emission under hot band resonant pump in highly-doped Nd:YAG ceramics, *Opt. Commun.* **195** (2001) 225–232.

[22] V. Lupei, A. Lupei, N. Pavel, I. Shoji, T. Taira, and A. Ikesue, Laser emission under thermally-activated resonant pump in concentrated Nd:YAG ceramics, *Appl. Phys. Lett.* **79** (2001) 590–592.

[23] V. Lupei, A. Lupei, N. Pavel, T. Taira, and A. Ikesue, Comparative investigation of spectroscopic and laser emission characteristics under direct 885-nm pump of concentrated Nd:YAG ceramics and crystals, *Appl. Phys. B* **73** (2001) 757–762.

[24] S. Goldring and R. Lavi, Nd:YAG laser pumped at 946 nm, *Opt. Lett.* **33** (2008) 669–671.

[25] H. Yagi, T. Yanagitani, and K. Ueda, $Nd^{3+}:Y_3Al_5O_{12}$ laser ceramics: flashlamp pumped laser operation with a UV cut filter, *J. Alloys Comp.* **421** (2006) 195–199.

[26] A. Ikesue, K. Kamata, and K. Yoshida, Synthesis of Nd^{3+}, Cr^{3+}-codoped YAG ceramics for high efficiency solid state lasers, *J. Am. Ceram. Soc.* **78** (1995) 2545–2547.

[27] H. Yagi, T. Yanagitani, H. Yoshida, M. Nakatsuka, and K. Ueda, Highly efficient flash-lamp pumped Cr^{3+} and Nd^{3+} co-doped $Y_3Al_5O_{12}$ ceramic lasers, *Jpn. J. Appl. Phys.* **45** (2006) 133–135.

[28] T. Saiki, M. Nakatsuka, and K. Imasaki, Highly efficient laser action of Nd^{3+} and Cr^{3+}-doped yttrium aluminium garnet ceramics based on phonon-assisted cross-relaxation using solar light source, *Jpn. J. Appl. Phys.* **49** (2010) 082702.

[29] V. Lupei, A. Lupei, and A. Ikesue, Transparent Nd and (Nd, Yb)-doped Sc_2O_3 ceramics as potential new laser materials, *Appl. Phys. Lett.* **86** (2005) 111118.

[30] V. Lupei, A. Lupei, C. Gheorghe, S. Hau, and A. Ikesue, Efficient sensitization of Yb^{3+} emission by Nd^{3+} in Y_2O_3 transparent ceramics and the prospect for high energy Yb lasers, *Opt. Lett.* **34** (2009) 2141–2143.

[31] V. Lupei, A. Lupei, C. Gheorghe, and A. Ikesue, Sensitization of Yb^{3+} emission in $(Nd,Yb):Y_3Al_5O_{12}$ transparent ceramics, *J. Appl. Phys.* **108** (2010) 123112.

[32] V. Petit, P. Camy, J.-L. Doualan, and R. Moncorge, Continuous-wave and tunable laser emission of Yb^{3+} in Nd:Yb:CaF_2, *Appl. Phys. Lett.* **88** (2006) 051111.

[33] A. A. Kaminskii, S. N. Bagaev, K. Ueda, A. Sirakawa, T. Tokurakawa, H. Yagi, T. Yanagitani, and J. Dong, Stimulated emission spectroscopy of fine-grained garnet ceramics Nd:YAG between 77–650 K, *Laser Phys. Lett.* **6** (2009) 682–687.

[34] W. Koechner, Absorbed pump power, thermal profile and stress in a cw pumped Nd:YAG crystal, *Appl. Opt.* **9** (1970) 1429–1434.

[35] V. Lupei, Directions for performance enhancement and power scaling of the Nd lasers, *Prog. Quantum Electron.* (to be published).

[36] D. Kracht, R. Wilhelm, M. Frede, K. Dupre, and L. Ackermann, 407 W end-pumped multi-segmented Nd:YAG laser, *Opt. Express* **13** (2005) 10140–10144.

[37] R. Wilhelm, M. Frede, and D. Kracht, Power scaling of end-pumped solid state rod lasers by longitudinal dopant concentration gradients, *IEEE J. Quantum Electron.* **44** (2008) 232–244.

[38] K. Yoshida, H. Ishii, T. Kumada, T. Kamimura, A. Ikesue, and T. Okamoto, All ceramic composite with layer by layer structure by advanced ceramic technology, *Proc. SPIE* **5647** (2005) 247–254.

[39] H. Yagi, K. Takeuchi, K. Ueda, Y. Yamasaki, T. Yanagitani, and A. A. Kaminskii, The physical properties of composite YAG ceramics, *Laser Phys.* **15** (2005) 1338–1344.

[40] D. Kracht, M. Frede, R. Wilhelm, and C. Fallnich, Comparison of crystalline and ceramic composite Nd:YAG for high-power diode end pumping, *Opt. Express* **13** (2005) 6212–6216.

[41] D. Kracht, D. Freiburg, R. Wilhelm, M. Frede, and C. Fallnich, Core-doped ceramic Nd:YAG laser, *Opt. Express* **14** (2006) 2690–2694.

[42] A. Strasser and M. Ostermeyer, Improving the brightness of side-pumped power amplifiers by using core doped ceramic rods, *Opt. Express* **14** (2006) 6687–6693.

[43] A. Ikesue, Y. L. Aung, T. Yoda, S. Nakayama, and T. Kamimura, Fabrication and laser performance of polycrystalline and single crystal Nd:YAG by advanced ceramic processing, *Opt. Mater.* **29** (2007) 1289–1294.

[44] D. Kouznetsov and J. F. Bisson, Role of undoped cap in the scaling of thin-disk lasers, *J. Opt. Soc. Am.* **B 25** (2008) 338–345.

[45] R. Wilhelm, D. Freiburg, M. Frede, and D. Kracht, End pumped Nd:YAG laser with longitudinal hyperbolic dopant profile, *Opt. Express* **16** (2008) 20106–20116.

[46] R. Wilhelm, D. Freiburg, M. Frede, D. Kracht, and C. Fallnich, Design and comparison of composite rod crystals for power scaling of diode end-pumped Nd:YAG lasers, *Opt. Express* **17** (2009) 8229–8236.

[47] B. Jiang, T. Huang, Y. Wu, W. Lai, and Y. Pan, Synthesis and microstructure analysis of composite Nd:YAG/YAG transparent ceramics, *Chin. Opt. Lett.* **7** (2009) 505–507.

[48] S. N. Bagayev, A. A. Kaminskii, Yu. L. Kopylov, I. M. Kotelyanskii, and V. B. Kravchenko, Simple method to join YAG ceramics and crystals, *Opt. Mater.* **34** (2012) 951–954.

[49] J. A. Wisdom, R. M. Gaume, and R. L. Byer, Laser gain scanning spectroscopy: a new characterization technique for dopant engineered media, *Opt. Express* **18** (2010) 18912–18921.

[50] J. P. Hollingsworth, J. D. Kuntz, F. J. Rierson, and Th. S. Soules, Nd diffusion in YAG ceramics, *Opt. Mater.* **34** (2011) 592–595.

[51] K. Marquardt, E. Petrishcheva, R. Abart, E. Gardes, R. Wirth, R. Dohmen, H.-W. Becker, and W. Heinrich, Volume diffusion of ytterbium in YAG: thin film experiments and combined TEM-RBS analysis, *Phys. Chem. Miner.* **37** (2010) 751–760.

[52] H. Furuse, J. Kawanaka, M. Myanaga, T. Saiki, K. Imasaki, M. Fujita, K. Takeshita, S. Ishii, and Y. Izawa, Zig-zag active mirror laser with cryogenic Yb:YAG/YAG composite ceramic, *Opt. Express* **19** (2011) 2448–2455.

[53] E. A. Khazanov, Investigation of Faraday isolator and Faraday mirror designs for multi-kilowatt power lasers, *Proc. SPIE* **4968** (2003) 115–126.

[54] H. Yoshida, K. Tsubakimoto, Y. Fujimoto, K. Mikami, H. Fujita, N. Miyanaga, H. Nozawa, H. Yagi, T. Yanagitani, Y. Nagata, and H. Kinoshita, Optical properties and

Faraday effect of ceramic terbium gallium garnet for a room temperature Faraday rotator, *Opt. Express* **19** (2011) 15181–15187.

[55] H. Lin, S. Zhou, and H. Tang, Synthesis of Tb$_3$Al$_5$O$_{12}$(TAG) transparent ceramics for potential magneto-optical application, *Opt. Mater.* **33** (2011) 1833–1836.

[56] R. Feldman, Y. Golan, Z. Burshtein, S. Lackel, I. Moshe, A. Meir, Y. Lumer, and Y. Shimony, Strengthening of polycrystalline (ceramic) Nd:YAG elements for high-power laser applications, *Opt. Mater.* **33** (2011) 695–701.

[57] J. Lu, M. Prabhu, J. C. Song, C. Li, J. Xu, K. Ueda, A. A. Kaminskii, H. Yagi, and T. Yanagitani, Optical properties and highly efficient laser oscillation of Nd:YAG ceramics, *Appl. Phys. B* **71** (2000) 469–473.

[58] I. Shoji, S. Kurimura, Y. Sato, T. Taira, A. Ikesue, and K. Yoshida, Optical properties and laser characteristics of highly Nd^{3+}-doped Y$_3$Al$_5$O$_{12}$ ceramics, *Appl. Phys. Lett.* **77** (2000) 939–941.

[59] V. Lupei, N. Pavel, and T. Taira, Efficient laser emission in concentrated Nd laser materials under pumping into the emitting level, *IEEE J. Quantum Electron.* **38** (2002) 240–245.

[60] H. F. Li, D. G. Xu, Y. Yag, Y. Y. Wu, R. Zhou, T. L. Zhang, X. Zhao, P. Wang, and J. Q. Yao, Experimental 511 W composite Nd:YAG ceramic laser *Chin. Phys. Lett.* **22** (2005) 2565–2567.

[61] R. Huss, R. Wilhelm, C. Kolleck, J. Neumann, and D. Kracht, Suppression of parasitic oscillations in a core-doped ceramic Nd:TAG laser by Sm:YAG cladding, *Opt. Express* **18** (2010) 13094–13101.

[62] S. J. McNaught, H. Komine, S. B. Weiss, R. Simpson, A. M. F. Johnson, J. Machan, C. P. Asman, M. Weber, G. C. Jones, M. W. Valley, A. Jankevics, D. Burchman, M. McClellan, J. Solee, J. Marmo, and J. Injeyan, 100 kW coherently combined slab MOPAs, *CLEO* 2009, Tech. Dig. Paper CThA1.

[63] H. Injeyan, Solid-state laser power scaling to 100 kW and beyond, *CLEO* 2010, paper CTuQQ1.

[64] A. Mandl and D. E. Klimek, Textron's J-HPSSL 100 kW ThinZag® laser program, *CLEO* 2010, paper JThH2.

[65] D. E. Klimek and A. Mandl, Nd:YAG ceramic ThinZag® laser development, in *High-Power Laser Handbook*, eds. H. Injeyan, S. Palese, and G. Goodno, McGraw Hill (2011), Chapter 9, pp. 207–224.

[66] C. Y. Li, Y. Bo, B. S. Wang, C. Y. Tian, Q. J. Peng, D. F. Cui, Z. Y. Xu, W. B. Liu, X. Q. Feng, and Y. B. Pan, A kilowatt level diode-side-pumped QCW Nd:YAG ceramic laser, *Opt. Commun.* **283** (2010) 5145–5148.

[67] R. M. Yamamoto, J. M. Parker, K. L. Allen, R. W. Allmon, K. F. Alviso, C. P. J. Barry, B. S. Bhachu, C. D. Booley, A. K. Bunham, R. L. Combs, K. P. Cutter, S. N. Fochs, S. A. Gonzales, R. L. Hurd, K. N. LaFortune, M. A. Mc Clelland, R. D. Merril, L. Molina, C. W. Parks, P. H. Paks, A. S. Posey, M. D. Rotter, B. M. Roy, A. M. Rubenchik, T. F. Soules, and D. E. Webb, Evolution of a solid state laser, *Proc SPIE* **6552** (2007), paper 655205.

[68] R. M. Yamamoto, B. S. Bachu, K. P. Cutter, S. N. Fochs, S. A. Letts, Ch. W. Parks, M. A. Rotter, and T. F. Soules, The use of large transparent ceramics in a high powered, diode pumped solid state laser, *OSA Adv. Solid-State Photonics*, Nara (2008), paper WC5.

[69] V. Lupei, N. Pavel, and T. Taira, 1064-nm laser emission of highly doped Nd:YAG active components under 885-nm diode laser pumping, *Appl. Phys. Lett.* **80** (2002) 4309–4311.

[70] M. Frede, R. Wilhelm, and D. Kracht, 250 W end-pumped Nd:YAG laser with direct pumping into the emitting laser level, *Opt. Lett.* **31** (2006) 3618–3619.

[71] H. Yagi, T. Yanagitani, K. Takaichi, K. Ueda, and A. A. Kaminskii, Characterization and laser performances of highly transparent Nd:$Y_3Al_5O_{12}$ laser ceramic, *Opt. Mater.* **29** (2007) 1257–1262.

[72] S. S. Zhang, G. P. Wang, X. Y. Zhang, Z. H. Cong, S. Z. Fan, Z. J. Liu, and W. J. Sun, Continuous-wave ceramic Nd:YAG laser at 1123 nm, *Laser Phys. Lett.* **6** (2009) 864–867.

[73] C. Y. Li, Q. J. Peng, B. S. Wang, Y. Bo, D. F. Cui, Z. Y. Xu, X. Q. Feng, and Y. B. Pan, QCW diode-side-pumped Nd:YAG ceramic laser with 247 W output power at 1123 nm, *Appl. Phys. B* **103** (2011) 285–289.

[74] Z. Wang, H. Liu, J. Wang, Y. Lv, Y. Sang, R. Lan, H. Yu, X. Xu, and Z. Shao, Passively Q-switched dual-wavelength laser output of LD-end-pumped ceramic Nd:YAG laser, *Opt. Express* **17** (2009) 12076–12081.

[75] J. Lu, J. Lu, T. Murai, K. Takaichi, T. Uematsu, J. Xu, K. Ueda, H. Yagi, T. Yanagitani, and A. A. Kaminskii, 36 W diode-pumped continuous-wave 1319 Nd:YAG ceramic laser, *Opt. Lett.* **27** (2002) 1120–1122.

[76] J. Lu, J. Lu, A. Shirakawa, K. Ueda, H. Yagi, T. Yanagitani, V. Gabler, H. J. Eichler, and A. A. Kaminskii, New highly efficient 1.3 μm CW generation in the $^4F_{3/2} \rightarrow$ $^4I_{13/2}$ channel for nanocrystalline Nd^{3+}:$Y_3Al_5O_{12}$ ceramic laser under diode pumping, *Phys. Status Solidi A* **189** (2002) R11–R13.

[77] S. G. P. Strohmaier, H. J. Eichler, J. F. Bisson, H. Yagi, K. Takaichi, K. Ueda, T. Yanagitani, and A. A. Kaminskii, Ceramic Nd:YAG laser at 946 nm, *Laser Phys. Lett.* **2** (2005) 383–386.

[78] L. Guo, R. Lan, H. Liu, H. Yu, Z. Zhang, J. Wang, D. Hu, S. Zhang, L. Chen, Y. Zhao, X. Xu, and Z. Wang, 1319 nm and 1338 nm dual wavelength operation of LD end pumped Nd:YAG ceramic laser, *Opt. Express* **18** (2010) 9098–9106.

[79] L. Chen, Z. Wang, H. Liu, S. Zhuang, H. Ye, L. Guo, R. Lann, J. Wang, and X. Xu, Continuous-wave tri-wavelength operation at 1064, 1319 and 1338 nm of LD end-pumped Nd:YAG ceramic laser, *Opt. Express* **18** (2010) 22167–22173.

[80] H. Cai, J. Chou, T. Feng, G. Yao, Y. Qi, Q. Lou, J. Dong, and Y. Wei, Dual wavelength competitive output in Nd:$Y_3Sc_{1.5}Al_{3.5}O_{12}$ ceramic disk laser, *Opt. Commun.* **281** (2008) 4401–4405.

[81] T. Saiki, K. Imasaki, S. Motokoshi, C. Yamanaka, M. Fujita, M. Nakatsuka, and Y. Izawa, Disk-type Nd/Cr:YAG ceramic laser pumped by arc-metal-halide lamp, *Opt. Commun.* **268** (2006) 155–159.

[82] H. Yagi, T. Yanagitani, H. Yoshida, M. Nakatsuka, and K. Ueda, The optical properties and laser characteristics of Cr^{3+} and Nd^{3+} co-doped $Y_3Al_5O_{12}$ ceramic, *Opt. Laser Technol.* **39** (2007) 1295–1300.

[83] T. Saiki, S. Motokoshi, K. Imasaki, M. Nakatsuka, C. Yamanaka, and K. Fujioka, Two-pass amplification of CW laser by Nd/Cr:YAG ceramic active mirror under lamp light pumping, *Opt. Commun.* **282** (2009) 936–939.

[84] T. Saiki, S. Motokoshi, K. Imasaki, K. Fujioka, H. Fujita, M. Nakatsuka, and C. Yamanaka, Laser pulse amplified by Nd/Cr:YAG ceramic amplifier using lamp and solar light sources, *Opt. Commun.* **282** (2009) 1358–1362.

[85] T. Ohkubo, T. Yabe, K. Yoshida, S. Uchida, T. Funatsu, B. Bagheri, T. Oishi, K. Daito, M. Ishioka, Y. Nakayama, N. Yasunaga, K. Kido, Y. Sato, C. Baasandash, K. Kato, T. Yanagitani, and Y. Okamoyo, Solar-pumped 80 W laser irradiated by a Fresnel lens, *Opt. Lett.* **34** (2009) 175–177.

[86] C. Greskovich and J. P. Chernoch, Polycrystalline ceramic lasers, *J. Appl. Phys.* **44** (1973) 4599–4606.

[87] C. Greskovich and J. P. Chernoch, Improved polycrystalline ceramic laser, *J. Appl. Phys.* **45** (1974) 4495–4502.

[88] J. Lu, T. Murai, K. Takaichi, T. Uematsu, K. Ueda, H. Yagi, T. Yanagitani, and A. A. Kaminskii, $Nd^{3+}:Y_2O_3$ ceramic laser, *Jpn. J. Appl. Phys.* **40** (2001) L1277–L1279.

[89] J. Lu, K. Takaiki, T. Uematsu, A. Shirakawa, M. Musha, K. Ueda, H. Yagi, T. Yanagitani, and A. A. Kaminskii, Promising ceramic laser material: highly transparent $Nd:Lu_2O_3$, *Appl. Phys. Lett.* **81** (2002) 4324–4326.

[90] Q. Yang, S. Lu, B. Zhang, H. Zhang, J. Zhou, Z. Yum, Y. Qi, and Q. Lou, Properties and laser performances of Nd-doped yttrium lanthanum oxide transparent ceramics, *Opt. Mater.* **33** (2011) 692–694.

[91] H. Kurokawa, A. Shirakawa, M. Tokurakawa, K. Ueda, S. Kuretake, N. Tanaka, Y. Kintaka, K. Kageyama, H. Takagi, and A. A. Kaminskii, Broadband-gain Nd^{3+}-doped $Ba(Zr, Mg, Ta)O_3$ ceramics for ultrashort pulse generation, *Opt. Mater.* **33** (2010) 667–669.

[92] H. Zhao, X. Sun, J. M. Zhang, Y. K. Zou, K. K. Li, Y. Wang, H. Jiang, P. L. Huang, and X. Chen, Lasing action and amplification in Nd doped electrooptic lanthanum lead zirconate titanate ceramics, *Opt. Express* **19** (2011) 2965–2971.

[93] T. T. Basiev, M. E. Doroshenko, V. A. Konyushkin, and V. V. Osiko, $SrF_2:Nd^{3+}$ laser fluoride ceramics, *Opt. Lett.* **35** (2010) 4009–4011.

[94] J. Akiyama, Y. Sato, and T. Taira, Laser ceramics with rare-earth-doped anisotropic materials, *Opt. Lett.* **35** (2010) 3598–3600.

[95] J. Akiyama, Y. Sato, and T. Taira, Laser demonstration of diode-pumped Nd^{3+} doped fluoroapatite anisotropic ceramics, *Appl. Phys. Express* **4** (2011) 022703.

[96] C. Y. Wang, J. H. Ji, Q. H. Lou, X. L. Zhu, and Y. T. Lu, Kilohertz electro-optic Q-switched Nd:YAG ceramic laser, *Chin. Phys. Lett.* **23** (2006) 1797–1799.

[97] Y. Qi, X. Zhu, Q. Lou, J. Ji, J. Dong, and R. Wei, High energy LDA side-pumped electrooptically Q-switched Nd-YAG ceramic laser, *J. Opt. Soc. Am. B* **24** (2007) 1042–1245.

[98] T. Omatsu, K. Nawaka, D. Sauder, A. Minassian, and M. J. Damzen, Over 40 W diffraction-limited Q-switched output from neodymium-doped YAG ceramic bounce amplifier, *Opt. Express* **14** (2006) 8198–8204.

[99] P. Li, R. Wang, X. Zhang, Y. Wang, and X. Chen, Compact and efficient diode-pumped actively Q-switched 1319 nm Nd:YAG ceramic laser, *Laser Phys.* **20** (2010) 1603–1617.

[100] C. Zhang, X. Y. Zhang, Q. P. Wang, Z. H. Kong, S. Z. Fan, X. H. Chen, Z. J. Liu, and Z. Zhang, Diode-pumped Q-switched 946 nm Nd:YAG ceramic laser, *Laser Phys. Lett.* **6** (2009) 521–525.

[101] A. Ikesue, K. Yoshida, and K. Kamata, Transparent Cr^{4+} doped YAG ceramics for tunable lasers, *J. Am. Ceram. Soc.* **78** (1995) 2545–2547.

[102] K. Takaichi, J. Lu, T. Murai, T. Uematsu, A. Shirakawa, K. Ueda, H. Yagi, T. Yanagitani, and A. A. Kaminskii, Chromium doped $Y_3Al_5O_{12}$ ceramic: a novel saturable absorption for passive self Q-switced 1 μm solid-state lasers, *Jpn. J. Appl. Phys.* **41** (2002) L96–L98.

[103] H. Yagi, K. Takaichi, K. Ueda, T. Yanagitani, and A. A. Kaminskii, Influence of annealing conditions on the optical properties of chromium-doped $Y_3Al_5O_{12}$, *Opt. Mater.* **29** (2006) 392–396.

[104] J. M. Degnan, Optimization of passively Q-switched lasers, *IEEE J. Quantum Electron.* **31** (1995) 1890–1901.

[105] Y. Kalisky, Cr^{4+} doped crystals: their use as lasers and passive Q-switches, *Prog. Quantum Electron.* **26** (2004) 249–303.

[106] H. Sakai, H. Kan, and T. Taira, 1 ΩM peak power single-mode high-brightness passively Q-switched Nd^{3+}:YAG microchip laser, *Opt. Express* **16** (2008) 19891–19899.

[107] K. Takaichi, J. Lu, T. Murai, T. Uematsu, A. Shirakawa, K. Ueda, H. Yagi, T. Yanagitani, and A. A. Kaminskii, Chromium doped $Y_3Al_5O_{12}$ ceramic: a novel saturable absorption for passive self Q-switced 1 μm solid-state lasers, *Jpn. J. Appl. Phys.* **41** (2002) L96–L98.

[108] Y. Feng, J. Lu, T. Takaichi, K. Ueda, H. Yagi, T. Yanagitani, and A. A. Kaminskii, Passively Q-switched ceramic Nd^{3+}:YAG/Cr^{4+}:YAG lasers, *Appl. Opt.* **43** (2004) 2944–2947.

[109] J. Kong, Z. L. Zhang, D. Y. Tang, G. Q. Xie, C. C. Chan, and Y. H. Shen, Diode-end-pumped passively Q-switched Nd:YAG ceramic laser with Cr^{4+}:YAG saturable absorber, *Laser Phys.* **18** (2008) 1508–1511.

[110] R. Lan, Z. Wang, H. Liu, H. Yu, L. Guo, L. Chen, S. Zhang, X. Xu, and J. Wang, Passively Q-switched Nd:YAG ceramic laser towards large pulse energy and short pulse width, *Laser Phys.* **20** (2010) 187–191.

[111] J. H. Yu, H. Zhang, Z. Wang, J. Wang, Y. Yu, X. Zhang, K. Lan, and M. Jiang, Dual wavelength neodymium doped yttrium aluminum garnet laser with chromium-doped yttrium aluminum garnet as frequency selector, *Appl. Phys. Lett.* **94** (2009) 041126.

[112] T. Saiki, J. Motokoshi, K. Imasaki, K. Fujioka, H. Yoshida, H. Fujita, M. Nakatsuka, and C. Yamanaka, Nd^{3+} and Cr^{3+} doped yttrium aluminum garnet pulse laser using Cr^{4+} doped yttrium aluminum garnet crystal passive Q-switch, *Jpn. J. Appl. Phys.* **48** (2009) 122501.

[113] V. Lupei, A. Lupei, V. Ionita-Manzatu, S. Georgescu, and F. Domsa, Combined mechanical color center passive Q-switching of neodymium lasers, *Opt. Commun.* **48** (1983) 203–206.

[114] T. Dascalu, G. Philipps, H. Weber, N. Pavel, T. Beck, and V. Lupei, Investigation on Nd:YAG laser, passive Q-switch, externally controlled, quasi-continuous or continuous pumped, *Opt. Eng.* **35** (1996) 1247–1251.

[115] N. Pavel, M. Tsunekame, and T. Taira, Composite, all-ceramic Nd:YAG/Cr^{4+}:YAG monolithic microlaser with multiple-beam output for engine ignition, *Opt. Express* **19** (2011) 9378–9384.

[116] T. Saiki, S. Motokoshi, K. Imasaki, K. Fujioka, H. Yoshida, H. Fujita, M. Nakatsuka, and C. Yamanaka, High repetition rate laser pulses amplification by Nd/Cr:YAG ceramic amplifier under CW arc-lamp-light pumping, *Opt. Commun.* **282** (2009) 2556–2559.

[117] T. Omatsu, A. Minassian, and M. J. Damzen, Passive Q-switching of a diode-side-pumped Nd doped 1.3 μm ceramic YAG bounce laser, *Opt. Commun.* **282** (2009) 4784–4788.

[118] P. Li, X. H. Chen, H. N. Zhang, and Q. P. Wang, Diode-end-pumped passively Q-switched 1319 nm Nd:YAG ceramic laser with V^{3+}:YAG saturable absorber, *Laser Phys.* **21** (2011) 1708–1711.

[119] J. Dong, K. Ueda, H. Yagi, and A. A. Kaminskii, Laser-diode pumped self-Q-switched microchip lasers, *Opt. Rev.* **15** (2008) 57–74.

[120] J. Dong, D. Y. Tang, J. Lu, K. Ueda, H. Yagi, and T. Yansagitani, Multi-pulse oscillations and instabilities in microchip self-Q-switched transverse mode laser, *Opt. Express* **17** (2009) 16980–16993.

[121] J. Li, Y. Wu, Y. Pan, H. Kou, Y. Shi, and J. Guo, Densification and microstructural evolution of Cr^{4+},Nd^{3+}:YAG transparent ceramics for self Q-switched lasers, *Ceram. Int.* **34** (2008) 1675–1679.

[122] G. Xie, D. Tang, J. Kong, and L. Qian, Diode-pumped passive Q-switching Nd:YAG ceramic laser with GaAs saturable absorber, *J. Opt. A: Pure Appl. Opt.* **9** (2007) 621–625.

[123] J. Zhao, S. Zhao, and G. Zhang, Optimization of passively Q-switched and mode-locked laser with Cr^{4+}:YAG saturable absorber, *Opt. Commun.* **284** (2011) 1648–1651.

[124] L. Yang, B. Feng, Z. Zhang, V. Grebier, B. Liu, and H. J. Eichler, Low modelocking saturation intensity of co-doped Nd^{3+}, Cr^{4+}:YAG crystals as saturable absorbers, *Opt. Mater.* **22** (2003) 59–63.

[125] W. D. Tan, F. Chen, R. J. Knize, J. Zhang, D. Tang, and J. Li, Passive mode-locking of ceramic Nd:YAG using (7, 5) semiconducting single walled carbon nanotubes, *Opt. Mater.* **33** (2011) 679–683.

[126] W. D. Tan, C. Y. Su, R. J. Knize, G. Q. Xie, L. J. Li, and D. Y. Tang, Mode locking of ceramic Nd:yttrium aluminum garnet with graphene as a saturable absorber, *Appl. Phys. Lett.* **96** (2010) 031106.

[127] L. Guo, W. Hou, H. B. Zhang, Z. P. Sun, D. F. Cui, Z. Y. Xu, Y. G. Wang, and X. Y. Ma, Diode-end-pumped passively mode-locked ceramic Nd:YAG laser with a semiconductor saturable mirror, *Opt. Express* **13** (2005) 4085–4089.

[128] J. A. Wisdom, D. S. Hum, M. J. F. Digonet, M. J. Fejer, and R. L. Byer, 2.6 watt average-power mode-locked ceramic Nd:YAG laser, *Proc. SPIE* **6469** (2007) 64690C.

[129] E. Martin-Rodriguez, P. Molina, A. Benayas, L. E. Bausa, J. Garcia-Sole, and D. Jaque, Suppression of Q-switched instabilities in a passively mode-locked $Nd:Y_3Al_5O_{12}$ ceramic laser, *Opt. Mater.* **31** (2009) 725–728.

[130] Y. Sato, J. Shirakawa, T. Taira, and A. Ikesue, Characteristics of Nd^{3+}-doped $Y_3ScAl_4O_{12}$ ceramic laser, *Opt. Mater.* **29** (2007) 1277–1282.

[131] H. Okada, M. Tanaka, H. Kiriyama, Y. Nakai, Y. Ochi, A. Sugiyama, H. Daido, T. Kimura, T. Yanagitani, H. Yagi, and N. Meichin, Laser ceramic materials for subpicocesond solid-state lasers using Nd^{3+} doped mixed scandium garnets, *Opt. Lett.* **35** (2010) 3048–3050.

[132] G. Q. Xie, D. Y. Tang, W. D. Tan, H. Luo, H. J. Zhang, H. H. Yu, and J. Y. Wang, Subpicosecond pulse generation from a Nd:CLNGG disordered crystal laser, *Opt. Lett.* **34** (2009) 103–105.

[133] G. Q. Xie, L. J. Lian, P. Yuan, D. Y. Tang, W. D. Tan, H. H. Yu, H. J. Zhang, and J. Y. Wang, Generation of 534 fs pulses from a passively mode-locked Nd: CLNGG-CNGG disordered crystal hybrid laser, *Laser Phys. Lett.* **7** (2010) 483–486.

[134] V. Lupei, A. Lupei, C. Gheorghe, L. Gheorghe, A. Achim, and A. Ikesue, Crystal field disorder in the optical spectra of Nd^{3+} and Yb^{3+}-doped calcium lithium niobium gallium garnet laser crystals and ceramics, *J. Appl. Phys.* **112** (2012) 063110.

[135] Q. Xie, D. Tang, J. Kong, and L. Qian, Passive mode-locking of a Nd:YAG ceramic laser by optical interference in a GaAs wafer, *Opt. Express* **15** (2007) 5360–5366.

[136] X. J. Chen, J. Q. Xu, M. J. Wang, B. X. Jiang, W. X. Zhang, and Y. B. Pan, Ho:YAG ceramic laser pumped by Tm:YLF laser at room temperature, *Laser Phys. Lett.* **7** (2010) 351–354.

[137] H. Chu, D. Chen, J. Zhang, H. Yang, D. Tang, T. Zhao, and X. Yang, In band pumped highly efficient Ho:YAG ceramic laser with 21 W output power at 2097 nm, *Opt. Lett.* **36** (2011) 1575–1577.

[138] G. A. Newburgh, A. Word-Daniels, A. Michael, L. D. Merkle, A. Ikesue, and M. Dubinskii, Resonantly diode-pumped $Ho^{3+}:Y_2O_3$ ceramic 2.1 μm laser, *Opt. Express* **19** (2011) 3604–3611.

[139] N. Ter-Gabrelian, L. D. Merkle, E. R. Knapp, G. L. Messing, and M. Dubinskii, Efficient resonantly pumped tape cast composite ceramic Er:YAG laser at 1645 nm. *Opt. Lett.* **35** (2010) 922–924.

[140] D. Y. Shen, H. Chen, X. Qin, J. Zhang, D. Tang, X. F. Yang, and T. Zhao, Polycrystalline ceramic Er:YAG laser in band pumped by a high power Er, Yb fiber laser at 1532 nm, *Appl. Phys. Express* **4** (2011) 052701.

[141] C. Zhang, D. Y. Shen, Y. Wang, L. J. Qinn, L. Zhang, X. P. Qin, D. Y. Tang, X. F. Yang, and T. Zhao, High power polycrystalline Er:YAG ceramic laser at 1617 nm, *Opt. Lett.* **36** (2011) 4767–4769.

[142] N. Ter-Gabrelian, L. D. Merkle, A. Ikesue, and M. Dubinskii, Ultralow quantum-defect eye safe $Er:Sc_2O_3$ laser, *Opt. Lett.* **33** (2008) 1524–1526.

[143] T. Sanamyan, J. Simmons, and M. Dubinskii, Er^{3+}-doped Y_2O_3 ceramic laser at 2.7 μm with direct pumping of the upper level, *Laser Phys. Lett.* **7** (2010) 206–209.

[144] T. Sanamyan, M. Kanskar, Y. Xiao, D. Kodlaya, and M. Dubinskii, High power diode pumped 2.7 μm $Er^{3+}:Y_2O_3$ laser with nearly quantum defect-limited efficiency, *Opt. Express* **19** (2011) A1082–A1087.

[145] X. J. Cheng, J. Q. Xu, W. X. Zhang, B. X. Jiang, and Y. B. Pan, End-pumped Tm:YAG ceramic slab laser, *Chin. Phys. Lett.* **26** (2009) 074204.

[146] W. H. Zhang, Y. B. Pan, J. Zhou, W. B. Liu, J. Li, B. X. Jiang, X. J. Cheng, and J. Q. Xu, Diode-pumped Tm:YAG ceramic laser, *J. Am. Ceram. Soc.* **92** (2009) 2434–2437.

[147] Y. W. Zhou, Y. D. Zhang, X. Zhong, Z. Y. Wei, W. X. Zhang, B. X. Jiang, and Y. B. Pan, Efficient Tm:YAG ceramic laser at 2 μm, *Chin. Phys. Lett.* **27** (2010) 074213.

[148] S. Zhang, M. Wang, L. Xu, Y. Wang, Y. Tang, X. Cheng, W. Chen, J. Xu, B. Jiang, and Y. Pan, Efficient Q-switced Tm:YAG ceramic slab laser, *Opt. Express* **19** (2011) 727–732.

[149] Y. Wang, D. Shen, H. Chen, J. Zhang, X. Qin, D. Tang, X. Yang, and T. Zhao, Highly efficient Tm:YAG ceramic laser resonantly pumped at 1617 nm, *Opt. Lett.* **36** (2011) 4485–4487.

[150] O. L. Antipov, S. Yu. Golovkin, O. N. Gorshkov, N. G. Zakharov, A. P. Zinoviev, A. P. Kasatkin, M. V. Kruglyov, M. O. Marychev, A. A. Novikov, N. V. Sakharov, and E. V. Chuprunov, Structure, optical and spectroscopic properties and efficient 2 μm lasing of new $Tm^{3+}:Lu_2O_3$ ceramics, *Quantum Electron.* **41** (2011) 863–866.

[151] S. Chenais, F. Druon, S. Forget, F. Balembois, and P. Georges, On thermal effects in solid-state lasers: the case of ytterbium-doped materials, *Prog. Quantum Electron.* **30** (2006) 89–153.

[152] K. Takaichi, H. Yagi, J. Lu, A. Shirakawa, K. Ueda, and T. Yanagitani, Yb^{3+}-doped $Y_3Al_5O_{12}$ ceramics – a new solid-state laser material, *Phys. Status Solidi A* **200** (2003) R5–R8.

[153] J. Dong, A. Shirakawa, K. Ueda, H. Yagi, T. Yanagitani, and A. A. Kaminskii, Efficient Yb^{3+}:$Y_3Al_5O_{12}$ ceramic microchip laser, *Appl. Phys. Lett.* **89** (2006) 091114.

[154] E. P. Ostby, R. A. Ackerman, J. C. Huie, and R. C. Gentilman, Ceramic Yb:YAG microchip laser, *Proc. SPIE* **6346** (2007) 63460V.

[155] S. Nakamura, Y. Matsubara, T. Ogawa, and S. Wada, High-power high-efficiency Yb^{3+} doped $Y_3Al_5O_{12}$ ceramic laser at room temperature, *Jpn. J. Appl. Phys.* **47** (2008) 2149–2151.

[156] S. Nakamura, H. Yoshioka, Y. Matsubara, T. Ogawa, and S. Wada, Efficient tunable Yb:YAG ceramic laser, *Opt. Commun.* **281** (2008) 4411–4414.

[157] Q. Hao, W. Li, H. Pan, X. Zhang, B. Jiang, Y. Pan, and H. Zeng, Laser-diode pumped 40 W Yb:YAG ceramic laser, *Opt. Express* **17** (2009) 17734–17738.

[158] A. Pirri, D. Alderighi, G. Toci, and M. Vannini, High-efficiency, high-power and low threshold Yb^{3+}:YAG ceramic laser, *Opt. Express* **17** (2009) 23344–23349.

[159] A. Pirri, D. Alderighi, G. Toci, and M. Vanini, A ceramic based Yb^{3+}:YAG laser, *Laser Phys.* **20** (2010) 931–935.

[160] H. Cai, J. Zhou, H. Zhao, Y. Qi, Q. Lou, and J. Dong, Continuous-wave and Q-switched performance of an Yb:YAG/YAG composite thin disk ceramic laser pumped with 970 nm laser diode, *Chin. Opt. Lett.* **6** (2008) 852–854.

[161] M. Tsunekame and T. Taira, High power operation of diode-edge-pumped, composite all-ceramic Yb:$Y_3Al_5O_{12}$ microchip laser, *Appl. Phys. Lett.* **90** (2007) 121101.

[162] A. Giesen and J. Speiser, Fifteen years of work on thin-disk lasers: results and scaling laws, *IEEE J. Quantum Electron.* **13** (2007) 598–609.

[163] W. P. Lathan, A. Lobad, T. C. Newell, and D. Stalnaker, 6.5 kW Yb:YAG ceramic thin disk laser, *AIP Conf. Proc.* **1278** (2010) 758–764.

[164] S. Nakamura, H. Yoshioka, T. Ogawa, and S. Wada, Broadly tunable Yb^{3+}-doped $Y_3Al_5O_{12}$ ceramic laser at room temperature, *Jpn. J. Appl. Phys.* **48** (2009) 060205.

[165] S. Nakamura, High power and high efficiency Yb:YAG ceramic lasers at room temperature, in *Frontiers in Guided Wave Optics and Optoelectronics*, ed. B. Pal, INTECH, Croatia (2010), pp. 513–528.

[166] J. Dong, A. Shirakawa, K. Ueda, H. Yagi, T. Yanagitani, and A. A. Kaminskii, Laser-diode pumped heavy-doped Yb:YAG ceramic laser, *Opt. Lett.* **32** (2007) 1890–1892.

[167] S. T. Fredrich-Thornton, C. Hirt, F. Tellkamp, K. Petermann, G. Huber, K. Ueda, and H. Yagi, Highly doped Yb:YAG thin-disk lasers: a comparison between single crystals and ceramic active media, *Tech. Dig. Adv. Solid-State Photon. Conf.*, Nara (2008), paper WB13.

[168] J. Dong, K. Ueda, H. Yagi, A. A. Kaminskii, and Z. Cai, Comparative study of the effect of Yb concentration on laser characteristics of Yb:YAG ceramics and crystals, *Laser Phys. Lett.* **6** (2009) 282–289.

[169] A. Pirri, G. Toci, D. Alderigi, and M. Vannini, Effect of the excitation density on the laser output of two differently doped Yb:YAG ceramics, *Opt. Express* **18** (2010) 17262–17272.

[170] V. I. Chani, G. Boulon, W. Zhao, T. Yanagida, and A. Yoshikawa, Correlation between segregation of the rare earth dopants in melt crystal growth and ceramic processing for optical applications, *Jpn. J. Appl. Phys.* **49** (2010) 075601.

[171] J. F. Bisson, D. Kouznetsov, K. Ueda, S T. Fredrich-Thornton, K. Petermann, and G. Huber, Switching of emissivity and photoconductivity in highly doped Yb^{3+}:Y_2O_3 and Lu_2O_3 ceramics, *Appl. Phys. Lett.* **90** (2007) 201901.

[172] C. Hirt, S. T. Fredrich-Thornton, F. Tellkamp, K. Petermann, and G. Huber, Photoconductivity measurements indicating a nonlinear loss mechanism in highly Yb-doped oxides, *Tech. Dig. Adv. Solid-State Photon. Conf.*, Nara (2008), paper MF1.

[173] U. Wolters, S. T. Fredrich-Thornton, F. Tellkamp, K. Petermann, and G. Huber, Photoconductivity in Yb-doped materials at high excitation densities and its effect on highly Yb-doped thin-disk lasers, *CLEO Europe–EQEC*, Munich (2009), paper CA9–2.

[174] C. Brandt, S. T. Fredrich-Thornton, K. Petermann, and G. Huber, Photoconductivity in Yb-doped oxides at high excitation densities, *Appl. Phys. B* **102** (2011) 765–768.

[175] C. W. Xu, D. W. Luo, J. Zhang, H. Yang, X. P. Qin, W. D. Tan, and D. Y. Tang, Diode pumped highly efficient Yb:Lu$_3$Al$_5$O$_{12}$ ceramic laser, *Laser Phys. Lett.* **9** (2012) 30–34.

[176] H. Furuse, J. Kawanaka, K. Takeshita, N. Miyanaga, T. Saiki, K. Imasaki, M. Fujita, and S. Ishii, Total-reflection active-mirror laser with cryogenic Yb:YAG ceramics, *Opt. Lett.* **34** (2009) 3439–3441.

[177] H. Furuse, J. Kawanaka, M. Myanaga, T. Saiki, K. Imasaki, M. Fujita, K. Takeshita, S. Ishii, and Y. Izawa, Zig-zag active mirror laser with cryogenic Yb:YAG/YAG composite ceramic, *Opt. Express* **19** (2011) 2448–2455.

[178] J. Lu, K. Takaichi, T. Uematsu, A. Shirakawa, M. Musha, K. Ueda, H. Yagi, T. Yanagitani, and A. A. Kaminskii, Yb^{3+}:Y$_2$O$_3$ ceramic – a novel solid state material, , *Jpn. J. Appl. Phys.* **41** (2002) L1373–L1375.

[179] J. Kong, J. Lu, K. Takaichi, T. Uematsu, K. Ueda, D. Y. Tang, D. Y. Shen, H. Yagi, T. Yanagitani, and A. A. Kaminskii, Diode-pumped Yb:Y$_2$O$_3$ ceramic laser, *Appl. Phys. Lett.* **82** (2003) 2556–2558.

[180] K. Takaichi, H. Yagi, J. Lu, J. F. Bisson, A. Shirakawa, K. Ueda, T. Yanagitani, and A. A. Kaminskii, Highly efficient continuous-wave operation at 1030 and 1075 nm wavelengths of LD-pumped Yb^{3+}:Y$_2$O$_3$ ceramic laser, *Appl. Phys. Lett.* **84** (2004) 317–319.

[181] J. Kong, D. Y. Tang, C. C. Chan, J. Lu, K. Ueda, H. Yagi, and T. Yanagitani, High-efficiency 1040 and 1078 nm laser emission of a Yb–Y$_2$O$_3$ ceramic laser with 976 nm diode pump, *Opt. Lett.* **32** (2007) 247–249.

[182] K. Takaichi, H. Yagi, A. Shirakawa, K. Ueda, S. Hosokawa, and A. A. Kaminskii, Lu$_2$O$_3$:Yb^{3+} ceramic – a novel gain material for high power solid state lasers, *Phys. Status Solidi A* **202** (2005) R1–R3.

[183] J. Sanghera, J. Frantz, W. Kim, G. Villalobos, C. Baker, B. Shaw, B. Sadovski, M. Hunt, F. Miklos, A. Lutz, and I. Aggarwal, 10% Yb^{3+}–Lu$_2$O$_3$ ceramic laser with 74% efficiency, *Opt. Lett.* **36** (2011) 576–578.

[184] A. Pirri, G. Toci, and M. Vannini, First laser oscillation and broad tenability of 1 at.% Yb-doped Sc$_2$O$_3$ and Lu$_2$O$_3$ ceramics, *Opt. Lett.* **36** (2011) 4284–4286.

[185] J. Lu, J. F. Bisson, K. Takaichi, T. Uematsu, A. Shirakawa, M. Musha, K. Ueda, H. Yagi, T. Yanagitani, and A. A. Kaminskii, Yb^{3+}:Sc$_2$O$_3$ ceramic laser, *Appl. Phys. Lett.* **83** (2003) 1101–1103.

[186] Q. Hao, W. Li, H. Zhang, Q. Yang, C. Dou, H. Zhou, and W. Lu, Low-threshold and broadly tunable laser of Yb^{3+} doped yttrium lanthanum oxide ceramics, *Appl. Phys. Lett.* **92** (2008) 211106.

[187] G. L. Bourdet, O. Casagrande, N. Deguil-Robin, and B. Le Garrec, Performances of cryogenic cooled lasers based on ytterbium doped sesquioxide ceramics, *J. Phys. Conf. Ser.* **112** (2008) 032054.

Ceramic lasers

[188] A. Lyberis, G. Patriarche, P. Gredin, D. Vivien, and M. Mortier, Origin of light scattering in yttrium doped calcium fluoride transparent ceramics for high power lasers, *J. Eur. Ceram. Soc.* **31** (2011) 1619–1630.

[189] T. T. Basiev, M. E. Doroshenko, and V. A. Konyushkin, Nd^{3+} and Yb^{3+} doped fluoride laser ceramics, *Tech. Dig. OSA Adv. Opt. Mater. Conf AIOM*, Istanbul (2011), paper AIThA3.

[190] T. T. Basiev, M. E. Doroshenko, P. P. Fedorov, V. A. Konyushkin, S. V. Kuznetsov, V. V. Osiko, and M. Sh. Akchurin, Efficient laser based on CaF_2–SrF_2–YbF_3 nanoceramics, *Opt. Lett.* **33** (2008) 521–523.

[191] J. Kawanaka, S. Tokita, N. Nishioka, K. Ueda, M. Fujita, T. Kawashima, H. Yagi, and T. Yanagitani, 42 mJ Q-switched active-mirror laser oscillator with a cryogenic Yb:YAG ceramic, *OSA Adv. Solid-state Photonics* (2007), paper MB2.

[192] J. Dong, A. Shirakawa, K. Ueda, H. Yagi, T. Yanagitani, and A. A. Kaminskii, Near-diffraction-limited passively Q-switched $Yb:Y_3Al_5O_{12}$ ceramic laser with peak power >150 kW, *Appl. Phys. Lett.* **90** (2007) 131105.

[193] J. Dong, A. Shirakawa, K. Ueda, H. Yagi, T. Yanagitani, and A. A. Kaminskii, Ytterbium and chromium doped composite $Y_3Al_5O_{12}$ ceramic for self Q-switched laser, *Appl. Phys. Lett.* **90** (2007) 191106.

[194] H. Yoshioka, S. Nakamura, T. Ogawa, and S. Wada, Diode-pumped mode-locked Yb:YAG ceramic laser, *Opt. Express* **17** (2009) 8919–8925.

[195] B. Zhou, Z. Y. Wei, Y. W. Zou, Y. D. Zhang, X. Zhang, G. L. Bourdet, and J. L. Way, High-efficiency diode-pumped femtosecond Yb:YAG ceramic laser, *Opt. Lett.* **35** (2010) 288–290.

[196] H. Yoshioka, S. Nakamura, T. Ogawa, and S. Wada, Dual-wavelength mode-locked Yb:YAG ceramic laser in single cavity, *Opt. Express* **18** (2010) 1479–1486.

[197] J. Saikawa, Y. Sato, T. Taira, and A. Ikesue, Passive mode-locking of a mixed $Yb:Y_3ScAl_4O_{12}$ ceramic laser, *Appl. Phys. Lett.* **85** (2004) 5845–5847.

[198] M. Tokurakawa, H. Kurokawa, A. Shirakawa, K. Ueda, H. Yagi, T. Yanagitani, and A. A. Kaminskii, Continuous-wave and mode-locked lasers on the base of partially disordered crystalline $Yb^{3+}:\{YGd_2\}[Sc](Al_2Ga)O_{12}$ ceramic, *Opt. Express* **18** (2010) 4390–4395.

[199] Yu. Senatsky, A. Shirakawa, Y. Sato, J. Hagiwara, J. Lu, K. Ueda, H. Yagi, and T. Yanagitani, Nonlinear refraction index of ceramic laser media and perspectives of their usage in a high-power laser driver, *Laser Phys. Lett.* **1** (2004) 500–506.

[200] A. Shirakawa, K. Takaichi, H. Yagi, M. Tanisho, J. F. Bisson, J. Lu, K. Ueda, T. Yanagitani, and A. A. Kaminskii, First modelocked ceramic laser: femtosecond $Yb:Y_2O_3$ ceramic laser, *Laser Phys.* **14** (2004) 1375–1381.

[201] J. Kong, D. Y. Tang, J. Lu, K. Ueda, H. Yagi, and T. Yanagitani, Passive mode-locking $Yb:Y_2O_3$ ceramic laser with a GaAs saturable absorption mirror, *Opt. Commun.* **237** (2004) 165–168.

[202] M. Tokurakawa, K. Takaichi, A. Shirakawa, K. Ueda, H. Yagi, S. Hosokawa, T. Yanagitani, and A. A. Kaminskii, Diode-pumped mode-locked $Yb^{3+}:Lu_2O_3$ ceramic laser, *Opt. Express* **14** (2006) 12832–12838.

[203] M. Tokurakawa, K. Takaichi, A. Shirakawa, K. Ueda, H. Yagi, T. Yanagitani, and A. A. Kaminskii, Diode pumped 188 fs mode-locked $Yb^{3+}:Y_2O_3$ ceramic laser, *Appl. Phys. Lett.* **90** (2007) 071101.

[204] G. Q. Xie, D. V. Tang, L. M. Zhao, L. J. Qian, and K. Ueda, High-power self-mode-locked $Yb:Y_2O_3$ ceramic laser, *Opt. Lett.* **32** (2007) 2741–2743.

[205] M. Tokurakawa, A. Shirakawa, K. Ueda, H. Yagi, T. Yanagitani, and A. A. Kaminskii, Diode pumped sub 100 fs Kerr mode-locked Yb^{3+}:Sc_2O_3 ceramic laser, *Opt. Lett.* **32** (2007) 3382–3384.

[206] M. Tokurakawa, A. Shirakawa, K. Ueda, S. Hosokawa, T. Yanagitani, and A. A. Kaminskii, Diode-pumped 65 fs Kerr-lens mode-locked Yb^{3+}:Lu_2O_3 and non-doped Y_2O_3 combined ceramic laser, *Opt. Lett.* **33** (2008) 1380–1382.

[207] M. Tokurakawa, A. Shirakawa, K. Ueda, H. Yagi, M. Noriyuki, T. Yanagitani, and A. A. Kaminskii, Diode-pumped ultrashort pulse generation based on Yb^{3+}:Sc_2O_3 and Yb^{3+}:Y_2O_3 ceramic multi-gain oscillator, *Opt. Express* **17** (2009) 3353–3361.

[208] S. Pearce, R. Yasuhara, A. Yoshida, J. Kawanaka, T. Kawashima, and H. Kan, Efficient generation of 200 mJ nanosecond pulses at 100 Hz repetition rate from a cryogenic cooled Yb:MOPA system, *Opt. Commun.* **282** (2009) 2199–2203.

[209] Y. Takeuchi, J. Kawanaka, A. Yoshida, R. Yasuhara, T. Kawashima, H. Kan, and N. Miyanaga, Sub-kHz cryogenic Yb:YAG regenerative amplifier by using a total-reflection active mirror, *Appl. Phys. B* **104** (2011) 29–32.

[210] G. A. Torchia, P. F. Meilan, A. Rodenas, D. Jaque, C. Mendez, and L. Roso, Femtosecond laser written surface wavequides fabricated in Nd:YAG ceramic, *Opt. Express* **15** (2007) 13266–13271.

[211] G. A. Torchia, A. Rodenas, A. Benayas, E. Cantelar, L. Roso, and D. Jaque, Highly efficient laser action in fs-written Nd:yttrium aluminum garnet ceramic waveguides, *Appl. Phys. Lett.* **92** (2008) 111102.

[212] A. Rodenas, G. Zhou, D. Jaque, and M. Gu, Direct laser writing of three-dimensional photonic structures in Nd:yttrium aluminum garnet laser ceramics, *Appl. Phys. Lett.* **93** (2008) 151104.

[213] A. Rodenas, G. A. Torchia, G. Lifante, E. Cantelar, J. Lamela, F. Jaque, L. Roso, and D. Jaque, Refractive index change mechanisms in femtosecond laser written ceramic Nd:YAG waveguides: micro-spectroscopy experiments and beam propagation calculation, *Appl. Phys. B* **95** (2009) 85–96.

[214] A. Benayas, W. F. Silva, C. Jacinto, E. Cantelar, J. Lamela, F. Jaque, J. R. Vasquez de Aldana, G. A. Torchia, L. Roso, A. A. Kaminskii, and D. Jaque, Thermally resistant waveguides fabricated in Nd:YAG ceramics by crossing femtosecond damage filaments, *Opt. Lett.* **35** (2010) 330–332.

[215] A. Benayas, W. F. Silva, A. Rodenas, C. Jacinto, J. Vasquez de Aldana, F. Chen, Y. Tan, R. R. Thomsom, N. D. Psaila, D. T. Reid, G. A. Torchia, A. K. Kar, and D. Jaque, Ultrafast laser writing of optical waveguides in ceramic Yb:YAG: a study of thermal and non-thermal regimes, *Appl. Phys. A* **104** (2011) 301–309.

[216] Y. Tan and F. Chen, Proton-implanted optical channel waveguides in Nd:YAG laser ceramics, *J. Phys. D: Appl. Phys.* **43** (2010) 075105.

[217] D. Xu, Y. Wang, H. Li, J. Yao, and Y. H. Tsang, 104 W high stability green laser generation by using diode laser pumped intracavity frequency doubled Q-switched composite ceramic Nd:YAG, *Opt. Express* **15** (2007) 3991–1997.

[218] A. A. Kaminskii, K. Ueda, H. J. Eichler, S. N. Bagaev, K. Takaichi, J. Lu, A. Shirakawa, H. Yagi, and T. Yanagitani, Observation of nonlinear lasing chi((3))-effects in highly transparent nanocrystalline Y_2O_3 and $Y_3Al_5O_{12}$ ceramics, *Laser Phys. Lett.* **1** (2004) 6–11.

[219] A. A. Kaminskii, S. N. Bagaev, H. J. Eichler, K. Ueda, K. Takaichi, A. Shirakawa, H. Yagi, T. Yanagitani, and H. Rhee, Observation of high order Stokes and anti-Stokes (3) generation in highly transparent laser-host Lu_2O_3 ceramics, *Laser Phys. Lett.* **3** (2006) 310–313.

[220] A. A. Kaminskii, H. Rhee, H. J. Eichler, K. Ueda, K. Takaichi, A. Shirakawa, and M. Tokurakawa, New nonlinear laser effects in crystalline fine-grained ceramics based on cubic Sc_2O_3 and Lu_2O_3 oxides: second and third harmonic generation and cascaded sum-frequency mixing in UV spectral region, *Laser Phys. Lett.* **5** (2008) 109–113.

[221] A. A. Kaminskii, S. N. Bagaev, K. Ueda, H. Yagi, H. J. Eichler, A. Shirakawa, M. Tokurakawa, H. Rhee, K. Takaichi, and T. Yanagitani, Nonlinear-laser-effects in novel garnet-type fine-grained ceramic host $\{YGd_2\}[Sc_2](Al_2Ga)O_{12}$ for Ln^{3+} lasants, *Laser Phys. Lett.* **6** (2009) 671–677.

[222] I. Shoji, Y. Sato, S. Kurimura, V. Lupei, T. Taira, A. Ikesue, and K. Yoshida, Thermal-birefringence-induced depolarization in Nd:YAG ceramics, *Opt. Lett.* **27** (2002) 234–236.

[223] I. Shoji, T. Taira, and A. Ikesue, Thermally-induced-birefringence effects of highly Nd^{3+}-doped $Y_3Al_5O_{12}$ ceramic lasers, *Opt. Mater.* **29** (2007) 1271–1276.

[224] E. A. Khazanov, Thermally induced birefringence in Nd:YAG ceramics, *Opt. Lett.* **27** (2002) 716–718.

[225] I. B. Mukhin, O. V. Palashov, E. A. Khazanov, A. Ikesue, and Y. L. Aung, Experimental study of thermally induced depolarization in Nd:YAG ceramics, *Opt. Express* **13** (2005) 5983–5987.

[226] A. Soloviev, I. Snetkov, V. Zelenogorski, I. Kozhevatov, O. Palashov, and E. Khazanov, Experimental study of thermal lens features in laser ceramics, *Opt. Express* **16** (2008) 21012–21021.

[227] R. Kawai, Y. Miyasaka, K. Otsuka, T. Narita, J. Y. Ko, I. Shoji, and T. Taira, Oscillation spectra and dynamic effects in a highly-doped microchip Nd:YAG ceramic laser, *Opt. Express* **12** (2004) 2293–2302.

[228] K. Otsuka, T. Narita, Y. Miyasaka, C. C. Lin, J. Y. Ko, and S. C. Chu, Non-linear dynamics in thin-slice Nd:YAG ceramic laser: coupled local-mode laser model, *Appl. Phys. Lett.* **89** (2006) 081117.

[229] K. Otsuka, Polarization properties of laser-diode-pumped microchip Nd:YAG ceramic lasers, in *Frontiers in Guided Wave Optics and Optoelectronics*, Ed. B. Pal, INTECHY, Croatia (2010) pp. 529–550.

[230] Q. L. Ma, Y. Bo, N. Zong, Q. J. Peng, D. F. Cui, Y. B. Pan, and Z. Y. Xu, 108 W Nd:YAG ceramic laser with birefringence compensation resonator, *Opt. Commun.* **283** (2010) 5183–5186.

[231] K. Otsuka and T. Ohtomo, Polarization properties of laser-diode-pumped micro-grained Nd:YAG ceramic laser, *Laser Phys. Lett.* **5** (2008) 659–663.

[232] O. Parriaux, J. F. Bisson, K. Ueda, S. Tonchev, E. Gagot, J. C. Pommier, and S. Teynaud, Polarization control of a Yb:YAG ceramic microchip laser by constructive-interference resonant grating mirror, *J. Mod. Opt.* **55** (2008) 1899–1912.

[233] I. Moshe, S. Jackel, Y. Lumer, A. Meir, R. Feldman, and Y. Shimony, Use of polycrystalline Nd:YAG rods to achieve pure radially and azimuthally polarized beams from high-average-power lasers, *Opt. Lett.* **35** (2010) 2511–2513.

[234] M. P. Thirugnasambandam, Y. Senatsky, and K. Ueda, Generation of radially and azimuthally polarized beams in Yb:YAG laser with intracavity lens and birefringent crystal, *Opt. Express* **19** (2011) 1905–1914.

[235] K. Tokunaga, S. C. Chu, H. Y. Hsiao, T. Ohtomo, and K. Otsuka, Spontaneous Mathieu–Gauss mode oscillations in micro-grained Nd:YAG ceramic lasers with azimuth laser-diode pumping, *Laser Phys. Lett.* **6** (2009) 635–638.

[236] M. P. Thirugnasambandam, Y. Senatsky, and K. Ueda, Generation of very high order Laguerre–Gaussian modes in Yb:YAG ceramic lasers, *Laser Phys. Lett.* **7** (2010) 637–643.

[237] M. P. Thirugnasambandam, Y. Senatsky, A. Shirakawa, and K. Ueda, Multi-ring modes generation in Yb:YAG ceramic laser, *Opt. Mater.* **33** (2011) 675–678.

[238] L. Li, D. Lin, L. X. Peng, K. Ueda, A. Sirakawa, M. Musha, and W. B. Chen, Passively Q-switched Nd:YAG ceramic microchip laser with azimuthally polarized light, *Laser Phys. Lett.* **6** (2009) 711–714.

[239] J. Huang, J. Dong, Y. Cao, W. Wang, H. Zheng, J. Li, F. Shi, Y. Ge, S. Dai, and W. Lin, Passively mode-locked radially polarized laser based on ceramic Nd:YAG rod, *Opt. Express* **19** (2011) 2120–2125.

[240] J. H. Nuckolls, Grand challenges of inertial fusion energy, *J. Phys. Conf. Ser.* **244** (2010) 012007.

[241] K. Mima, V. Tikhonchuk, and M. Perlado, Inertial fusion experiments and theory, *Nucl. Fusion* **51** (2011) 094004.

[242] E. I. Moses, T. Diaz de la Rubia, J. F. Latkowski, J. C. Farmer, R. P. Abbott, K. J. Kramer, P. F. Peterson, H. F. Shaw, and R. F. Lehman II, A sustainable fuel cycle based on laser inertial fusion energy (LIFE), *Fusion Sci. Technol.* **56** (2009) 566–572.

[243] E. I. Moses, Ignition of the National Ignition Facility: a path towards inertial fusion energy, *Nuclear Phys.* **49** (2009) 104022.

[244] Ch. Lyon, The LMJ project: an overview, *J. Phys. Conf. Ser.* **244** (2010) 012003.

[245] J. Kawanaka, N. Miyanaka, T. Kawashima, K. Tsubakimoto, Y. Fijiomoto, H. Kubomura, S. Matsuoka, T. Ikeyama, Y. Suzuki, N. Tschiya, T. Jitsumo, H. Frukawa, T. Kanabe, H. Fujita, K. Yoshida, N. Nakano, J. Nishimae, M. Nakatsuka, K. Ueda, and K. Tomabechi, New concept for laser fusion energy driver by using cryogenically-cooled Yb:YAG ceramic, *J. Phys. Conf. Ser.* **112** (2008) 032058.

[246] A. Yoshida, S. Tokita, J. Kawanaka, T. Yanagitani, H. Yagi, F. Yamamura, and T. Kawashima, Numerical gain estimation of cryogenic Yb:YAG ceramic for IFE reactor driver, *J. Phys. Conf. Ser.* **112** (2008) 032042.

[247] R. Yasuhara, R. Katai, J. Kawanaka, T. Kawashima, H. Miyajima, and H. Kan, 1 J × 100 Hz cryogenic Yb:YAG laser development for feasibility demonstration of GENBU main laser, *Rev. Laser Eng.* **36** (2008) 1092–1093.

[248] J. C. Chanteloup, D. Albach, A. Lucianetti, K. Ertel, S. Bannerjee, P. D. Mason, C. Hernandez-Gomez, J. L. Collier, J. Hein, M. Wolf, J. Korner, and B. J. LeGarrec, Multi kJ level laser concepts for HiPER facility, *J. Phys. Conf. Ser.* **244** (2010) 012010.

[249] S. Jacquemot *et al.*, Studying ignition schemes on European laser facilities, *Nucl. Fusion* **51** (2011) 094025.

[250] E. A. Khazanov and A. M. Sergeev, Petawatt lasers based on optical parametric amplifiers: their state and prospects, *Physics-Uspekhi* **51** (2008) 969–974.

[251] E. A. Khazanov and A. M. Sergeev, Concept study of a 100 PW femtosecond laser based on laser ceramics doped with Chromium ions, *Laser Phys.* **17** (2007) 1398–1403.

[252] S. Bakhramov, Sh. Payziyev, and A. Kazimov, Feasibility of creation of ceramic disk laser pumped by concentrated solar flux of big solar furnace, *J. Renew. Sust. Energ.* **1** (2009) 063103.

[253] T. Yabe, M. S. Mohamed, S. Uchida, C. Baasandash, Y. Sato, M. Tsuji, and Y. Mori, Noncatalithic dissociation of MgO by laser pulses towards sustainable energy cycle, *J. Appl. Phys.* **101** (2007) 123106.

Index

absorption, 18, 22
  coefficient, 22, 55, 58, 220, 221, 225, 386
  distribution, 59, 398
  efficiency, 55, 390, 393
  rate, 19
amplification by stimulated emission, 22
amplified spontaneous emission (ASE), 54, 78, 81,
    329, 386, 399, 402
artificial tooth, 294

back scattering image (BEI), 200, 205
birefringence
  thermal, 87, 422
Bixbyite, 109, 162, 163, 165, 166
bonding interface, 181, 243, 245, 246, 249, 250
Bridgman method, 204

capsule free, 143, 153, 156, 158
cation vacancy, 216
Ce-doped ceramics, 321
ceramic fiber laser, 246, 255, 265
ceramic granular structure, 104, 358
  effect on laser beam, 422
  grain boundaries, 28, 105, 301, 311
  pores, 104
  randomly oriented grains, 104
ceramic jewelry, 297
ceramic laser components, 401
ceramic monolithic composite materials, 109, 402
ceramic raw material
  granulation (spray drying), 105
  solid-state synthesis (reaction), 105
  wet (soft) chemistry synthesis, 105
ceramic sintering process
  grain growth, 105, 311
  liquid-phase mass transport, 311
  sintering aid, 105, 301, 310, 317
  solid-state mass transport, 311
  solute drag effect, 105, 311
ceramics
  coarse-grained, 105, 358
  fine-grained, 106

  mechanical properties, 108
  nanoceramics, 106
charge compensation, 92, 101, 110, 312, 316, 344
clad–core, 219, 242, 243, 246, 248, 252, 264
Coble, 7, 121, 122, 267
cold isostatic pressing (CIP), 123, 138, 169
composite ceramic, 270, 404, 413, 419, 422
concentration quenching, 165, 195, 329, 335, 365
conversion of excitation
  down-conversion, 40, 42, 346, 365, 393
  quantum splitting, 365
  upconversion, 40, 346, 351, 368, 393
Cr:ZnSe laser, 277
Cr^{2+} doped II–VI compound laser, 275, 277
Cr^{2+}-doped ceramics, 342
Cr^{3+}-doped ceramics, 343
Cr^{4+}-doped ceramics, 255, 259, 284, 344, 409
crystal field interaction, 29, 31, 32, 35
crystal growth, 102
crystal symmetry, 28
cubic fluorides, 110
cubic sesquioxides, 109, 302, 320
  structure, 109, 303
Czochralski method, 2, 14, 102, 107, 161, 187, 200,
    204, 231, 253, 286, 308, 316, 345

de-excitation processes
  global, 64, 395
  spatial distribution, 82, 397
defective sites, 28
diffusion coefficient, 122, 216
doping
  charge compensation, 33, 313, 314
  concentration, 28, 33, 108, 319, 348, 359, 391
  coordination, 28, 34, 300
  crystallographic site, 28, 300, 363
  dead sites, 60, 108, 308
  defective centers, 309
  distribution, 33, 35, 358
  distribution, average-distance model, 33, 43, 49
  distribution, continuous, 33, 44, 50
  distribution, correlated, 33

Printed in the United States
by Baker & Taylor Publisher Services